T0317667

Automotive Power Transmission Systems

Automotive Series

Series Editor: Thomas Kurfess

Automotive Power Transmission Systems	Zhang and Mi	September 2018
Hybrid Electric Vehicles: Principles and Applications with Practical Perspectives, 2nd Edition	Mi and Masrur	October 2017
Hybrid Electric Vehicle System Modeling and Control, 2nd Edition	Liu	April 2017
Thermal Management of Electric Vehicle Battery Systems	Dincer, Hamut and Javani	March 2017
Automotive Aerodynamics	Katz	April 2016
The Global Automotive Industry	Nieuwenhuis and Wells	September 2015
Vehicle Dynamics	Meywerk	May 2015
Vehicle Gearbox Noise and Vibration: Measurement, Signal Analysis, Signal Processing and Noise Reduction Measures	Tůma	April 2014
Modeling and Control of Engines and Drivelines	Eriksson and Nielsen	April 2014
Modelling, Simulation and Control of Two-Wheeled Vehicles	Tanelli, Corno and Savaresi	March 2014
Advanced Composite Materials for Automotive Applications: Structural Integrity and Crashworthiness	Elmarakbi	December 2013
Guide to Load Analysis for Durability in Vehicle Engineering	Johannesson and Speckert	November 2013

Automotive Power Transmission Systems

Yi Zhang
University of Michigan-Dearborn
USA

Chris Mi
San Diego State University
USA

This edition first published 2018

Registered Offices
John Wiley & Sons, Inc., 111 River Street, Hoboken, NJ 07030, USA
John Wiley & Sons Ltd, The Atrium, Southern Gate, Chichester, West Sussex, PO19 8SQ, UK

Editorial Office
The Atrium, Southern Gate, Chichester, West Sussex, PO19 8SQ, UK

For details of our global editorial offices, customer services, and more information about Wiley products visit us at www.wiley.com.

Wiley also publishes its books in a variety of electronic formats and by print-on-demand. Some content that appears in standard print versions of this book may not be available in other formats.

Library of Congress Cataloging-in-Publication Data

Names: Zhang, Yi, 1962 September 1- author. | Mi, Chris, author.
Title: Automotive power transmission systems / by Yi Zhang, Chris Mi.
Description: Hoboken, NJ : John Wiley & Sons, 2018. | Includes
bibliographical references and index. |
Identifiers: LCCN 2018013178 (print) | LCCN 2018028719 (ebook) | ISBN
9781118964910 (pdf) | ISBN 9781118964903 (epub) | ISBN 9781118964811 (cloth)
Subjects: LCSH: Automobiles–Transmission devices.
Classification: LCC TL262 (ebook) | LCC TL262 .Z43 2018 (print) | DDC 629.2/44–dc23
LC record available at https://lccn.loc.gov/2018013178

Cover design by Wiley
Cover images: Background: © solarseven/Shutterstock; Left: Courtesy of Yi Zhang and Chris Mi;
Middle: © Dong liu/Shutterstock; Right: © Scharfsinn86/Getty Images

Contents

Series Preface

Automotive power transmission systems are critical elements of any automobile. The ability to transmit power from the engine of a vehicle to the rest of the drive train is of primary importance. Furthermore, the design of power transmission systems is of critical importance to the overall vehicle system performance, as it affects not only performance characteristics such as torque and acceleration, but it also directly affects fuel efficiency and emissions. The power transmission system also presents one of the most complex design tasks in the overall automotive systems design and integration because it must interface with a variety of power plants such as internal combustion, electric, and hybrid plants. This is further complicated by the fact that engineers must consider a variety of transmission designs such as manual, automatic, and continuously variable systems. Furthermore, all of these elements must be condensed into the smallest, lightest package possible while functioning under significant loads over long periods of time.

Automotive Power Transmission Systems presents a thorough discussion of the various concepts that must be considered when designing a power transmission system. The book begins with an excellent discussion of how a transmission is designed by matching the engine output and the vehicle performance via proper transmission ratio selection. It then proceeds to discuss the basics of manual transmission and the analysis and design of essential transmission subsystems and components such as the gears, torque converter, and clutches. The authors then discuss more advanced transmission types such as dual clutch transmissions, continuously variable transmissions and automatic transmissions. In the final chapters, advanced control concepts for transmissions are presented, leading to the final chapters on electric and hybrid powertrains. This powerful combination of concepts results in a text that has both breadth and depth that will be valued as both a classroom text and a reference book.

The authors of *Automotive Power Transmission Systems* have done an excellent job in providing a thorough technical foundation for vehicle power transmission analysis and control. The text includes a number of clearly presented examples that are of significant use to the practicing engineer, resulting in a book that is an excellent blend of practical applications and fundamental concepts. The strength of this text is that it links a number of fundamental concepts to very pragmatic examples, providing the reader with significant insights into modern automotive power transmission technology. The authors have done a wonderful job in clearly and concisely bringing together the significant breadth of technologies necessary to successfully implement a modern power transmission system, providing a fundamentally grounded book that thoroughly explains power transmissions. It is well written, and is authored by recognized experts in a field that is critical to the automotive sector. It provides an excellent set of pragmatic and fundamental perspectives to the reader and is an excellent addition to the Automotive series.

Thomas Kurfess
January 2018

Preface

Automotive power transmission systems deliver output from the power source, which can be an internal combustion engine or an electric motor or a combination of them, to the driving wheels. There are many valuable books and monographs published for internal combustion engines (ICE), but only a few can be found in the public domain, as referenced in this book, that are specifically written for automotive transmissions. Technical publications by the Society of Automotive Engineers (SAE) in transmissions are mostly for conventional ICE vehicles and are basically collections of research papers that are aimed at readers with high expertise in transmission sub-areas. The purpose of this book is to offer interested readers, including undergraduate or graduate students and practicing engineers in the related disciplines, a systematic coverage of the design, analysis, and control of various types of automotive transmissions for conventional ICE vehicles, pure electric vehicles, and hybrid vehicles. The aim is that this book can be used either as a textbook for students in the field of vehicular engineering or as a reference book for engineers working in the automotive industry.

The authors have taught a series of courses on powertrain systems for both ICE and electric-hybrid vehicles over many years in the graduate programs of mechanical engineering, electrical engineering, and automotive systems engineering at the University of Michigan-Dearborn. The lecture notes of these courses form the framework for the book chapters, the main topics of which are highlighted below.

The book starts with automotive engine matching in Chapter 1, which covers the following technical topics: output characteristics of internal combustion engines, vehicle road loads and acceleration, driving force (or traction) and power requirements, vehicle performance dynamics and fuel economy, and transmission ratio selection for fuel economy and performance. The formulation and related analysis in Chapter 1 on road loads, performance dynamics, and powertrain kinematics are applicable to all vehicles driven by wheels and will be used throughout the book.

Chapter 2 covers manual transmissions, focusing on gear layouts, clutch design, synchronizer design, and synchronization analysis. Detailed analysis is provided on the operation principles of synchronizers and on the synchronization process during gear shifts. Example production transmissions are used as case studies to demonstrate principles and approaches that are then generally applicable.

For readers' convenience, Chapter 3 provides the basics of the theory of gearing and gear design with specific application to manual transmissions (MT). With example transmissions, the chapter details geometry design, gear load calculation, and gear strength and power ratings for standard and non-standard gears using existing equations or formulae from AGMA standards. The chapter also includes a separate section on the kinematics of planetary gear trains which are widely applied in automatic transmissions (AT). Readers are strongly recommended to read this section before reading Chapters 5 and 6.

Chapter 4 covers the structure, design, and characteristics of torque converters, focusing on torque converter operation principles, functionalities, and input–output characteristics. Methods for the determination of engine–converter joint operation states are presented in detail. The chapter also deals with the modeling of the combined operation of the entire vehicle system that consists of the engine, torque converter, automatic transmission, and the vehicle itself.

Chapters 5 and 6 can be considered as the core of the book, as these two chapters present the design, analysis, and control of conventional automatic transmissions (AT) which are typically designed with planetary gear trains. Chapter 5 focuses on how multiple gear ratios are achieved by different combinations of clutches and planetary gear trains. A systematic method will be presented in this chapter for the design and analysis on the gear ratios and clutch torques of automatic transmissions. The chapter also gives an in-depth analysis of the dynamics of automatic transmissions during gear shifts and the general vehicle powertrain dynamics in a systematic approach, using an eight-speed production automatic transmission as the example in the case study.

Chapter 6 concentrates on hardware and software technologies of both component and system levels which are applied in the control systems for the implementation of transmission functionalities. The chapter begins with the functional descriptions of the hardware components, including hydraulic components, electronic sensors, and solenoids. The chapter then presents transmission control system configurations and the related design guidelines. Examples based on the production transmissions of previous and current generations are used to demonstrate the operation logic and functions of the control systems. A specific section is devoted to present concurrent transmission control technologies commonly applied in the automotive industry. This focuses on the accurate clutch torque control during gearshifts and torque converter clutch actuation. The chapter ends with the identification of control variables and control system calibration.

Chapter 7 mainly presents the design and control of belt type continuously variable transmissions (CVT), starting with the structural layouts of CVT systems and key components, including the basic CVT kinematics and operation principles. Topics are concentrated on force analysis during the CVT's operations, and the mechanisms for torque transmission and ratio changes. The chapter provides details of control system design and the analysis of the control of ratio changing processes. CVT system control strategies, including continuous ratio control, stepped ratio control, and system pressure control are also presented.

The design and control of dual clutch transmissions (DCT) are covered in Chapter 8. The chapter concentrates on the dynamic modeling and analysis of DCT operations, including DCT vehicle launch and shifts. DCT control system design, and shift and launch control processes are included here, and the chapter also dedicates a specific section to DCT clutch torque formulation during launch and shifts, using an electrically actuated dry DCT as the example in the case study.

Chapter 9 covers power train systems for pure electric vehicles (EV). It includes several key technical topics: design optimization and control of electric machines for EV applications, power electronics for electric power transmission and inverter design, and system control under various operation modes. The chapter also includes a section on mechanical transmissions with a fixed ratio or two ratios which are specifically designed for pure electric vehicles. Two-speed or multi-speed automated gear boxes enable EV driving motors to operate within the speed range for optimized efficiency and performance.

Finally, hybrid powertrain systems are discussed in Chapter 10, which presents various hybrid powertrain configurations including series, parallel, and complex architectures. It provides detailed analysis of the operation modes and operation control for hybrid vehicle powertrain systems. Production hybrid vehicles are used as case studies in mode analysis and operation control.

As highlighted above, each chapter of the book is dedicated to a specific transmission, and readers may choose the chapter of interest to read. If the book is used as a textbook, the course syllabus can follow the order of the chapters. If the book is used as a reference, readers with transmission expertise may just choose the chapter of interest, and those readers without broad expertise may wish to first read Chapters 1 and 4 and then read the chapter of interest.

The authors would like to express their hearty thanks for the help received from friends and colleagues in preparing the manuscript. We would like to thank especially Prof. Qiu Zhihui of Xian Jiaotong University and Prof. He Songping of Huazhong University of Science and Technology for their help in drawing the pictures for this book. We also want to thank the publisher, John Wiley & Sons, for giving us the opportunity to publish this book, and we dedicate our deep appreciation to Ms Ashmita Rajaprathapan for her invaluable contributions in editing and finalizing the book. Lastly and most importantly, the authors would like to express their thanks to engineers, scholars, and researchers who have contributed to the technologies of vehicular power transmission systems and whose work may or may not have been specifically acknowledged in the reference lists.

Yi Zhang and Chris Mi

1

Automotive Engine Matching

1.1 Introduction

Internal combustion engines have been the primary power source for automotive vehicles since the beginning of the automotive industry. Although automobiles powered by electric motors have entered the automotive market and are likely to grow in market share, the vast majority of vehicles will still be powered by internal combustion engines in the foreseeable future. This is partly due to the bottleneck in the development of key technologies for electric vehicles, such as battery energy density, durability and charging time, and the lack of infrastructure and facilities necessary for the daily use of electric vehicles. On the other hand, proven crude oil reserves can still fuel internal combustion engines for decades to come.

Modern internal combustion engines are sophisticated systems that integrate synergistically mechanical, electrical, and electronic subsystems. Engine technologies are subjects of study in great breadth and depth in the areas of combustion, heat transfer, mechanical design and manufacturing, material engineering, and electronic control [1,2,3,4]. However, this book does not cover engines themselves and is concerned only with how the engine outputs are transmitted to the driving wheels. Readers interested in engine topics are directed to the books referenced here or other related books. The engine outputs, in terms of power and torque, fuel economy, and emissions, are considered as given throughout the text of this chapter and indeed the whole book. Note that engine mapping data are highly proprietary and is usually not available in the public domain. Figures and plots pertaining to engine data in this book are mainly for illustration purposes and may not show the precise data of production engines.

The main topic of this chapter is the matching between the engine outputs and vehicle performance through the selection of transmission ratios. The chapter specifically covers: output characteristics of internal combustion engines; vehicle road loads and acceleration; driving force (or traction) and power requirements; vehicle performance dynamics and fuel economy; and transmission ratio selection. These topics are interconnected and are described in sequential order.

Although the chapter concerns automotive engine matching, as the title indicates, the formulation and related analysis of road loads, performance dynamics, and powertrain kinematics are applicable for all ground vehicles driven by wheels. The equations derived in this chapter will be referenced throughout the book wherever needed by the text.

Automotive Power Transmission Systems, First Edition. Yi Zhang and Chris Mi.

1.2 Output Characteristics of Internal Combustion Engines

The output of an internal combustion engine depends on its design, control, and calibration. Although computer simulation can be used to analyse engine output, engine mapping is the only experimental approach to obtain reliable engine output data. For a given production engine, its output data are provided in terms of power and torque, as well as specific fuel consumption and emissions.

1.2.1 Engine Output Power and Torque

The operation status of an internal combustion engine is defined by its crankshaft rotational speed and the output torque from its crankshaft. The output torque and power depend on the throttle opening and the engine speed, i.e. the crankshaft rotational speed in RPM. It should be noted that the output torque and output power are not independent since power is the product of torque and angular velocity. The torque map of a typical IC engine is shown in Figure 1.1, where the two horizontal axes are respectively the throttle opening as a percentage of the wide open throttle (WOT) or as a degree of throttle angle and the engine speed in RPM. The vertical axis shows the engine output torque in foot pounds in the imperial standard or in newton-meters in the international standard (SI). Without considering the engine transient behavior, the engine static output torque can be found from Figure 1.1, usually by numerical interpolation, for a given set of engine speed and throttle opening. This is the torque as a load at which the engine reaches dynamic equilibrium at the specified engine speed and the throttle opening.

In practice, the engine output torque is often plotted as a curve against the engine speed for specific throttle openings as shown in Figure 1.2, where the throttle opening for each torque curve is represented as a percentage of the wide open throttle angle. Clearly, the engine output torque is a function of engine RPM for a given throttle opening and there is a torque vs RPM curve for each throttle opening. Figures 1.1 and 1.2 provide the same output torque data and are just drawn for convenience of reference.

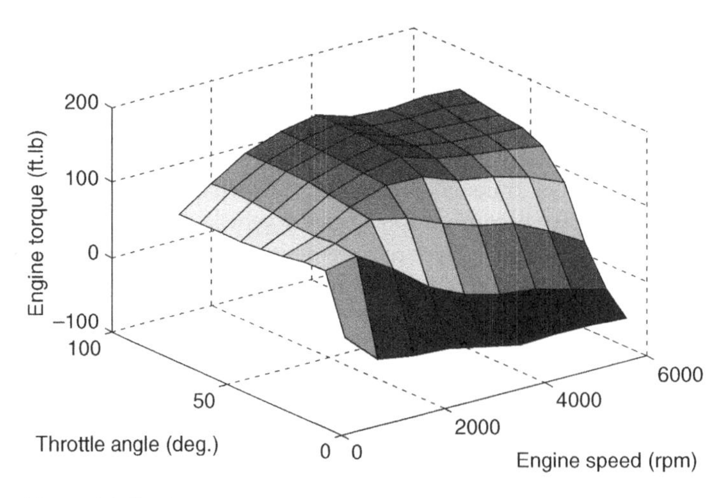

Figure 1.1 Engine output torque map.

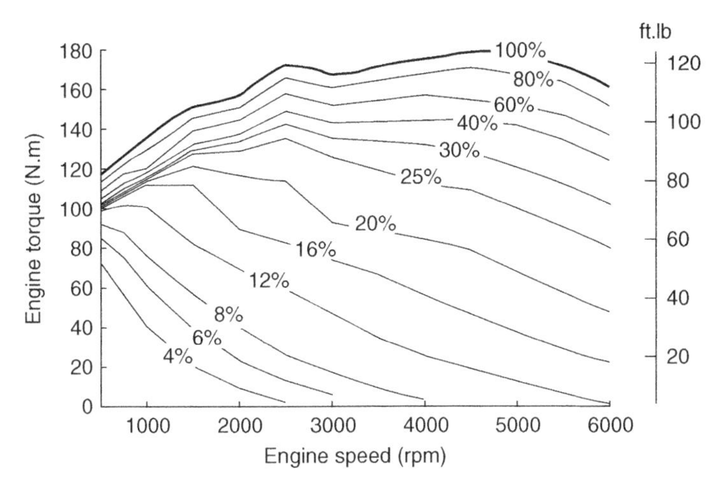

Figure 1.2 Engine torque curves for various throttle openings.

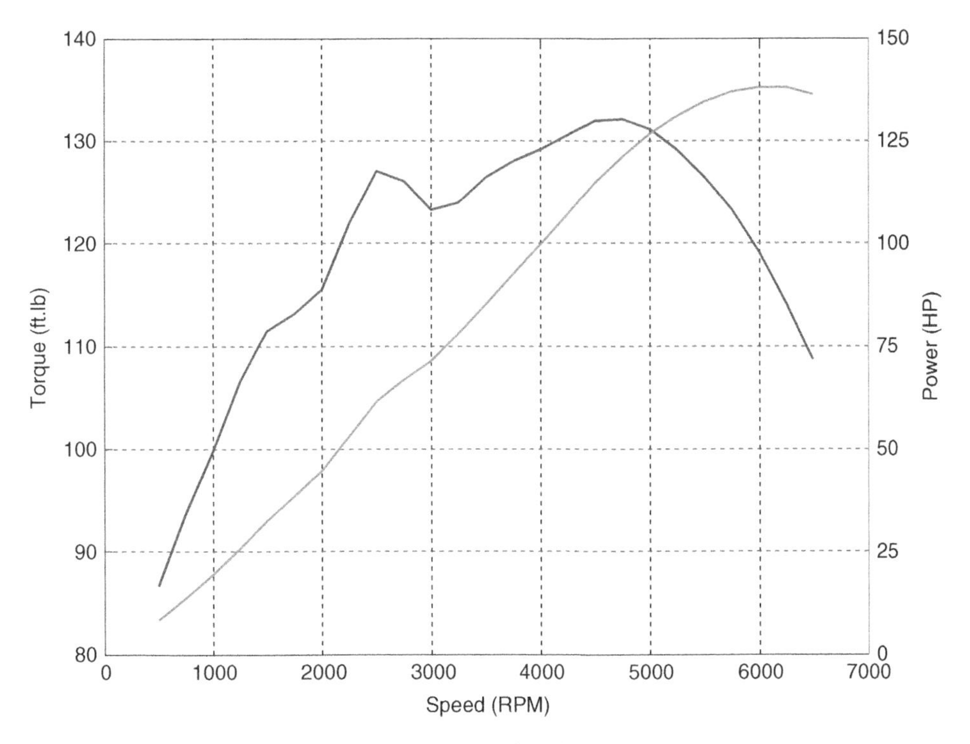

Figure 1.3 Engine torque and power at wide open throttle.

Apparently, the engine capacity torque output or the power output is achieved at the wide open throttle (WOT), as shown in Figure 1.3. It should be noted that the maximum engine torque and the maximum engine power occur only at two separate RPM values on the WOT torque and power curves. The two popularly referred engine performance

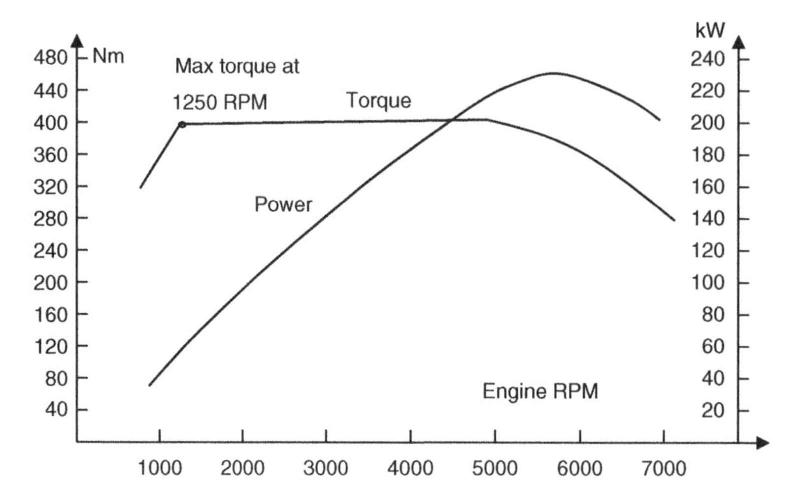

Figure 1.4 Typical torque curve of turbo engines.

specifications, engine power and engine torque, are actually the peak values for the power and torque on the WOT output plot. As can be observed in Figure 1.3, internal combustion engines provide stable power output within a range of engine rotational speed, defined by the so-called idle RPM and redline RPM. Below the idle, the engine does not run stably but stalls, without being able to provide any usable output. On the other side, running the engine beyond the redline speed may cause excessive damage to the engine.

The shape of the torque curve in the operation range defined by the idle and redline is characteristic of the IC engine, depending on its design, fuel injection method, control, and calibration. As an example, the torque curve in Figure 1.3 has a local peak at around 2500 RPM and the maximum engine torque occurs around 4800 RPM. The engine power increases from the idle point almost linearly up to a peak at around 6100 RPM.

In general, the torque curves for naturally aspirated engines can be categorized as rising and buffalo shaped [5], while for turbo-charged engines, the torque curves are flat from a certain low RPM up to a relatively high RPM, as shown in Figure 1.4. Using turbo technology, the maximum output torque can be increased by more than 50% for the same engine displacement. To make things better, this maximum torque becomes available at a much lower RPM in comparison to naturally aspirated engines and stays flat up to a high RPM. This provides the vehicle with much better acceleration performance, especially at low vehicle speed. Turbo engines with small displacements provide outputs in torque and power equivalent to those of naturally aspirated engines of much larger displacements but consumes less fuel. Because of these advantages, vehicles powered by turbo engines are increasingly popular and represent the trend in the automotive industry.

1.2.2 Engine Fuel Map

Engine fuel efficiency is a top performance specification in today's automotive industry. The fuel consumption data of internal combustion engines are indispensable for the design, operation and control of vehicle powertrain systems. These data are experimentally obtained by intensive engine mapping and are usually provided for a production

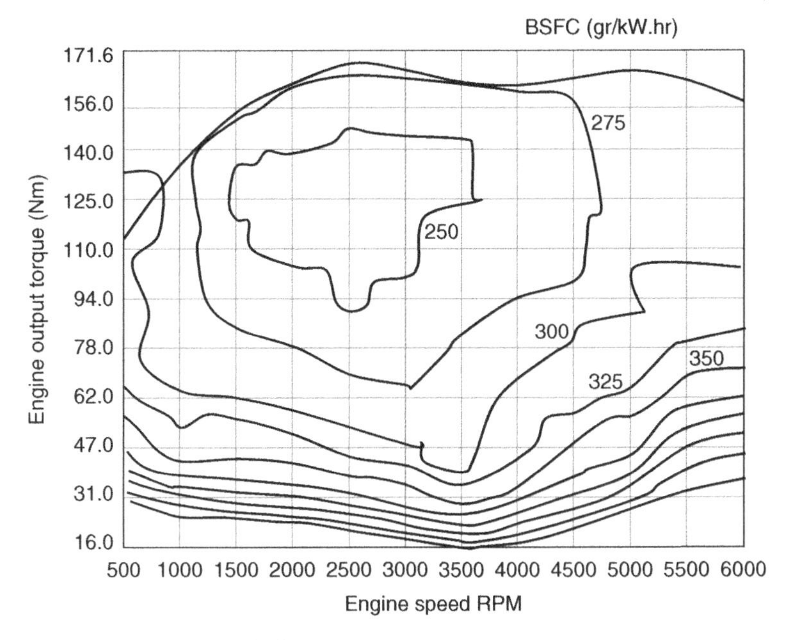

Figure 1.5 Engine specific fuel consumption map.

engine as a fuel map, which indicates in contour plots the specific fuel consumption of the engine at a given operation status, as shown in Figure 1.5. The specific fuel consumption is the amount of fuel that the engine needs to burn in order to do one horsepower hour of work, either in litres, grams, or pounds of weight.

The horizontal and vertical axes in Figure 1.5 are respectively the engine RPM and the engine output torque that define the engine operation status. The numbers by the contours are the specific fuel consumption in gram per kilowatt-hour (grams of fuel the engine consumes for it to do one kilowatt-hour of work). For example, if the engine runs stably at 4000 RPM with an output torque of 94 Nm, the specific fuel consumption is 275 gr/kW.hr. At the operation status defined by the RPM and the torque, the engine power is 39.37 kW. If the engine runs at the status continuously for one hour, the fuel consumed by the engine will then be 10.82 kg. The engine fuel consumption along a contour is the same even though the operation status is different, so the hourly fuel consumption of the engine is also 10.82 kg if it runs stably at 4740 RPM with an output torque of 125 Nm. Apparently, the engine will be more fuel efficient if it runs along the contour with 250 gr/kW.hr. It can also be observed that the most fuel efficient operation status is near the point with 2500 RPM and 125 Nm. When a vehicle is driven on a road at constant speed, the engine operation status depends on the road load, the vehicle speed, and the transmission gear ratio, the last determining the engine RPM and torque at a given vehicle speed and thus largely affects the vehicle fuel economy.

1.2.3 Engine Emission Map

When internal combustion engines generate power to propel ground vehicles, unwanted pollutants, harmful to the environment and to human health, are emitted in the process of combustion. These pollutants include CO, CO_2, NO_X, and other harmful gasses or particulate matter. The standard on emission control is increasingly stringent in the automotive industry due to environmental and human health concerns. Engine technologies, especially the technologies for combustion control and after-combustion treatment, are the key to minimizing the emission of pollutants. Transmissions also contribute to lowering vehicle emission levels by keeping the engine running in more efficient and less polluting operating ranges.

Engine emission maps are even more difficult to obtain than fuel maps because the quantity of a pollutant under various operation conditions is hard to measure. Computer simulation can be used to analyse engine emissions, but reliable emission data can only be obtained experimentally through extensive tests. Engine emission maps are provided for a given engine in formats similar to engine fuel maps. The specific quantity of a particular pollutant emitted by the engine is interpolated from the emission contour for a given engine operation status. Using the emission maps of the engine, the amount of emission of a pollutant can be simulated for a specified drive range.

1.3 Road Load, Driving Force, and Acceleration

Various forces are applied to a vehicle when it travels on a road surface. These forces include gravity, wheel–road contacting forces, road load, and driving force, which is also termed traction. Road load is against the motion of the vehicle, while traction force, or driving force, propels the motion of the vehicle. The driving force of a vehicle originates from the engine via the transmission and is fundamentally limited by the road traction limit. The total road load is the resultant of three separate road loads: rolling resistance, grade load, and air resistance. Figure 1.6 is the free body diagram of a vehicle of weight W that is being accelerated uphill.

In the free body diagram, v and a are the vehicle speed and acceleration respectively. R_A is the air drag or air resistance. The air drag is a distributed load, but for simplicity, it is assumed to be a point load acting at height h_A. R_F and R_R are the rolling resistance from

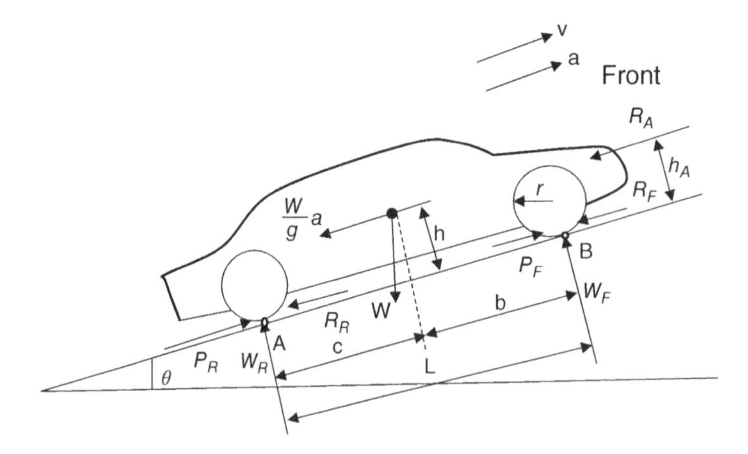

Figure 1.6 Free body diagram of a vehicle accelerated uphill.

the front and rear wheels respectively. P_F and P_R are the driving force from the front and rear wheels respectively. W_F and W_R are the axle loads, which are respectively the contact force between the front wheels and the road surface and between the rear wheels and the road surface. A and B denote the points of contact between the wheels and the road surface. θ is the grade angle of the slope and r is the rolling radius of the tire. The height of the center of gravity and the height of the air resistance are respectively denoted as h and h_A. For passenger cars, these two heights are assumed to be the same. The vehicle wheelbase is L and the longitudinal position of the gravity center is determined by b and c, which is the distance from the gravity center to the front axle and rear axle respectively. Unless otherwise stated, the US customary unit system will be used in the equations, where forces are in pounds, linear dimensions are in feet, speed is in ft/s, and acceleration is in ft/s^2. It should be noted that the inertia force $\gamma^W/_g\, a$ in the free body diagram is in the opposite direction to the acceleration, based on the D'Alembert's principle. γ is the equivalent mass factor that is introduced to account for the mass moments of inertia of all rotational components in the powertrain, including transmission input and output shafts, gears in the power flow path, drive shaft, differential, wheels, etc. The value of γ can be accurately determined based on the total vehicle equivalent kinetic energy as follows,

$$\frac{1}{2}\gamma\frac{W}{g}v^2 = \frac{1}{2}\frac{W}{g}v^2 + \sum_{i=1}^{n}\frac{1}{2}J_i\omega_i^{\,2} \tag{1.1}$$

In the equation above, J_i is the mass moment of inertia of each rotational component and n is the total number of rotational components in the powertrain. The equivalent mass factor is then determined as,

$$\gamma = 1 + \sum_{i=1}^{n}\frac{gJ_i}{W}\left(\frac{\omega_i}{v}\right)^2 \tag{1.2}$$

For a given vehicle, the ratio $\frac{\omega_i}{v}$ is a constant for each rotational component that depends on the transmission gear ratios. Empirical formula and tables are available for the approximation of the equivalent mass factor [6]. For passenger cars, the value of γ is small and can be considered to be equal to one for vehicle acceleration analysis and transmission ratio selections.

1.3.1 Axle Loads

The forces in the free body diagram (Figure 1.6) form a system of equilibrium, and three scalar equations can be written based on the condition of equilibrium. As shown below, the first two equations are based on the conditions that the sum of moments made by all forces about point A and point B must be equal to zero. The third equation is that the sum of all forces, including the inertia force, must be equal to zero in the direction of vehicle motion.

$$\begin{aligned} \sum M_A &= 0 \\ \sum M_B &= 0 \\ \sum F &= 0 \end{aligned} \tag{1.3}$$

These equations can be arranged to express the axle loads and the inertia force as follows:

$$W_F = \frac{1}{L}\left(Wc\cos\theta - R_A h_A - \gamma\frac{W}{g}ah - Wh\sin\theta\right) \tag{1.4}$$

$$W_R = \frac{1}{L}\left(Wb\cos\theta + R_A h_A + \gamma\frac{W}{g}ah + Wh\sin\theta\right) \tag{1.5}$$

$$\gamma\frac{W}{g}a = P_F + P_R - R_F - R_R - R_A - W\sin\theta \tag{1.6}$$

The first two equations determine the dynamic axle weights for the vehicle. During acceleration, there is a weight transfer equal to the magnitude of the inertia force from the front axle to the rear axle, as shown in Eqs (1.4) and (1.5). The static axle weights on level ground are obtained from the equations by making the slope angle θ, the air drag R_A, and the acceleration a equal to zero. It should be noted that tractions are available from both front and rear wheels only for a four wheel drive vehicle. P_R is zero for front wheel drive and P_F is equal to zero for rear wheel drive. Total driving force and rolling resistance from both front wheels and rear wheels are:

$$\begin{aligned} P &= P_F + P_R \\ R &= R_F + R_R \end{aligned} \tag{1.7}$$

The rolling resistance depends on many factors, such as tire material, texture, tread, inflation, speed, etc. Accurate calculation of the rolling resistance is very difficult, indeed impractical. For simplicity, it is common practice in the automotive industry to calculate the rolling resistance by:

$$R = fW\cos\theta \approx fW \tag{1.8}$$

where f is the rolling resistance coefficient and is approximately equal to 0.02. By rearranging Eqs (1.4–1.7) with the assumption that $h \approx h_A$, the axle loads can be solved in the following form:

$$P - fW = \gamma\frac{W}{g}a + R_A + W\sin\theta \tag{1.9}$$

$$W_F = \frac{Wc}{L} - \frac{h}{L}(P - fW) \tag{1.10}$$

$$W_R = \frac{Wb}{L} + \frac{h}{L}(P - fW) \tag{1.11}$$

where $\frac{c}{L}$ and $\frac{b}{L}$ are the weight distribution factors. The term $\frac{h}{L}(P - fW)$ is the dynamic weight transfer. Eqs (1.10) and (1.11) represent the dynamic axle weights in terms of the static axle weights and the weight transfer. The dynamic axle weight on the driving axle determines the maximum traction available for the vehicle under a given road condition.

1.3.2 Road Loads

There are three kinds of road loads that are against vehicle motion when the vehicle travels on a road surface: rolling resistance, air drag, and grade load, as shown in Figure 1.6.

The rolling resistance is calculated by Eq. (1.8). The grade load is the component of gravity on the slope direction and is equal to $W \sin \theta$. At level ground, only rolling resistance and air drag exist. At high vehicle speed, the air drag becomes more significant than the rolling resistance.

There are two causes for the generation of air resistance: friction between the air and the vehicle body surface; and air turbulence formed around the vehicle body [6]. The latter is the main cause of air drag for ground vehicles. Factors affecting the magnitude of the air drag include the shape and finish of vehicle body, the vehicle frontal projected area, air density and atmospheric condition, and most importantly, the vehicle's speed. It is very challenging to exactly determine the air drag by analytical means. In the standard of the Society of Automotive Engineers (SAE), the air resistance or air drag is calculated by the following formulation [6]:

$$R_A = 0.26 C_D A \left(\frac{v}{10}\right)^2 \tag{1.12}$$

where C_D is the unit less air drag coefficient that mainly depends on vehicle body shape and body surface smoothness. The air drag coefficient can be determined with high accuracy by wind tunnel testing. Modern passenger cars with streamlined body can have an air drag coefficient as low as 0.26. A is the vehicle frontal projected area in ft^2 that mainly depends on the vehicle size. This is the area of the vehicle body that confronts the air flow in the direction perpendicular to vehicle motion. To determine this area, a flat board can be held perpendicular to the road surface behind the parked vehicle and a flashlight is then used to beam the vehicle body horizontally in front of the vehicle. The area of the shadow casted on the board is the frontal projected area. As shown in the formulation, the air drag is proportional to the square of the vehicle speed v relative to the wind. With the speed v in mph, the formulation determines the air drag as a force in pounds. For the analysis and calculations of vehicle dynamics, the vehicle speed and acceleration are often in ft/s and ft/s^2, then the formulation for air drag will be used in the following form:

$$R_A = 0.00118 C_D A v^2 \tag{1.13}$$

The equation above is directly transformed from Eq. (1.12) by considering that one mph is equal to 1.467 ft/s. The resultant road is the sum of the rolling resistance, grade load and air drag, as expressed below,

$$R = 0.00118 C_D A v^2 + fW + W \sin\theta \tag{1.14}$$

1.3.3 Powertrain Kinematics and Traction

There are various layouts for vehicle powertrain systems. In this section, the powertrain of a rear wheel drive vehicle with a manual transmission (MT), in the layout shown in Figure 1.7, is used as the example for demonstration. The analysis on powertrain kinematics and related equations derived in the example are applicable to all other powertrain layouts.

The engine output torque and output angular velocity are denoted as T_e and ω_e respectively, the transmission output torque and angular velocity are denoted as T_t and ω_t, the

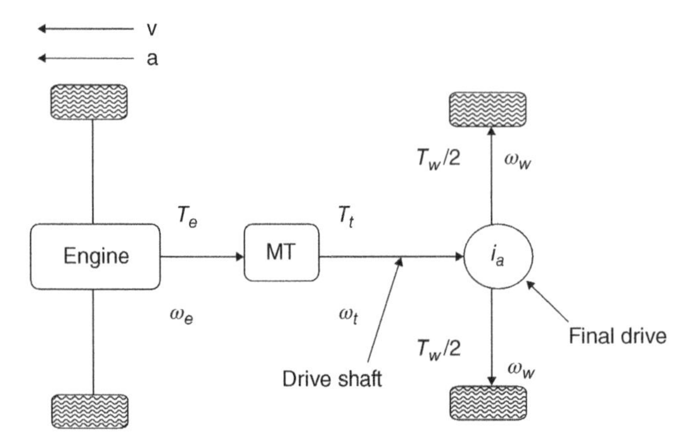

Figure 1.7 Layout of RWD manual transmission powertrain.

angular velocity of the driving wheel is denoted as ω_w, and the torque on each of two wheels on the driving axle is denoted as $\frac{T_w}{2}$. As shown in Figure 1.7, the engine output is transmitted through the transmission to the drive shaft, which then transmits the transmission output to the final drive. The final drive assembly contains a pair of spiral bevel or hypoid gears that multiplies the transmission output torque by the final drive ratio i_a and transmits the rotation of the drive shaft to the wheels between the two perpendicular axes. The outputs of the engine and the transmission are related by the following equation:

$$T_t = \eta_t i_t T_e \tag{1.15}$$

$$\omega_t = \frac{\omega_e}{i_t} \tag{1.16}$$

where i_t is the transmission ratio defined as the division of the input angular velocity by the output angular velocity and η_t is the transmission efficiency. The transmission ratio is a stepped variable for manual transmissions. For a five speed transmission, the ratio varies between the highest value in the first gear and the lowest value in the fifth gear. Note that the accurate determination of the transmission efficiency is experimental by nature, and the efficiency is assumed to be a constant value throughout the text of this book.

The transmission output is further transmitted by the final drive to the driving wheels. The torque on the two driving wheels and the angular velocity of the driving wheels are related to the engine output torque and engine angular velocity by the following equations:

$$T_w = \eta_a i_a T_t = \eta_a \eta_t i_a i_t T_e \tag{1.17}$$

$$\omega_w = \frac{\omega_t}{i_a} = \frac{\omega_e}{i_a i_t} \tag{1.18}$$

where i_a is the final drive ratio and η_a is the final drive efficiency. When a vehicle travels on road, there is always a small amount of slippage between the tire and the road surface. The amount of tire spillage is determined by the tire slip rate defined as:

$$\delta = \frac{\omega r - v}{\omega r} \tag{1.19}$$

Under normal driving conditions, the slip rate is very small ($\delta \leq 0.02$) and can be neglected for the selection of gear ratios in engine–transmission matching. If the slippage is not considered, then vehicle motion and driving wheel rotation are related as,

$$\begin{aligned} \omega_w &= \frac{v}{r} \\ \alpha_w &= \frac{a}{r} \end{aligned} \tag{1.20}$$

where α_w is the angular acceleration of the driving wheels. Based on Eqs (1.17), (1.18), and (1.20), the engine output torque, the engine angular velocity, the torque on the driving wheels, and the vehicle speed are related by the following equations,

$$\begin{cases} T_w = \eta i_a i_t T_e \\ \omega_e = i_a i_t \dfrac{v}{r} \end{cases} \tag{1.21}$$

where η is the overall powertrain efficiency. The second of these equations relates the engine angular velocity and the vehicle speed without considering tire slippage. Knowing the torque on the driving wheels, the driving force that originates from the engine and propels the vehicle is then determined by the following equation for the rear wheel drive (RWD) vehicle in the example:

$$P = P_R = \frac{\eta i_a i_t T_e}{r} \tag{1.22}$$

The maximum value of the driving force for ground vehicles driven by wheels is limited by road–tire contact conditions and is commonly termed as the traction limit. For the RWD vehicle in question, the traction is limited by the following inequality:

$$P = P_R \leq \mu W_R \tag{1.23}$$

where μ is the so-called traction coefficient, which depends on the road condition and the tire properties. On a standard highway surface (road surface with skid number 81), the value of the traction coefficient is equal to 0.81. The traction force or driving force available from the powertrain cannot exceed the traction limit, or the driving wheels will slip excessively. W_R is the dynamic rear axle load, which is determined by Eq. (1.11). It is emphasized here that Eqs (1.15–1.23) are derived for the RWD vehicle in the example, and are applicable for all other types of powertrain systems as mentioned previously. With the driving force determined by Eq. (1.22), the vehicle acceleration can then be determined from Eq. (1.16) by:

$$a = \frac{P - (fW + W\sin\theta + 0.00118 C_D A v^2)}{\gamma \frac{W}{g}} \tag{1.24}$$

Eq. (1.9), or Eq. (1.24), is actually the equation of motion of the vehicle when it runs on a straight path. The vehicle acceleration can be calculated by Eq. (1.24) at any given vehicle speed if the engine throttle opening and the road condition are provided. The equivalent mass factor γ in Eq. (1.24) is approximately equal to one for passenger vehicles. The following example demonstrates how the equation series is used for the calculation of vehicle acceleration and fuel economy.

Example 1.1 A manual transmission used for a vehicle with data given below has six forward speeds and the gear ratios are: 1st gear (3.72), 2nd gear (2.31), 3rd gear (1.51), 4th gear (1.07), 5th gear (0.81), 6th gear (0.63). The engine WOT output is as given in Figure 1.3 and the fuel map is as given in Figure 1.5. The vehicle has the following data:

Front axle weight: 1750 lb	Rear axle weight: 1550 lb
Center of gravity height: 17 in.	Wheelbase: 104 in.
Air drag coefficient: 0.32	Frontal projected area: 22.90 sq.ft
Tire radius: 11.70 in.	Roll resistance coefficient: 0.018
Powertrain efficiency: 0.94	Traction coefficient: 1.0
Max. power @6000 RPM: 138 HP	Max. torque @4500 RPM: 132 ft.lb

a) The engine RPM drops by 662 (RPM) when a 4–5 upshift is made at a vehicle speed of 45 mph. Determine the final drive ratio of the vehicle.
b) Determine the engine torque and the engine power when the vehicle is cruising at a constant speed of 65 mph on level ground in the 6th and 5th gears respectively.
c) Determine the fuel economy in mpg or litres per 100 km for the conditions in (b).
d) Determine the maximum acceleration that the vehicle can achieve in 4th gear at a speed of 65 mph on a 2% slope.

Solution:

a) $$v = 45\,\text{mph} = 1.46(45) = 65.7\,\text{ft/s}$$

$$\omega = \frac{v}{r} = \frac{65.7}{11.70/12} = 67.38\,\text{rad/s} = 643.5\,\text{RPM}$$

The engine RPM drops by 662 in a 4–5 upshift, so $i_4 i_a \omega_w - i_5 i_a \omega_w = 662$, and:

$$i_a = \frac{662}{\omega_w(i_4 - i_5)} = \frac{662}{643.5(1.07 - 0.81)} = 3.96$$

b) $$v = 65\,\text{mph} = 65(1.46) = 94.9\,\text{ft/s} = 104\,\text{km/h}$$

$$R = fW + 0.00118 c_D A v^2 = 0.018(1750 + 1550) + 0.00118(.32)(22.9)(94.9)^2 = 137.3\,\text{lb}$$

$$\omega_w = \frac{v}{r} = \frac{94.9}{11.7/12} = 97.33\,\text{rad/s}$$

Since the vehicle speed is constant, engine power is the same for both gears and is equal to:

$$\frac{Rv}{\eta} = \frac{137.3(94.9)}{.94} = 13861.6\,\text{ft.lb/s} = \frac{13861.6}{550} = 25.2\,\text{HP} = 18.8\,\text{kW}$$

Engine torque depends on the gear ratio and is calculated respectively for the 6th and 5th gears,

$$\text{6th gear}: T_e^{(6)} = \frac{Rr}{\eta i_6 i_a} = \frac{137.3(11.7/12)}{.94(.63)(3.96)} = 57.1\,\text{ft.lb} = 76.5\,\text{Nm}$$

$$\text{5th gear}: T_e^{(5)} = \frac{i_6}{i_5} T_e^{(6)} = \frac{.63}{.81}(57.1) = 44.4\,\text{ft.lb} = 59.5\,\text{Nm}$$

c) The angular velocity of the engine for the condition in question (b) is respectively,

$$\omega_e^6 = i_6 i_a \omega_w = (.63)(3.96)(97.33) = 242.8 \text{ rad/s} = 2319 \text{ RPM}$$

$$\omega_e^5 = i_5 i_a \omega_w = (.81)(3.96)(97.33) = 312.2 \text{ rad/s} = 2981 \text{ RPM}$$

So the engine operation status is defined respectively as (2319, 76.5) and (2981, 59.5). From the fuel map (Figure 1.5), the specific fuel consumption is 275 gr/(kW.hr) and 285 gr/(kW.hr) respectively. The fuel consumed per hour can then be calculated as:

$$\text{6th gear}: 18.8(275) = 5170 \text{ gram} = 6.89 \text{ litres};$$

$$\text{5th gear}: 18.8(285) = 5358 \text{ gram} = 7.14 \text{ litres}$$

Fuel consumption in litres per 100 km:

$$\text{6th gear}: \frac{100(6.89)}{104} = 6.63; \text{5th gear}: \frac{100(7.14)}{104} = 6.87$$

Fuel consumption in mpg:

$$\text{6th gear}: \frac{65}{5.17/3.03} = 38 \text{ mpg}; \text{5th gear} = \frac{65}{5.36/3.03} = 36.7 \text{ mpg}$$

d) $$\omega_e = i_4 i_a \omega_w = 1.07(3.96)(97.73) = 412.41 \, {}^{\text{rad}}/_{\text{s}} = 3938 \text{ RPM}$$

At 3938 RPM, the WOT engine torque T_e is found from Figure 1.3 and is equal to 129 ft.lb.

$$P = \frac{\eta i_4 i_a T_e}{r} = \frac{0.94(1.07)(3.96)(129)}{11.7/12} = 526.98 \text{ lb}$$

$$R = fw + 0.00118 c_D A v^2 + 0.02 \text{ W} = 137.366 + 203.3 \text{ lb}$$

$$a = \frac{P - R}{W/g} = \frac{526.98 - 203.3}{3300/32.2} = 3.16 \text{ ft/s}^2$$

1.3.4 Driving Condition Diagram

As can be observed from Eqs (1.21) and (1.22), the driving force available from the engine at a given throttle position can be calculated for a given vehicle speed. This is because the engine angular velocity is related to the vehicle speed via the transmission and final drive ratios, and the engine output torque is a function of engine speed at a given throttle. Apparently, the capacity propulsive force available from the engine is obtained when the engine is operating at wide open throttle (WOT) and this capacity driving force is a function of vehicle speed as determined by Eqs (1.21) and (1.22). Since the transmission has stepped gear ratios, each gear of the transmission provides a function or relationship between the engine torque and the vehicle speed. Similarly, the road load is also a function of vehicle speed, as shown in Eq. (1.14). Therefore, the net force for acceleration at engine capacity can be found by subtracting the road load from the driving force available under WOT condition at any vehicle speed. The driving force and road load can be plotted over the speed range of the vehicle in the so-called driving condition diagram, as shown in Figure 1.8. In this figure, the horizontal axis is for the vehicle speed and the vertical axis is for the driving force and road load. In the US customary unit, the driving force and road load are in pounds, and the vehicle speed is in ft/s, while in International standard, the traction (i.e. driving force) and road load are in N and the vehicle speed is in m/s.

The driving condition diagram in Figure 1.8 is for an example vehicle equipped with a five-speed manual transmission. There are five separate traction curves, one for each gear. There must be a certain amount of slippage between the engine output and transmission input at vehicle launch because the engine cannot provide output torque below the idle RPM. Each driving force curve or traction curve covers the vehicle speed range corresponding to the engine speed range from idle to redline. For example, the vehicle speeds at the starting point and end point on the first gear driving force or traction curve are determined as:

$$\begin{aligned} v_{1i} &= \frac{\pi\, n_e^{idle} i_a i_1 r}{30} \\ v_{1r} &= \frac{\pi\, n_e^{redline} i_a i_1 r}{30} \end{aligned} \tag{1.25}$$

where v_{1i} and v_{1r} are the vehicle speed in the first gear corresponding to the engine idle speed n_e^{idle} and redline speed $n_e^{redline}$ respectively. For any other point along the first gear traction curve, we can first divide the interval $[v_{1i}, v_{1r}]$ evenly and then pick up a vehicle speed from the interval. This speed is then used to determine the engine angular velocity from Eq. (1.21). Knowing the engine angular velocity, the engine torque can then be found from the engine WOT torque output. The driving force is finally determined by using Eq. (1.22).

In the driving condition diagram, the road load is plotted against vehicle speed on different grades, starting from level ground. The road load curves on different grades are parallel parabolic curves. In the SAE standard, the steepness of a slope is expressed by the grade percentage defined as follows:

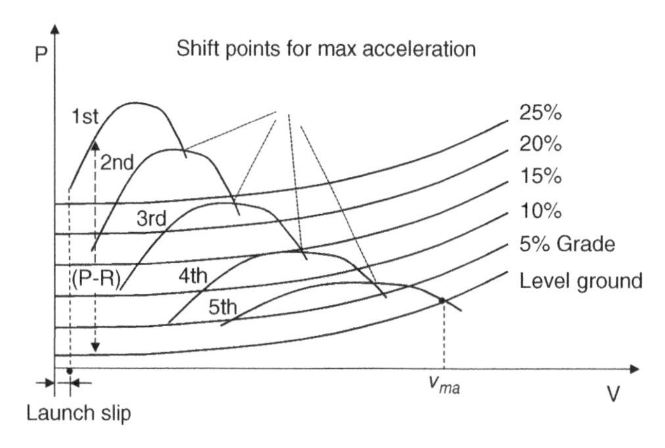

Figure 1.8 Driving condition diagram.

$$\frac{G}{100} = \tan\theta \approx \sin\theta \tag{1.26}$$

where θ is the slope angle and G is the grade number. If the slope angle θ is small, the slope percentage can be approximated as $\sin\theta$. Thus the grade load is often calculated as $\frac{G}{100}W$.

The following data are required to complete the driving condition diagram for a vehicle: engine output data, the efficiency and ratios of the transmission and final drive, vehicle data such as weight, tire rolling radius, frontal projected area and air drag coefficient. The driving condition diagram provides graphically technical data on vehicle dynamic performance. For example, performance related data in the following list can be obtained by observing the driving condition diagram:

- Net driving force available for vehicle acceleration in different gears and on different grades over the whole vehicle speed range. This net force is the difference between the traction curve and the road load curve.
- Maximum vehicle speed based on powertrain capacity. As shown in Figure 1.8, the maximum speed on level ground occurs at the intersection between the traction curve of the fifth gear and the level ground road load curve for the vehicle in the example. It is also possible to find the maximum vehicle speed on other grades and to find the gear at which the maximum speed is achieved.
- Shifting points for maximum vehicle acceleration. To reach the maximum vehicle speed from standstill in the shortest time, the driver must make gear shifts at points that give the largest net driving force for acceleration. For the driving condition diagram shown in Figure 1.8, the shift points should be at the points of intersection of the two adjacent traction curves. If the traction curves do not intersect, then the shift point for maximum acceleration should be at the redline point.
- Grade on which the vehicle can be started at a given gear. Theoretically, the vehicle can be started on a grade as long as the traction curve is above the road load curve. The higher the traction curve above the road load curve, the easier it is to start the vehicle.

The list above shows some of the vehicle performance related data that can be directly obtained from the driving condition diagram. It should be pointed out that the driving condition diagram can be stored as a computer data file for easy reference or interpolation.

1.3.5 Ideal Transmission

The ideal transmission is the one that can vary the gear ratio continuously and is therefore called continuously variable transmission (CVT). Since the transmission ratio can be varied continuously, the engine speed can be controlled at the value that optimizes a selected objective at any given vehicle speed. This optimized objective can be the best fuel economy or the maximum power available from the powertrain. If the maximum power is the objective, then the CVT gear ratio is controlled to keep the engine speed corresponding to the peak power for the whole vehicle speed range. This leads to the following equation for the traction curve of CVT:

$$Pv = \eta(Power)_{max} = C \tag{1.27}$$

where $(Power)_{max}$ is the engine maximum output power, η is the efficiency of the whole powertrain including the CVT and the final drive, and C is a constant. It should be noted here that the efficiency of the CVT itself is actually lower than that of a manual stepped ratio transmission due to the higher friction loss. The fuel economy advantage of CVT is from its capability that the engine operation status can be always controlled at the most efficient range. As shown in Figure 1.9, Eq. (1.27) represents a hyperbola on the driving condition diagram for the traction curve of a CVT powertrain. The hyperbolic traction curve of a CVT powertrain is the envelope to the piecewise traction curves of a stepped ratio transmission. If the same overall powertrain efficiency is assumed, the hyperbola will touch the piecewise traction curves at the point corresponding to the maximum engine power. Several important observations can be made from Figure 1.9:

- For a stepped ratio transmission, the maximum engine power can only be useful at one vehicle speed in each gear. There are always losses of engine power potential due to the mismatch between the vehicle speed and engine status.
- The more gears a stepped ratio transmission has, the more the piecewise traction curves and the closer these curves get to the hyperbola (the ideal CVT traction curve). Generally, more transmission gears provide better matching to the engine and make the engine output power available over a wider vehicle speed range. For example, a semi-truck may have as many as 16 gears in the transmission to fully utilize the engine power potential over a wide speed range.
- Similarly, more transmission gears provide better engine matching in terms of fuel economy because they allow the engine to run closer to the status corresponding to the most fuel efficient range.
- Figure 1.9 provides a judgment on how well the transmission matches the engine. A good match must have those piecewise traction curves closely enveloped by the hyperbola.

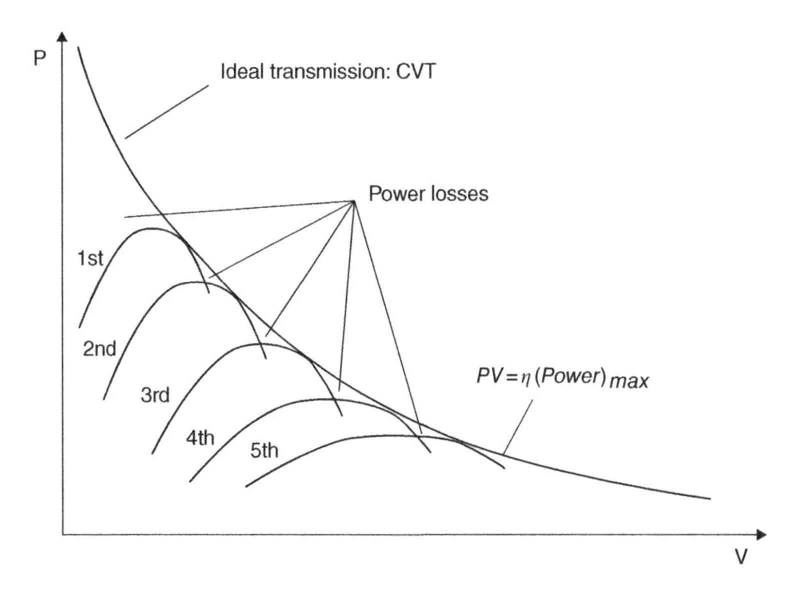

Figure 1.9 Traction curve of the ideal transmission.

A complete driving condition diagram that includes the CVT traction curve provides important technical data on vehicle performance as described previously. In addition, it tells us, to a certain degree, how the transmission ratios match the engine output.

1.3.6 Power–Speed Chart

When a vehicle is driven on a road surface, the power available from the powertrain, $(Power)_A$, and the power of the road load, $(Power)_R$, are respectively determined by the following equations:

$$(Power)_A = Pv \tag{1.28}$$

$$(Power)_R = Rv \tag{1.29}$$

where the driving force P and the road load R are respectively represented by Eqs (1.22) and (1.14). Obviously, the power available from the powertrain must be higher than the road load power for acceleration, and the two powers are balanced when the vehicle cruises at constant vehicle speed. In general, vehicle power requirements depend on its operation conditions. In a power–speed chart, the power available from the powertrain and the road load power to be overcome are both plotted against the vehicle speed, as shown in Figure 1.10. In this chart, the horizontal axis is the vehicle speed, in either ft/s or m/s, depending on the units used. The vertical axis represents the power in horsepower or kilowatts available from the powertrain in each gear, the maximum engine power, and the road load power. The power available from the powertrain is plotted for each gear separately and the road load power is plotted on different grades. The maximum engine power is a horizontal line in the power–speed chart and is realized only by the ideal transmission (CVT) for all vehicle speeds. While the driving condition diagram tells the net driving force for vehicle acceleration, the power–speed chart provides a graphic quantification on the net power reserve under all vehicle operation conditions. The combination of the driving condition diagram and the power–speed chart gives a complete picture of the availability of driving force and power for acceleration or

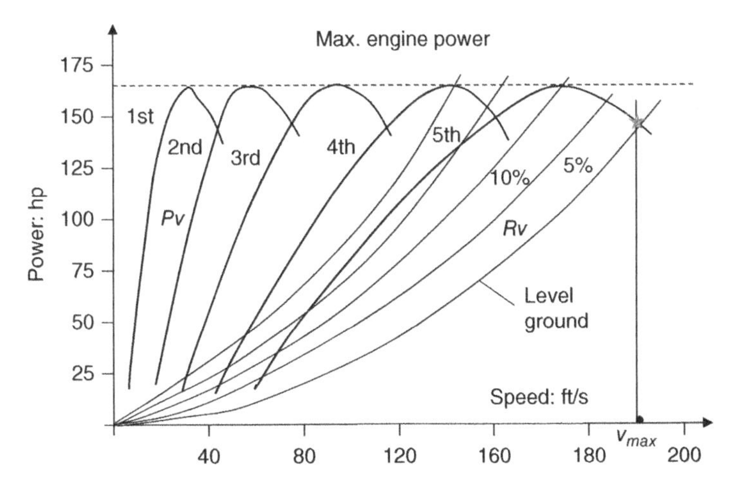

Figure 1.10 Power–speed chart.

gradability. For example, the maximum vehicle speed on level ground can also be found on the power–speed chart. As shown in Figure 1.10, the power curve for the highest gear intersects with the road load power curve on level ground. The maximum vehicle speed is achieved at this intersection because no extra power from the powertrain would be available to overcome the additional road power resulting from any further speed increase.

1.4 Selection of Gear Ratios

The principal considerations for the selection of transmission gear ratios are vehicle performance and fuel economy. The process of gear ratio selection is trial and error by nature, in which experience and data on previous vehicles play an important role. There is no closed form formula that would provide the precise values for the gear ratios that match the engine output for the best results in dynamic performance and fuel efficiency. However, approximate gear ratio values can be calculated analytically by equations derived in this section. These approximate ratios can be used as the starting values for the finalization of transmission ratios.

1.4.1 Highest Gear Ratio

The highest gear ratio is that for the top gear and has the lowest value. This ratio is usually selected such that the vehicle will achieve the maximum speed on level ground that is allowed by the maximum engine power. As discussed previously, vehicle maximum speed occurs at the intersection between the available power curve for the highest gear and the road load power curve on level ground (Figure 1.10). There exists a unique gear ratio for the highest gear such that the available power curve defined by this ratio, the line of maximum power, and the road load power curve on level ground intersect at the same point, which is point H as shown in Figure 1.11. Obviously, the vehicle speed corresponding to point H is the very maximum vehicle speed as allowed by the engine capacity power. At point H, the maximum engine power is fully matched by the highest gear ratio to balance the road load power at the maximum vehicle speed.

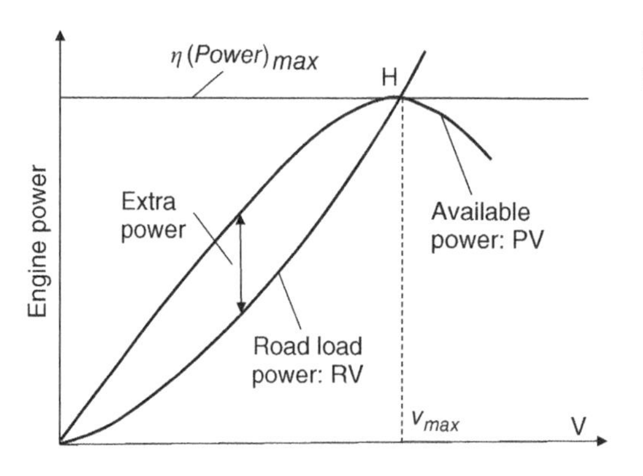

Figure 1.11 Matching maximum engine power for maximum vehicle speed.

As shown in Figure 1.3, the maximum engine power only occurs at a unique engine speed, and this speed corresponds to the maximum vehicle speed calculated as follows:

$$v_{max} = \frac{\pi n_{maxp}}{30(i_{th}i_a)} r \tag{1.30}$$

where n_{maxp} is the engine RPM at which the engine output power is the maximum, r is the tire rolling radius, and $(i_{th}i_a)$ is the overall powertrain ratio. Since the maximum engine power is balanced by the road load power, the following equation becomes apparent,

$$Rv_{max} = \eta(Power)_{max} \tag{1.31}$$

where R is the road load determined by Eq. (1.14) with the slope angle θ is equal to zero, η is the overall powertrain efficiency that is assumed to be a constant, and $(Power)_{max}$ is obtained directly from the WOT engine output plot. An equation of the following form can be derived by plugging Eqs (1.14) and (1.30) into Eq. (1.31):

$$\frac{\pi n_{maxp}}{30(i_{th}i_a)} r \left[0.00118 C_D A \left(\frac{\pi n_{maxp}}{30(i_{th}i_a)} r \right)^2 + fW \right] = \eta(Power)_{max} \tag{1.32}$$

This equation contains only one unknown $(i_{th}i_a)$, which is the overall powertrain ratio. Through some simple manipulations, this equation can be transformed to a cubic equation in terms of the unknown $(i_{th}i_a)$:

$$\eta(Power)_{max}(i_{th}i_a)^3 - fW \frac{\pi n_{maxp}}{30} r (i_{th}i_a)^2 - 0.00118 C_D A \left(\frac{\pi n_{maxp}}{30} r \right)^3 = 0 \tag{1.33}$$

For a given vehicle the only unknown in the equation above is overall powertrain ratio $(i_{th}i_a)$, which can be solved by a simple iteration procedure. The transmission ratio in top gear, i_{th}, can then be easily determined if the final drive ratio i_a is given.

1.4.2 First Gear Ratio

The first gear ratio has the largest value and is designed such that the vehicle will be capable of negotiating the theoretical maximum grade that is allowed by the maximum engine output torque. When the vehicle negotiates the maximum grade in the first gear at a constant low speed, the vehicle acceleration is zero and the air drag can be neglected in the calculation of road load. Thus, the road load on the maximum grade at low vehicle speed is approximated as:

$$R = fW\cos\theta_{max} + W\sin\theta_{max} \approx fW + \frac{WG_{max}}{100} \tag{1.34}$$

The equation for the road load above uses the designation for the grade defined in Eq. (1.26). For the vehicle to be able to negotiate the maximum grade, the driving force in the first gear when the engine torque is the maximum must satisfy the following inequality,

$$\frac{T_{emax}(i_{t1}i_a)\eta}{r} \geq fW + \frac{WG_{max}}{100} \tag{1.35}$$

Therefore,

$$(i_{t1}i_a) = \beta \frac{Wr\left(f + \frac{G_{max}}{100}\right)}{\eta T_{emax}} \tag{1.36}$$

where β is a reservation factor that should be larger than one. The engine maximum output torque T_{emax} is obtained directly from the engine WOT output torque plot (Figure 1.3). The powertrain based gradability is usually designed to reach the gradability allowed by the traction limit. That is to say, the value of $\frac{G_{max}}{100}$ in Eq. (1.36) can be determined based the traction condition when the vehicle is running uphill. If rolling resistance and air drag are neglected in calculation, then the traction based gradability for a RWD vehicle is limited by the following inequality:

$$W \sin\theta_{max} \leq \mu W_R = \mu \frac{W}{L}(b\cos\theta_{max} + h\sin\theta_{max}) \tag{1.37}$$

where W_R is the real axle weight determined by Eq. (1.5) when the acceleration and air resistance are zero, and μ is the traction coefficient. Solving Eq. (1.37) for $\tan\theta_{max}$ and using the definition for grade percentage in Eq. (1.26), the maximum traction based gradability for a RWD vehicle can be approximated by:

$$G_{max} = 100\tan\theta_{max} \approx (100)\frac{\mu\frac{b}{L}}{1-\mu\frac{h}{L}} \tag{1.38}$$

Finally, the overall first gear ratio for the RWD vehicle can be found by plugging Eq. (1.38) into Eq. (1.36). The first gear ratio thus determined enables the vehicle with the maximum gradability allowed by the engine torque capacity. Note that the maximum vehicle gradability $\frac{G_{max}}{100}$ is approximately equivalent to a vehicle acceleration of $\frac{G_{max}}{100}g$ on level ground. This can be observed from Eq. (1.24) by dropping the air drag and rolling resistance in the equation. As can be observed in Eq. (1.36), the value of the first gear ratio heavily depends on the vehicle weight and the engine maximum output torque. For vehicles that have a high ratio between total vehicle weight and engine output torque, such as a heavy duty truck, there needs to be a very large first gear ratio for gradability and acceleration capability during launch.

1.4.3 Intermediate Gear Ratios

The low and high gears define the transmission ratio range, and the gap between low and high gears must be bridged by a number of intermediate gears. These ratios affect the engine RPM range under various vehicle operation statuses. Generally, internal combustion engines have a certain RPM range within which the engine output torque is close to the maximum. The lowest specific fuel consumption usually falls within this range. For a given engine, the low and high bounds of this range can be selected from the engine output torque plot. Apparently, for the best results in acceleration performance and fuel efficiency, the intermediate gear ratios should be designed such that the engine RPM is kept within this range while the vehicle operates in different gears. As shown in Figure 1.12, the low bound and high bound of the RPM range are denoted by L and M respectively.

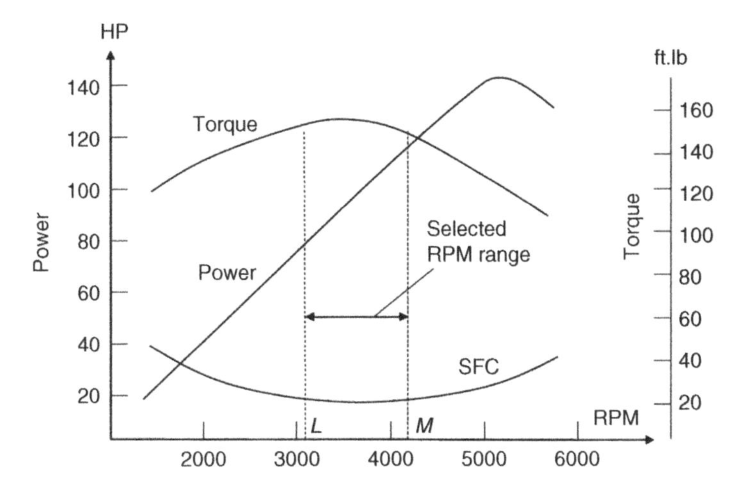

Figure 1.12 Engine RPM range.

The intermediate gear ratios are designed such that the engine RPM is maintained within the interval $[L, M]$ when the transmission shifts gears. Suppose the vehicle is being accelerated in the current gear, or a lower gear. As the vehicle speed increases, the engine speed also increases toward the high bound M. To keep the engine speed within the range, an upshift must be made when the high bound M is reached. Immediately after the upshift, the engine speed should be down to the low bound L, while the vehicle speed remains almost unchanged because the shift only lasts a short time. If the current (lower) overall gear ratio is i_L, then the vehicle speed corresponding to the high bound M is,

$$v = \frac{\pi M}{30 i_L} r \tag{1.39}$$

After the upshift, the engine speed drops to the low bound L, but the vehicle speed remains almost the same and is related to the next (higher) overall gear ratio i_H as:

$$v = \frac{\pi L}{30 i_H} r \tag{1.40}$$

These two equations result in the following relation for the current and the next overall powertrain ratios:

$$i_H = \frac{L}{M} i_L = c i_L \tag{1.41}$$

where c is a constant, since the lower and upper bounds are specified. After the upshift, the vehicle will be driven with the gear ratio i_H. As the vehicle speed increases, the engine speed will reach the high bound M again. Then an upshift is made again to keep the engine RPM within the range. By similar analysis, the gear ratio after the next upshift will be $c i_H$ or $c^2 i_L$. Thus it can be observed by deduction that the gear ratios should form a geometric progression if the engine RPM is to be kept within the speed range $[L, M]$. For example, if designed in a geometric progression, the gear ratios of a five speed manual transmission will be related by the following equation,

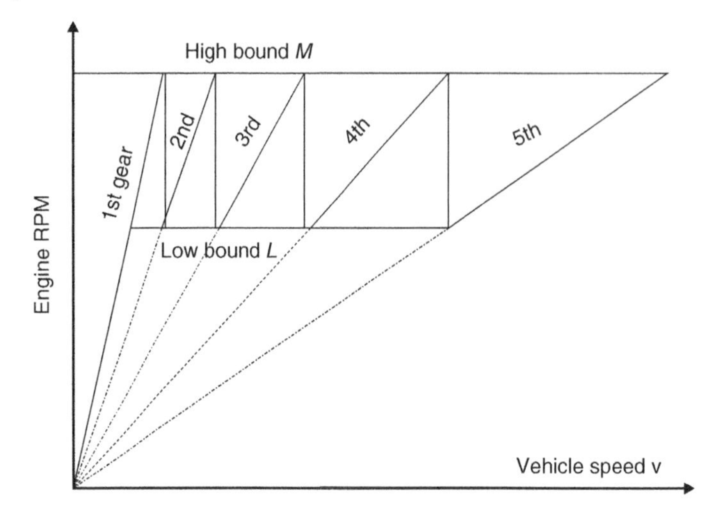

Figure 1.13 Engine RPM vs vehicle speed for gear ratios in geometric progression.

$$\frac{i_5}{i_4} = \frac{i_4}{i_3} = \frac{i_3}{i_2} = \frac{i_2}{i_1} = \frac{L}{M} = c \tag{1.42}$$

If all upshifts are made at the same engine RPM, then the engine RPM will drop by the same amount, as shown in Figure 1.13. Clearly, the engine speed is kept between the low bound and the high bound.

The number of gears needed to match the engine output largely depends on the shape of the engine output torque curve and on the type of vehicle. For an engine with sharply peaked torque curve, the ratio between the low bound and high bound $c = \frac{L}{M}$ is higher than that for an engine with a flatter torque curve. This means that more gear ratios are needed to bridge the gap between the low gear and the high gear. The higher the value of $\frac{L}{M}$, the more the transmission gears needed. In other words, the narrower the RPM range within which the engine is kept to operate, the more gear ratios are needed. If a relatively small engine is used as the power plant for a large vehicle, such as the case for a semi-truck, it is highly desirable to keep the engine RPM nearest to the peak output, thus the ratio $\frac{L}{M}$ is close to one and a large number of gears are required in the transmission.

Summarizing this analysis, the first gear ratio $i_{t1}i_a$ for the whole powertrain is designed for maximum gradability or acceleration, the highest gear ratio $i_{th}i_a$ is designed for the maximum vehicle speed allowed by the maximum engine power, and the intermediate gear ratios fall in a geometric progression such that the engine RPM will be kept within the same range for optimized engine performance in torque and fuel efficiency. The final drive ratio i_a is always a constant and is realized separately from the transmission. Therefore the first gear ratio i_{t1} and the highest gear ratio i_{th} of the transmission become known after the final drive ratio is selected. If the number of gears is pre-determined, then the intermediate gear ratios can be calculated by the following equations:

$$c = \sqrt[(n-1)]{\frac{i_{th}}{i_{t1}}} \tag{1.43}$$

$$i_{t2} = c i_{t1};\ i_{t3} = c^2 i_{t1}\ , \ldots\ldots, i_{t(n-1)} = c^{(n-2)} i_{t1} \tag{1.44}$$

where n is the number of gears in the transmission. If the engine RPM range is specified, that is, the value of $c = \frac{L}{M}$ is predetermined, then the number of gears can be calculated firstly by using:

$$n = 1 + \frac{\ln \frac{i_{th}}{i_{t1}}}{\ln c} \tag{1.45}$$

The answer from this equation must be rounded up to the nearest integer. Then Eqs (1.43) and (1.44) are used to calculate all the intermediate gear ratios. It can be observed from Eq. (1.45) that if the first gear ratio i_{t1} is large, as in the case of heavy duty truck, the value of n, i.e. the number of gears, will also be large. Similarly, if c gets closer to one, the number of gears will become larger, as mentioned previously.

1.4.4 Finalization of Gear Ratios

The gear ratio values calculated above may not be the final gear ratios used for the transmission, but these values serve as a good starting point in the transmission ratio selection. After these starting values are calculated, the driving condition diagram and the power–speed chart, Figures 1.8 and 1.10, can then be plotted for the vehicle to judge how good the engine–transmission match is. With the selected gear ratios, the vehicle acceleration performance and fuel economy can also be simulated under various drive ranges. Based on the simulation results, necessary modifications of the gear ratios can be made for priorities in acceleration performance or fuel economy or for the optimized trade-off between the two.

Note that, in addition to the geometric progression method described in this section, transmission gear ratios can also be designed in progressive steps [5]. When transmission gear ratios are in progressive steps, the difference between two adjacent gears becomes smaller toward the high gears. This means that the engine RPM will not drop the same amount during upshifts, as shown in Figure 1.13. Instead, the engine RPM drop decreases for upshifts in higher gears. Gear ratios for passenger car transmissions often have low gear ratios close to a geometric series and high gear ratios with the characteristics of progressive steps.

Example 1.2 A five-speed manual transmission is used for a FWD car with the following data:

Front axle weight: 1820 lb	Rear axle weight: 1700 lb
Center of gravity height: 14 in.	Wheelbase: 105 in.
Air drag coefficient: 0.31	Frontal projected area: 21.50 sq.ft
Tire radius: 12.0 in.	Roll resistance coefficient: 0.02
Powertrain efficiency: 0.94	Road adhesion coefficient: 1.0
Max. power @ 5500 RPM: 198 HP	Max. torque @3850 RPM: 199 ft.lb.

The torque output curve of the engine is given in Figure 1.14. Design the overall powertrain ratio of the highest gear for the highest vehicle speed and the ratio of the lowest gear for the maximum gradability, using a reservation factor $\beta = 1.35$.

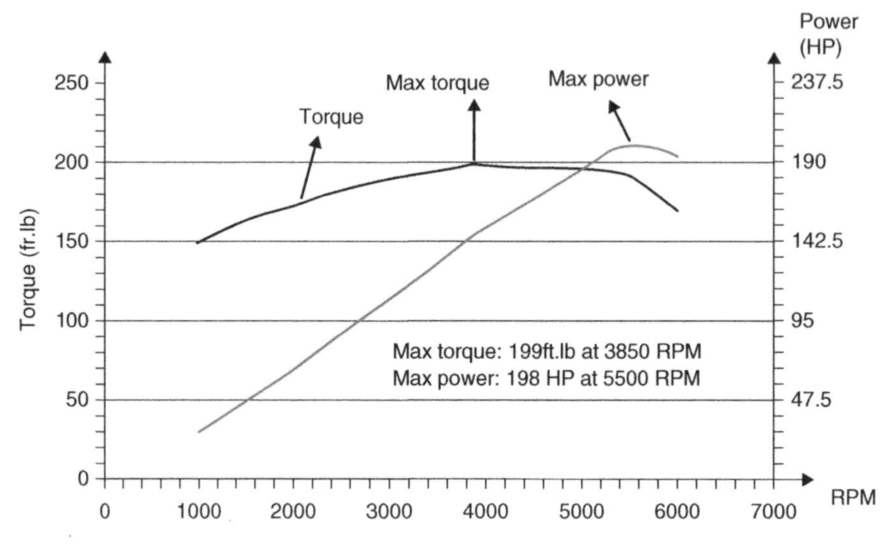

Figure 1.14 Engine WOT output for Example 1.2.

a) Determine all the gear ratios assuming they fall into a geometric progression.
b) Determine the maximum grade the vehicle can negotiate in 4th gear at a constant speed of 65 mph.
c) What will be the maximum vehicle speed if the overall powertrain ratio of 5th gear is designed as 2.15?

Note: Show how the data are taken from the plot.

Solution:

a) Fifth gear ratio:
Plugging the relevant data given in the problem into Eq. (1.33), we can obtain a cubic equation with $(i_a i_5)$ as the unknown:

$$102366(i_a i_5)^3 - 40527.2(i_a i_5)^2 - 1500368.7 = 0$$

Solving the equation above by iteration, $(i_a i_5) = 2.57$
First gear ratio:
For an FWD vehicle, the traction based maximum gradability is determined as:

$$W \sin\theta_{max} \le \mu_o \left(\frac{c}{L} W \cos\theta_{max} - \frac{h}{L} \sin\theta_{max} \right)$$

$$G_{max} = 100 \tan\theta_{max} = 100 \frac{\mu_o \frac{c}{L}}{1 + \mu_o \frac{h}{L}} = 100 \frac{1.0\left(\frac{1820}{3520}\right)}{1 + 1.0\left(\frac{14}{105}\right)} = 45.6$$

$$(i_a i_1) = \beta \left[\frac{Wr\left(f + \frac{G_{max}}{100}\right)}{\eta T_{emax}} \right] = 1.35 \left[\frac{0.02(3520) + 0.456(3520)}{199(.94)} \right] = 12.10$$

Intermediate gear ratios:

Since the gear ratios are in a geometric series, $i_5 = c^4 i_1$, so, $c = \sqrt[4]{\frac{2.57}{12.10}} = 0.679$, the other ratios are:

$$i_a i_2 = 0.679(12.10) = 8.22$$
$$i_a i_3 = 0.679(8.22) = 5.58$$
$$i_a i_4 = 0.679(5.58) = 3.79$$

b) Maximum grade:

$$v = 65\,\text{mph} = 1.467(65) = 95.36\,\text{ft/s}$$
$$R = 0.02 + (3520) + 0.00118(.31)(21.5)(95.36)^2 + \sin\theta = 140.25 + 3520\sin\theta$$
$$\omega_e = i_4 i_a \frac{v}{r} = 3.79\left(\frac{95.36}{1.0}\right) = 361.41\,\text{rad/s} = 3451\,\text{RPM}$$

From the torque plot, the engine torque T_e is found to be 192 ft.lb at 3451 RPM, so

$$P = \frac{0.94(3.79)(192)}{1.0} = 684\,\text{lb}$$

At maximum grade, traction and road load are balanced, so

$$140.25 + 3520\sin\theta_{max} = 684$$
$$G_{max} \approx 100\sin\theta_{max} = 15.2$$

c) When the vehicle reaches its maximum speed, the available power from the power-train is fully balanced by road load power, i.e. $Rv_{max} = \cong \eta(Power)$ or,

$$v_{max}\left[(0.02(3520) + 0.00118(0.31)(21.5)v_{max}^2\right) = 0.94(\text{Power})$$
$$74.89v_{max} + 0.008367v_{max}^3 = Power$$

Note that *Power* is a function of engine RPM. The solution for v_{max} is based on iteration.

First iteration: assuming $v_{max} = \dfrac{3.14(5500)(1.0)}{30(2.57)} = 223.99\,\text{ft/s}$, then,

$$\text{Road load power}: 74.89(223.99) + 0.008367\left(223.99^3\right) = 110802\,\text{ft.lb/s} = 201\,\text{HP}$$
$$\text{RPM} = 2.15\left(\frac{30}{\pi}\right)\frac{v}{r} = 2.15\left(\frac{30}{\pi}\right)\left(\frac{223.99}{1.0}\right) = 4598.7$$

At 4598 RPM, the engine power is about 165 HP and is below 201 HP. So $v_{max} < 223.99\,\text{ft/s}$

Second iteration: assuming $v_{max} = 190\,\text{ft/s}$, then

Road load power: $74.87(190) + 0.008367(190)^3 = 129\,\text{HP}$

$$\text{RPM} = 2.15\left(\frac{190}{1.0}\right)\frac{30}{\pi} = 3900$$

At 3900 RPM, the engine power is 152 HP and is higher than 129 HP. So $v_{max} > 190\,\text{ft/s}$

Third iteration: assuming $v_{max} = 210\,\text{ft/s}$, then

Road load power: $74.89(210) + 0.008367(210)^3 = 168\,\text{HP}$

$$\text{RPM} = 2.15\left(\frac{210}{1.0}\right)\frac{30}{\pi} = 4311$$

At 4311 RPM, the engine power is about 162 HP, so $v_{max} \approx 210\,\text{ft/s} = 143\,\text{mph}$

References

1 Stone, R.: *Introduction to Internal Combustion Engines*, Second Edition, Society of Automotive Engineers, 1993. ISBN 1-56091-390-8.
2 Taylor, C.F.: *The Internal Combustion Engine in Theory and Practice, Vol. 1: Thermodynamics, Fluid Flow Performance*, Second Edition, MIT Press, 1985. ISBN 13:978-0262700269.
3 Taylor, C.F.: *The Internal Combustion Engine in Theory and Practice, Vol. 2: Combustion, Fuels Design*, Second Edition, MIT Press, 1985. ISBN 13:978-0262700276.
4 Isermann, R.: *Engine Modelling and Control*, Springer-Verlag, 2014, ISBN: 978-3-642-39934-3.
5 Lechner, G.; Naunheimer, H.: *Automotive Transmissions: Fundamentals, Selection, Design and Application*, Springer, 1999. ISBN 3-540-65903.
6 Gillespie, T.D.: *Fundamental Vehicle Dynamics*, Society of Automotive Engineers, 1992. ISBN 978-1-56091-199-9.

Problem

A FWD vehicle has the following data:

Front axle weight: 1750 lb
Rear axle weight: 1200 lb
Center of gravity height: 15 in.
Wheelbase: 105 in.
Air drag coefficient: 0.30
Frontal projected area: 22.0 square feet
Tire radius: 11.40 in.
Roll resistance coefficient: 0.02
Max. power @6000 RPM: 138 HP
Max. torque @4500 RPM: 132 ft.lb
Powertrain efficiency: 0.96
Final drive ratio: 3.143

A six-speed manual transmission is used for the vehicle and the gear ratios from 1st to 4th gears are: 1st gear (3.92), 2nd gear (2.76), 3rd gear (1.85), 4th gear (1.35). The engine WOT output plot is as given in Figure 1.3.

a) The vehicle runs in the 5th gear at a speed of 55 mph with the engine speed at 2450 RPM. The driver then makes a 5–6 upshift and the engine RPM drops by 500 RPM immediately after the shift. Determine the 5th and the 6th gear ratios.

b) Determine the engine torque and work done by the engine when the vehicle cruises for 1.5 miles at a constant speed of 65 mph on level ground in the 6th and 5th gears respectively.
c) The driver floors the gas pedal and simultaneously makes a 6–5 downshift when the vehicle runs on a 3% slope at a speed of 65 mph. Determine the vehicle acceleration immediately after the 6–5 downshift.
d) What is the steepest percentage slope the vehicle can negotiate at a speed of 70 mph?

2

Manual Transmissions

2.1 Introduction

The gear ratios discussed in Chapter 1 that match the engine outputs for optimized vehicle performance and fuel economy are realized by different types of transmissions. Manual transmissions (MT) are the oldest type and have a history as old as the automotive industry [1]. In a manual transmission, engine power is transmitted by gear pairs on fixed axes from input to output, and gear shifts are manually made by the driver. Although the basic structure and operation principles have remained almost the same ever since the advent of automobiles, manual transmissions have undergone an evolution of changes aimed at improving ease of operation and shift smoothness. The earliest manual transmissions used sliding gears for gear shifts [2]. To make a shift, the driver would separate a gear pair by pulling one of the gears out of mesh, and then pushing and sliding another gear into mesh. It was very difficult to make shifts this way and gear grinding was unavoidable during shifts. Later versions had constant mesh gear design in which gears responsible for shifts had a dogtooth ring attached. During a shift, the driver would push a sleeve with internal spline teeth on the transmission shaft and slide it into mesh with the dogtooth ring. Tooth grinding was still unavoidable since the speed of the sleeve and the dogtooth ring were different during shifts. It was in the 1930s that synchronizers were widely applied in manual transmissions [1,2]. The use of synchronizers greatly improved vehicle drivability and made driving much easier and more pleasant. To a certain degree, synchronized manual transmissions contributed to the wide spread of automobiles in the daily life of the populace.

The design of synchronized manual transmissions has not changed much for many decades, but advances in material and manufacturing technologies have made these products more durable and reliable. In comparison with automatic transmissions (AT) that offer better driver convenience, manual transmissions are less costly and generally more fuel efficient. For some drivers, manual transmission vehicles are the preferred choice because they offer sportier driving techniques and experience than the automatic counterparts. In addition, manual transmissions with multiple gear ratios are more suitable for heavy duty trucks due to their advantages in cost and fuel economy. For these reasons, it can be said that manual transmissions are a mature and everlasting product in the automotive industry.

The market share of manual transmissions varies with vehicle type and marketplace [3]. For passenger vehicles, manual transmissions have less than 20% market share in

Automotive Power Transmission Systems, First Edition. Yi Zhang and Chris Mi.

North America, but the number is closer to 80% in Europe. In the Chinese market, which is currently the world's largest, manual transmission market share is about 60%. For trucks, including pickups and heavy duty vehicles, manual transmission share is well above 50% in all markets. For the world market as a whole, the manual transmission market share floats around 58%. These numbers indicate the significant status of manual transmissions in today's automotive market.

The overall vehicle powertrain system includes the engine, transmission, transfer case for four wheel drive (4WD), drive shaft, final drive and differential, half shafts, universal or constant velocity joints, and driving wheels. The transmission is the core in the driveline from the engine output to the wheels. The focus of this chapter is on manual transmissions. Readers are recommended to study publications that provide technical details on other driveline components, such as transfer case, CV joints, and universal joints [4,5]. A manual transmission is a mechanical system that consists of dozens of components, including clutch, gears, shafts, synchronizers, bearings, and seals. This chapter starts with the general layouts of vehicle powertrains and the basic structures of manual transmissions, followed by the analysis on the power flow and transmission ratios. At the component level, the chapter focuses on the design and analysis of clutches and synchronizers which are specifically developed for automotive transmissions. A dynamic model will be introduced for the analysis of manual transmission shift dynamics. Detailed analysis on the synchronization process and synchronizer design are covered based on the model. Transmission gear design will be covered separately in the next chapter. It should be noted that this chapter uses example transmissions to demonstrate general principles and approaches, which are applicable for the design and analysis of all other manual transmissions.

2.2 Powertrain Layout and Manual Transmission Structure

Vehicle powertrain system layouts depend on whether the vehicle has front wheel drive (FWD), real wheel drive (RWD), or four wheel drive (4WD). For FWD vehicles, the transmission and the final drive, which contains the differential, are integrated into the same assembly. FWD vehicles usually have transversely mounted engines, with the engine crankshaft parallel to the drive axle, as shown in Figure 2.1a. A FWD vehicle can also have the engine mounted longitudinally as shown in Figure 2.1b, then a pair of spiral bevel gears or hypoid gears must be used to transmit the power to the driving wheels from the transmission output shaft which is perpendicular to the axle. In comparison with RWD layout, FWD layout offers lower manufacturing cost and better passenger room due to its compactness. It also offers somewhat better traction in cold weather conditions because the front axle has a larger portion of the weight distribution. Most passenger cars or vans today have front FWD layout because of these advantages.

RWD vehicles always have longitudinally mounted engines. The transmission and the final drive are separate assemblies, as shown in Figure 2.2a, with a drive shaft connecting the transmission output and the final drive input. Universal joints (Hooke joints) are used at the two ends of the drive shaft to accommodate assembly condition and drive line flexibility. The final drive assembly in an RWD vehicle contains a pair of spiral bevel gears or hypoid gears that provide the final drive ratio and the differential that allows a speed

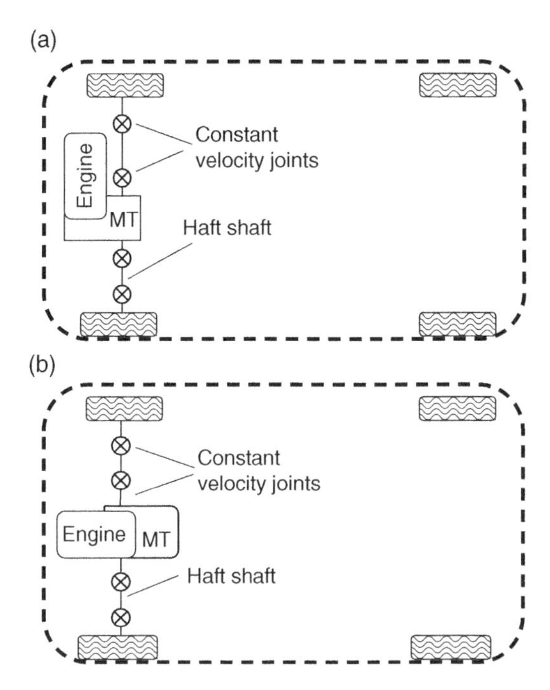

Figure 2.1 Alternative front wheel drive layouts.

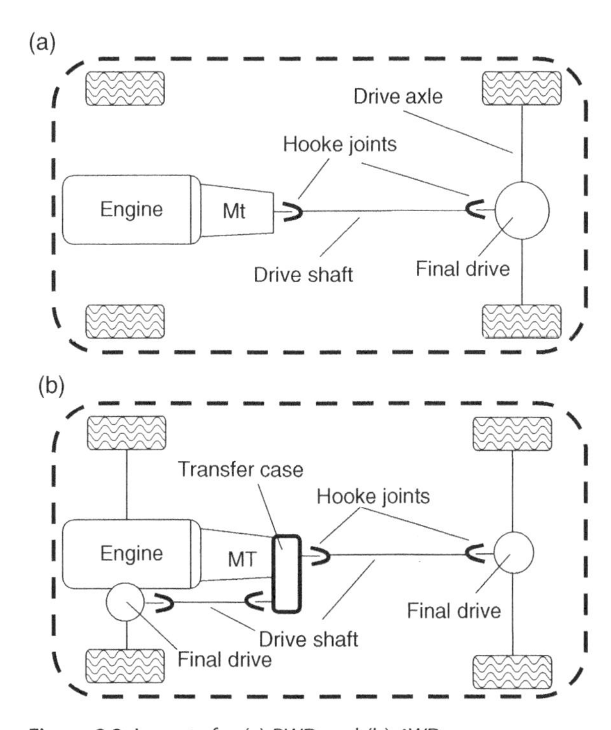

Figure 2.2 Layouts for (a) RWD and (b) 4WD.

difference between the two driving wheels. The weight distribution of RWD cars can be designed close to a perfect 50/50 between the front and rear axles. This optimized weight distribution leads to improved drivability and handling in comparison with that of FWD vehicles. There are drivers who prefer RWD cars because of the better drivability (at least as perceived by them) in steering and cornering. This may be one of the reasons why RWD is used for most of the luxury and sports cars. The RWD layout in Figure 2.2a is also used for the powertrains of light to middle duty trucks, such as pickups and delivery trucks. This layout can be conveniently modified to fit a 4WD system, as shown in Figure 2.2b. For 4WD vehicles, there are two final drive and differential assemblies, one for each drive axle. Other 4WD layouts may originate from an FWD configuration with transversely mounted engine. In a 4WD powertrain shown in Figure 2.2b, the transmission output is split by the transfer case to the final drives on the front and rear axles. The distribution of transmission output torque between the front and rear axles depends on the design and control of the transfer case.

The vast majority of non-commercial passenger vehicles have powertrain layouts as shown in Figures 2.1 and 2.2, perhaps with very few exceptions for fancy sports cars. Other powertrain system layouts can also be adopted to meet the requirements of different vehicle types and functions. For example, Figure 2.3a shows the typical powertrain layout for a semi-truck with two drive axles in series. Standard cargo boxes are hooked to the semi-trailer on top of the double axles that provide propulsion together. Spiral bevel

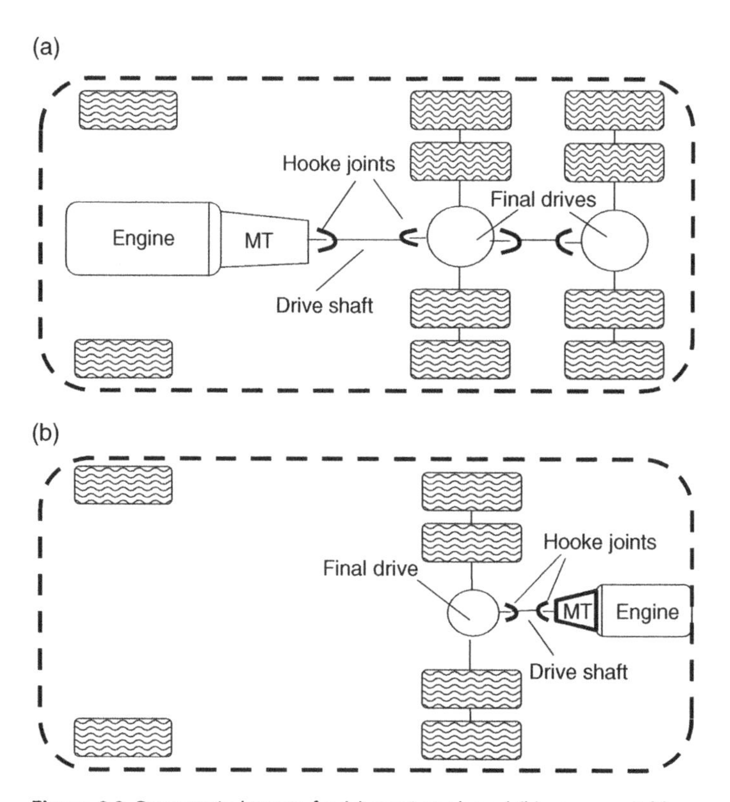

Figure 2.3 Powertrain layouts for (a) semi-truck and (b) commercial bus.

gears or hypoid gears can be used in the final drives for the transmission of power from the longitudinal drive shaft to the axles. Both spiral bevel gears (with the pinion and gear axes intersected) and hypoid gears are used in the final drive of heavy duty trucks or buses. A pair of helical gears is sometimes used to provide an additional reduction ratio in the final drive assembly. The hypoid gear sets used in the layout shown in Figure 2.3a have the pinion axis offset above the axle, allowing the assembly of the differential at the center of the hypoid gear. Figure 2.3b shows the layout for a large passenger bus that has the engine longitudinally mounted in the rear. It should be noted here that either a manual transmission or an automatic transmission can be fitted to the same powertrain system layout of a vehicle.

Most of the manual transmissions used for passenger cars and vans today have five or six forward gears and one reverse. Only a few sports cars or high end models are equipped with transmissions with six or more forward speeds for fuel economy and performance. Manual transmissions are a mature product in the automotive industry that shares similar structures. Figure 2.4 shows the section view of a typical five-speed FWD manual transmission that fits to the powertrain layout in Figure 2.1a for passenger cars. Key components in the manual transmission are marked in the drawing. The gear pairs for different speeds and synchronizers are laid out on the input shaft and the output shaft. All gears except the reverse gears are cylindrical helical gears. There are six gears on each shaft since there are five forward and one reverse gears. Both the input and output shafts are supported on cylindrical roller bearings on the engine side and on deep groove ball bearings on the outside. The whole output shaft and a section of the input shaft are hollow to reduce weight and mass moment of inertia. The 1st gear and the 5th gear are on the two ends of the shaft, close to the bearings, to minimize shaft deflection. The synchronizer for the first and second gears is on the output shaft. The 3–4 synchronizer and the 5th synchronizer are mounted on the input shaft. The reverse gear is realized by a sliding idler between the input and output shafts. This requires a little more effort than shifting into a forward gear, but has little effect on drivability since reverse gear is always engaged when the vehicle is at rest. The output gear for reverse is attached to the 1–2 synchronizer assembly and is machined on the sleeve of the 1–2 synchronizer. The input gear of the final drive is machined on the output shaft, and the ring gear (final drive output) is bolted to the differential carrier, which is supported on both sides by tapered roller bearings that resist thrusts. Each of the two side gears of the differential is connected to the driving wheel via a haft shaft, which is coupled to the side gear and the wheel hub by CV joints. The transmission housing consists of two pieces bolted together with sealing and is attached to the engine assembly by bolts. The housing is designed with an empty space on the input side, called clutch well, for the clutch assembly. Note that the axes of the shafts, the final drive, and the sliding idler are actually not within a plane, but at angles with respect to each other for assembly and compactness considerations.

A typical five-speed manual transmission for an RWD passenger vehicle is shown in Figure 2.5. This transmission fits the powertrain layout in Figure 2.2a. There are three shafts in this transmission design. The input shaft is supported on the housing by a deep groove ball bearing and at the center of the engine flywheel by a pilot bearing. To minimize shaft deflection, both the counter shaft and the output shaft are supported at three locations. The counter shaft carries all of the gears on it and is supported by a cylindrical roller bearing on the engine side, a double row roller bearing on the supporting wall of

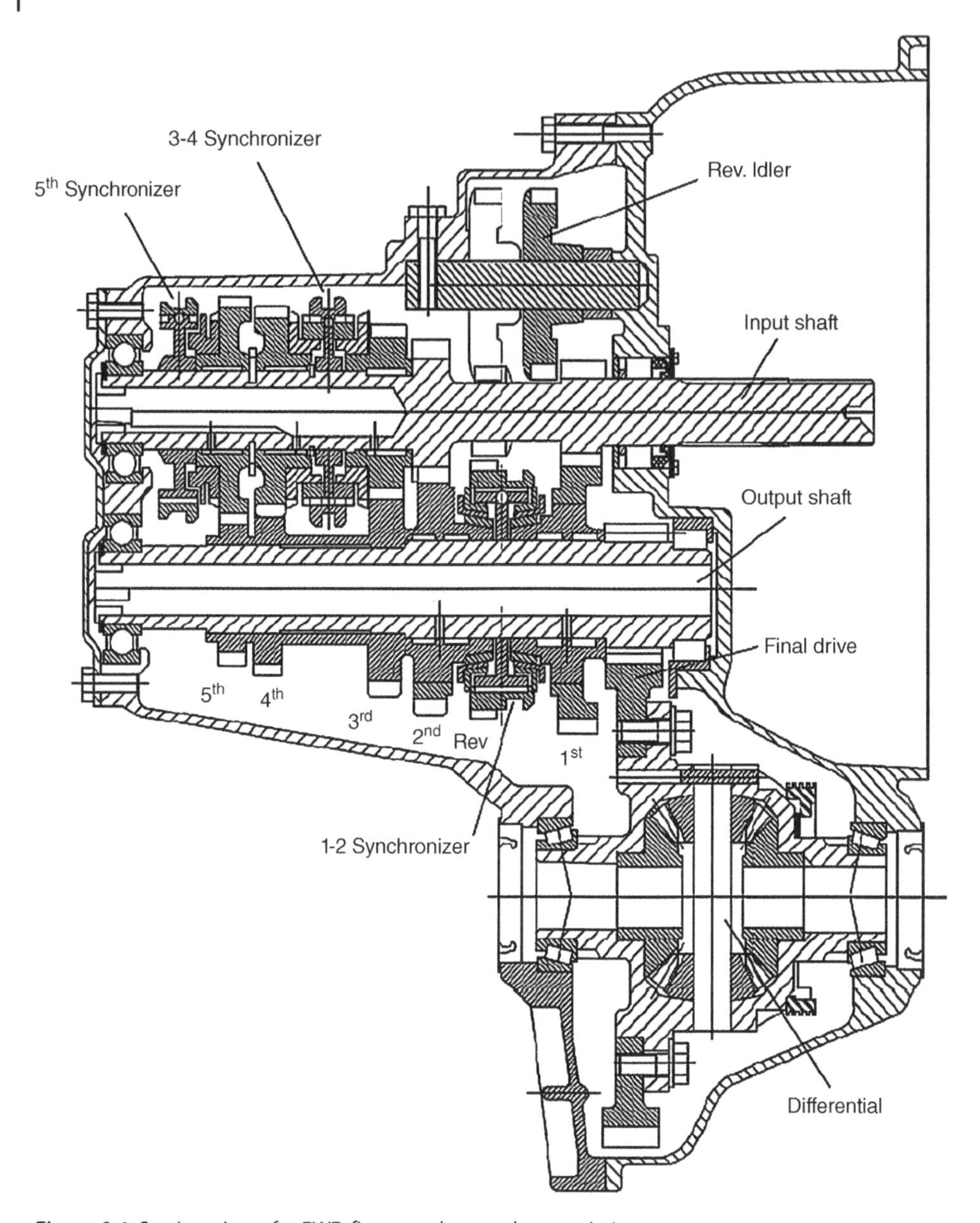

Figure 2.4 Section view of a FWD five-speed manual transmission.

the housing, and a ball bearing at the other side. The output shaft is supported by ball bearings on the supporting wall and on the output side, as well as by a roller bearing at the center of the input gear. The first, second, and 5th gears are located by the bearing to minimize shaft deflections. The reverse gears are constantly meshed and the reverse gear is engaged by the R-5 synchronizer. All three synchronizers are mounted on the output shaft and thus all gears freewheel on the output shaft unless engaged. The housing is designed with an extra space on the output side for the assembly of the shifting

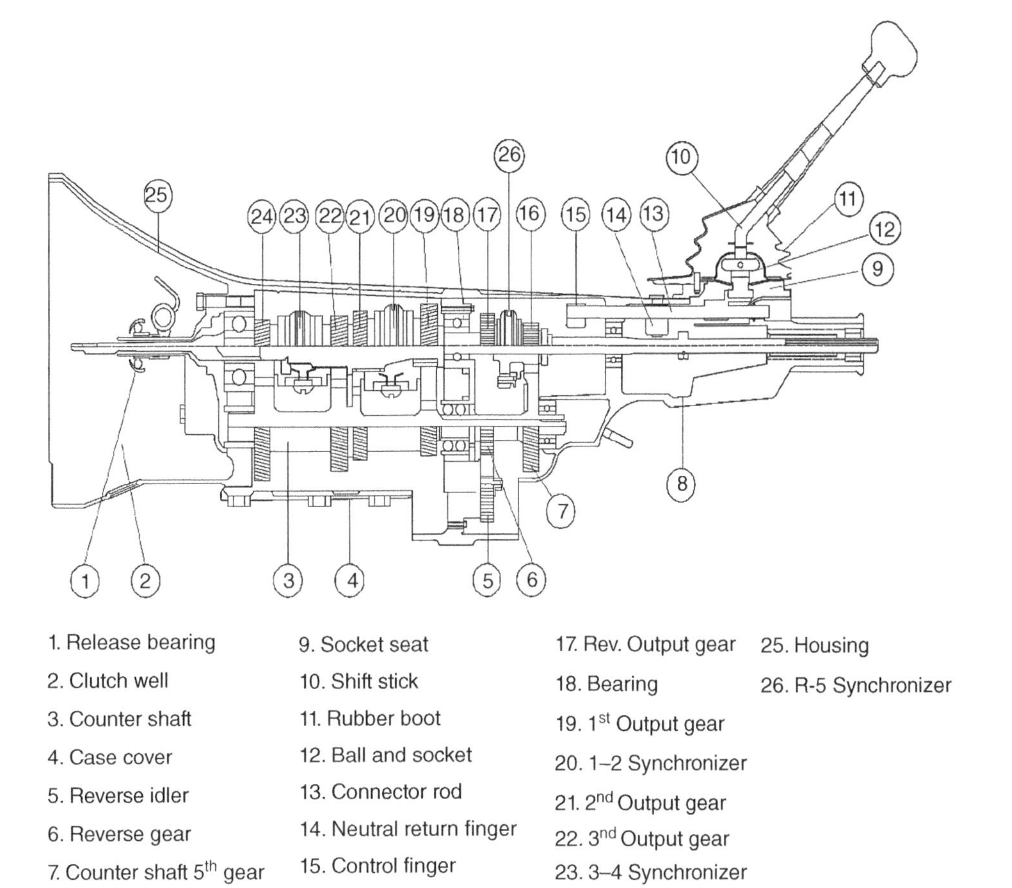

Figure 2.5 Section view of an RWD five-speed manual transmission.

mechanism and speedometer gear or speed sensor. There are variations of RDW manual transmissions with five or six speeds based on improvements from the one shown in Figure 2.5. In the newer versions [6], the supporting wall in the transmission housing and the bearings on it are eliminated because sufficient rigidity and strength are guaranteed by new materials and manufacturing technologies for the housing.

The structure of a manual transmission, or any other mechanical systems in general, can be illustrated by a simple and intuitive sketch, commonly termed as "stick diagram" in the automotive industry. A stick diagram uses a set of symbols to represent various components and interconnections between them in a mechanical system or machinery. The symbols used in a stick diagram are easily identifiable and are used consistently to represent powertrain components throughout this book. A list of symbols for powertrain components is shown in Figure 2.6.

The stick diagrams for the five-speed FWD manual transmission in Figure 2.4 and the five-speed RWD manual transmission in Figure 2.5 are shown in Figures 2.7a and 2.7b, respectively. Clutches are not shown in the stick diagrams for clarity of the drawing.

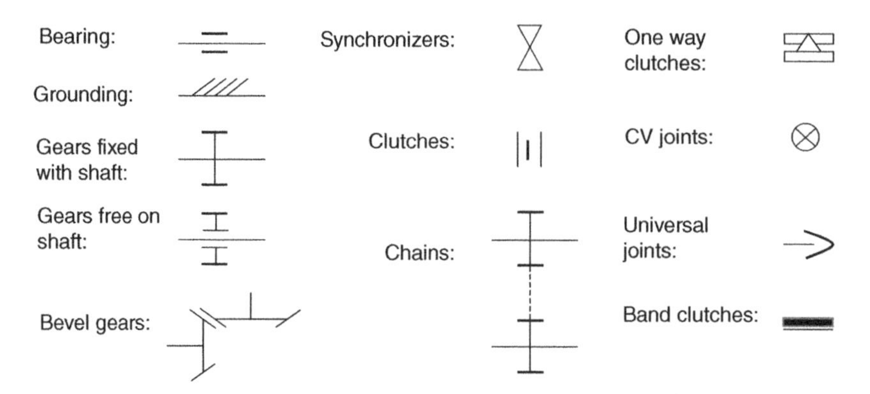

Figure 2.6 Symbols for common powertrain components.

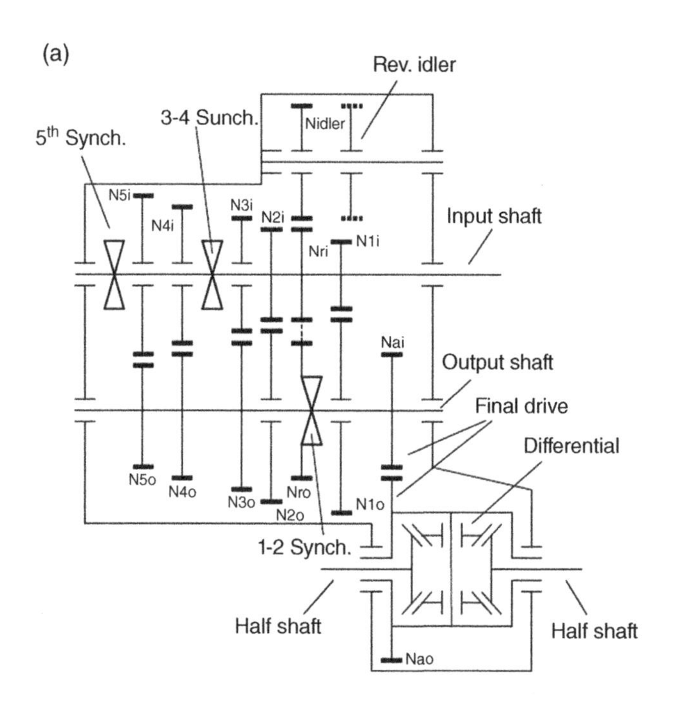

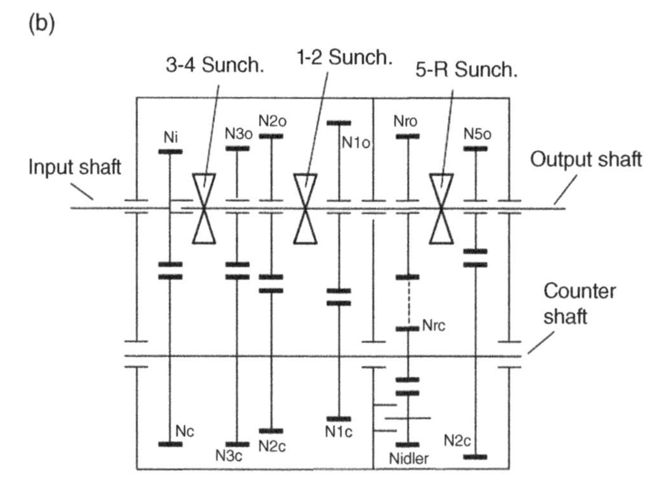

Figure 2.7 Stick diagrams for (a) five-speed FWD MT and (b) five-speed RWD MT.

These two stick diagrams provide a clear picture for the structural layouts of the two manual transmissions and are easy to understand because of the intuitive feature of the symbols. The numbers of teeth for each gear are also labeled in the stick diagrams for the identification of power flows and gear ratios, which are discussed in the next section. The first of the two subscripts indicates the speed and the second indicates the shaft that the gear is on. For example, N_{4i} is the number of teeth of the 4th gear on the input shaft. For the FWD MT, the integrated final drive is a pair of helical gears labeled as N_{ai} and N_{ao}. The RWD MT has two stages of gears for each speed, which shares a common input gear pair labeled as N_i and N_c.

2.3 Power Flows and Gear Ratios

It is fairly straightforward to figure out the power flow in each gear for a manual transmission. With the numbers of teeth labeled with the subscripts as shown in Figure 2.7, the power flow can be identified directly from the input to the output. For example, the power flow in each forward gear for the FWD MT in Figure 2.7a is shown as follows:

1st gear: Input shaft (gear N_{1i}) → Output shaft (gear N_{1o} and gear N_{ai}) → Final drive output (gear N_{ao})
2nd gear: Input shaft (gear N_{2i}) → Output shaft (gear N_{2o} and gear N_{ai}) → Final drive output (gear N_{ao})
3rd gear: Input shaft (gear N_{3i}) → Output shaft (gear N_{3o} and gear N_{ai}) → Final drive output (gear N_{ao})
4th gear: Input shaft (gear N_{4i}) → Output shaft (gear N_{4o} and gear N_{ai}) → Final drive output (gear N_{ao})
5th gear: Input shaft (gear N_{5i}) → Output shaft (gear N_{5o} and gear N_{ai}) → Final drive output (gear N_{ao})

It should be noted that, in each gear, only the concerned gear in the power flow is coupled to the shaft by the synchronizer, and gears not in the power flow freewheel on their shafts. The power flow in reverse gear differs from forward gears and involves an idler gear between the input and output shafts, shown in the following:

$$\begin{aligned}\text{Rev. gear}: \text{ Input shaft (gear}N_{ri}) &\rightarrow \text{Idler gear } (N_{idler}) \rightarrow \text{Output} \, shaft \, (N_{ro}) \\ &\rightarrow \text{Final drive output (gear}N_{ao})\end{aligned}$$

In all manual transmissions, the reverse gear is realized by the reverse idler. Due to their being one more external mesh than the forward gears, the reverse idler gear will reverse the direction of rotation of the output shaft without contributing to the gear ratio. When the vehicle travels in a straight path, the final drive output, i.e. the ring gear of the final drive, will rotate at the same angular velocity as the two driving wheels. The differential allows speed difference for the two driving wheels on the driving axle when the vehicle travels on curves. The power flow for the RWD five-speed MT shown in Figure 2.7b can be found in a similar fashion. All forward gears, except the 4th gear, involves two stages of gearing, as follows:

1st gear: Input shaft (gear N_i) → Counter shaft (gear N_c and gear N_{1c}) → Output shaft (gear N_{1o}) → Final drive
2nd gear: Input shaft (gear N_i) → Counter shaft (gear N_c and gear N_{2c}) → Output shaft (gear N_{2o}) → Final drive

3rd gear: Input shaft (gear N_i) → Counter shaft (gear N_c and gear N_{3c}) → Output shaft (gear N_{3o}) → Final drive
5th gear: Input shaft (gear N_i) → Counter shaft (gear N_c and gear N_{5c}) → Output shaft (gear N_{5o}) → Final drive

The 4th gear is a direct drive, with the output shaft coupled to the input shaft gear N_i by the 3–4 synchronizer. All gears freewheel in the 4th gear. The reverse gear power flow involves an idler gear as follows:

$$\text{Rev. gear: Input shaft (gear } N_i) \rightarrow \text{Counter shaft} \rightarrow \text{Idler gear } (N_{idler}) \rightarrow \text{output shaft (gear } N_{1o}) \rightarrow \text{Final drive}$$

Similar to all other gearboxes, the gear ratios of an automotive transmission are defined as the ratio between the input angular velocity to the output angular velocity, by the following general formula:

$$i_t = \frac{\omega_{input}}{\omega_{output}} = (-1)^n \frac{\textit{numbers of teeth of driven gears}}{\textit{numbers of teeth of driving gears}} \tag{2.1}$$

where n is the number of external meshes in the power flow path. Clearly, if n is odd, then the output angular velocity will be opposite to the input. For the FWD MT in Figure 2.7a, the value of transmission ratio in each gear and the final drive gear ratio are calculated by the following equations:

$$i_1 = \frac{N_{1o}}{N_{1i}}; i_2 = \frac{N_{2o}}{N_{2i}}; i_3 = \frac{N_{3o}}{N_{3i}}; i_4 = \frac{N_{4o}}{N_{4i}}; i_5 = \frac{N_{5o}}{N_{5i}}; i_r = -\frac{N_{ro}}{N_{ri}}; i_a = \frac{N_{ao}}{N_{ai}} \tag{2.2}$$

It should be noted here that the transmission ratio in a FDW transmission refers to the ratio between the input angular velocity and the output angular velocity that is also the input angular velocity of the final drive. For the transmission in Figure 2.7a, there are two external meshes in the power flow of forward gears from the transmission input shaft to the final drive output, so the direction of rotation for the input shaft and the driving wheels is the same. But in the reverse gear, there are three external meshes in the power flow, so the direction of rotation of the driving wheels is opposite to that of the input shaft and the vehicle moves backward, as illustrated in Figure 2.8.

For the RWD MT in Figure 2.7b, there are more than one pairs of gears involved in the power flows, with the exception of the 4th gear, as mentioned previously. Using Eq. (2.1), the five forward ratios and the reverse ratio are determined in terms of the tooth numbers of involved gears in the following:

$$i_1 = \frac{N_c N_{1o}}{N_i N_{1c}}; i_2 = \frac{N_c N_{2o}}{N_i N_{2c}}; i_3 = \frac{N_c N_{3o}}{N_i N_{3c}}; i_4 = 1; i_5 = \frac{N_c N_{5o}}{N_i N_{5c}}; i_r = -\frac{N_c N_{ro}}{N_i N_{rc}} \tag{2.3}$$

The minus sign for the reverse gear ratio indicates that the rotational direction of the output shaft in reverse gear is opposite to that of the forward gears. In this RWD MT, the 4th gear is a direct drive with a ratio of one and the 5th gear is an overdrive with a ratio below one. In the automotive industry today, some passenger vehicles are also equipped with FWD or RWD six-speed manual transmissions with typical layouts shown in Figure 2.9.

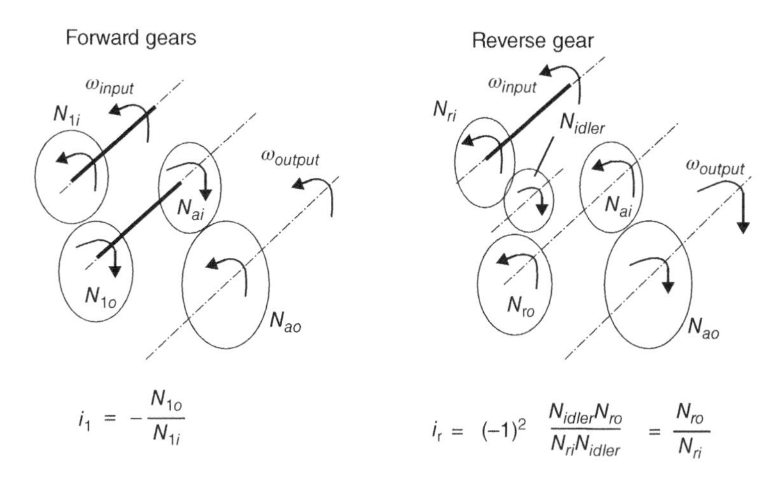

Figure 2.8 Reverse idler gear.

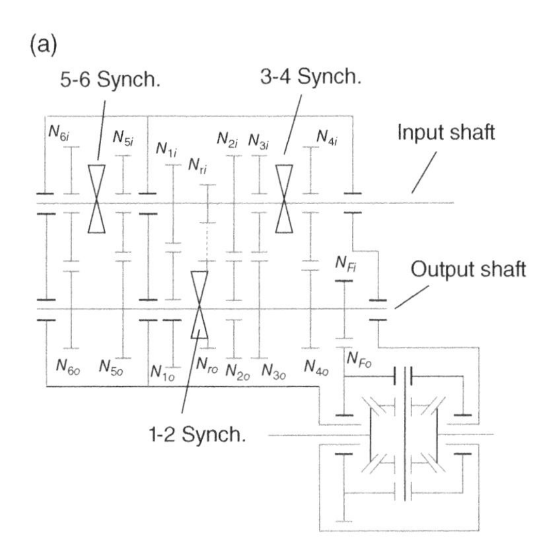

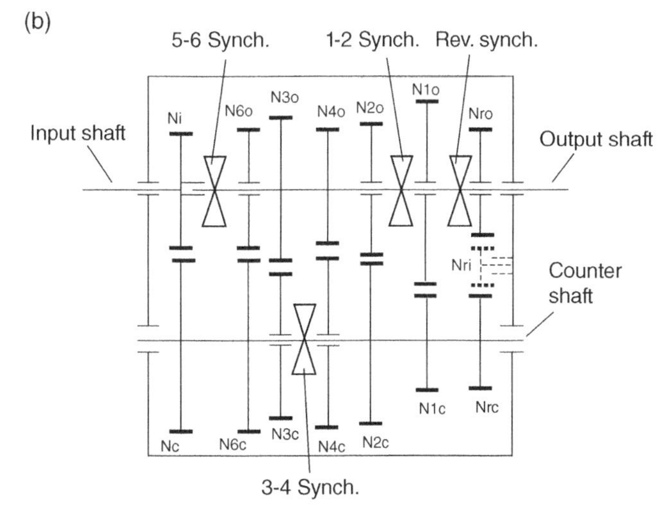

Figure 2.9 **Six**-speed MTs for (a) FWD and (b) RWD.

2.4 Manual Transmission Clutches

As mentioned in Chapter 1, internal combustion engines cannot provide stable output torque below the idle speed. When a vehicle is being launched from rest, the vehicle speed is zero but the engine speed must be above the idle. This means that there must be a launch device that allows slippage while transmitting the engine torque to the transmission input during vehicle launch. The slippage ends at the time when the vehicle speed has gradually increased to the value that satisfies Eq. 1.21. In a manual transmission vehicle, the clutch functions as the launch device that bridges the gap between the engine RPM and the vehicle speed. The clutch has another important function in a manual transmission vehicle. When disengaged, it provides a disconnection between the engine output and the transmission input. During a shift, the clutch is briefly disengaged to cut off power supply to the transmission so that the disengagement of the current gear and the engagement of the next gear can be carried out smoothly by the driver. Most of the clutches in manual transmissions are actuated manually through clutch pedal depression by the driver, but some are automated hydraulically or electrically for ease of operation

2.4.1 Clutch Structure

Manual transmission clutches are friction devices that require a clamping force or normal force on friction surfaces for the generation of friction. The clamping force is generated by the compression of springs. There are two types of springs used in manual transmission clutches, coil springs and Belleville or diaphragm springs. Coil spring clutches have higher torque capacity and are thus mainly used for trucks. Belleville clutches are mainly used for passenger vehicles due to structural simplicity and compactness. The structure of a coil spring clutch is illustrated by a section view in Figure 2.10.

As shown in the Figure 2.10, the clutch assembly consists of the friction disk, pressure plate, clutch cover, release lever, coil springs, and release bearing. Friction is generated on the two faces of the friction disk that is splined to the transmission input shaft and sandwiched between the pressure plate and the engine flywheel which is machined with a flat contacting surface. The friction disk is assembled with a torsional spring damper that is designed to cushion the dynamic impacts caused by clutch actuation and engine output harmonics. The pressure plate is attached to the clutch cover by flexible links as shown in Figure 2.10. The flexible links carry the pressure plate in rotation with the clutch cover that is bolted to the flywheel and meanwhile allow the pressure plate to move slightly in the axial direction. A set of coil springs is installed circumferentially between the pressure plate and the clutch cover. Compression in the coil springs is generated in assembly when the clutch cover is bolted to the flywheel. This compression is adjustable by the tightness of the assembly bolts. The clutch is normally engaged since sufficient magnitude of clamping force is generated by the coil spring at assembly. This clamping force is applied by the pressure plate to the friction disk. Engine torque is transmitted by the friction generated on the two contacting faces of the friction disk – one between the disk and the flywheel, and the other between the disk and the pressure plate – to the transmission input shaft. The disengagement of the clutch and the control of the clamping force (i.e. the clutch torque) are realized through symmetrically mounted

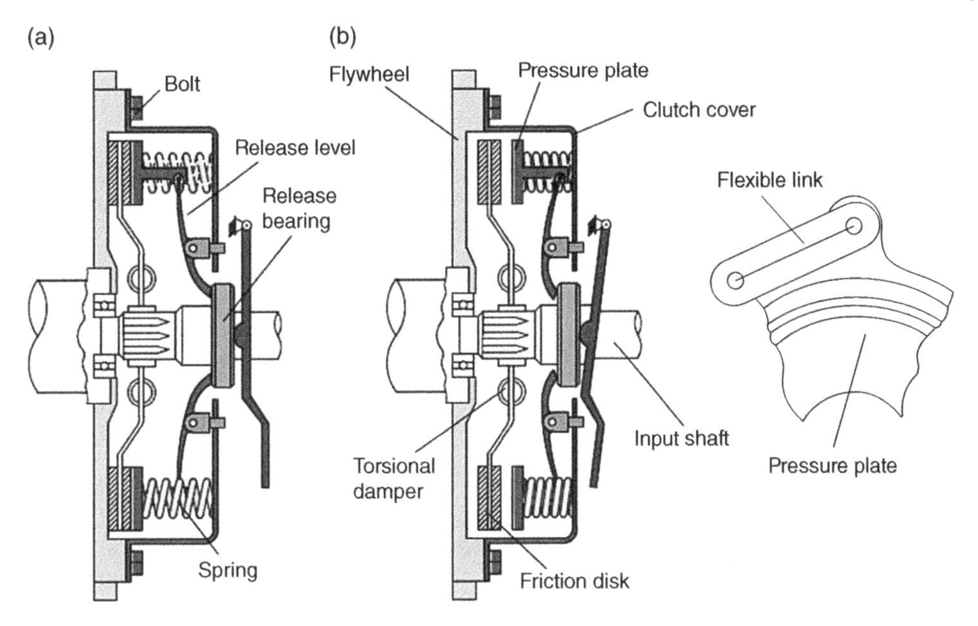

Figure 2.10 Structure of coil spring clutch (a) engaged (b) disengaged.

release levers, each of which is on the central line between two coil springs, pivoting about a pin joint on the clutch cover. The upper end of the release lever contacts the pressure plate and the lower end contacts the release bearing. The engaged position and the disengaged position of the clutch are shown in Figures 2.10a and 2.10b respectively. Before the vehicle is launched, the driver will first depress the clutch pedal fully to disengage the clutch through the clutch actuation linkage. The release bearing moves leftward and the pressure plate moves rightward as the clutch pedal is depressed, placing the clutch in the disengaged position as shown in Figure 2.10b. As the vehicle is being launched, the driver will gradually lift the clutch pedal so that the release bearing moves rightward due the spring force and the pressure plate moves leftward until the clutch is engaged as shown in Figure 2.10a. Clutch operation during transmission shifts is the same as for vehicle launch. The clutch in Figure 2.10 uses one friction disk. To increase the torque transmission capacity, two or more friction disks may be used, with an extra pressure plate between the two for additional friction faces.

As shown in Figure 2.11, the structure of the Belleville or diaphragm clutch is similar to that of coil spring clutches, but only differs in the spring that generates the clamping force. In a Belleville clutch, the conically shaped diaphragm spring serves as both the clamping force generator and the release lever. Multiple slots are machined on the diaphragm spring so that its inner side forms touching fingers to contact the release bearing. These slots are also conducive to uniform deformation of the inner side along the axial direction when pushed by the release bearing. The diaphragm spring is attached to the clutch cover by pins that fit into the holes on the upper side of the slots. Two circular rings are placed on the rivet pins, on each side of the diaphragm spring. These two circular rings retain the position of the diaphragm spring and also allow it to pivot about the pins. The outer side of the diaphragm spring is attached to the pressure plate. This

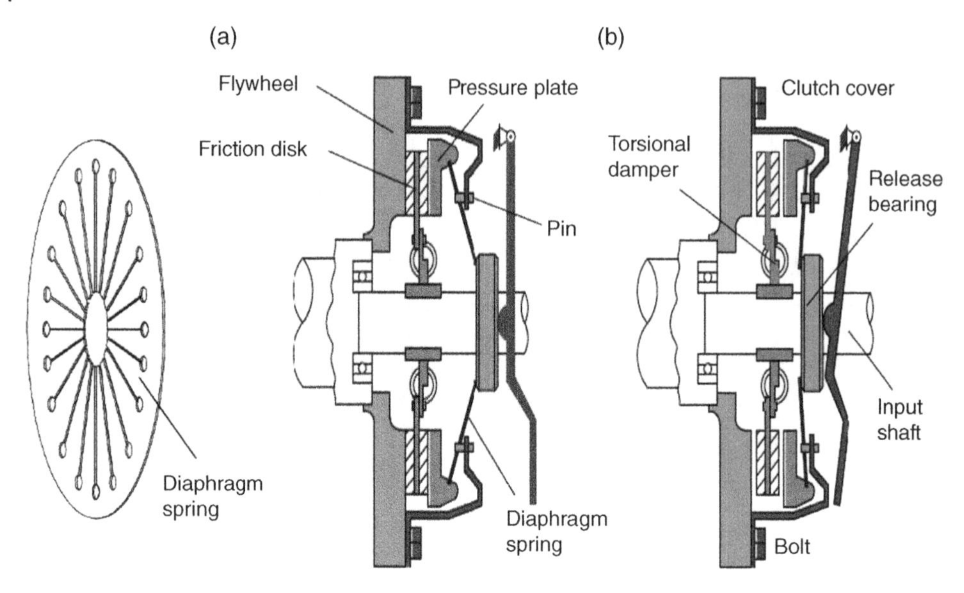

Figure 2.11 Diaphragm spring and structure of Belleville clutch (a) engaged (b) disengaged.

attachment allows for a slight relative motion between the diaphragm spring and the pressure plate. Before the clutch cover is bolted to the engine flywheel, the diaphragm spring is not deformed and there is a small gap between the clutch cover and the flywheel even though the flywheel, the friction disk, and the pressure plate touch each other. When the clutch cover is bolted to the engine flywheel, the small gap is eliminated and the diaphragm spring is therefore deformed in the axial direction, generating the clamping force on the pressure plate. This results in bending in the upper portion of the diaphragm spring and an axial deflection at the pivoting point that is equal to the gap. By design, this deflection of the diaphragm spring generates sufficient clamping force to fully engage the clutch at assembly. The friction disk is then clamped tightly between the pressure plate and the flywheel. When the clutch needs to be released, the clutch actuation linkage pushes the release bearing to the left, carrying the diaphragm spring inner side to the left also. The leftward deflection at the inner side reduces the clamping force on the pressure plate due to the lever effect of the diaphragm spring itself. If the release bearing moves to the left sufficiently, then the combined effect of spring deformation and leverage will pull the pressure plate completely out of contact with the friction disk, fully disengaging the clutch. Note that the axial stiffness of the diaphragm spring – i.e. the relation between the axial spring deflection and the axial force applied by the release bearing on the spring inner side – can be analytically formulated and validated by experiments.

The clutch actuation mechanism is fairly simple in structure. The release bearing is actuated by the clutch release fork which pivots about a joint attached to the clutch well of the transmission housing. This fork straddles over the groove on the release bearing and is connected to the clutch pedal by either a cable or a linkage. In some designs, the clutch fork is connected to the clutch pedal by a hydraulic hose with a master cylinder on the driver side and a slave cylinder on the clutch side. The slave cylinder can also be mounted with the release bearing on the transmission input shaft for compactness, with the piston pushing the release bearing directly. A mechanical advantage is designed between the clutch pedal and the clutch release fork to reduce the effort of the driver's foot on the clutch pedal.

2.4.2 Clutch Torque Capacity

The clutch for a manual transmission must be capable of transmitting the maximum engine torque; in other words, the clutch torque capacity must be higher than the maximum engine torque. As mentioned previously, clutch torque is generated by the friction on the friction disk due the clutch clamping force. It is therefore clear that the clutch torque capacity depends on several factors: the friction coefficient of the friction disk, the clamping force, and the dimensions of the disk. In formulating the clutch torque capacity, an assumption must be made about the distribution of the normal pressure generated by the clamping force on the disk face. One assumption is that the clamping force is evenly distributed over the disk. The other assumes that the clamping force is distributed in such a way that the power of friction per disk area is a constant from the inner radius to the outer radius. This assumption is equivalent to uniform wear on the disk face and is termed accordingly. The following equation quantifies the uniform wear assumption:

$$\mu p_n v = C_f \tag{2.4}$$

where μ is the friction coefficient of the friction disk, p_n is the distribution of the clamping force or the normal pressure in $^{N}/_{m^2}$ or psi. v is the speed at a point on the disk friction face. C_f is the friction power per unit area that is a constant with a unit of $^{W}/_{m^2}$ or $^{ft.lb.s^{-1}}/_{ft^2}$. Here, W is watts. Based on this assumption, the clamping force distribution or the normal pressure is then described by:

$$p_n = \frac{C_f}{\mu v} = \frac{C_f}{\mu \omega r} = \frac{C}{r} \tag{2.5}$$

where r is the radius of a point on the disk friction face, as shown in Figure 2.12. C is also a constant since the term $\frac{C_f}{\mu\omega}$ is constant for a given disk at a given angular velocity ω. Therefore, the maximum normal pressure p_{max} occurs at the inner radius and the constant C is equal to $\frac{d}{2}p_{max}$. The distribution of the clamping force is thus expressed as $p_n = \frac{1}{2}p_{max}\frac{d}{r}$. The clamping force and the friction torque on one friction face are then determined by integration through the following equations:

$$F = 2\pi \int_{\frac{d}{2}}^{\frac{D}{2}} p_n r dr = 2\pi \int_{\frac{d}{2}}^{\frac{D}{2}} \frac{1}{2} d p_{max} dr = \frac{\pi}{2} p_{max}\, d(D-d) \tag{2.6}$$

$$T = 2\pi\mu \int_{\frac{d}{2}}^{\frac{D}{2}} \mu p_n r^2 dr = 2\pi \int_{\frac{d}{2}}^{\frac{D}{2}} \frac{1}{2}\mu d p_{max} r dr = \frac{\pi}{8}\mu p_{max}\, d\left(D^2 - d^2\right) = \mu F \frac{D+d}{4} \tag{2.7}$$

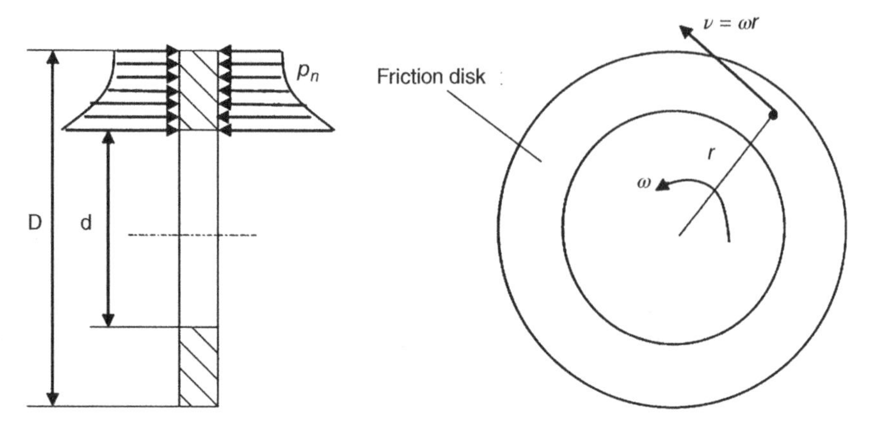

Figure 2.12 Distribution of clamping force on disk face.

where d and D are the inner and outer diameters of the friction disk respectively. Since each disk has two faces, the clutch torque for a clutch with n disks is determined by:

$$T_C = n\frac{D+d}{2}\mu F \tag{2.8}$$

For a given clutch, $n\frac{D+d}{2}$ is a constant. The friction coefficient depends on the lining material and can be considered as a given value in clutch design. For ceramic lining, the friction coefficient is about 0.25, and for an organic lining, it is about 0.30. In real world applications, the value of friction coefficient is affected by clutch temperature to some degree and gradually fades away with the clutch service life. The dominating parameter in the clutch torque is the clamping force F generated by the clutch spring. Note that Eq. (2.8) will also be used to model the torque in wet clutches in automatic transmissions where the clamping force F is generated by hydraulic pistons.

2.4.3 Clutch Design

The clutch for a manual transmission must be capable of transmitting the maximum engine torque. This means that the clutch torque capacity must be designed to be higher than the maximum engine torque with some reservation, i.e. $T_C = (1.15 \sim 1.20)T_{emax}$.As can be observed in Eq. (2.8), there are only three design parameters, which are the inner and outer diameters of the friction disk and the clamping force. The friction coefficient is considered as a constant and its value depends on the lining material, as already mentioned. The inner and outer diameters must guarantee both sufficient friction surface area and enough room in the radial direction for the assembly of the spring damper. In addition, the outer diameter is also limited by the centrifugal stress at the disk perimeter at high RPMs and the availability of assembly room in the clutch well. The magnitude of the clamping force depends on the maximum contact stress allowed for the disk lining material. The contact stress is in magnitude equal to the normal pressure p_n. After the normal pressure p_n is determined based on friction disk material properties, the disk dimension, namely the inner and outer diameters can then be designed according to the required torque capacity and assembly conditions. Note that the clutch spring has to be further deformed at the fully disengaged position. The spring force at the fully disengaged position is usually designed to be about 15% larger than the spring force F at the engaged position. This means that the spring stiffness is equal to $\frac{0.15F}{\delta}$, with δ as the pressure plate travel between the engaged and disengaged positions. Knowing the spring forces at the two clutch positions and the spring stiffness, clutch designers can then select the springs from product inventories of spring suppliers. In the automotive industry today, manual transmission clutches are mature products and can be readily supplied by clutch manufacturers according to the required torque capacity.

2.5 Synchronizer and Synchronization

Gear shifting in a transmission involves coupling together two components or assemblies that turn at different angular velocities in the same direction. This is dynamically similar to the case when an object moving linearly is collided behind by another object moving collinearly at a higher speed. If the rotational speeds of the two components during a gear shift are not brought to the same value, i.e. not synchronized, before they are coupled together, a collision will occur between them, resulting sharp noises, gear grinding, and even component damage. It is possible to make synchronized shifts, or nearly synchronized shifts, in a manual transmission without synchronizers if the driver is highly skilled. But for an average driver, it is a very difficult (if not impossible) job to operate a vehicle equipped with a non-synchronized manual transmission. The vast majority of manual transmissions today are equipped with synchronizers, with a few exceptions for heavy duty trucks. Note that the synchronization issue during shifts is also important for automatic transmissions. In an automatic transmission, synchronization during shifts is realized by controlling the slippage in hydraulically actuated clutches, as discussed in Chapter 6.

2.5.1 Shift without Synchronizer

In the following qualitative analysis, the five-speed MT in Figure 2.7b is used as the example to demonstrate the 3–4 upshift and the 4–3 downshift processes in the absence of synchronizers. Other upshifts or downshifts are similar in nature, so the example can be extended to all other manual transmissions for generality.

3–4 Upshift: The 3–4 upshift involves the disengagement of the 3rd gear and the engagement of the 4th gear. This is realized by moving the 3–4 shifter or the 3–4 synchronizer sleeve, illustrated in Figure 2.13, from the 3rd gear position to the 4th gear position along the output shaft so that the internal teeth on the 3–4 shifter engage the dog teeth ring on the input gear (the 4th gear) N_i. When the vehicle is driven in 3rd gear at speed v, the 3rd gear N_{3o} is engaged by the 3–4 shifter to the output shaft and turns with it at the angular velocity determined by:

$$\omega_{3o} = \omega_{3-4\,shifter} = \omega_{out} = i_a \frac{v}{r} \tag{2.9}$$

where i_a is the final drive ratio and r is the tire rolling radius. The 4th gear, i.e. the input gear N_i on the input shaft, turns at an angular velocity higher than the output, that is,

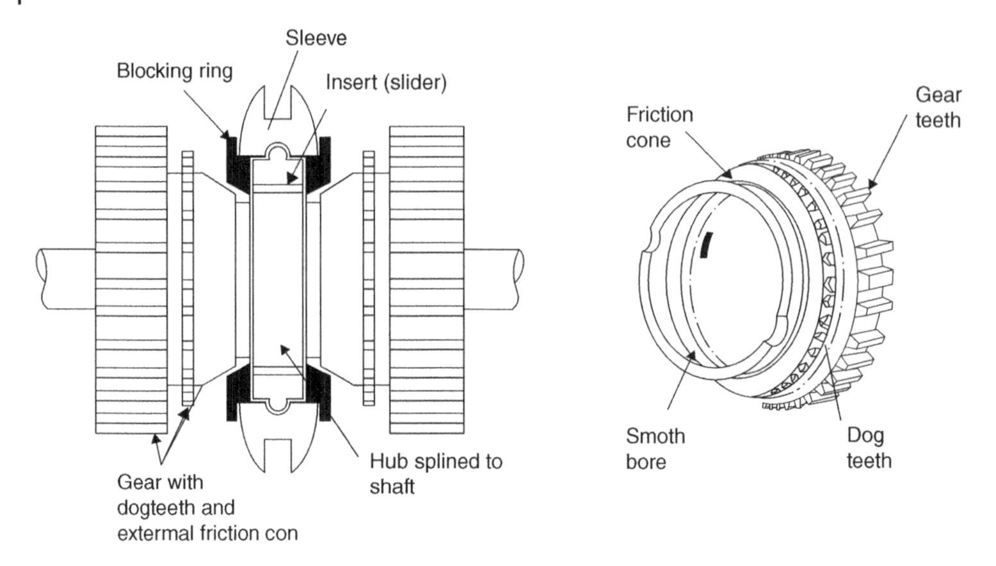

Figure 2.13 Synchronizer assembly and gear with dog teeth.

$\omega_i = i_3\omega_{out} > \omega_{3-4\,shifter}$ since $i_3 > 1$. If the driver decides to make a 3–4 upshift, the clutch is first disengaged to cut off the engine power and then gear N_{3o} is decoupled from the output shaft by the 3–4 shifter, which is now at the neutral position. While the 3–4 shifter stays at neutral, both angular velocities, ω_i and $\omega_{3-4\,shifter}$, decrease since there is no power supplied to the transmission. However, ω_i decreases at a much higher rate than $\omega_{3-4\,shifter}$ because the 3–4 shifter turns with the output shaft that is coupled to the whole vehicle mass. Therefore, the two angular velocities, ω_i and $\omega_{3-4\,shifter}$, will become equal themselves at some time. The driver can then actuate the shift stick to engage the 4th gear at or near that time, making a synchronized shift without the help of a synchronizer. Clearly, to make this happen, the driver needs high skill and must know the right time to engage the target gear.

4–3 Downshift: In downshifts, the next gear turns at higher speed than the current gear, contrary to upshifts. For the five-speed MT in the example, a 4–3 downshift involves the disengagement of the 4th gear N_i and the engagement of the 3rd gear N_{3o} through the 3–4 shifter. When the vehicle is driven in 4th gear at speed v, gear N_i is coupled by the 3–4 shifter to the output shaft and both turn at the output angular velocity determined by:

$$\omega_i = \omega_{3-4\,shifter} = \omega_{out} = i_a\frac{v}{r} \tag{2.10}$$

The 3rd gear N_{3o} turns at a lower angular velocity determined by:

$$\omega_{3o} = \frac{i_4}{i_3}i_a\frac{v}{r} = \frac{1}{i_3}i_a\frac{v}{r} < \omega_{3-4\,shifter} \tag{2.11}$$

When a 4–3 downshift is initiated, first the clutch is disengaged and then the 4th gear N_i is decoupled by the 3–4 shifter, which now stays at the neutral position for a fraction of a second. While the 3–4 shifter stays at the neutral position, both angular velocities,

$\omega_{3-4\,shifter}$ and ω_{N3o}, decrease since power is cut off. However, $\omega_{3-4\,shifter}$ decreases at a much lower rate than ω_{3o} for the same reason as mentioned previously in the analysis for the 3–4 upshift. This makes it impossible for the two angular velocities to become equal by themselves because $\omega_{3-4\,shifter}$ is already higher than ω_{3o} before the 4–3 shift is initiated and drops at a much lower rate than ω_{3o} at the neutral position. Therefore synchronized downshift will not happen without driver's intervention. In this case, the driver can use a so-called "double declutching shift" technique to make synchronized or near synchronized downshifts. In double declutching, the driver will briefly engage the clutch and step on the gas pedal while the shift stick is at neutral position. This action will quickly increase angular velocity ω_{3o} to be above $\omega_{3-4\,shifter}$ as the engine RPM flares. The driver then disengages the clutch again. Now that ω_{3o} is higher than $\omega_{3-4\,shifter}$ but drops at a much higher rate than $\omega_{3-4\,shifter}$ while the 3–4 shifter is still in the neutral position, the two angular velocities will become equal at some point of time as in the 3–4 upshift example. An experienced driver would then be able to make a synchronized downshift by actuating the shifting stick at the right time.

2.5.2 Shift with Synchronizer

It can be seen from this analysis that even though synchronized shifts are possible for transmissions without synchronizers for well-experienced drivers, driving such vehicles will still be a very difficult and unpleasant endeavor for an average driver. This is the reason why manual transmissions for today's passenger vehicles are all synchronized. To understand how synchronizers work, it is helpful to first take a look at the structure of a typical synchronizer, as shown in Figures 2.13 and 2.14.

Figure 2.13 illustrates the assembly of a typical synchronizer with a gear on each side. This assembly consists of several key components: synchronizer hub, shifting sleeve, two blocking rings (one on each side of the hub), and inserts (or sliders) that are located between the synchronizer hub and the shifting sleeve. The hub has both internal splines and external splines. The internal splines couple the hub to the shaft so the hub and the

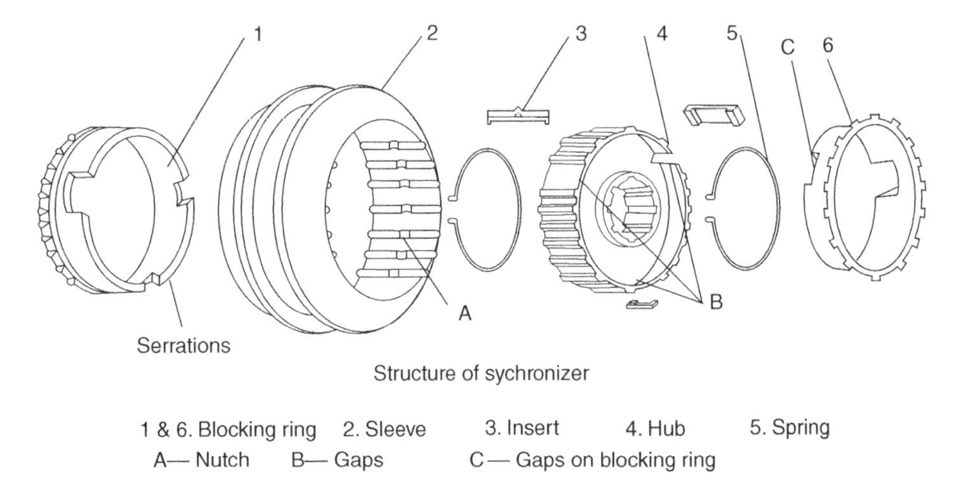

Figure 2.14 Exploded view of a synchronizer.

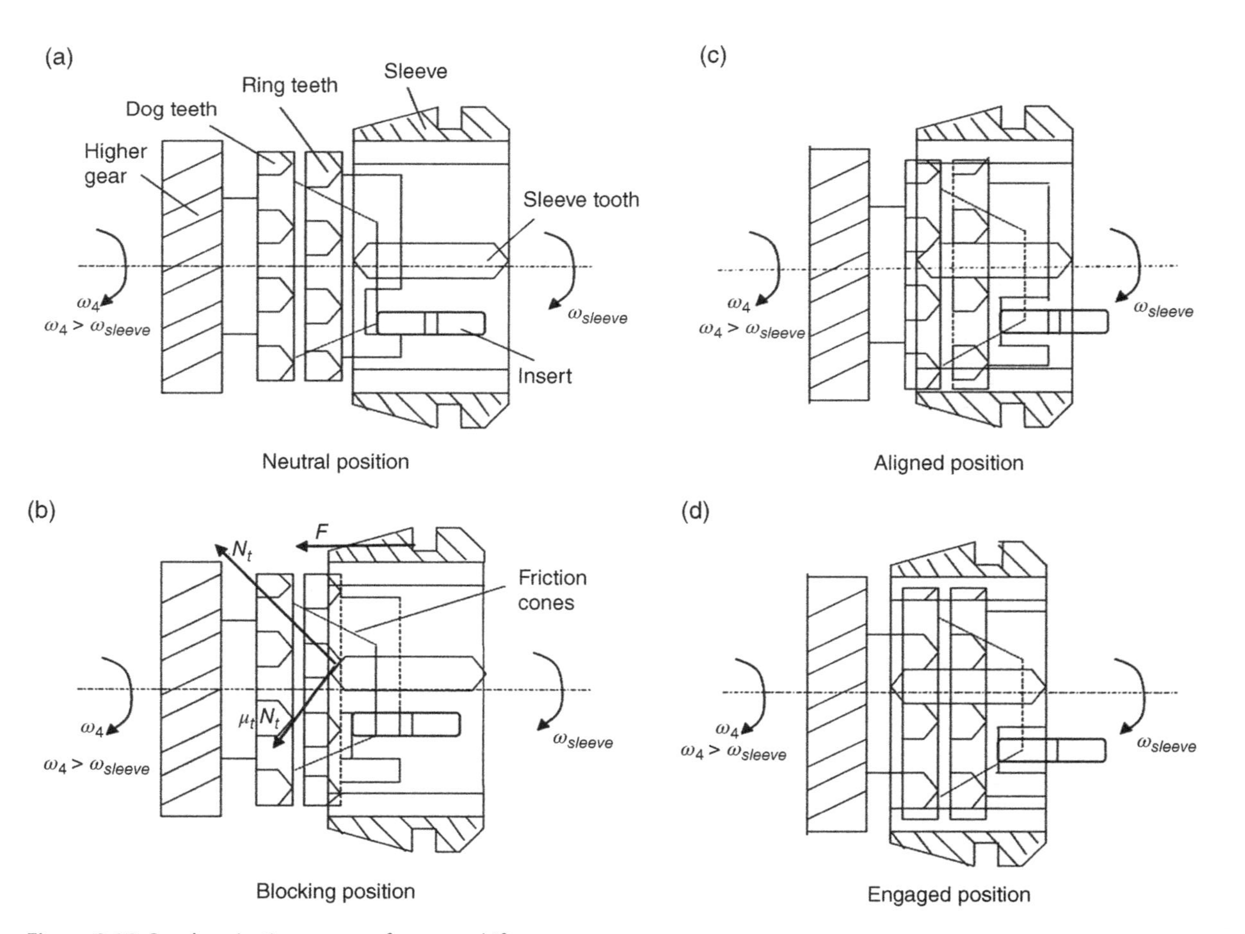

Figure 2.15 Synchronization process for an upshift.

shaft always turn together. The external spline teeth of the hub mesh with the internal spline teeth of the shifting sleeve. This allows the shifting sleeve to slide over the hub during a shift, while turning with it. Gears are located on each side of a synchronizer and always freewheel on the shaft unless engaged by the shifting sleeve. For the example transmission, we can imagine that the 4th gear is on the left side of the synchronizer and the 3rd gear is on the right side, as shown in Figure 2.13. The engagement of a gear is through the mesh between the dog teeth on it and the internal teeth of the sleeve. The dog teeth form a ring with the same pitch radius as the internal spline teeth on the sleeve. Power is transmitted by the regular gear teeth once the related gear is engaged by the shifting sleeve. The two blocking rings, placed between the gear and the hub on each side, both have a ring of teeth with the same pitch radius and other tooth element proportions as the internal spline teeth of the shifting sleeve. Clearly, the shifting sleeve has to slide over the blocking ring to engage a gear. The blocking rings are usually made of bronze to enhance wear resistance and are both machined with an internal conical surface that contacts the external conical surface on the gear. It is the friction generated between the two contacting conical surfaces that realizes the synchronization during a shifting process. Double friction cones can be designed to minimize synchronization time for quicker shifts, as in sports car applications. Without the blocking rings, the assembly is just a plain gear shifter that disengages and engages gears during shifts without synchronization.

The exploded view in Figure 2.14 offers a better understanding on the structure and operation of synchronizer. As shown in the figure, three symmetrically arranged gaps are machined on the external spline of the hub (4). The three inserts (3) have the same width as these gaps and are placed in them. There is a small ridge at the middle of each insert that fits into a notch on the internal spline of the sleeve (2). Two ring-shaped springs (5), one on each end of the insert, fit into the corners of the three inserts and push them outward against the sleeve internal spline. There are also three symmetrically arranged gaps on each of the blocking rings (1). The three inserts (3) are longer than the gaps on the hub, therefore the two ends of these inserts extend into the gaps on the two blocking rings in the assembly. The blocking ring gaps are slightly wider than the width of the inserts, so a small relative rotation is allowed between the blocking ring (1) and the inserts (3). The teeth on the two blocking rings (1) have the same pitch radius and tooth element proportions as the internal spline teeth of the sleeve (2). The tooth space on the blocking ring and the tooth on the sleeve internal spline are aligned by design only when the inserts are exactly located at the middle of the gaps on the blocking ring. This is the only relative position between the blocking ring and the sleeve that will allows the sliding of the sleeve over the blocking ring. When the vehicle is driven in any particular gear, the insert contacts one side of the gap on the blocking ring by the gear that is not engaged and carries the blocking ring in rotation with the synchronizer assembly and the shaft.

3-4 Upshift: When the vehicle is driven in 3rd gear, the sleeve of the 3–4 synchronizer, i.e. the 3–4 shifter, engages the 3rd gear N_{3o} and couples it to the output shaft. The inserts contact the two blocking rings at the lower end of the gap on it, as shown in Figure 2.15a. The blocking rings are thus carried by the inserts to rotate with the output shaft and the 3–4 synchronizer assembly at angular velocity ω_{sleeve} that is equal to ω_{out}, as determined by Eq. (2.9). When a 3–4 upshift is initiated, the clutch is firstly disengaged to cut off engine power and meanwhile the 3rd gear N_{3o} is disengaged by the sleeve, or the 3–4 shifter. The sleeve is now temporarily at the middle position, i.e. the neutral position, and the inserts still contact the blocking ring at the same point as in 3rd gear operation, as shown in Figure 2.15a. As the driver pushes or pulls the shift stick continuously, the sleeve is forced to move leftward by the shifting force F and tends to carry the inserts in this motion due to the combined effects of the ridge on the inserts and the friction between the inserts and the addendum of the spline teeth. This quickly leads the left end of the inserts to contact the blocking ring and applies a force on it in the axial direction. Contact force and friction with it are then generated between the friction cones of the 4th gear and the blocking ring.

This friction then turns the blocking ring quickly through a small angle in the direction as carried by the 4th gear since it turns faster than the 3rd gear and as allowed by the width difference between the gap and the inserts, until the inserts contact the other side of the gap on the blocking ring, as shown in Figure 2.15b. At this position, there is no relative rotation between the blocking ring and the sleeve, but the sleeve keeps moving leftward as pushed by the shift force F until the teeth on the sleeve internal spline contact the teeth on the blocking ring as shown in Figure 2.15b. Since the inserts are located at the end side of gaps on the blocking ring, the tooth space on the blocking ring is not aligned with the tooth space of the sleeve internal spline. As a result, the teeth of the blocking ring and the teeth of the sleeve internal spline contact against each other at the blocked position shown in Figure 2.15b. Note that the blocking ring must turn quickly enough so that the insert reaches the other side of the gap on the blocking ring before the sleeve slides to the blocked position. The ends of the blocking ring teeth and the sleeve internal spline are machined with round-up surfaces that form an arrow shaped tip, so that at the contact point between the blocking ring teeth and sleeve internal spline teeth, contact force N_t and friction force $\mu_t N_t$ will be generated on an inclined side surface on the tooth end as the sleeve is pushed by the shifting force F.

Meanwhile, the friction cone on the blocking ring will be pressed against the friction cone on the 4th gear, and friction is generated due the existence of the normal contact force on the conical surfaces and the relative motion between the two cones. At the blocked position, the blocking ring is subject to the friction between the friction cones of the 4th gear and the blocking ring itself, and at the contact between itself and the sleeve, the contact force N_t and friction force $\mu_t N_t$. The resultant of contact force N_t and friction force $\mu_t N_t$ has a component in the tangential direction of the blocking ring as shown in Figure 2.15b and Figure 2.16b, forming a torque about the axis of the synchronizer that tends to rotate the blocking ring to make way for the sleeve to slide over. This effect is resisted by a resistant torque generated by the friction between the two friction cones on the blocking ring. This resistant torque is reactive and is by design larger than or equal to the torque formed by the resultant of contact force N_t and friction force $\mu_t N_t$. Therefore, the sleeve is blocked by the blocking ring at the blocked position.

While the sleeve is at the blocked position, the friction between the two friction cones acts against the 4th gear and decelerates its angular velocity ω_4. As time goes by, the angular velocity of the 4th gear decreases to be equal to that of the sleeve, i.e. $\omega_4 = \omega_{sleeve}$. When these two angular velocities are synchronized, the friction between the two friction cones disappears due to the lack of relative motion. The shift force F, contact force N_t and friction force $\mu_t N_t$ still exist because of the driver's actuation on the shifting stick. Now that there is no friction between the two friction cones to resist the torque formed by the resultant of contact force N_t and friction force $\mu_t N_t$ in the tangential direction, the blocking ring will turn slightly so that the inserts will be positioned

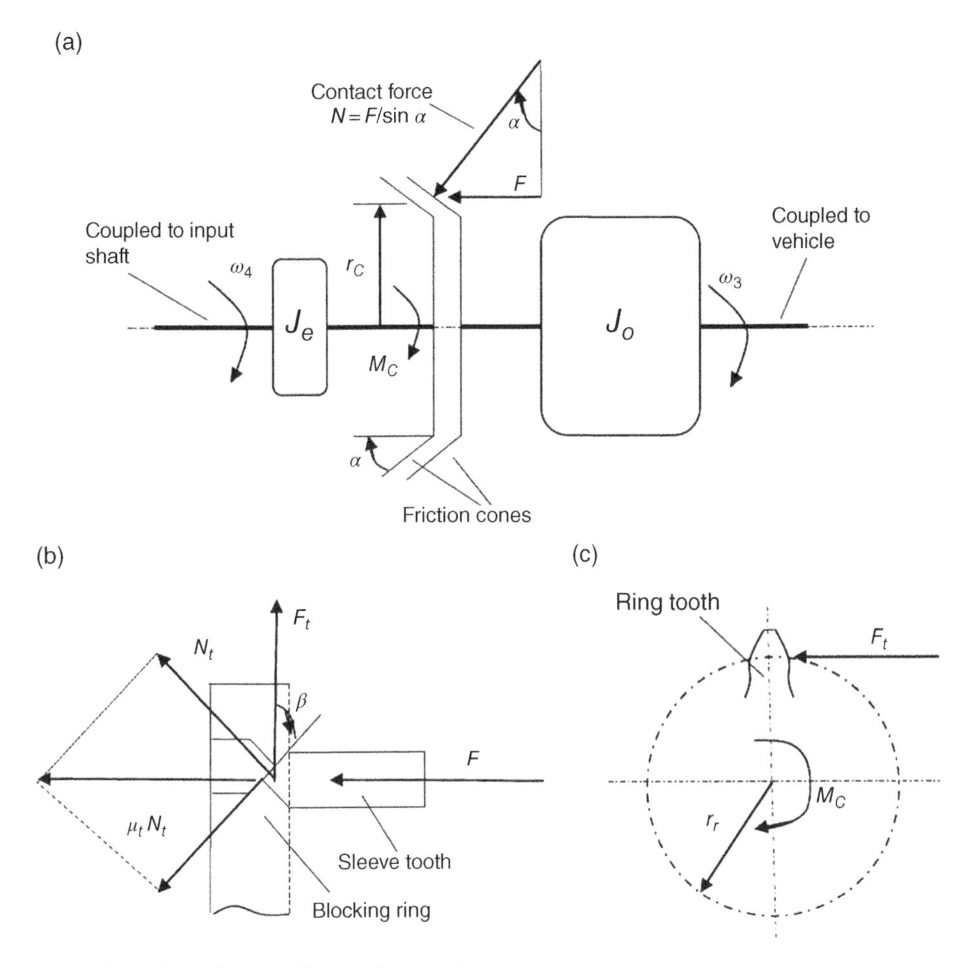

Figure 2.16 Dynamic model for synchronization process.

at the middle of the gap to align the tooth space on the blocking ring and the internal sleeve spline teeth, allowing the sleeve to slide over the blocking ring, as shown in Figure 2.15c. After the sleeve slides over the blocking ring, its internal spline teeth then contact the dog teeth on the 4th gear at the round-up tooth ends. Now that the angular velocity of the 4th gear and the sleeve are the same, the dog teeth can be engaged quickly and smoothly, as shown in Figure 2.15d.

4–3 Downshift: The 4–3 downshift, or any other downshifts, are basically similar to the 3–4 upshift analysed above. There are some general differences between downshifts and upshifts. Before a 4–3 downshift is initiated, the 2–3 synchronizer sleeve engages the 4th gear N_i, as shown in Figure 2.15d. In the 4–3 downshift process, the 4th gear is firstly disengaged and the sleeve is then at the neutral position, as illustrated in Figure 2.15a. Unlike the 3–4 upshift, the inserts remain at the position in the blocking ring gap as shown in Figure 2.15a during the synchronization process. The friction between the friction cones of the 3rd gear and the blocking ring on the right side (not shown in Figure 2.15) of the 3–4 synchronizer speeds up the angular velocity of the 3rd gear

N_{3o} on the output shaft and synchronizes it with the angular velocity of 3–4 synchronizer sleeve. The 4–3 downshift is otherwise the same as the 3–4 upshift.

2.6 Dynamic Modeling of Synchronization Process

A lumped mass model is shown in Figure 2.16a for the dynamic analysis of the synchronization process during manual transmission shifts. The model is based on a 3–4 upshift for the example 5-speed RWD MT, shown again in Figure 2.17 for readers' convenience, but is generally applicable for all shifts in other manual transmissions. As shown in Figure 2.16, F is the shifting force transferred from the shifting stick, N is the contact force between the two friction cones, and M_c is the friction torque generated by the friction cones about the shaft. The geometry of the friction cones is defined by the mean radius r_c and the cone angle α. The pitch radius of the sleeve internal spline and the blocking ring are the same, denoted as r_r in Figure 2.16b. The to-be-synchronized component in the 3–4 upshift is the input gear or 4th gear N_i and its angular velocity is denoted as ω_4. In the 3–4 upshift, ω_4 is to be reduced by the friction torque M_c to be synchronized with ω_3, which is the angular velocity of the 3–4 synchronizer sleeve. It should be noted that the 3–4 synchronizer sleeve and the output shaft rotate together at the same angular velocity since the 3–4 synchronizer is coupled with the output shaft.

When a manual transmission makes a shift, the angular velocity of the output shaft is related to the vehicle speed and is almost unchanged because the vehicle has a huge inertia in comparison with the to-be-synchronized component. The angular velocities of synchronizers that are splined to the output shaft and gear pairs rotating with the output shaft are also almost unchanged. All other components, including gears, shafts, and synchronizers that rotate separately from the output shaft, will change angular velocity and need to be synchronized as a lumped mass. As shown in Figure 2.16a, the equivalent mass moment of inertia of this lumped mass is denoted as J_e, and the equivalent mass moment of inertia of the vehicle and components coupled to the output shaft is denoted as J_o.

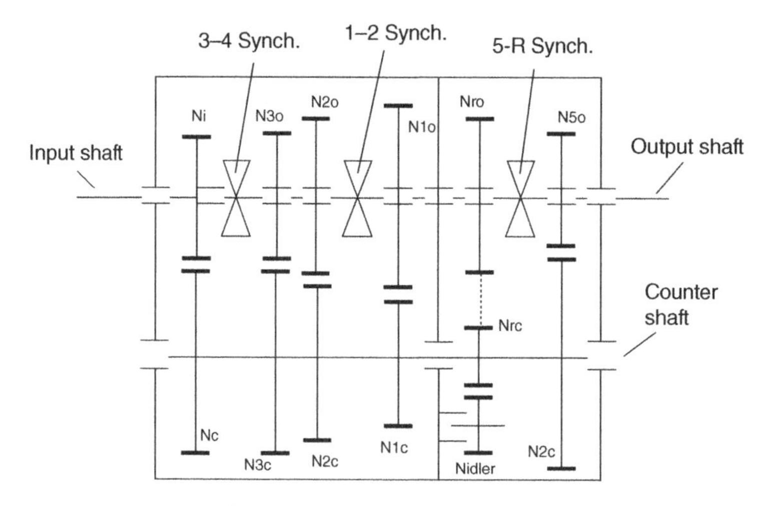

Figure 2.17 Example five-speed RDW MT.

2.6.1 Equivalent Mass Moment of Inertia

Note that the components that rotate separately from the output shaft are always the same during all shifts for the same transmission. **To find the equivalent mass moment of inertia, the first step is to identify those components in the transmission whose angular velocities change during shifts.** These components can be easily identified by observing the stick diagram. For the example transmission in Figure 2.17, the output shaft and the three synchronizers on it are coupled to the vehicle inertia and therefore do not change angular velocity. All other components, including the input shaft with the clutch friction disk on it, the counter shaft with all gears on it, the reverse idler and reverse gear that freewheels on the output shaft, and all gears that freewheel on the output shaft during shifts, do change angular velocity and need to be accounted for in the equivalent mass moment of inertia. **The second step in finding the equivalent mass moment of inertia is to identify the component that is to be synchronized directly by the synchronizer.** The directly synchronized component varies for different shifts. For the example MT in Figure 2.17, the component to be synchronized directly is the input shaft with gear N_i in a 3–4 upshift. But for a 4–5 upshift, the component to be synchronized directly is gear N_{5o}. The equivalent mass moment of inertia to be synchronized is lumped on the component that is to be synchronized directly by the synchronizer. The determination of the equivalent mass moment of inertia is based on the condition that the kinetic energy of the equivalent mass moment of inertia must be same as the sum of kinetic energies of all individual components that change angular velocity during a shift, i.e.

$$\frac{1}{2}J_e\omega_s^2 = \frac{1}{2}\sum_{i=1}^{n} J_i\omega_i^2 \tag{2.12}$$

$$J_e = \sum_{i=1}^{n} J_i\left(\frac{\omega_i}{\omega_s}\right)^2 \tag{2.13}$$

where ω_s is the angular velocity of the component to be synchronized directly, ω_i and J_i are respectively the angular velocity and the mass moment of inertia of each component that changes angular velocity during the shift. The summation is for all those components that need to be synchronized as lumped masses. In summary, the equivalent mass moment of inertia can be determined for any shift by following steps:

- Assuming the transmission output speed remains the same, identify all the components in the transmission that will change angular velocities during shifts. These components are the same for the same transmission in all shifts.
- Identify the component, which can be a gear or a shaft, to be synchronized directly in the shift of interest.
- Use Eq. (2.13) to find the equivalent mass moment of inertia for the shift.

For the 3–4 upshift in the example transmission shown in Figure 2.17, ω_s is the angular velocity of the 4th gear N_i, i.e. ω_4 as shown in Figure 2.16a. ω_i is the angular velocity of each individual component that changes angular velocity, including the input shaft, the counter shaft with all gears on it, the reverse idler, and all gears that freewheel on the output shaft during shifts. The summation is for all of these individual components. The angular velocity ratios in Eq. (2.13) depend on the tooth numbers of related gears.

For the 3–4 upshift in the example, the equivalent mass moment of inertia is determined by:

$$J_e = J_{Ni}\left(\frac{\omega_4}{\omega_4}\right)^2 + J_{CS}\left(\frac{\omega_{CS}}{\omega_4}\right)^2 + J_{idler}\left(\frac{\omega_{idler}}{\omega_4}\right)^2 + J_{N1o}\left(\frac{\omega_{1o}}{\omega_4}\right)^2 + J_{N2o}\left(\frac{\omega_{2o}}{\omega_4}\right)^2 + J_{N3o}\left(\frac{\omega_{3o}}{\omega_4}\right)^2 + + J_{N5o}\left(\frac{\omega_{5o}}{\omega_4}\right)^2 + J_{Nro}\left(\frac{\omega_{ro}}{\omega_4}\right)^2 \quad (2.14)$$

$$J_e = J_{Ni} + J_{CS}\left(\frac{N_i}{N_C}\right)^2 + J_{idler}\left(\frac{N_i N_{Rc}}{N_C N_{idler}}\right)^2 + J_{N1o}\left(\frac{N_i N_{1C}}{N_C N_{1o}}\right)^2 + J_{N2o}\left(\frac{N_i N_{2C}}{N_C N_{2o}}\right)^2 + J_{N3o}\left(\frac{N_i N_{3C}}{N_C N_{3o}}\right)^2 + J_{N5o}\left(\frac{N_i N_{5C}}{N_C N_{5o}}\right)^2 + J_{NRo}\left(\frac{N_i N_{rC}}{N_C N_{ro}}\right)^2 \quad (2.15)$$

In Eq. (2.15), J_{CS} is the mass moment of inertia of the counter shaft assembly that includes all of the gears that rotate with it. All other mass moments of inertia are for each individual gear as indicated by the subscripts. The equivalent mass moment of inertia for other shifts can be found similarly by replacing the angular velocity of the component to be synchronized directly, ω_s in Eq. (2.13) accordingly. For example, for 4–5 shift, the component to be synchronized directly will be the fifth gear N_{5o} and ω_s is replaced by ω_{5o} in Eq. (2.13) to find the equivalent mass moment of inertia in a 4–5 shift. Note that if the component to be directly synchronized is the same, then the equivalent mass moments of inertia are also the same regardless of shifts. For example, the equivalent mass moment of inertia for 3–4 upshift and 5–4 downshift is the same as determined by Eq. (2.15).

The location of the synchronizers affects the number of components whose angular velocities will change during shifts and the equivalent mass moment of inertia. For example, the 3–4 synchronizer is located on the counter shaft for the six-speed RWD MT in Figure 2.9. This arrangement will reduce the number of components to be synchronized during shifts. In this case, the counter shaft is the component to be synchronized directly by the 3–4 synchronizer in any shift whose target gear is 3rd or 4th. In general, putting a synchronizer on the counter shaft for an RWD MT reduces the equivalent mass moment of inertia, but increases the angular velocity difference to be synchronized during shifts.

The equivalent mass moment of inertia coupled to the vehicle, J_O, can be determined in similar fashion. For the 3–4 shift in the example, J_O is determined on the output shaft as follows:

$$J_o = J_{3-4\,Synch.} + J_{1-2\,Synch.} + J_{5-R\,Synch.} + \frac{1}{2}\frac{W}{g}\left(\frac{r}{i_a}\right)^2 \gg J_e \quad (2.16)$$

The equivalent mass moment of inertia coupled to the vehicle is far larger than the mass moment of inertia to be synchronized. Therefore, it is reasonable to assume that the angular velocity of those components that are coupled the output shaft is unchanged during synchronization.

2.6.2 Equation of Motion during Synchronization

Referring to Figure 2.17, the equation of motion for the equivalent mass moment of inertia during the 3–4 upshift is represented in the following based on Newton's second law:

$$M_C = J_e \frac{d\omega_4}{dt} \tag{2.17}$$

Since the synchronization time is fairly short, the equation above can be approximated by its finite form as:

$$M_C = J_e \frac{\Delta\omega_4}{\Delta t} \tag{2.18}$$

where $\Delta\omega_4$ is the change of angular velocity before and after shift of the input shaft with the fourth gear N_i, which is the component to be synchronized directly in the 3–4 upshift, and Δt is the synchronization time. Clearly, larger equivalent mass moment of inertia, bigger change in the angular velocity during shift, and shorter synchronization time correspond to a larger friction torque being required for synchronization. The angular velocity difference in the 3–4 upshift can be calculated by:

$$\Delta\omega_4 = \Delta\omega_4{}^{(A)} - \Delta\omega_4{}^{(B)} = \frac{v}{r} i_a i_4 - \frac{v}{r} i_a\, i_3 = \frac{v}{r} i_a\,(1 - i_3) \tag{2.19}$$

As can be observed in this equation, the angular velocity difference that needs to be synchronized is proportional to the difference of the gear ratios between the current and next gears. It is also proportional to the vehicle speed v at which the shift is made. In addition, the angular velocity difference to be synchronized in upshifts, such as the 3–4 upshift in the example, is always negative. This means that the component to be synchronized directly in upshifts is always slowed down. As shown in Figure 2.17, the contact force between the friction cones generated by the shift force F is equal to $F/_{\sin\alpha}$. The friction torque M_C generated in the friction cones is thus determined by:

$$M_C = \mu F r_c / \sin\alpha \tag{2.20}$$

where μ is the friction coefficient of the friction cones and $\mu \approx 0.1$. For a given synchronizer, the friction torque M_C depends on only the shift force F. As can be observed from Eq. (2.18), the magnitude of the shift force directly affects the synchronization time Δ_t, which should be below half a second so that a shift, including actuations for disengagement, engagement, and synchronization, can be completed within a second. Shorter synchronization time requires a larger shift force F, which originates from the driver's hand pushing or pulling the shifting stick. Although a mechanical advantage is designed into the shift mechanism between the driver's hand and the shift sleeve, the shift force F should be kept within an appropriate range since a large mechanical advantage needs a large travel of driver's hand on the shifting stick. To reduce the effort and enhance the shifting feel of the driver, the shift force on the synchronizer sleeve should be below 100 N for passenger cars, and below 300 N for trucks. Before synchronization is achieved, the friction torque M_C has two crucial effects for the synchronization process, one of which is to reduce the angular velocity of the fourth gear and the other is to prevent the blocking ring from turning by the tangential resultant F_t of the contact force N_t and friction force $\mu_t N_t$ at the contact point between the blocking ring and the sleeve, as shown in Figures 2.16a and 2.16b. The friction torque M_C exists only when there is relative motion between the two friction cones. At the synchronization point, the relative angular velocity $\Delta\omega_4$ is equal to zero and the friction torque becomes zero also, as observed from Eq. (2.18), even though the shift force F still exists because the driver pulls or pushes the shifting stick continuously during the shift.

2.6.3 Condition for Synchronization

As shown in Figure 2.16b, the sleeve of the 3–4 synchronizer is blocked by the blocking ring and is in equilibrium before the synchronization is achieved, therefore,

$$N_t \cos\beta + \mu_t N_t \sin\beta = F \tag{2.21}$$

$$N_t = \frac{F}{\cos\beta + \mu_t \sin\beta} \tag{2.22}$$

where β is the contact angle at the tooth tips of the blocking ring and the sleeve internal spline, as shown in Figure 2.16b. Note that the contact force N_t and the friction force $\mu_t N_t$ are at the contact between the blocking ring teeth and the sleeve internal spline teeth, and the contact force N and friction force μN are at the contact between the two friction cones. All of these forces originate from the shift force F. The tangential component F_t is the sum of the projections of the contact force N_t and the friction force $\mu_t N_t$ in the tangential direction of the blocking ring, determined as:

$$F_t = N_t \sin\beta - \mu_t N_t \cos\beta = \left(\frac{\sin\beta - \mu_t \cos\beta}{\cos\beta + \mu_t \sin\beta}\right) F \tag{2.23}$$

As discussed previously, the tangential force component F_t in the equation above tends to turn the blocking ring in the direction to yield for the sleeve to slide over it. The critical condition for achieving synchronization is that the blocking ring remains at the blocked position before the synchronization point. To satisfy this condition, the friction torque generated by the friction cones must be larger than the torque generated by the tangential force component on the blocking ring, i.e.

$$M_C \geq F_t r_r \tag{2.24}$$

$$\mu r_c / \sin\alpha \geq \left(\frac{\sin\beta - \mu_t \cos\beta}{\cos\beta + \mu_t \sin\beta}\right) r_r \tag{2.25}$$

The inequality (2.25) is the condition for guaranteeing that the blocking ring does not turn before synchronization is achieved and is therefore the condition for synchronization. The shift force F does not appear in this inequality since it is a common factor for both M_C and F_t in inequality (2.24) and cancels itself out. It can be observed from inequality (2.25) that the friction coefficient μ_t at the contact between the blocking teeth and the sleeve internal spline teeth makes the right-hand side smaller. Neglecting μ_t makes the right-hand side larger and strengthens the synchronization condition. The strengthened synchronization condition is then represented by the following inequality:

$$\tan\beta \leq \frac{\mu r_c}{r_r \sin\alpha} \tag{2.26}$$

To satisfy the synchronization condition, the cone angle α should be as small as possible to increase the contact force N for a larger friction torque M_C. But it has to be above a minimum value to avoid the locking of the friction cones. This minimum cone angle depends on the friction coefficient of the friction cones and is equal to the friction angle, i.e. $\alpha_{min} = \tan^{-1}\mu$. The friction coefficient μ is about 0.1 as mentioned previously, therefore the minimum cone angle to avoid friction cone locking is about 6°. In synchronizer design, the cone angle is usually above 6° for a factor of safety. The ratio between the friction cone mean radius and the blocking ring pitch radius is about 0.85. Thus the contact angle β is in the neighborhood of 35°. It should be noted that the friction coefficient μ depends on many factors, such as temperature and lubrication, and is impossible to be accurately determined. The synchronization condition and the analysis in this paragraph provide only a guideline for synchronizer design, so experimental data and experience of existing production synchronizers are crucial to the selection of the related design parameters. Synchronizers are mature products in today's automotive industry and can usually be ordered from suppliers from existing inventories based on transmission specifications.

As long as the synchronization condition is satisfied, the angular velocity of the to-be-synchronized component during a shift, ω_4 in the 3–4 shift of the example transmission, will be synchronized to be the same as the coupling component, the 3–4 synchronizer sleeve in the example, as described by Eq. 2.17. As soon as synchronization is achieved, the friction torque M_C disappears and the blocking ring is then turned slightly by the tangential force component F_t, aligning the tooth space of the blocking ring with the teeth of the sleeve internal spline. The shift force F further pushes the sleeve over the blocking ring and eventually engages the sleeve with the dog teeth of the fourth gear.

Example 1.1 The stick diagram of a five-speed FWD MT is shown in Figure 2.7a. The tire radius of the vehicle is R (ft). The numbers of teeth are labeled in the drawing.

a) Determine the equivalent mass moment of inertia to be synchronized in a 1–2 upshift and a 5–4 downshift respectively, in terms of the gear tooth numbers and mass moments of inertia of the involved parts.
b) A 1–2 upshift is to be made at a vehicle speed of v (mph). Assuming the synchronization time is Δt and the angular velocity change is uniform during synchronization, determine the friction torque to be generated by the synchronizer friction cones.
c) Determine the work done by the friction in the process of synchronization for the shift in (b).
d) The synchronizers used in the transmission are the same, and the angular velocity change during synchronization is assumed to be uniform. Find the ratio between the magnitude of the shift force in a 4–5 upshift, F_{45}, and the magnitude of the shift force in a 5–4 downshift, F_{54}. Both shifts are made at the same engine speed and within the same synchronization time.

Solution:

a) For this transmission, the following components need to be synchronized during shifts: the input shaft assembly that includes gears N_{1i}, N_{2i}, N_{ri} and the 3–4 and 5th synchronizers, gears N_{1o} and N_{2o}. In a 1–2 shift, gear N_{2o} is directly synchronized by the 1–2 synchronizer that turns with the output shaft. In a 5–4 shift, the input shaft is directly synchronized by the 3–4 synchronizer to rotate at the same angular velocity as gear N_{4i}. The equivalent mass moments of inertia for the 1–2 shift and 5–4 shift are determined respectively in the following by using Eq. 2.13.

$$J_{12} = J_{N2o} + J_{input}\left(\frac{N_{2o}}{N_{2i}}\right)^2 + J_{N1o}\left(\frac{N_{2o}N_{1i}}{N_{2i}N_{1o}}\right)^2$$

$$J_{54} = J_{input} + J_{N2o}\left(\frac{N_{2i}}{N_{2o}}\right)^2 + J_{N1o}\left(\frac{N_{1i}}{N_{1o}}\right)^2$$

b) $$\omega_{output} = 1.467 i_a \frac{V}{R} = 1.467 \frac{N_{ao}}{N_{ai}} \frac{v}{R} \left({}^1/_s\right)$$

The angular velocity of gear N_{2o} before and after the 1–2 shift is:

$$\omega_{12}^B = \frac{i_1}{i_2}\omega_{output} = 1.467 \frac{N_{1o}}{N_{1i}} \frac{N_{2i}}{N_{2o}} \frac{N_{ao}}{N_{ai}} \frac{v}{R}$$

$$\omega_{12}^A = \omega_{output} = 1.467 \frac{N_{ao}}{N_{ai}} \frac{v}{R}$$

With J_{12} determined in (a), the friction torque can be determined using Eq. (2.18) as:

$$M_{12} = J_{12}\frac{\Delta\omega}{\Delta t} = J_{12}\frac{\omega_{12}^A - \omega_{12}^B}{\Delta t} = 1.467 \frac{J_{12}}{\Delta t} \frac{N_{ao}}{N_{ai}} \frac{v}{R}\left(1 - \frac{N_{1o}}{N_{1i}} \frac{N_{2i}}{N_{2o}}\right) \text{ (ft.lb)}$$

c) Friction work is equal to the change of kinetic energy. Therefore:

$$W_{12} = \frac{1}{2}J_{12}\left(\omega_{12}^A\right)^2 - \frac{1}{2}J_{12}\left(\omega_{12}^B\right)^2$$
$$= \frac{1}{2}J_{12}\left[\left(1.467\frac{N_{ao}}{N_{ai}}\frac{v}{R}\right)^2 - \left(1.467\frac{N_{1o}}{N_{1i}}\frac{N_{2i}}{N_{2o}}\frac{N_{ao}}{N_{ai}}\frac{v}{R}\right)^2\right] \text{ (ft.lb)}$$

d) The equivalent mass moment of inertia for the 4–5 upshift and 5–4 downshift is the same since the component to be synchronized in both shifts is the same. As observed in Eq. (2.18) and Eq. (2.20), the shift forces in the 4–5 upshift and the 5–4 downshift are proportional to the angular velocity differences to be synchronized in the two shifts.

$$\text{For the 4} - \text{5 upshift: } \Delta\omega_{45} = \frac{\omega_e}{i_4} i_5 - \omega_e = \omega_e\left(\frac{i_5}{i_4} - 1\right)$$

$$\text{For the 4} - \text{5 downshift: } \Delta\omega_{54} = \frac{\omega_e}{i_5} i_4 - \omega_e = \omega_e\left(\frac{i_4}{i_5} - 1\right)$$

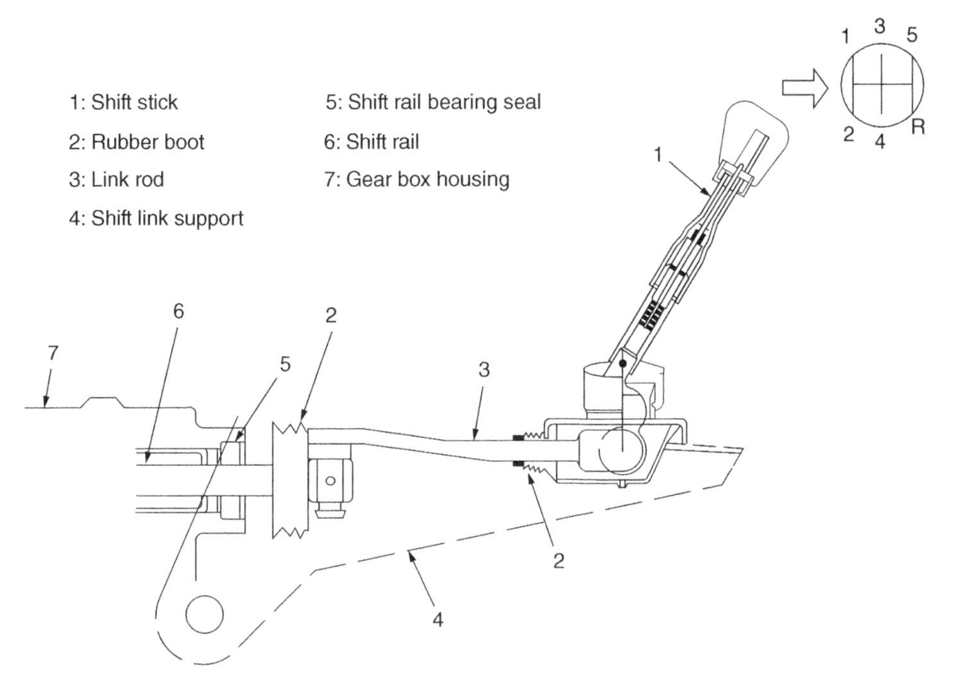

Figure 2.18 Shifting stick and shift pattern.

Since the shifts are made at the same engine speed, the shift force ratio is: $\left|\frac{F_{54}}{F_{45}}\right| = \left|\frac{\frac{i_4}{i_5}-1}{\frac{i_5}{i_4}-1}\right| = \frac{i_4}{i_5} = \frac{N_{4o}\,N_{5i}}{N_{4i}\,N_{5o}}$

2.7 Shifting Mechanisms

The shifting mechanism of a manual transmission must fulfill the following functional requirements:

a) Fully engage the synchronizer sleeve internal spline teeth with the gear dog teeth and keep the gear engaged without dropping out.
b) Prevent different gears from being engaged at the same time.
c) Differentiate the engagement of reverse gear from that of forward gears.
d) Be able to produce an appropriate mechanical advantage between the shift stick and the synchronizer sleeve to reduce the driver's effort during shifts.

The functions above are realized through the external shifting mechanism and internal shifting mechanism which are interconnected in kinematics by either a connection rod or cables. Shifts in the vast majority of, if not all, manual transmissions today are realized through the shifting stick that has two motions if actuated by the driver, a horizontal motion for gear selection and a fore and aft motion for gear engagement in a so-called H pattern. The internal shifting mechanism may have multi-rail design or single (or main) rail design. The following text uses a main-rail design as the example for analysis. The design and operation of multi-rail the shifting mechanism is similar in principle.

The external shifting mechanism shown in Figure 2.18 is typical for RWD manual transmissions, as shown in Figures 2.7b and 2.9b. The low end of the shifting stick (1) is connected with the main shifting rail (6) (or the single shifting rail if designed so) in the internal shifting mechanism, shown in Figure 2.19, by the shift link rod (3). The shifting stick pivots on a ball and socket joint mounted on the gear shifter support (4) rigidly attached to the transmission housing. Force and motion originated from the driver's hand are transmitted to main shift rail through the shift link rod (3). The joints at the two ends of the shift link rod allow the main shifting rod to rotate and move axially. The shift stick is blocked from moving into the reverse gear position unless it is lifted by the driver at the neutral position to release the blocking first. The joints on both ends of the shift link rod (3) are enclosed by rubber boots (2) and are lubricated at assembly. The ball and socket for the shifting stick is also lubricated at assembly and enclosed in rubber root that serves both sealing and cosmetic purposes.

The internal shifting mechanism is shown in Figure 2.19, including an internal shift rod (2) which is used to support the 3–4 shifter fork (19) and the 5-R shifter fork (4). The 1–2 shifter fork (3) is supported on the main shift rail (18) itself. The shift finger (11) and shift pin support (5) are rigidly attached to the main shift rail (18) by bolts. Spring (15) is

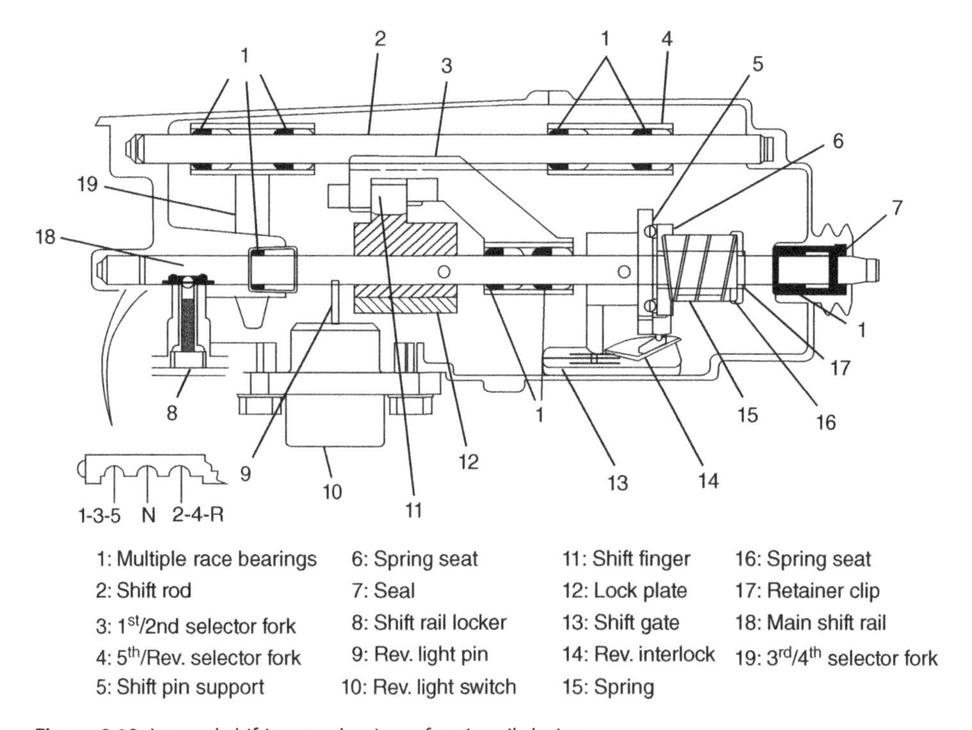

Figure 2.19 Internal shifting mechanism of main-rail design.

mounted on the main shift rail between the spring seat (16), which is retained axially by a snap ring (17) and spring support (6) that is mounted on the housing. Note that the 3–4 shifter and the 5-R shifter can also be mounted on the main shift rail in a single rail design that does not need the auxiliary shift rod (2). In either design, the shifter forks are supported on the shift rail by multiple race ball bearings (1) which minimize friction for linear motion. The shifter fork straddles over the groove of a synchronizer and maintains a fixed angular position as designed. The main shift rail is supported on both ends also by multiple race bearings. During a shift, the main shift rail has two motions: rotation for selection and axial motion for gear engaging. When the main shift rail rotates in the selection motion, the shift finger (11) on it will be aligned with the receiving notch on the selected shifter or selector fork. Since the angular position of the shifter is fixed, there is only one rotational angle for the main shift rail to align its shift finger with the receiving notch of the shifter. Meanwhile, only at this position, will the shift pin on the shift pin support (5) be aligned with the shift gate (13) for the selected gear. The shift gate guarantees that only the selected gear will be engaged and also provides the effect of guiding the engaging motion of the shifting stick and keeping the angular position of the main shift rail at an engaged position. Once a gear is selected during a shift – that is, the shift finger (11) is received and connected with the selected shifter and the shift pin is aligned with the gate for the selected gear – the driver will push or pull the shifting stick to actuate an axial motion for the main shift rail to engage the selected gear.

The axial positions of the main shift rail (18), namely the left position for 1–3–5 gears and the right position for 2–3–R gears, as shown in Figure 2.19, are locked by the shift rail locking mechanism (8). Three shallow ball-shaped or V-shaped races, are machined on the left end of the main shift rail which fit into a hole in the transmission housing. The center distances between these three races are equal to the travel of the shifters to fully engage the gear dog teeth. A ball is seated on top of a spring inside a hole drilled on the transmission housing, as shown in Figure 2.19. This ball is jacked up by the spring and fits into the races when the main shift rail is located at the three positions corresponding to the synchronizer sleeve or the shifter, locking the main shift rail in these positions unless the driver pulls or pushes the shifting stick, and therefore, keeping gears engaged without dropping out. In addition, the locking mechanism, designed appropriately, will yield a "clicking" or "sucking in" feel when a gear is disengaged or engaged. The spring (15) is not deformed at the neutral position and helps return the main shift rail to the neutral position. It also provides a damping feel for the driver while moving the shifting stick. In reverse gear, an actuating pin (9) will close the reverse warning light circuit to signal the driver and the pedestrians around the vehicle. The locking plate (12) and reverse interlock (14) double guarantee that only one gear can be engaged at a time.

Other designs for the shifting mechanism are also applied for manual transmission, depending on assembly and cost considerations. For example, three shift rails can be used, each for two gear positions and each with a locking mechanism. In this design, the shifters are fixed to the rails and move with the rails axially during shifts. Interlocks using balls and pins are placed between the three shifting rails, allowing only one rail to move axially at a time. The three receiving notches on the rail ends are arranged in a row with a distance between each. The shift finger on the lower end of the shifting stick is directly received and connected by a receiving notch on the rail for the selected gear. Other aspects for the shifting mechanism and operations are similar to those already discussed. Note that in manual transmissions for FWD vehicles, shift cables are often used instead of a shift rod, due to the peculiar assembly conditions. Usually, two cables contained in steel tubes are used, one for selection motion and the other for engaging motion. Since the cables are contained in steel tubes, both push and pull motions during shifts will be transmitted from the shifting stick to the shifting levers attached to the transmission housing. The shifting levels then complete the gear selection and engage for the intended shift. The details of these shifting mechanisms can be found in technicians' manuals for most passenger vehicles [6] in the market, which can be referred to for further studies on the subject.

References

1 Gott, P.G.: *Changing Gears: The Development of the Automotive*, Society of Automotive Engineers, 1991, ISBN 1-56091-099-2.
2 Manual transmission – Wikipedia https://en.wikipedia.org/wiki/Manual_transmission.
3 Transmission type market share – global car production 2017 https://www.statista.com/statistics/204123/transmission-type-market-share-in-automobile-production-worldwide/.
4 Fenton, J.: *Handbook of Automotive Powertrain and Chassis Design*, Professional Engineering Publishing, 1998, ISBN 1-86058-075 0.
5 Lechner, G.; Naunheimer, H.: *Automotive Transmissions: Fundamentals, Selection, Design and Application*, Springer, 1999. ISBN 3-540-65903.
6 Ford Parts and Service Division, Technical Training: *Aerostar Electronic Four-Wheel Drive and Scorpio MT-75 Manual Transmission*, 1989, Ford Motor Company.

Problems

1 The stick diagram of a six-speed RWD MT is shown Figure 2.9b. The tire radius of the vehicle is R (ft) and the final drive ratio i_a is 3.25. The numbers of teeth are labeled in the drawing. Some gear ratios are given as: 1st gear (3.92), 2nd gear (2.76), 3rd gear (1.91), 4th gear (1.41). The number of teeth of gears N_i and N_c are 19 and 23 respectively.
   a) Determine the equivalent mass moment of inertia to be synchronized in a 1–2 upshift and a 5–4 downshift respectively, in terms of the mass moments of inertia of the involved parts and the labels for the tooth numbers.
   b) A 2–3 upshift is to be made at a vehicle speed of V (mph). Assuming the synchronization time is Δt and the angular velocity change is uniform during synchronization, determine the friction torque to be generated by the synchronizer friction cones.
   c) Determine the work done by the friction and the angle of rotation of gear N_{4c} in the process of synchronization for the shift in (b).
   d) The synchronizers used in the transmission are the same and the angular velocity change during synchronization is assumed to be uniform. If the magnitude of the shift force in a 2–3 upshift, F_{23}, is 10 N, determine the magnitude of the shift force in a 1–2 upshift, F_{12}. Both shifts are made at the same engine angular velocity ω_e and within the same synchronization time Δt.

Note: In your answers for (a), (b), and (c), only the equivalent mass moments of inertia are allowed to contain labels for tooth numbers.

2 The stick diagram of a six-speed FWD MT is shown in Figure 2.9a. The tire radius of the vehicle is R (ft). The numbers of teeth are labeled in the drawing for clarity. The gear ratios from the 1st to the 4th gears are: 1st gear (3.92), 2nd gear (2.76), 3rd gear (1.85), 4th gear (1.35). The numbers of teeth of the final drive gears are: $N_{Fi} = 21$, $N_{Fo} = 66$.

a) Determine the equivalent mass moment of inertia to be synchronized in a 1–2 upshift and a 4–3 downshift respectively, in terms of the mass moments of inertia of the involved parts.
b) A 2–3 upshift is to be made at a vehicle speed of V (mph). Assuming the synchronization time is Δt and the angular velocity change is uniform during synchronization, determine the friction torque to be generated by the synchronizer friction cones.
c) Determine the work done by the friction and the angle of rotation of gear N_{4i} in the process of synchronization for the shift in (b).
d) The synchronizers used in the transmission are the same and the angular velocity change during synchronization is assumed to be uniform. If the shift force in a 2–3 upshift, F_{23}, is 10 N, determine the shift force in a 1–2 shift, F_{12}. Both shifts are made at the same engine speed and within the same synchronization time.

Note: Your answers must not contain the gear tooth labels.

3

Transmission Gear Design

3.1 Introduction

Gears are one of the most important components for mechanical systems. Although gears are applied in all machinery, the automotive industry is by far the largest user of gears because of the huge number of vehicle sales. In automotive powertrain systems, gears are indispensable, regardless what type of transmission is used for the vehicle. Different gear designs are applied for different types of transmissions. Lay-shaft gears are applied mostly for manual transmissions and sometimes also for automatic transmissions, where all gears rotate about fixed axes and multiple transmission ratios are achieved by multiple gear pairs rotating about parallel shafts, as discussed in Chapter 2. Most of the conventional automatic transmissions use planetary gear trains to transmit power and achieve multiple gear ratios, with multiple clutches to control the power flows. Different from lay-shaft gear systems, not all gears rotate about fixed axes in planetary gear systems. Continuously variable transmissions do not need as many as gears as manual transmissions or conventional automatic transmissions, but they still use gears in the final drive to increase the overall drive train ratio.

The history of gears dates back thousands of years in human civilization [1], but it was not until the eighteenth century that the gear industry started to take shape. The vast majority of gears in the gear industry today are involute gears (i.e. gears with involute tooth profiles), which were originally designed by Leonhard Euler [2]. Gears with cycloid tooth profiles are also used for some applications, such as mechanical clocks and apparatus, which mainly involve the transmission of motion. Many types of gears have been developed for various applications. In vehicle powertrain systems, spur gears, helical gears, and bevel gears (including straight, spiral, and hypoid bevel gears) are mostly applied for the transmission of power. Because of the large volume and wide application, gears are standardized mechanical components in the industry.

Gear design standard and practice is based on the principles and approaches of the theory of gearing. For a comprehensive study of gear design fundamentals, see *Analytical Mechanics of Gears* by E. Buckingham [3], *Gear Geometry and Applied Theory* by F.L. Litvin [4], and *Handbook of Practical Gear Design* by D.W. Dudley [5]. The contents of this chapter will be focused on gear design using existing equations or formulas available in AGMA standards [6,7] rather than on general gear design theory. Following the introduction section, the chapter will first provide a summary of gear design fundamentals on conjugate motions, characteristics of involute gearing, and key definitions and

Automotive Power Transmission Systems, First Edition. Yi Zhang and Chris Mi.

terminologies. The chapter will then present the design of tooth element proportions for spur and helical involute gears according to the related AGMA standards. The design of non-standard gears, including the long-short addendum system and the general non-standard gears will be highlighted in the chapter since the design of these gears is very useful for vehicle transmission applications. The chapter will also cover the calculation of gear forces in the power flows of automotive transmissions. A separate section will be included on the strength design and power rating of gear pairs based on AGMA standard formulation. Note here that the design of hypoid gears used in the final drives of RWD and 4WD vehicles is quite different from cylindrical gears and readers are referred to the related AGMA standard [8] for the design of these gear drives.

The chapter ends with a section on the kinematics of planetary gear trains which are widely applied in automatic transmissions. The section specifically covers the structure and kinematics of three types of planetary gear trains – the simple planetary gear train, dual-planet planetary gear train, and Ravigneaux planetary gear train – since it is these three types of planetary gear trains that are widely applied in automatic transmissions. The kinematic characteristics and torque relations of these three types of planetary gear trains will be analysed and formulated in this section, which will be applied for the design and analysis of automatic transmissions in Chapter 5.

3.2 Gear Design Fundamentals

3.2.1 Conjugate Motion and Definitions

In planar gearing, a pair of gears transmits rotational motion and power from one axis to another through conjugate motion between the tooth profiles of the two gears. As shown in Figure 3.1, the two gears are centered at point O_1 and O_2, with E as the center distance. The two conjugate tooth profiles, Σ_1 and Σ_2, rotate with the respective gear about the gear axis at point O_1 and O_2 at angular velocities ω_1 and ω_2. According to the Lewis theorem [4], the common normal at the point of contact (point of tangency) between the two conjugate profiles in the process of conjugate motion must intersect the line O_1O_2 connecting the centers of rotation and divide the line O_1O_2 into two segments O_1I and O_2I that are related by the following equation:

$$\frac{O_2I}{O_1I} = \frac{\omega_1}{\omega_2} = i \tag{3.1}$$

where i is the gear ratio. If i is a constant, then the position of point I remains the same in the process of conjugate motion. Thus, to transmit uniform rotary motion from one shaft to another by means of gear teeth, the common normal to the profiles of the gear teeth at all points of contact must pass through a fixed point in the line connecting the centers of the two gears. Point I is in kinematics the instantaneous center of rotation between the two gears and is called the pitch point, denoted as P, in gear design. As shown in Figure 3.1, the two circles centered at point O_1 and O_2 with radius r_1 and r_2 equal to O_1P and O_2P are called the pitch circles. The relative motion between the two gears is equivalent in kinematics to the relative motion between the two circular disks with the pitch radii rolling against each other without sliding. Therefore, the pitch circles are the centrodes of the two gears in kinematics.

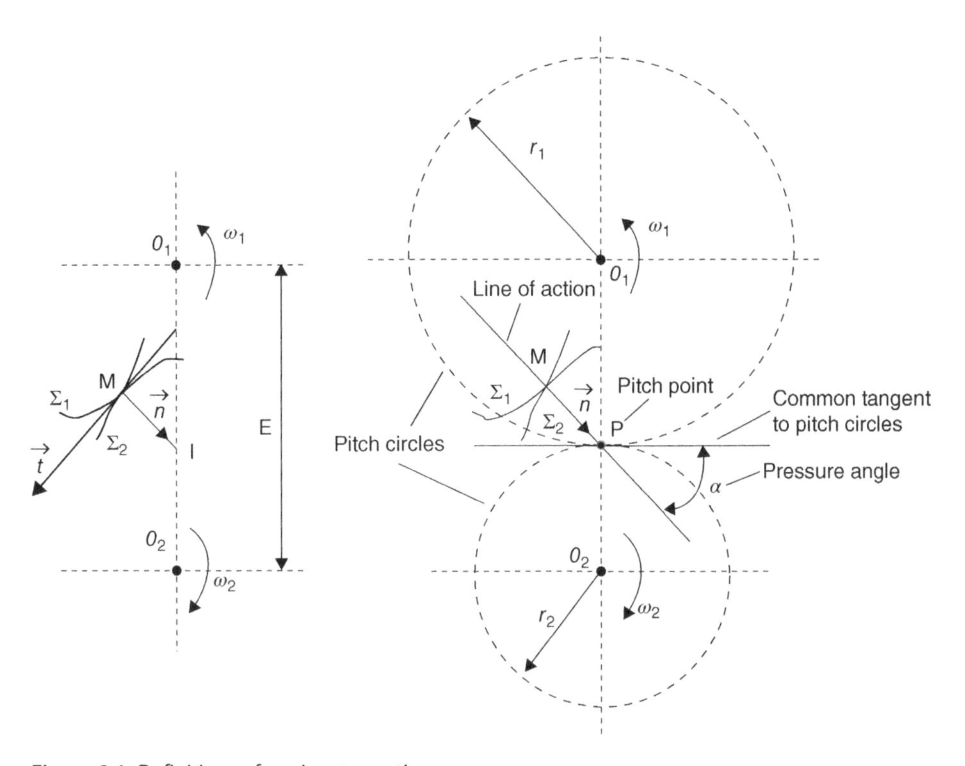

Figure 3.1 Definitions of conjugate motion.

During the process of the conjugate motion between a pair of gear tooth profiles, the points of contact at different instants form a line in the fixed frame, which is called the line of action. In other words, the two gear tooth profiles only contact at points along the line of action. The curvilinear nature of the line of action depends on the conjugate profiles. At a point of contact along the line of action, the common normal to the two conjugate profiles passes through the pitch point P and forms angle α with the common tangent to the two pitch circles, as shown in Figure 3.1. This angle α is called the pressure angle in the theory of gearing and is a very important gear design parameter. In kinematics, the pressure angle affects the transmission of force from the input gear to the output gear and the loading condition of the gears, shafts, and bearings.

3.2.2 Property of Involute Curves

As mentioned in the introduction, involute curves are used as the profiles for the vast majority gears in power transmission applications. To understand why this is the case, it is necessary to take a look at the geometry of involute curves and how these curves are generated. A circular involute curve is uniquely defined by its base circle. As shown in Figure 3.2, an involute curve starts from point M_o on the base circle with radius r_b. The straight line AM with end point M rolls on the base circle without sliding and is always in tangency to the base circle. Initially, this line is in tangency to the base circle at point M_o, and as it rolls on the base circle, the end point M will trace out the involute curve. The generation of the involute curve can also be considered as the result of unwinding a line

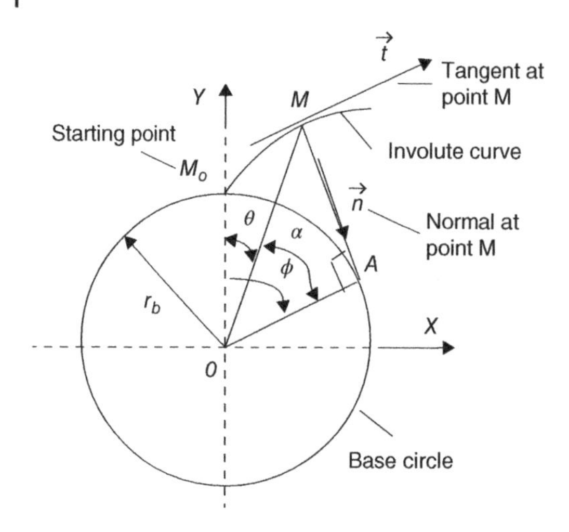

Figure 3.2 Generation of involute curve.

with end point M_o that initially wraps around the base circle. As the line unwinds from the base circle, it is always kept taut so that the portion of the line away from the base circle is always in tangency to the base circle. The end point of the line will then trace out the involute curve as the line unwinds from the base circle.

It is obvious from the generation of involute curves that the normal of an involute curve at any point along it is in tangency to the base circle. Since line AM rolls on the base circle without sliding, the arc length AM_o, $r_b\phi$, is equal to the length of line AM. As shown in Figure 3.2, the Cartesian coordinates of point M on the involute curve can be determined by the following equations:

$$x = r_b(\sin\phi - \phi\cos\phi) \tag{3.2}$$

$$y = r_b(\cos\phi + \phi\sin\phi) \tag{3.3}$$

where ϕ is the so-called unwinding angle of the involute curve and must be in radians. The curvilinear equation of the involute curve can also be written in the following form:

$$x = R\sin\theta \tag{3.4}$$

$$y = R\cos\theta \tag{3.5}$$

$$R = \frac{r_b}{\cos\alpha}; \theta = \phi - \alpha \tag{3.6}$$

where angle α is called the pressure angle that is formed between line OM and line OA and is a variable along the involute curve. For triangle ΔOAM, $\tan\alpha = {}^{AM}/_{r_b} = \phi$, therefore, $\theta = \tan\alpha - \alpha$. In gear design, the relation $\theta = \tan\alpha - \alpha$ is defined as the involute function denoted as:

$$\text{Inv}(\alpha) = \tan\alpha - \alpha \tag{3.7}$$

3.2.3 Involute Curves as Gear Tooth Profiles

Involute spur gears, i.e. gears with straight teeth, have symmetric involute tooth profiles with the same base circle. The tooth profiles on one side of the tooth can be imagined as

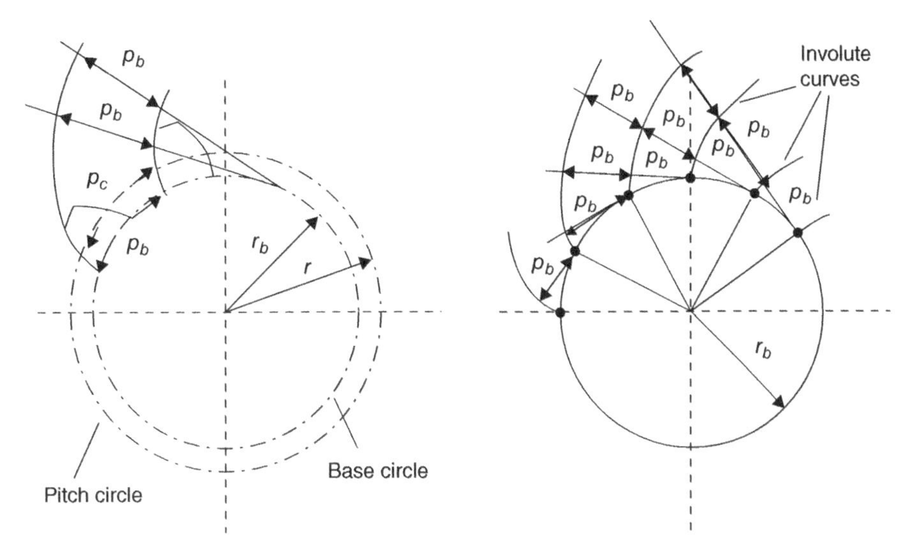

Figure 3.3 Formation of involute gear teeth.

the curves formed by evenly spaced knots on a string that is unwinding from the base circle, as shown in Figure 3.3. The distance between two neighboring tooth profiles measured along the normal is constant and is called the normal pitch or the base pitch, denoted as p_b. Apparently, the base pitch is the even distribution of the base circle circumference on the number of gear teeth N:

$$p_b = \frac{2\pi r_b}{N} \tag{3.8}$$

The arc distance between two neighboring tooth profiles measured on the pitch circle is called the circular pitch or simply pitch, denoted as p_c or just p without the subscript. Similar to the base pitch, the circular pitch is the even distribution of the pitch circle circumference on the number of teeth, i.e. $p = {}^{2\pi r}/_{N}$, with r being the pitch circle radius. It can be observed from Eqs (3.6) and (3.8) that the pitch circle radius, the base circle radius, the circular pitch and the base pitch are related to the pressure angle α by the following equations:

$$r = \frac{r_b}{\cos\alpha}; p_b = p\cos\alpha \tag{3.9}$$

3.2.4 Characteristics of Involute Gearing

Because of the geometric properties of involute curves, gears with involute profiles have unique features that are advantageous for applications in power transmission and for manufacturability. The characteristics of involute gears described in the following list warrant that these gears are almost exclusively applied in the gear industry.

a) The line of action for a pair of involute gears is a straight line, as shown in Figure 3.1. As mentioned previously, the normal at any point along an involute curve is always tangent to the base circle. This means that the common normal at any point of contact between the tooth profiles of an involute gear pair must be in tangency to the two base circles. There are only two common tangents to the two base circles in a gear pair, one for each pair of contacting tooth branches. It follows immediately that the tooth profiles only contact each other along a common tangent to the base circles. Thus, the common tangent to the two base circles is the line of action by definition. Furthermore, the line of action, or the common tangent to the two base circles, intersects the gear center line O_1O_2 at the pitch point P.

b) The pressure angle of involute gears is a constant in the process of conjugate motion. This follows directly from (a) since the position and orientation of the common tangent to the two base circles do not change as the gears rotate. This feature is advantageous for the loading conditions on the gear teeth, bearings, and shafts since the contact force is always applied in a constant direction.

c) Involute tooth profiles are mutually conjugative. This means that if the tooth profile of a gear is involute, then the tooth profile of the other gear in the pair is also involute. This feature can be rigorously proven by the theory of gearing [4]. Thanks to this feature, involute gears possess excellent exchangeability and manufacturability.

d) If the pitch diameter of one gear is given and the pitch diameter of the other is infinitely large, then the conjugate motion is between a gear and a rack. If the gear tooth profile is involute, then the rack tooth profile is a straight line. Vice versa, it is also true. In this case, the line of action passes through the pitch point and is tangent to the gear base circle. This feature is very important for gear manufacturing because all involute gears can be cut by a rack cutter – or basic rack as commonly termed in gear theory – with a straight blade. In production, gears are mostly manufactured by a hob in the hobbing process, as shown in Figure 3.4. The hob is a cylindrical worm with multiple straight slots. These slots form the cutting edges for the hob and allow the removal of cutting chips from the gear blank. In the axial section of the hob, the profile of the worm teeth is the same as that of the basic rack. In kinematics, the turning of the hob is equivalent to the translation of a rack with straight teeth. The angular velocities of the hob and the gear blank are controlled at a constant ratio according to the tooth number of the gear being cut and the pitch of the hob. The hob also moves along the gear axis as the feed motion to cut the whole length of the gear teeth.

e) The change of center distance does not change the gear ratio of a pair of involute gears. The gear ratio of a pair of involute gears solely depends on the ratio between the base circle radii. Once the gears are cut, the base circle radii are determined and do not change at whatever center distance the two gears are assembled. As shown in Figure 3.5, the common tangent to the two base circles defines the line of action when the gears are assembled at center distance E', shown on the right side of Figure 3.5, which is not the same as the standard center distance $E = r_1 + r_2$, with r_1 and r_2 being the pitch radii of the two gears, as shown on the left side of the Figure 3.5. Point I is the instantaneous center of rotation of the two gears. Since the gears are not assembled at the standard center distance, point I is not the point of tangency between the two pitch circles with pitch radius r_1 and r_2 respectively. The two circles with radius r_1' and r_2', equal to O_1I and O_2I respectively, intersect at point I and are called the operating pitch circles. The respective radius r_1' and r_2' are called the operating pitch radius. In kinematics, the motion between the two gears assembled at center distance

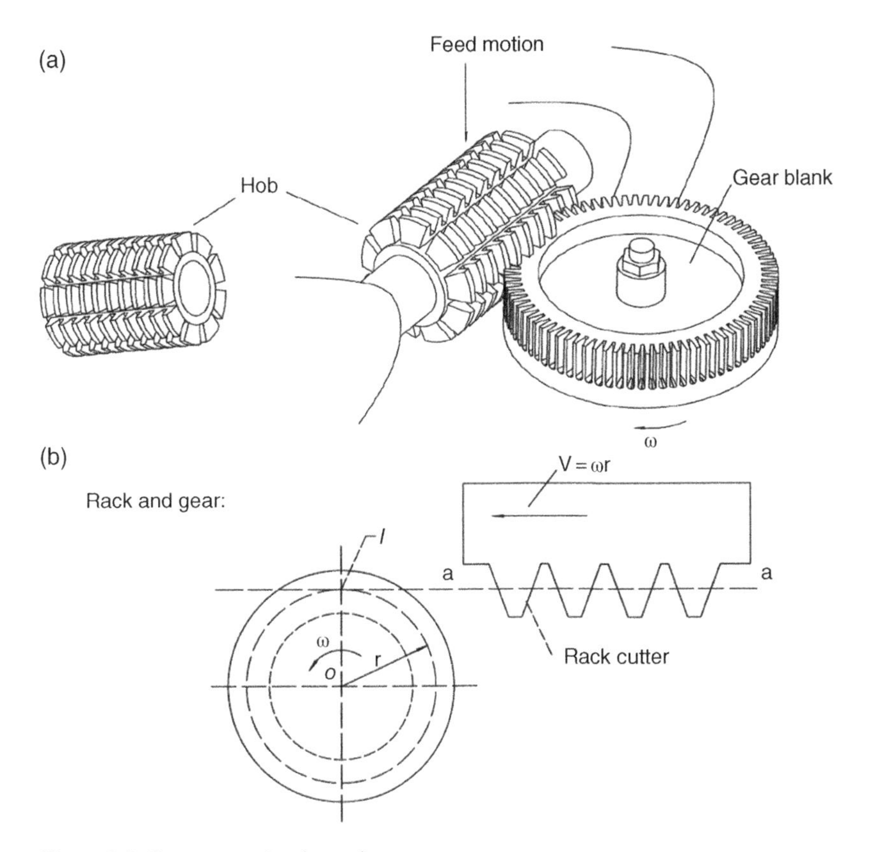

Figure 3.4 Gear generation by rack cutter.

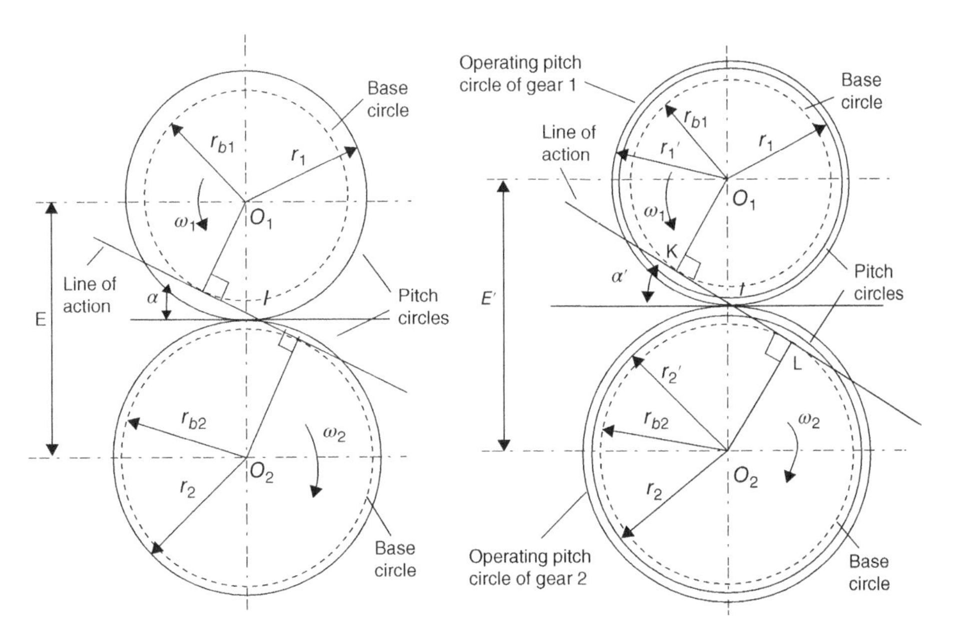

Figure 3.5 Involute gears assembled at non-standard center distance.

E' is equivalent to the rolling of the two operating pitch circles against each other without sliding.

f) Therefore, the operating pitch circles are actually the centrodes of the two gears. Based on the Lewis theory and Eq. (3.1), the gear ratio for the gear pair is $i = \frac{\omega_1}{\omega_2} = \frac{O_2I}{O_1I} = \frac{r_2'}{r_1'}$. Since the two triangles, ΔO_1KI and ΔO_2LI, are similar, the ratio between the base circle radii is equal to the ratio between the operating pitch radii, i.e. $\frac{r_2'}{r_1'} = \frac{r_{b2}}{rb_1}$. It is apparent then that the gear ratio remains the same since $i = \frac{\omega_1}{\omega_2} = \frac{r_2'}{r_1'} = \frac{r_{b2}}{rb_1} = \frac{r_2}{r_1}$. Note that when the two gears are assembled at a non-standard center distance, the operating pitch radii r_1' and r_2', and the operating pressure angle α' as shown in Figure 3.5 are different from the standard values.

3.3 Design of Tooth Element Proportions of Standard Gears

3.3.1 Gear Dimensional and Geometrical Parameters

The gear pitch circle is the design reference for gear dimension and geometry. The design parameters of the rack cutter and the corresponding parameters of the gear are shown in Figure 3.6. The geometry and dimension of a gear are determined by the design parameters, or the tooth element proportions:

N: number of teeth
P: diametral pitch
α_c: cutter pressure angle
a: addendum
b: dedendum
c: clearance ($c = b - a$)
r: pitch radius
r_a: addendum radius
r_d: dedendum radius
r_b: base circle radius
p_c: circular pitch
p_b: base pitch
t: tooth thickness
w: tooth space
p_n: cutter normal pitch ($p_b = p_n$)
α: gear pressure angle (for standard gears, $\alpha = \alpha_c$)

3.3.2 Standardization of Tooth Dimensions

In the AGMA Standard [6,7], the tooth element proportions are designed based on the diametral pitch, which is denoted as P and is defined as:

$$P = \frac{\pi}{p_c} = \frac{N}{d} \tag{3.10}$$

As defined in Eq. (3.10), the diametral pitch can be considered as the distribution of the number of teeth on the gear pitch diameter, as the term itself implies. The unit of the diametral pitch is the reverse of inch, i.e. $^1/_{\text{in}}$. Once the diametral pitch of a gear is specified, the pitch circle diameter and the circular pitch are then determined as:

$$p_c = \frac{\pi}{P}; d = \frac{N}{P} \tag{3.11}$$

Cutter geometry:

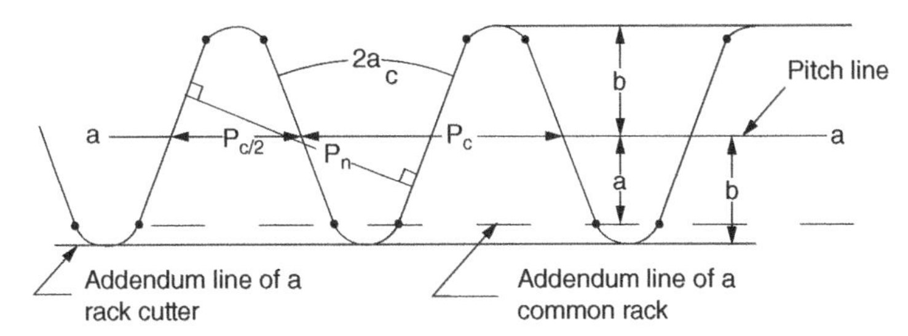

Gear geometry:

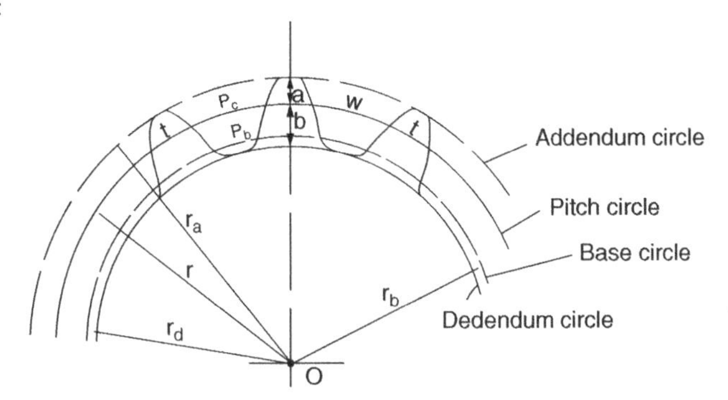

Figure 3.6 Gear tooth element proportions.

The diametral pitch is standardized in the gear industry and its values are divided into two groups as listed in the following:

a) Coarse pitch: 1, $1\frac{1}{4}$, $1\frac{1}{2}$, $1\frac{3}{4}$, 2, $2\frac{1}{2}$, $2\frac{3}{4}$, 3, $3\frac{1}{2}$, 4, 5, 6, 7, 8, 9, 10, 12, 14, 16, 18

b) Fine pitch: 20, 22, 24, 26, 28, 30, 48, 64, 72, 80, 96, 120

The fine pitch gears are mainly used for the transmission of motions in applications with low loads, such as instruments and control mechanisms. For gears used in automotive powertrain applications, the diametral pitches are in the series 6, 7, 8, 9, 10, 12. For passenger car transmission gears, diametral pitches 7, 8, 9, 10, 12 are usually used. For truck transmission gears, the diametral pitches are usually selected from 6, 7, 8, 9, 10.

In SI units, the gear tooth element proportions are designed based on the module, denoted as m and with milimeters (mm) as the unit. The module is the reverse of the diametral pitch numerically, that is, $= {}^{25.4}/_{p}$. For passenger vehicle transmission gears, the module is usually in the series 2, 2.5, 2.75, 3, 3.5, and for truck transmission gears, the module is usually selected between 3 and 5.5.

The gear pressure angle α on the pitch circle is the other standardized parameter in the gear industry. For spur gears, the standard pressure angle is 20° or 16°. The standardization of the gear pressure angle is directly related to the cutter pressure angle α_c, as shown in Figure 3.6. For standard gears, the gear pressure angle α is the same as the cutter pressure angle α_c. It is obvious that the diametral pitch and the pressure angle for a pair of gears in mesh must be the same.

3.3.3 Tooth Dimensions of Standard Gears

According to the AGMA standard [6,7], the tooth element proportions, mainly the addendum and the dedendum, are designed by slightly different formulae for coarse pitch and fine pitch, as represented in the following:

$$\text{Coarse Pitch}: a = \frac{1}{P}; b = \frac{1.250}{P} \tag{3.12}$$

$$\text{Fine Pitch}: a = \frac{1}{P}; b = \frac{1.250}{P} + 0.002 \tag{3.13}$$

For both diametral pitch series, the tooth thickness and the tooth space on the pitch circle are the same, equal to half of the circular pitch:

$$t = w = \frac{p_c}{2} = \frac{\pi}{2P} \tag{3.14}$$

Following the AGMA standard, other gear dimensional parameters are calculated by the following equations:

$$\text{Pitch circle diameter}: d = \frac{N}{P} \tag{3.15}$$

$$\text{Addendum diameter}: d_a = d + 2a \tag{3.16}$$

$$\text{Dedendum diameter}: d_d = d - 2b \tag{3.17}$$

$$\text{Base circle diameter}: d_b = d \cos\alpha \tag{3.18}$$

$$\text{Circular pitch}: p_c = p = \frac{\pi}{P} \tag{3.19}$$

$$\text{Tooth thickness}: t = \frac{p_c}{2} = \frac{\pi}{2P} \tag{3.20}$$

$$\text{Tooth space}: w = t = \frac{p_c}{2} = \frac{\pi}{2P} \tag{3.21}$$

3.3.4 Contact Ratio

A gear pair is designed to transmit motion and power from one gear to another continuously. For the gear pair in Figure 3.7, gear 1 (upper gear) is the driving member and gear 2 is the driven member. Line N_1N_2 is the line of action that is tangent to the two base circles. Point B_2 is the intersection between the addendum circle of gear 2 and the line of action, and point B_1 is the intersection of the addendum circle of gear 1 and the line of action. It is obvious that B_2 is the starting point and B_1 is the end point of the mesh cycle. Three cases are shown in Figure 3.7 to demonstrate the effect of the length of line segment B_1B_2 on the continuity of mesh. In case (a), B_1B_2 is just equal to the base pitch p_b. Mesh continuity is barely maintained in this case since when the current tooth pair ends contact at point B_1, the next pair just starts contact at point B_2. In case (b), B_1B_2 is longer

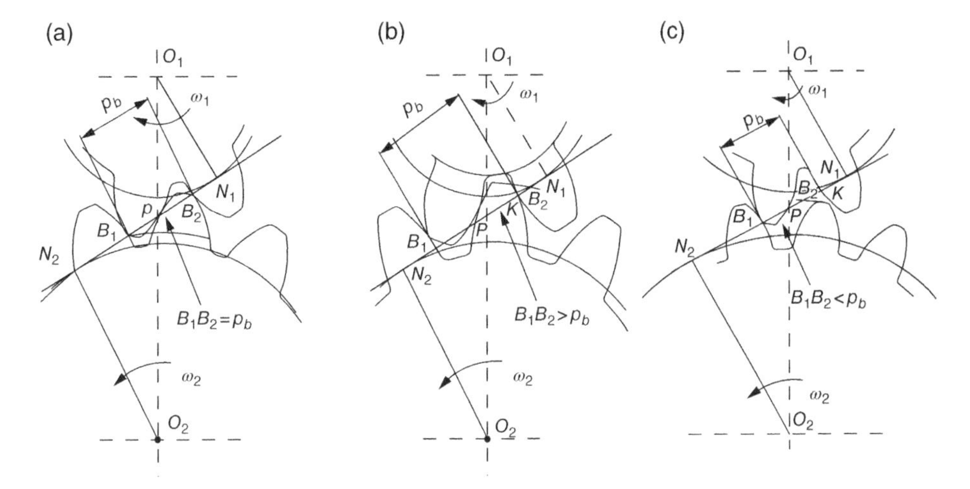

Figure 3.7 Gear mesh process and definition of contact ratio.

than the base pitch p_b. Mesh continuity is well maintained since the next pair of teeth is already in contact before the current pair comes out of contact. This guarantees that at least one pair of teeth will be in contact during the mesh process. In case (c), B_1B_2 is shorter than the base pitch p_b. In this case, mesh continuity cannot be maintained since when the current pair of teeth comes out contact, the next pair is not yet at contact. In gear design, mesh continuity is quantified by the contact ratio, defined as the ratio between the length of line segment B_1B_2 and the base pitch: $\frac{B_1B_2}{p_b}$.

The contact ratio of a pair of involute gears can be determined using Figure 3.8. By considering gear 1 as the driving member and gear 2 as the driven member, the starting point and end point of mesh are at B_1 and B_2, which are respectively the intersection between the line of action and the addendum circles of gear 2 and gear 1. Points K and L are the points of tangency between the line of action and the two base circles. For ΔO_1KB_2 and ΔO_2LB_1, the lengths of line segments KB_2 and B_1L can be determined as:

$$\left.\begin{aligned} KB_2 &= \sqrt{r_{a1}^2 - r_{b1}^2} \\ B_1L &= \sqrt{r_{a2}^2 - r_{b2}^2} \end{aligned}\right\} \tag{3.22}$$

where r_{a1} and r_{a2} are the respective addendum radii of gear 1 and gear1, r_{b1} and r_{b2} are the respective base circle radii. Length KL is related to the center distance E and is determined as:

$$KL = O_1I \sin\alpha + O_2I \sin\alpha = E \sin\alpha \tag{3.23}$$

It is apparent that $KB_2 + B_1L = KL + B_1B_2$. Therefore the contact ratio is determined by the following equation:

$$m_c = \frac{B_1B_2}{P_b} = \frac{KB_2 + B_1L - KL}{p_b} = \frac{P}{\pi \cos\alpha}\left[\sqrt{r_{a1}^2 - r_{b1}^2} + \sqrt{r_{a2}^2 - r_{b2}^2} - E \sin\alpha\right] \tag{3.24}$$

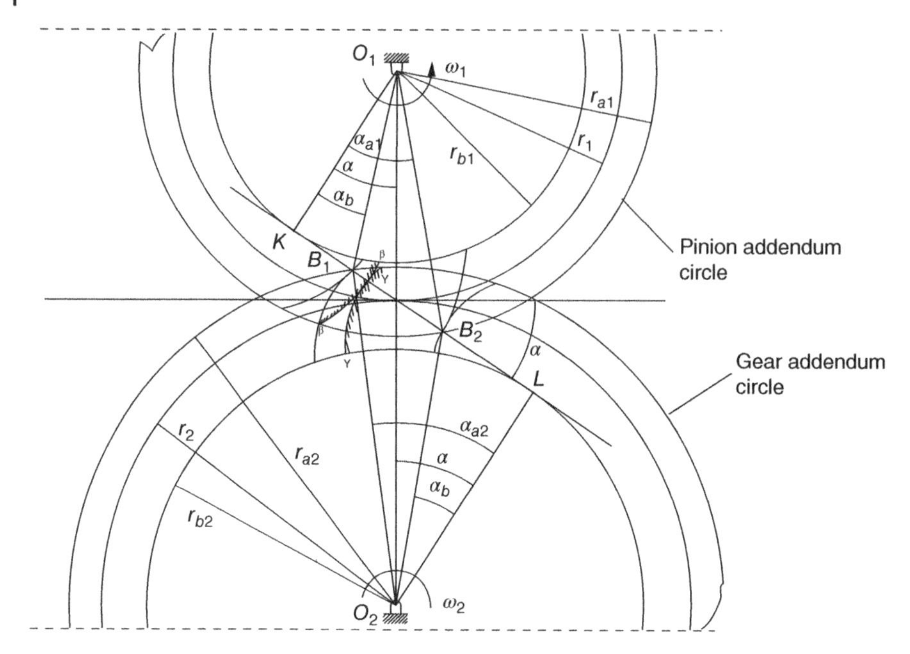

Figure 3.8 Determination of contact ratio.

In gear design, the contact ratio must be higher than 1.0. Higher contact ratio is beneficial for the sharing of contact load and also for the reduction of gear noise. It is observed from the equation above that larger addendum radii yield higher contact ratio. In gear design practice, non-standard gears with modified addendum radii are sometimes used to increase the contact ratio, as discussed later in the design of non-standard gears.

3.3.5 Tooth Thickness and Space along the Tooth Height

The tooth thickness is very important for gear design and manufacturing. If the tooth thickness is not controlled with high accuracy in gear manufacturing, the gear teeth may not be properly meshed. There will be either interference or backlash between the teeth of the two gears. The tooth thickness calculated by Eq. (3.20) is measured on the pitch circle. Along the tooth height, the tooth thickness varies from the largest at the base circle to the smallest at the addendum. The tooth thickness at different radii from the gear center is crucial to the mesh condition of non-standard gears, as discussed in the next section. In the following, a set of equations formulated in reference [4] are represented for the calculation of tooth thickness of involute gears at any radius from the gear center.

As shown in Figure 3.9, arc length $\widehat{AA'}$ is the standard tooth thickness t on the pitch circle with the pitch radius r, and arc length $\widehat{BB'}$ is the tooth thickness t_x on the circle with radius r_x. Angles β and β_x are respectively the angle between line OA and the middle line of the tooth and between line OA_o and the middle line of the tooth. The tooth thickness on the pitch circle is equal to $2\beta r$. The angle between line OA_o and line OA is determined

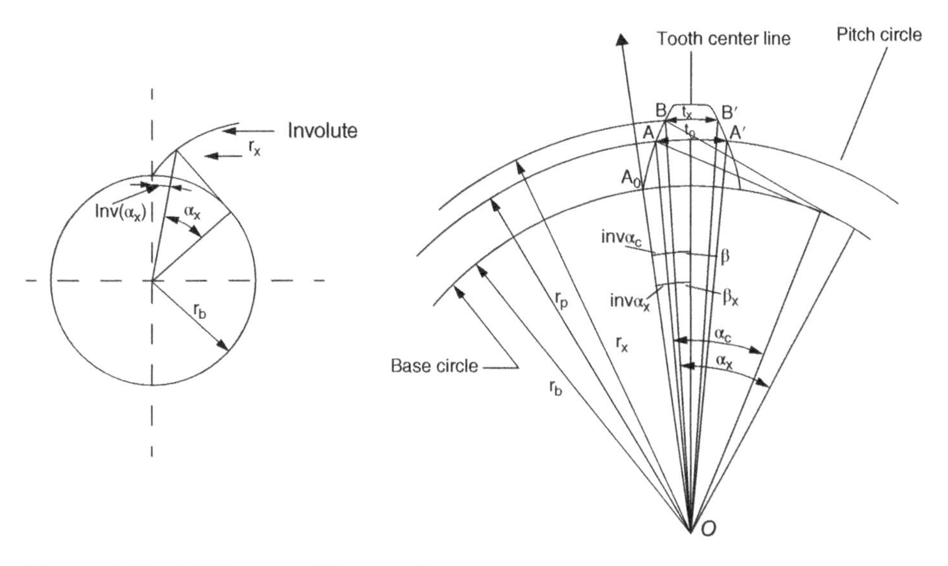

Figure 3.9 Tooth thickness at different radius.

by Eq. 3.7 and is equal to $\text{Inv}(\alpha_c)$, and the angle between line OA_o and line OB is equal to $\text{Inv}(\alpha_x)$. The following equations can be obtained by observing Figure 3.9:

$$r_b = r_x \cos\alpha_x = r\cos\alpha_c = \frac{N}{2P}\cos\alpha_c \tag{3.25}$$

$$\text{Inv}(\alpha_c) + \beta = \text{Inv}(\alpha_x) + \beta_x \tag{3.26}$$

$$\beta_x = \text{Inv}(\alpha_c) + \beta - \text{Inv}(\alpha_x) \tag{3.27}$$

where α_c is the cutter pressure angle and is the same as the pressure angle α for standard gears. The tooth thickness at radius r_x and the tooth thickness at pitch radius are equal to $2\beta_x r_x$ and $2\beta r$ respectively. Therefore:

$$t_x = 2\beta_x r_x = 2r_x[\text{Inv}(\alpha_c) + \beta - \text{Inv}(\alpha_x)] = 2r_x\left[\text{Inv}(\alpha_c) + \frac{t}{2r} - \text{Inv}(\alpha_x)\right] \tag{3.28}$$

The tooth space w_x at radius r_x is equal to the circular pitch on the circle with radius r_x minus the tooth thickness t_x, and is determined by the following equation:

$$w_x = \frac{2\pi r_x}{N} - t_x = \frac{2\pi r_x}{N} - 2r_x\left(\frac{t}{2r}\right) + 2r_x\text{Inv}(\alpha_x) - 2r_x\text{Inv}(\alpha_c)$$
$$= \frac{wr_x}{r} + 2r_x\text{Inv}(\alpha_x) - 2r_x\text{Inv}(\alpha_c) = 2r_x\left[-\text{Inv}(\alpha_c) + \frac{w}{2r} + \text{Inv}(\alpha_x)\right] \tag{3.29}$$

Summarizing these derivations, the tooth thickness and the tooth space at a given radius r_x for an involute gear are determined by the following set of equations:

$$\alpha_x = \cos^{-1}\left(\frac{N\cos\alpha_c}{2Pr_x}\right) \tag{3.30}$$

$$\text{Inv}(\alpha_x) = \tan\alpha_x - \alpha_x \tag{3.31}$$

$$t_x = 2r_x\left[\frac{\pi}{2N} + \text{Inv}(\alpha_c) - \text{Inv}(\alpha_x)\right] \tag{3.32}$$

$$w_x = 2r_x\left[\frac{\pi}{2N} - \text{Inv}(\alpha_c) + \text{Inv}(\alpha_x)\right] \tag{3.33}$$

The tooth thickness and tooth space change along the tooth height and are critical for the proper mesh of general non-standard gears discussed in the next section. For the derivation of mesh condition for general standard gears, Eqs (3.32) and (3.33) can be rewritten as:

$$\left.\begin{aligned} \frac{t_x}{r_x} &= \frac{t}{r} + 2[\text{Inv}(\alpha_c) - \text{Inv}(\alpha_x)] \\ \frac{w_x}{r_x} &= \frac{w}{r} - 2[\text{Inv}(\alpha_c) - \text{Inv}(\alpha_x)] \end{aligned}\right\} \tag{3.34}$$

3.4 Design of Non-Standard Gears

3.4.1 Standard and Non-Standard Cutter Settings

Non-standard gears differ from standard gears in the cutter settings that are used to cut them. As mentioned previously, gears can be cut by a rack cutter or "basic rack" as shown in Figures 3.4 and 3.6. In the cutting process, the cutter pitch line is in tangency to the pitch circle of the gear being cut. When a standard gear is cut, the middle line of the rack cutter coincides with the pitch line and is in tangency with the gear pitch circle, as shown in Figure 3.10. In this standard cutter setting, the distance between the middle line of the rack cutter and the gear center is equal to the gear pitch radius. However, the rack cutter can be displaced either toward or away from the gear center. If displaced away from the

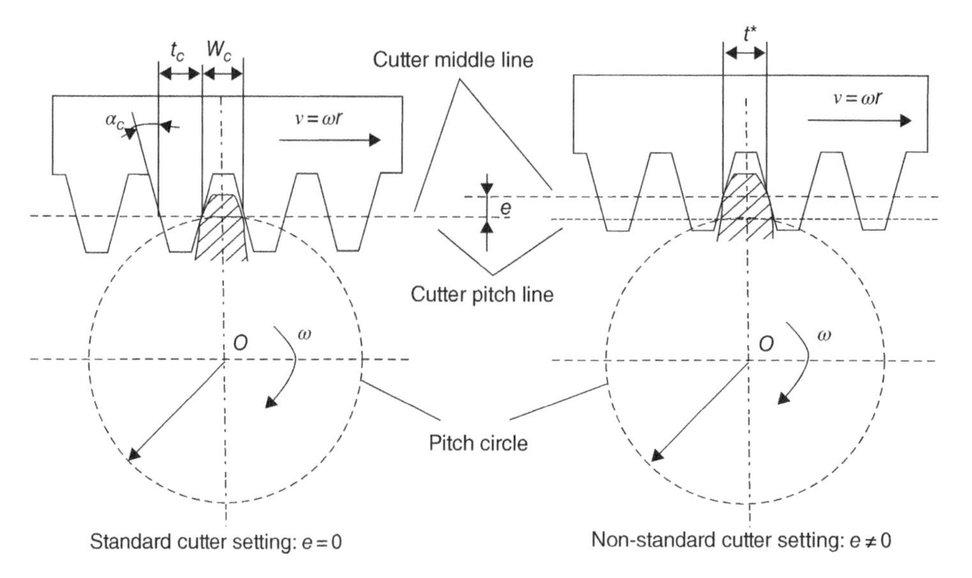

Figure 3.10 Standard and non-standard cutter settings.

gear center, the cutter displacement, denoted as e in Figure 3.10, is defined to be positive, otherwise, the displacement is negative. Gears that are cut when the cutter displacement is not zero are non-standard gears. Obviously, the cutter displacement for standard gears is equal to zero.

The tooth thickness t_c and tooth space w_c of the rack cutter on the middle line are designed as the same and are both equal to half a pitch, i.e. $\pi/_{2P}$ as shown in Figure 3.10. In the cutting process, the rack cutter travels a distance equal to the pitch when the gear rotates through a circular pitch, which is also equal to $\pi/_{2P}$. Therefore, for standard gears, the tooth thickness and tooth space of the gear are respectively equal to the tooth space and tooth thickness of the rack cutter on its middle line:

$$t = w_c = \frac{\pi}{2P}, \; w = t_c = \frac{\pi}{2P} \tag{3.35}$$

For non-standard gears, the cutter middle line does not coincide with cutter pitch line, that is in tangency with the gear pitch circle. For the positive cutter displacement shown in Figure 3.10, the tooth space on its pitch line becomes wider, while the tooth thickness becomes narrower, resulting in an increase in the tooth thickness and a decrease in the tooth space on the gear pitch circle, as determined by the following equations:

$$\left.\begin{aligned} t^* &= t_c + 2e\tan\alpha_c = \frac{\pi}{2P} + 2e\tan\alpha_c \\ w^* &= w_c - 2e\tan\alpha_c = \frac{\pi}{2P} - 2e\tan\alpha_c \end{aligned}\right\} \tag{3.36}$$

where α_c is the cutter pressure angle and e is the cutter displacement, shown as positive in Figure 3.10. It is obvious from Eq. (3.36) that when the cutter displacement is positive, the gear tooth thickness on the pitch circle is increased and the tooth space is decreased. It is also obvious that the gear dedendum and dedendum radius will be changed by the cutter displacement as follows:

$$\left.\begin{aligned} b^* &= b - e \\ r_d^* &= r - b + e \end{aligned}\right\} \tag{3.37}$$

where r and b are the gear pitch radius and the standard gear dedendum respectively. For positive cutter displacement, the dedendum will be decreased by an amount equal to the cutter displacement, while the dedendum circle radius will be increased by the cutter displacement.

3.4.2 Avoidance of Tooth Undercutting and Minimum Number of Teeth

Tooth undercutting is a phenomenon that occurs in the gear cutting process. When it occurs in the cutting process of involute gears, the tooth profile is not the involute curve over the whole tooth, but has an undercut portion in the root area. Figure 3.11 shows the comparison between a gear tooth with normal involute profile and an undercut gear tooth. When there is no undercutting, the envelope of the cutter traces at different positions during the cutting process forms the involute profile for the whole tooth. However, when undercutting occurs, the profile on the tooth root is not the involute profile required for the proper mesh of gears. It is obvious that undercutting must be avoided

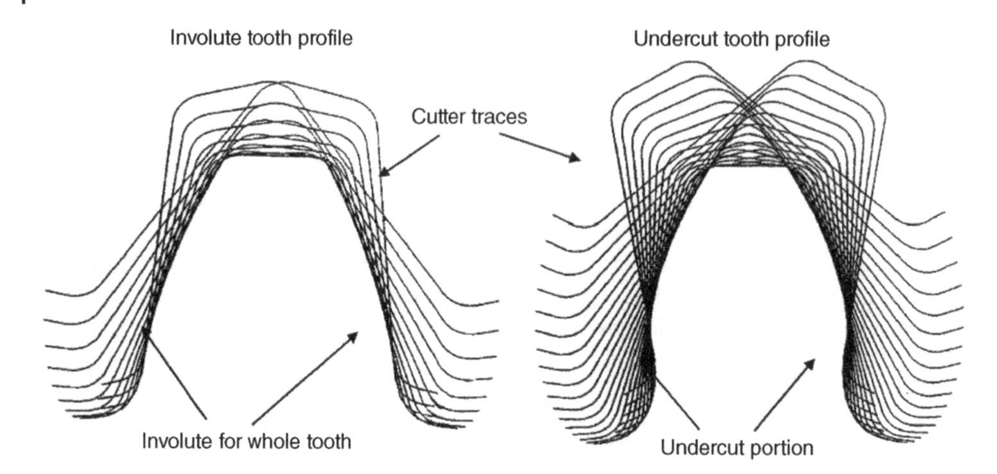

Figure 3.11 Undercutting of involute gears.

in gear design since an undercut gear cannot be meshed properly and the undercut root makes the gear strength much lower. The problem of undercutting can be solved in general by methods in the theory of gearing; for the case of involute gears, it can be solved based on involute curve geometry and the cutter setup [4], as shown in Figure 3.12.

The regular tooth profile of an involute gear must be above the base circle since involute curves start from the base circle. As shown in Figure 3.12, the rack cutter is displaced with an offset e. The cutter addendum is a and the cutter clearance is c. The cutter pressure angle and pitch are denoted as α_c and p_c. The cutter and the gear tooth profiles are denoted as Σ_1 and Σ_2 respectively. Point G is the starting point of the gear tooth profile

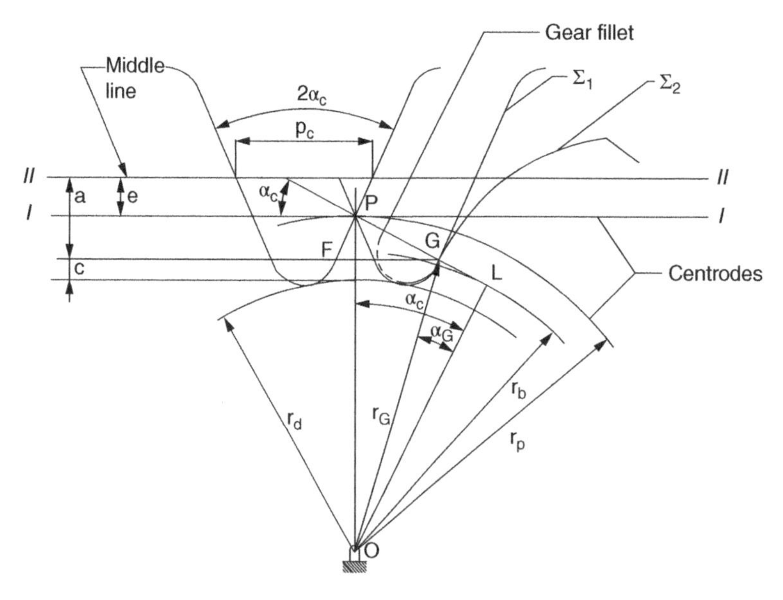

Figure 3.12 Undercutting avoidance for involute gears.

and α_G is the pressure angle at point G. Clearly, $r_G \geq r_b$ and $\alpha_G \geq 0$. Line PL is the line of action between the rack cutter and the gear and is in tangency to the gear base circle at point L. Point G is on the line of action since it is a normal contact point between the rack cutter profile and the gear tooth profile. For triangle ΔGOL, the following inequality can be derived:

$$\tan\alpha_G = \frac{PL - PG}{r_b} = \frac{GL}{r_b} = \frac{r_b \tan\alpha_c - \dfrac{a-e}{\sin\alpha_c}}{r_b} = \tan\alpha_c - \frac{a-e}{r_b \sin\alpha_c} \geq 0 \tag{3.38}$$

For standard gears, the cutter displacement e is zero, therefore,

$$\sin^2\alpha_c \geq \frac{a\cos\alpha_c}{r_b} = \frac{a}{r} = \frac{2}{N}; N \geq \frac{2}{\sin^2\alpha_c} = N_{min} \tag{3.39}$$

Equations (3.9) and (3.12) are used in the derivation of the inequality just given. It follows directly that to avoid undercutting of standard involute gears, the number of teeth of the gear must be larger than $\left({}^{2}/_{\sin^2\alpha_c}\right)$. This quantity is rounded up to the integer and is called the minimum number of teeth, denoted as N_{min}, for the avoidance of undercutting.

For non-standard gears, the cutter displacement e is not zero and the inequality in Eq. (3.38) can be transformed as follows:

$$\frac{a-e}{\sin\alpha_c} \leq r_b \tan\alpha_c = \frac{N}{2P}\sin\alpha_c \tag{3.40}$$

$$\text{N}\sin^2\alpha_G - 2(1 - Pe) \geq 0 \tag{3.41}$$

$$Pe \geq 1 - \frac{N\sin^2\alpha_c}{2} = 1 - \frac{N}{N_{min}} = \frac{N_{min-N}}{N_{min}} \tag{3.42}$$

where P is the diametral pitch, Pe is unitless and is denoted as ζ, which is called the modification coefficient for non-standard gears. The condition for non-undercutting can then be represented as $\zeta \geq {}^{(N_{min}-N)}/_{N_{min}}$. In gear design practice, there are two cases involving the use of modification coefficient ζ. In the first case, $N < N_{min}$ and $\zeta > 0$, i.e. the number of teeth of the gear is smaller than the minimum number of teeth for the avoidance of undercutting. In this case, the cutter must be displaced away from the gear center and the minimum cutter displacement is determined by:

$$\zeta_{min} = \frac{N_{min-N}}{N_{min}}, \; e_{min} = \frac{\zeta_{min}}{P} \tag{3.43}$$

In the second case, $N > N_{min}$ and $\zeta < 0$. The cutter setting can be either standard or the cutter can be displaced toward the gear center to change the tooth thickness for non-standard gears.

3.4.3 Systems of Non-standard Gears

Non-standard gears are designed for several purposes:

- To avoid undercutting of a gear with fewer teeth than the minimum number of teeth, N_{min}.

- To enlarge the tooth thickness of the pinion, the smaller gear, for the balance of strength in a gear pair. This is possible when the sum of the numbers of teeth in the gear pair is larger than $2N_{min}$, as will be shown later.
- To accommodate the assembly of gear pairs at non-standard center distance. This is often useful for gear designs that involve multiple gear pairs on the parallel shafts, such as in manual transmissions.
- To increase the contact ratio by increasing the tooth height.
- To optimize the tooth element proportions for specific applications, such as gear pumps and meters.

There are two systems for the design of non-standard gears in AGMA Standard [6,7], as categorized by the sum of cutter displacements for the pinion and for the gear, i.e. $\left(e_p + e_g\right)$:

1) Long-short addendum gear system: $e_p + e_g = 0$
2) General non-standard gear system: $e_p + e_g \neq 0$

3.4.4 Design of Long-Short Addendum Gear System

There must be no tooth undercutting for each of the two gears in the pair, so the cutter displacement, e_p for the pinion and e_g for the gear, must satisfy the inequality expressed by Eq. (3.42):

$$e_p \geq \frac{1}{P}\left(\frac{N_{min-N_p}}{N_{min}}\right) \tag{3.44}$$

$$e_g \geq \frac{1}{P}\left(\frac{N_{min-N_g}}{N_{min}}\right) \tag{3.45}$$

Since $e_p + e_g = 0$ for the long-short (L-S) addendum system, it follows directly that for the design of non-standard gears in this system, the following condition must be satisfied:

$$N_p + N_g \geq 2N_{min} = \frac{4}{\sin^2 \alpha_c} \tag{3.46}$$

In the design of L-S addendum gears, the first step is to check whether the inequality in Eq. (3.46) is satisfied. The cutter displacements must then be chosen according to the design priority. Usually, the pinion cutter displacement e_p is chosen to be positive for an increase in the pinion tooth thickness. The gear cutter displacement e_g has the same magnitude as e_p but is opposite in sign. Obviously, both e_p and e_g must satisfy Eqs (3.44) and (3.45). Once the cutter displacements are chosen, the tooth element proportions are then calculated by the following standard equations:

$$\text{Dedendum}: b_p = \frac{1.25}{P} - e_p;\ b_g = \frac{1.25}{P} - e_g \quad \text{(for coarse pitch)} \tag{3.47}$$

$$b_p = \frac{1.25}{P} + 0.002 - e_p;\ b_g = \frac{1.25}{P} + 0.002 - e_g \quad \text{(for fine pitch)} \tag{3.48}$$

$$\text{Addendum}: a_p = \frac{1}{P} + e_p;\ a_g = \frac{1}{P} + e_g \tag{3.49}$$

$$\text{Tooth thickness}: t_p = \frac{\pi}{2P} + 2e_p \tan\alpha_c;\ t_g = \frac{\pi}{2P} + 2e_g \tan\alpha_c \tag{3.50}$$

$$\text{Center distance}: E = \frac{1}{2P}\left(N_p + N_g\right) \tag{3.51}$$

It can be seen from Eq. (3.49) that the addendum of the pinion becomes longer and the addendum of the gear becomes shorter in the so-called L-S addendum system. It should be emphasized here that non-standard gears in the L-S addendum system are assembled at the standard center distance and the pressure angle is the same as the standard pressure angle. In summary, the characteristics of non-standard gears in the L-S addendum system are:

- The sum of cutter displacements is zero. Usually, the pinion cutter displacement is positive and the gear displacement is negative.
- The center distance is the same as the standard center distance E.
- The operation pressure angle of the non-standard gears in the L-S addendum system at assembly is the same as the standard pressure angle.
- The tooth thickness differs from that of the standard gears. Usually, the pinion tooth thickness is increased due to the positive cutter displacement for strength balance in the gear pair. The gear tooth thickness is reduced by the same amount.
- The addendum and dedendum differ from the standard gears.
- The operation pitch circles are the same as those in standard gears.

3.4.5 Design of General Non-Standard Gear System

As mentioned in Section 3.2, the change of center distance does not change the gear ratio of a pair of involute gears. However, if a pair of involute gears is assembled at a non-standard center distance, the operating pitch circles, i.e. the centrodes in kinematics, and the operating pressure angle are different from the standard values, as shown in Figure 3.13. It is obvious that the non-standard center distance E' is the sum of the operating pitch radius of the pinion and the gear, as expressed by the following equations:

$$r'_p = \frac{r_{bp}}{\cos\alpha'} = \frac{r_p \cos\alpha_c}{\cos\alpha'};\ r'_g = \frac{r_{bg}}{\cos\alpha'} = \frac{r_g \cos\alpha_c}{\cos\alpha'} \tag{3.52}$$

$$E' = r'_p + r'_g = \frac{\cos\alpha_c}{\cos\alpha'}\left(r_p + r_g\right) = \frac{N_p + N_g}{2P}\frac{\cos\alpha_c}{\cos\alpha'} = E\frac{\cos\alpha_c}{\cos\alpha'} \tag{3.53}$$

In the meshing process, the two operating pitch circles roll against each other without sliding. There cannot be either backlash or interference between the meshing teeth on the operating pitch circle. Using Eq. (3.36), the pinion tooth thickness and the gear tooth space on the respective pitch circles for the non-standard gear pair can be determined as follows:

$$t_p^* = \frac{\pi}{2P} + 2e_p \tan\alpha_c \tag{3.54}$$

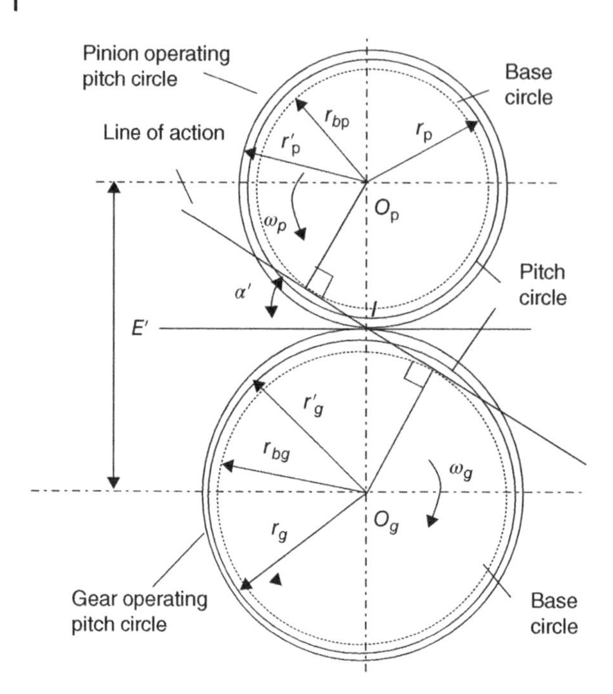

Figure 3.13 Operating pitch circles and operating pressure angle.

$$w_g^* = \frac{\pi}{2P} - 2e_g \tan\alpha_c \tag{3.55}$$

The pinion tooth thickness and the gear tooth space on the respective operating pitch circles, t'_p and w'_g, can be found by Eq. (3.34) with the substitution of r_x by the operating pitch radius, r'_p and r'_g respectively and the substitution of α_x by α', i.e.

$$t'_p = r'_p\left[\frac{t_p^*}{r_p} + 2[\text{Inv}(\alpha_c) - \text{Inv}(\alpha')]\right] = r'_p\left[\frac{\frac{\pi}{2P} + 2e_p \tan\alpha_c}{r_p} + 2[\text{Inv}(\alpha_c) - \text{Inv}(\alpha')]\right] \tag{3.56}$$

$$w'_g = r'_g\left[\frac{w_g^*}{r_g} - 2[\text{Inv}(\alpha_c) - \text{Inv}(\alpha')]\right] = r'_g\left[\frac{\frac{\pi}{2P} - 2e_g \tan\alpha_c}{r_g} - 2[\text{Inv}(\alpha_c) - \text{Inv}(\alpha')]\right] \tag{3.57}$$

For the proper mesh of the non-standard gears, the following equation must be observed:

$$\frac{t'_p / r'_p}{w'_g / r'_g} = \frac{N_g}{N_p} \tag{3.58}$$

On the left side of the equation above, the numerator is the pinion angle of rotation corresponding to the tooth thickness and the denominator is the gear angle of rotation corresponding to the tooth space on the respective operation pitch circle. It follows directly from Eqs (3.56–3.58) that the mesh condition for non-standard gears is represented by the following equation:

$$\text{Inv}(\alpha') = \frac{2(e_p + e_g)P\tan\alpha_c}{N_p + N_g} + \text{Inv}(\alpha_c) \tag{3.58a}$$

This equation is the **mesh condition of general non-standard gears** and must be satisfied in the design of general non-standard involute gears. For production gears, there is always a certain amount of tooth backlash which is controlled by tooth thickness tolerance specifications [9,10]. In manual transmissions, there are multiple pairs of gears on parallel shafts. In other words, multiple gear pairs are assembled at the same center distance. Since the standard center distance depends on the sum of teeth, which are integers and must satisfy the transmission ratio requirement, and the diametral pitch, which is selected in the standardized series, it is often impossible to have all the gear pairs with the same standard distance. In such applications, general non-standard gears can be designed to accommodate the center distance constraints, as described in the following steps:

Step 1: The operating pressure angle α' is determined from Eq. (3.53) as:

$$\alpha' = \cos^{-1}\left(\frac{N_p + N_g}{2PE'}\cos\alpha_c\right) \tag{3.59}$$

where E' is the non-standard distance at which the gears are to be assembled, i.e. the distance between the two gear shafts; α_c is the cutter pressure angle or the standard gear pressure angle; $(N_p + N_g)$ is the sum of teeth for the gear pair that must provide the specified gear ratio; and P is the diametral pitch selected in the standard series.

Step 2: The sum of the cutter displacement $(e_p + e_g)$ is then calculated based on the mesh condition for general non-standard gears represented by Eq. (3.58a), as follows,

$$(e_p + e_g) = \frac{N_p + N_g}{2P\tan\alpha_c}[\text{Inv}(\alpha') - \text{Inv}(\alpha_c)] \tag{3.60}$$

The sum of cutter displacements determined above is then distributed between the pinion and the gear in the pair based on considerations of strength and of avoidance of undercutting and interference [7,11,12].

Step 3: The tooth element proportions of the general non-standard gears are then calculated by the following equations:

$$\text{Tooth thickness}: t_p = \frac{\pi}{2P} + 2e_p\tan\alpha_c;\ t_g = \frac{\pi}{2P} + 2e_g\tan\alpha_c \tag{3.61}$$

$$\text{Dedendum}: b_p = \frac{1.25}{P} - e_p;\ b_g = \frac{1.25}{P} - e_g \ \text{(for coarse pitch)} \tag{3.62}$$

$$b_p = \frac{1.25}{P} + 0.002 - e_p;\ b_g = \frac{1.25}{P} + 0.002 - e_g \ \text{(for fine pitch)} \tag{3.63}$$

$$\text{Dedendum radius}: r_{dp} = r_p - \frac{1.25}{P} + e_p;\ r_{dg} = r_g - \frac{1.25}{P} + e_g \ \text{(for coarse pitch)} \tag{3.64}$$

$$r_{dp} = r_p - \frac{1.25}{P} - 0.002 + e_p;\ r_{dg} = r_g - \frac{1.25}{P} - 0.002 + e_g \ \text{(for fine pitch)} \tag{3.65}$$

$$\text{Addendum radius}: r_{ap} = E' - r_{dg} - c;\ r_{ag} = E' - r_{dp} - c \tag{3.66}$$

$$\text{Tooth height}: h = h_o - \left[\left(e_p + e_g\right) - \Delta E\right] \tag{3.67}$$

where h_o is the standard tooth height, which is equal to the sum of addendum and dedendum, i.e. $h_o = a + b$; ΔE is the difference between the center distance at which the gears are assembled and the standard center distance, and is equal to $(E' - E)$. The addendum radii determined by Eq. (3.66) are designed to provide the standard clearance that is equal to $(b - a)$. Obviously, the tooth height of general standard gears determined by Eq. (3.67) is different from the standard value and is shortened by $\left[\left(e_p + e_g\right) - \Delta E\right]$. After all tooth element proportions are calculated, Eq. (3.24) is then used to check the contact ratio of the designed general non-standard gears. Note here that when Eq. (3.24) is used for the contact ratio of general non-standard gears, the operating pressure angle α' and the non-standard center distance E' must be used in the equation instead of the standard pressure angle α and the standard center distance E.

3.5 Involute Helical Gears

Involute helical gears have an involute tooth profile on the transverse section, which is the cross-section perpendicular to the gear axis. The surface of an involute helical gear is a helicoid which is formed by an involute curve on the transverse section in a screw motion along the gear axis, as shown in Figure 3.14. Intuitively, an involute helical gear can be thought as the result of a spur involute gear having undergone a uniform twist.

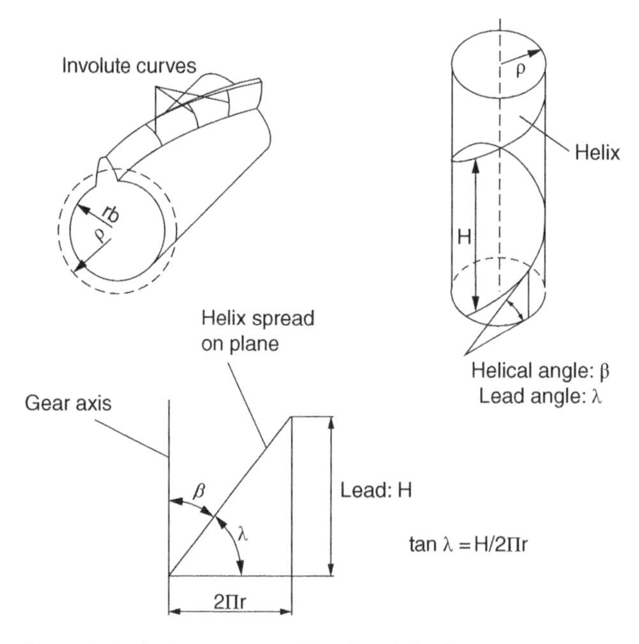

Figure 3.14 Basic geometry of involute helical gears.

The base circles for all the involute tooth profiles on the transverse sections form the so-called base cylinder with base radius r_b. Similarly, the pitch circles on the transverse sections form the pitch cylinder with pitch radius r. The tooth surface of an involute helical gear is between the base cylinder and the addendum cylinder with radius r_a. The intersection between the tooth surface and any cylinder with radius between r_b and r_a is a helical line or helix. The angle between the tangent to the helix on the pitch cylinder and the gear axis is called the helical angle, denoted as β. This helical angle can also be defined by spreading the helix on the pitch cylinder on a plane as shown in Figure 3.14. A helical gear can be right handed or left handed, depending the direction of the helix. The gear shown in Figure 3.14 is right handed. For helical gears with parallel axes, the two gears in mesh have the same magnitude for the helical angle, but opposite hands of the helix in external gearing. For internal gearing, the helical angle and helix direction are the same.

3.5.1 Characteristics of Involute Helical Gearing

Involute helical gears have all the characteristics of involute gears, plus additional features highlighted in the following:

a) On the transverse section, involute helical gears just behave as spur involute gears. Because of this, all of the equations and conclusions for the design and analysis of spur involute gears are valid if the mesh for helical involute gears is considered on the transverse section.
b) The involute tooth profiles along the gear axes come in and out of mesh at different time, i.e. at different rotational positions. This provides an extra contact ratio, called lengthwise contact ratio, in addition to the contact ratio on the transverse section. Because of the enhanced contact ratio, involute helical gears are stronger and quieter than spur involute gears.
c) Helical and spur involute gears are cut by the same machine tools and cutters. The manufacturing costs for helical and spur gears are about the same.
d) The tooth geometry design of helical gears is based on the tooth form on the normal section, which is the intersection between the tooth surface and the plane normal to the helix on the pitch cylinder. The normal diametral pitch P_n and the normal pressure angle α_n are standardized for helical gears. For the proper mesh of a pair of helical gears, the normal diametral pitch and the normal pressure angle must be the same respectively, i.e. $P_{n1} = P_{n2}$ and $\alpha_{n1} - \alpha_{n2}$.
e) Helical gears have axial loads, or thrusts. This is the disadvantage of helical gears.

Because of the advantages in strength and quietness as compared with spur gears, involute helical gears are widely applied for passenger vehicle transmissions. Having an axial load is a shortcoming of involute helical gears. When there are two gears on the same shaft in the power flow of a gear box, one driven for a gear pair and one driving for another gear pair, the axial load can be partially cancelled in gear design so as to minimize the thrust on the bearing.

3.5.2 Design Parameters on the Normal and Transverse Sections

As mentioned in the design of spur involute gears, the tooth geometry of involute spur gears is based on the rack cutter with a straight tooth, or the so-called basic rack shown in

Figure 3.15a. Similarly, the geometry of involute helical gears is based a basic rack shown in Figure 3.15b. The tooth surface of the rack cutter, or basic rack, is an inclined plane for involute helical gears. On the pitch plane, that is tangent to the pitch cylinder, the rack tooth profile is a straight line that forms the helical angle β with the lengthwise direction. The distance between two neighboring teeth measured on the transverse section is the transverse pitch p_t. The normal pitch p_n is measured perpendicular to the tooth profile on the pitch plane, as shown in Figure 3.15c. The transverse pressure angle α_t and the normal pressure angle α_n are respectively defined on the transverse section and the normal section of the basic rack, as shown in Figure 3.15d.

It can be observed from Figure 3.15c that the transverse pitch and the normal pitch are related to the helical angle by $p_t = \dfrac{p_n}{\cos\beta}$. Reversing both sides of this relation leads to the relation between the transverse diametral pitch P_t and the normal diametral pitch P_n. From Figure 3.15d, the relation between the transverse pressure angle α_t and the normal pressure angle α_n can be derived since the tooth height is the same whether it is measured on the transverse section or the normal section. In summary, the normal diametral pitch and the normal pressure angle are converted to the transverse diametral pitch and the transverse pressure angle by the following equation:

$$P_t = P_n \cos\beta \tag{3.68}$$

$$\tan\alpha_t = \frac{\tan\alpha_n}{\cos\beta} \tag{3.69}$$

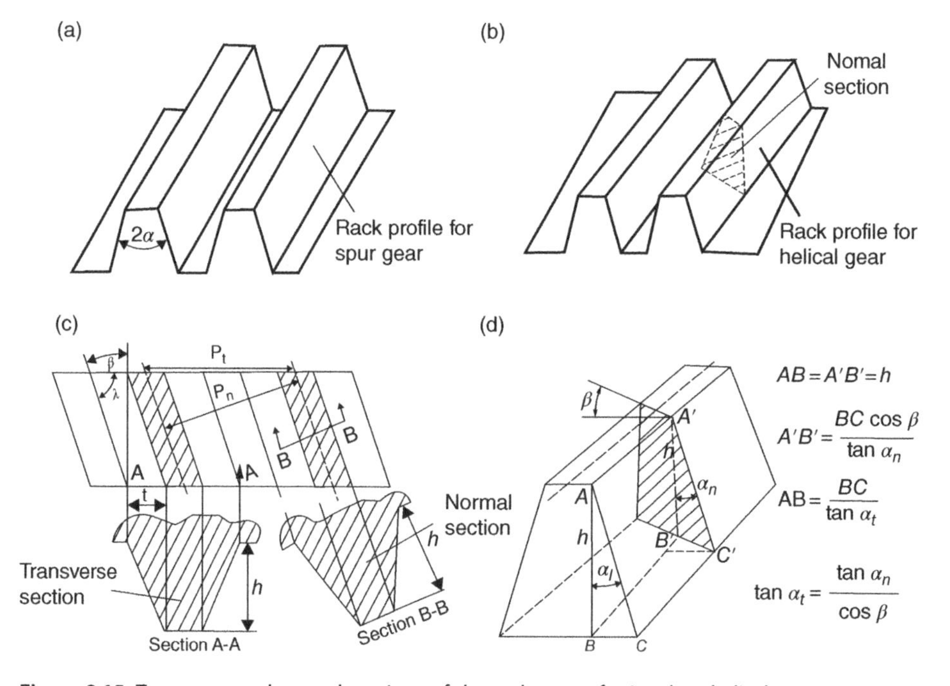

Figure 3.15 Transverse and normal sections of the rack cutter for involute helical gears.

3.5.3 Tooth Dimensions of Standard Involute Helical Gears

As mentioned previously, the normal pressure angle α_n and the normal diametral pitch P_n for involute helical gears are standardized in the gear industry. The tooth element proportions of involute helical gears are based on the tooth form on the normal section. According to the AGMA standard [7,12], the addendum and the dedendum are designed by the following equations:

$$a = \frac{1}{P_n}; b = \frac{1.250}{P_n} \tag{3.70}$$

The basic design parameters of an involute helical gear are its normal pressure angle α_n, normal diametral pitch P_n, helical angle β, and number of teeth N. The dimensional parameters are determined by the following equations:

$$\text{Pitch radius}: r = \frac{N}{2P_t} = \frac{N}{2P_n \cos\beta} \tag{3.71}$$

$$\text{Base radius}: r_b = r\cos\alpha_t = \frac{N\cos\alpha_t}{2P_n \cos\beta} \tag{3.72}$$

$$\text{Dedendum radius}: r_d = r - b = \frac{N - 2.5\cos\beta}{2P_n \cos\beta} \tag{3.73}$$

$$\text{Addendum radius}: r_a = r + a = \frac{N + 2\cos\beta}{2P_n \cos\beta} \tag{3.74}$$

$$\text{Transverse pitch}: p_c = p_t = \frac{\pi}{P_t} \tag{3.75}$$

$$\text{Normal pitch}: p_n = \frac{\pi}{P_n} \tag{3.76}$$

$$\text{Transverse tooth thickness}: t = \frac{p_t}{2} = \frac{\pi}{2P_t} \tag{3.77}$$

$$\text{Transverse tooth space}: w = t = \frac{\pi}{2P_t} \tag{3.78}$$

3.5.4 Minimum Number of Teeth for Involute Helical Gears

Since involute helical gears have the same characteristics as spur involute gears on the transverse section, the condition for the avoidance of undercutting represented by Eq. (3.39) can be applied directly if the parameters on the transverse section are used. Therefore, the following inequality must be observed to avoid undercutting of standard involute helical gears,

$$\sin^2\alpha_t \geq \frac{a\cos\alpha_t}{r_b} = \frac{a}{r} = \frac{2\cos\beta}{N}; N \geq \frac{2\cos\beta}{\sin^2\alpha_t} = N_{min} \tag{3.79}$$

As can be observed from Eq. (3.79), involute helical gears can be designed with fewer teeth than spur gears without the issue of undercutting. The higher the helical angle, the fewer the number of teeth the gear can have without undercutting. For example, if the normal pressure angle is 20° and the helical angle is 45°, the minimum number of teeth for an involute helical gear for the avoidance of undercutting is as few as seven.

3.5.5 Contact Ratio of Involute Helical Gears

The contact ratio of involute helical gears has two components: a transverse contact ratio and a lengthwise contact ratio. The transverse contact ratio is also termed the involute contact ratio, which is determined by Eq. (3.24). For involute helical gears, the transverse diametral pitch P_t and the transverse pressure angle α_t are used in Eq. (3.24), since these gears behave as spur involute gears on the transverse section. The lengthwise contact ratio comes from the fact that the tooth profiles on the two ends of the tooth length do not come in and out of mesh at the same time. Figure 3.16 shows the planar spread of the pitch cylinder and the tooth helixes on it, with β and L as the helical angle and the tooth length. Line AB shows the position of a tooth that just enters mesh. For spur gears, the tooth lines on the planar spread are straight lines parallel to the gear axis and separated by the circular pitch. When a spur gear rotates through a circular pitch, points A and B on tooth ends move to points A' and B', which indicate the end of the mesh for the tooth. However, in the case of helical gears, only the mesh on the end section with point A is finished. When the gear rotates through a circular pitch, the same tooth is still in mesh elsewhere along the tooth length. For the whole tooth to be out of mesh, the gear has to rotate additionally until the other end of the tooth helix, point B, moves through on the pitch circle an arc length $\widehat{B'B''}$ that is equal to $L \tan\beta$. The lengthwise contact ratio m_l is then defined as the ratio between the length of line segment $B'B''$ and the circular pitch on the transverse section. The total contact ratio of helical gears is the sum of the transverse contact ratio and the lengthwise contact ratio as follows:

$$m_l = \frac{L \tan\beta}{p_t} = \frac{P_n L \sin\beta}{\pi} \tag{3.80}$$

$$m_c = m_t + m_l = \frac{P_n L \sin\beta}{\pi} + \frac{P_t}{\pi \cos\alpha_t}\left[\sqrt{r_{a1}^2 - r_{b1}^2} + \sqrt{r_{a2}^2 - r_{b2}^2} - E \sin\alpha_t\right] \tag{3.81}$$

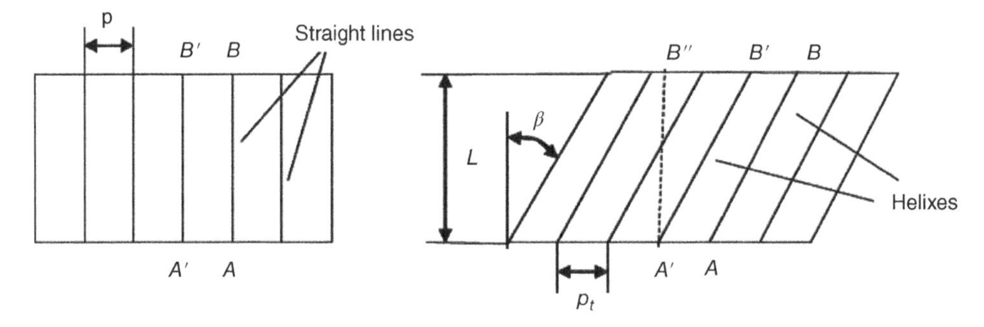

Figure 3.16 Lengthwise contact ratio for involute helical gears.

3.5.6 Design of Non-standard Involute Helical Gears

The design of non-standard involute helical gears is similar to the design of spur gears, as described in Section 3.4. Firstly, the transverse diametral pitch P_t and the transverse pressure angle α_t are converted from the normal diametral pitch P_n and normal pressure angle α_n by Eqs (3.68) and (3.69). Then the non-standard center distance E' and the mesh condition on the transverse section are used to determine the operating transverse pressure angle α_t' and the sum of the cutter displacements by the following equations:

$$\alpha_t' = \cos^{-1}\left(\frac{N_p + N_g}{2P_t E'}\cos\alpha_t\right) \tag{3.82}$$

$$\left(e_p + e_g\right) = \frac{N_p + N_g}{2P_t \tan\alpha_t}\left[\mathrm{Inv}(\alpha_t') - \mathrm{Inv}(\alpha_t)\right] \tag{3.83}$$

The equations for the design of non-standard spur gears, Eqs (3.61–3.67), are also applicable for the design of non-standard involute helical gears if the transverse pressure angle α_t is used. It is noted here that the normal diametral pitch P_n is used for the calculation of addendum and dedendum in these equations. The contact ratio of non-standard involute helical gears is calculated by Eq. (3.81), where the standard center distance E and the standard transverse pressure angle α_t are replaced by the non-standard center distance E' and non-standard transverse pressure angle α_t'.

3.6 Gear Tooth Strength and Pitting Resistance

3.6.1 Determination of Gear Forces

Figure 3.17 shows a pair of helical gears with the pinion as the driving member. The driving torque and the pinion angular velocity are in the same direction. Since the pinion is

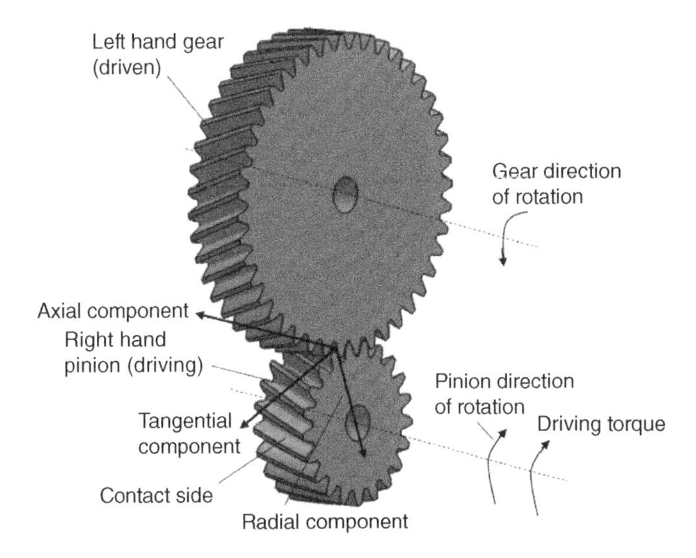

Figure 3.17 Directions of gear force components.

the driving member, the pinion tooth side that contacts the gear tooth is indicated as shown in the figure. The contact force between the pinion and the gear has three components: radial, tangential, and axial. The directions of the three force components applied to the pinion are shown in Figure 3.17. The radial component is always toward the center, the tangential component always forms a torque about the center to resist the driving torque, and the direction of the axial component or thrust depends on the hand of the helix and can be readily determined once the contact side of the driving member is indicated. The three force components applied by the pinion to the gear member are just in the opposite directions.

The magnitudes of the three force components shown in Figure 3.17 are related to the normal pressure angle α_n and the helical angle β. As shown in Figure 3.18, the contact force N acts on the normal section that is perpendicular to the helix on the pitch cylinder. The angle between the contact force and the intersection between the normal section and the pitch cylinder is the normal pressure angle α_n. The contact force is decomposed to two components on the normal section: the radial component F_r which is always toward the gear center and the tangential component on the normal section F_{nt}. F_{nt} is further decomposed into the tangential component F_t on the transverse section and axial component F_a. It is obvious that the driving torque T_d is equal to the tangential component F_t times the pitch radius r_p of the driving member, which is the pinion as shown in Figure 3.17, i.e. $T_d = r_p F_t$. By observing Figure 3.18, the gear contact force and the three gear force components are determined by the following equations:

$$F_r = N \sin\alpha_n \tag{3.84}$$

$$F_t = N \cos\alpha_n \cos\beta \tag{3.85}$$

$$F_a = N \cos\alpha_n \sin\beta \tag{3.86}$$

$$N = \frac{T_d}{r_p \cos\alpha_n \cos\beta} \tag{3.87}$$

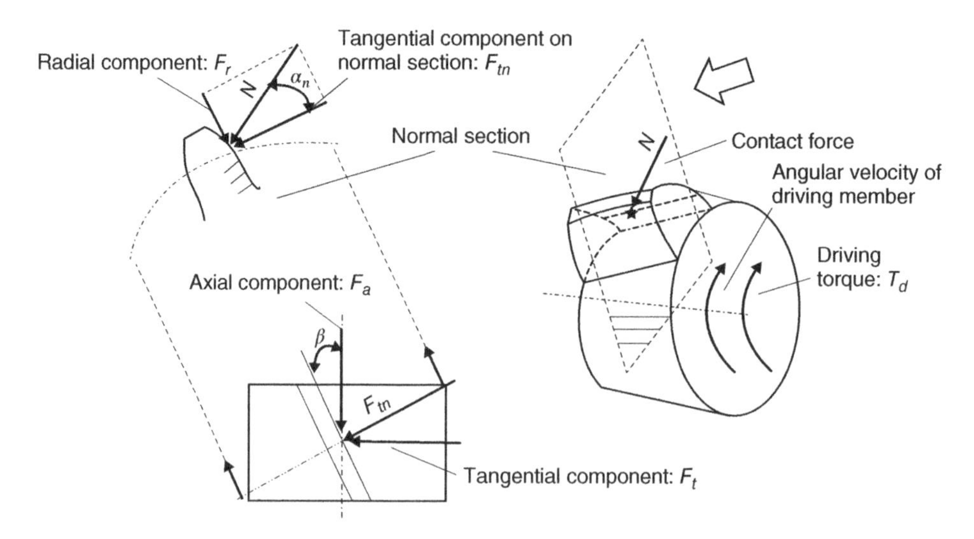

Figure 3.18 Determination of gear force components.

Note that Eqs (3.84–3.87) are also valid for spur gears. To determine the gear force components of spur gears, the normal pressure angle α_n is replaced by the pressure angle α of the spur gears and the helical angle β is zero. Obviously, the axial load for spur gears is equal to zero.

3.6.2 AGMA Standard on Bending Strength and Pitting Resistance

Gear strength and durability are rated on bending stress and contact stress in the AGMA standard [11,13]. For gears used in automotive transmissions, durability, which is related to pitting resistance, and tooth strength, which is related to fracture and fatigue resistance, are the most important design considerations. The AGMA standard *Fundamental Rating Factors and Calculation Methods for Involute Spur and Helical Gear Teeth* provides the formula for bending stress calculations based on the modification of the Lewis method for gear bending stress, and the formula for contact stress based on the modification of the Hertzian contact theory. Although these formulations are highly empirical by nature and may yield very different results since a large number of modifiers or factors are used to account for variations in materials, manufacturing, and applications, they are based on experiences, expertise, and real data from the whole gear industry and they have been proved to be valuable and practical in the design and application of gear transmission systems. The following text highlights the AGMA formulations on gear tooth bending strength and pitting resistance and their applications in gear design practice.

3.6.3 Pitting Resistance

In the AGMA fundamental pitting resistance formulation for involute spur and helical gears, the contact stress is calculated by the following equation:

$$s_c = C_p \sqrt{\frac{F_t C_a}{C_v} \frac{C_s}{dF} \frac{C_m, C_f}{I}} \tag{3.88}$$

In this formulation five quantities carry a unit:

s_c: This is the calculated contact stress in lb/in^2 or MPa.

C_p: Elastic coefficient in $[\text{lb/in}^2]^{\frac{1}{2}}$ or $[\text{MPa}]^{\frac{1}{2}}$, C_p is tabulated in the AGMA standard in terms of materials.

F_t: Tangent component of the contact force in Eq. (3.85) in lb or N.

d: Operating pinion pitch diameter, in in. or mm.

F: Net face width of the narrowest member, in in. or mm, the same as tooth length L in Eq. (3.82).

The other six quantities are unitless factors named as follows:

C_a: Application factor for pitting resistance.
C_s: Size factor for pitting resistance.
C_m: Load distribution factor for pitting resistance.
C_f: Surface condition factor for pitting resistance.
I: Geometry factor for pitting resistance.

The contact stress calculated by Eq. (3.88) must be lower than the allowable contact stress, which is modified by a another set of factors, as shown below:

$$s_c \leq s_{ac}\frac{C_L C_H}{C_T C_R} \tag{3.89}$$

The quantities in this formulation are as follows:

s_{ac}: Allowable contact stress in lb/in^2 or MPa.
C_L: Life factor for pitting resistance.
C_H: Tooth hardness ratio for pitting resistance.
C_T: Temperature factor for pitting resistance.
C_R: Reliability factor for pitting resistance.

3.6.4 Bending Strength

In the AGMA's fundamental bending strength formulation for involute spur and helical gears, the bending stress is calculated by the following equation:

$$s_t = \frac{F_t K_a}{K_v}\frac{P_t}{F}\frac{K_s K_m K_B}{J} \tag{3.90}$$

In this formula, s_t is the calculated bending stress in lb/in^2 or MPa. F_t and F are the same as in the formulation for pitting resistance. P_t is the transverse diametral pitch. The six unitless modification factors are as follows:

K_a: Application factor for bending strength.
K_B: Rim thickness factor for bending strength.
K_s: Size factor for bending strength.
K_m: Load distribution factor for bending strength.
K_v: Dynamic factor for bending strength.
J: Geometry factor for bending strength.

The bending stress calculated by Eq. (3.90) must be lower than the allowable bending stress modified as follows:

$$s_t \leq s_{at}\frac{K_L}{K_T K_R} \tag{3.91}$$

In this formulation, the allowable bending stress s_{at} is modified by three factors:

K_L: Life factor for bending strength.
K_T: Temperature factor for bending strength.
K_R: Reliability factor for bending strength.

In the AGMA standard for gear pitting resistance and bending strength, the allowable contact stress s_{ac} and the allowable bending stress s_{at} are tabulated in terms of gear materials, grades, heat treatment, and hardness. These allowable stress values are based on 10^7 cycles with 99% reliability. There are a large number of factors in the AGMA formulation, as shown in Eqs (3.88–3.91). The values of these factors have to be chosen carefully from plots or formulations recommended by the AGMA standard. The modification factors that share the same name (i.e. the same subscripts, for example, C_a and K_a) in the formulation have the same values and share the same plot or formula as provided in the standard. A separate AGMA standard [13,14] needs to be used for the values of the two geometry factors I and J.

3.7 Design of Automotive Transmission Gears

Involute helical gears are widely applied for automotive transmission gears. The design of transmission gears is based on the same principle and theory as for other applications and should take into considerations the vehicle operation conditions and requirements. Generally, durability, compactness, cost, and quietness are the design priorities for automotive gears, especially so for passenger vehicles. Because of the maturity of the automotive industry, the gear design for new transmission development can rely on the experience and data gained from existing products that have been proved in the industry. As the crystallization of these experiences and data, the AGMA standards on gear design provide guidelines in the highly empirical transmission gear design and development process. For the design of transmission gears, a series of AGMA gear design standards should be referenced or followed, as listed below:

AGMA 933-B03: Basic Gear Geometry
AGMA 913-A98: Method for Specifying the Geometry of Spur and Helical Gears
AGMA 2000-A88: Gear Classification and Inspection Handbook
AGMA 2002-B88: Tooth Thickness Specification and Measurement
AGMA 2001-D04: Fundamental Rating factors and Calculation Methods for Involute Spur and Helical Gear Teeth
AGMA 6002-B93: Design Guide for Vehicle Spur and Helical Gear Drives
AGMA 908-B89: Information Sheet – Geometry Factors for Determining the Pitting Resistance and Bending
Strength for Spur, Helical and Herringbone Gear Teeth, Parts 1 & 2
AGMA 918-A93: Summary of Numerical Examples Demonstrating the Procedures for Calculating Geometry
Factors for Spur and Helical Gears

For the design of transmission gears, the following data are usually known or specified:

- Maximum engine torque and power
- Vehicle weight and dimensional data
- Transmission ratios
- Space envelope for the transmission assembly
- Expected service life

The design of transmission gearing is a process that does not have a direct closed-form solution. Experiences and data on existing products in the industry play an important role in the selections of initial design parameters. The following summarizes the steps highlighting the gear design procedure for two example manual transmissions, one FWD and the other RWD, shown in Figures 2.4 and 2.5. These manual transmissions are typical for FWD passenger cars and for RWD commercial vehicles such as pickup truck and cargo vans. The stick diagrams are shown again in Figure 3.19 for convenience.

Step 1: Selection of normal pressure angle α_n, helical angle β and normal diametral pitch P_n

The normal pressure angle α_n is usually chosen as the standard value of 16° or 20°. The helical angle β is a very important design parameter that affects center distance, axial load, and gear contact ratio. For passenger car transmission gears, β can be initially chosen in the range $25° \ll \beta \leq 38°$. For commercial vehicles, the helical angle should be chosen at smaller values to reduce the axial load and can be chosen initially in the range $18° \ll \beta \leq 30°$. The final value for the helical angle depends on center distance and load

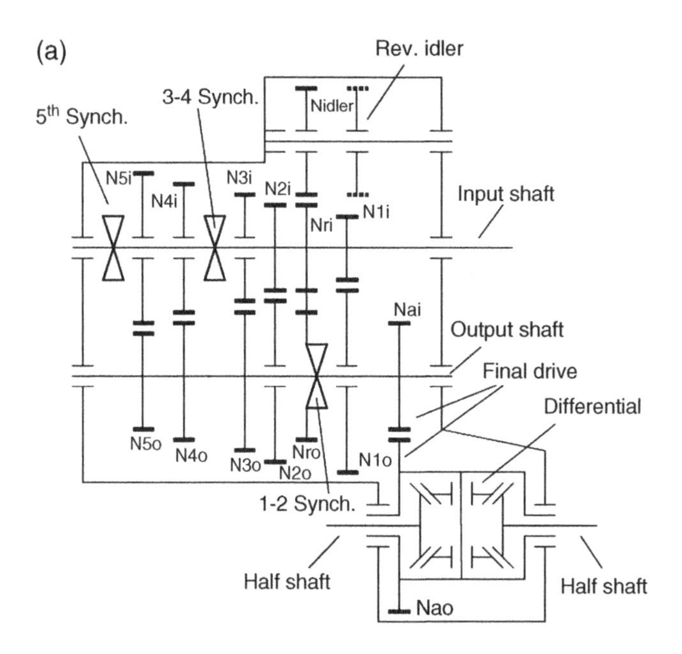

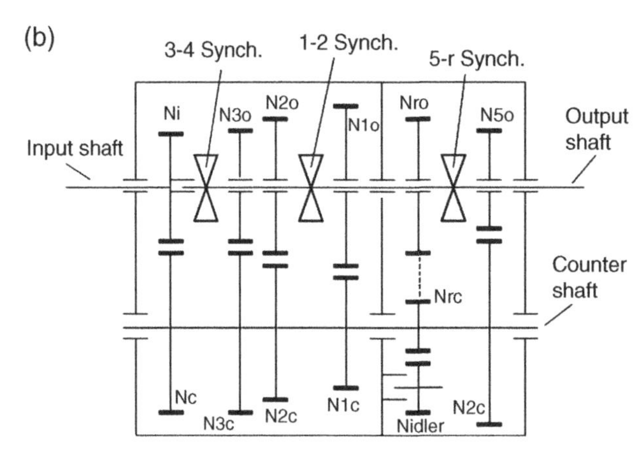

Figure 3.19 Transmission gear design examples.

constraints. Note that the helical gears on the output shaft in the MT shown in Figure 3.19a, and the helical gears on the counter shaft in the MT shown in Figure 3.19b, should have the same hands of helix so that the axial loads at the two gear meshes on these two shafts will partially cancel each other out. The gears on the input shaft and the final drive ring gear N_{1o} are right handed so that the axial load on the final drive ring gear will push the gear box against the engine. The normal diametral pitch P_n directly affects the dimension of the gears and the whole transmission size. It is recommended that the normal diametral pitch be selected according to the engine output power P_e as listed below:

$$\begin{aligned} \text{Passenger cars}: P_n &= 12 \quad \text{for} \quad P_e \ll 200\,\text{HP} \\ P_n &= 10 \quad \text{for} \quad P_e > 200\,\text{HP} \end{aligned}$$

$$\begin{aligned} \text{Commercial vehicles}: P_n &= 8 \quad \text{for} \quad P_e \ll 200\,\text{HP} \\ P_n &= 7 \quad \text{for} \quad 200 < P_e \le 250\,\text{HP} \\ P_n &= 6 \quad \text{for} \quad P_e > 250\,\text{HP} \end{aligned}$$

If the SI unit is used, then these diametral pitch values are converted to the module values by $\frac{25.4}{P_n}$. This value is then rounded to the closest module value in the standard series. For example, a normal diametral pitch $P_n = 10$ is converted to a normal module $m_n = 2.5$ mm. In many cases, a vehicle may be powered by optional engines with different output powers. Then the selection of the normal diametral pitch should fit the engine option with the highest power. For the FWD MT shown in Figure 3.19a, the final drive gears are subjected to the highest torque, and if necessary the normal diametral pitch for the final drive may be chosen as the coarser value next to the diametral pitch selected for all other gears.

Note that if strength and endurance requirements are satisfied, finer diametral pitch or smaller normal module are preferred. With the same size, gears can be designed with more teeth if a finer pitch is used. Gear pairs with more teeth and finer pitch in general run smoother and more quietly. For compact cars, the transmission gears are often designed with a normal module of 2 mm or a normal diametral pitch of 12.

Step 2: Selection of the numbers of teeth for different gears

The numbers of teeth must firstly guarantee that the specified gear ratios be achieved accurately, usually up to the second digit after the decimal point of the gear ratio value. Secondly, the numbers of teeth must also satisfy the constraints on the center distance. For the FWD MT in Figure 3.19a, the numbers of teeth and the center distances of gear pairs on the input and output shafts are constrained by the following equations:

$$E_{io} = \frac{1}{2} \frac{N_{ji} + N_{jo}}{P_n \cos\beta_j} \tag{3.92}$$

$$i_j = \frac{N_{jo}}{N_{ji}} \tag{3.93}$$

where the subscript j is 1, 2, 3, 4, 5, and r respectively. Apparently, the shaft distance or center distance, E_{io} is the same for all the six gear pairs. The value of E_{io} is usually specified within a narrow range due to assembly constraints. With the normal diametral pitch P_n and the helical angle β_j chosen in Step 1, the numbers of teeth for each gear pair,

N_{ji} and N_{jo}, can be determined by combining Eqs (3.92) and (3.93). But the solutions for N_{ji} and N_{jo} are generally not integers and must be rounded to the closest integers. These rounded integer numbers of teeth are then plugged back into Eq. (3.92) to solve for the helical angles. It is possible to design standard helical gears for all forward gears in this way. The reverse gears in the FWD MT shown in Figure 3.19a can be designed as a non-standard spur gear since the center distance of the spur reverse gears only depends on the numbers of teeth once the diametral pitch is chosen.

For the FWD MT shown in Figure 3.19a, the first gear N_{1i}, the second gear N_{2i}, and reverse gear N_{ri}, are machined on the input shaft. The numbers of teeth of these three gears can be chosen as: $N_{1i} = 12 \sim 14$, $N_{2i} = 17 \sim 19$, $N_{ri} = 12\tilde{}14$. The numbers of teeth for gears N_{1o}, N_{2o}, and N_{ro} are then determined by the specified ratios. The numbers of teeth for the first, second, and reverse gears and the selected normal diametral pitch are then plugged in Eq. (3.92) to solve the helical angles.

For the RWD MT in Figure 3.19b, the numbers of teeth and the center distances between the output shaft that is coaxial with the input shaft and the counter shaft must satisfy the following equations:

$$E_{io} = \frac{1}{2}\frac{N_i + N_c}{P_n \cos\beta_i} = \frac{1}{2}\frac{N_{jc} + N_{jo}}{P_n \cos\beta_j} \tag{3.94}$$

$$i_j = \frac{N_c N_{jo}}{N_i N_{jc}} \tag{3.95}$$

where the subscripts j is 1, 2, 3, 4, 5, and r respectively. In the design, there is an input gear pair, N_i and N_c, which is in the power flows of all gears except the 4th gear. Except for the direct 4th gear, the transmission ratios are the multiplication of the common input gear ratio and the other ratio designated for each gear. The input gear ratio, $\frac{N_c}{N_i}$, is usually designed to be around $1.25 \sim 1.35$, with $N_c = 17 \sim 19$ and $N_i = 22 \sim 24$. This allows for a near balanced strength for the input gears which undergo the longest service life under load. The numbers of teeth for other gear pairs are then chosen based on the gear ratio requirements and center distance constraints. The selection for the number of teeth of the first gear on the counter shaft, N_{1c}, is very important because the first gear pair determines the radial dimension of the transmission housing and has the largest strength imbalance between the pinion and the gear. In general, N_{1c} is chosen to be $14 \sim 16$. After the numbers of teeth for the first gears are chosen, the numbers of teeth for all other gears can then be chosen by using Eqs (3.94) and (3.95). As a rule of thumb, the counter shaft gear of the next higher gear is designed with $4 \sim 6$ more teeth. For example, N_{2c} is chosen to be $18 \sim 20$. The center distance constraints can be satisfied by choosing different helical angles for each gear pair. If the helical angle determined based on the center distance constraint is way off the range recommended in Step 1, then general non-standard gears can be designed to accommodate the center distance with the helical angle in the recommended range.

Step 3: Selection of the tooth length (face width) for gear pair

The tooth length L affects the contact area, and therefore the load capacity of transmission gears. The ratio between the tooth length and the normal diametral pitch is usually in the range: $L = (7 \sim 8.5)^1/_{P_n}$. For transmission gears with a normal diametral pitch of 12, the tooth length can be designed as $0.6 \sim 0.7$ inch.

Step 4: Tooth element proportions and tolerance specifications

After Step 3, the tooth element proportions can then be calculated by equations in Sections 3.3, 3.4, and 3.5. The contact ratio for each gear pair is calculated using Eqs (3.24), (3.80), and (3.81). Gears for automotive transmissions usually fall in class 9, 10, or 11 in the AGMA standard, depending on the category of the vehicle using the transmission. Gear tooth tolerances are specified in AGMA standards [9,10] based on the selected gear class.

Step 5: Gear strength and pitting resistance check

The gear tooth strength and pitting resistance are rated respectively using Eqs (3.90) and (3.91), and Eqs (3.90) and (3.91). Firstly, the gear force components are calculated by Eqs (3.84–3.87). The driving torque in Eq. (3.87) is the maximum engine output torque of the engine. For the FWD MT in Figure 3.19a, the first gears, 5th gears and the final drive gears should be the focus of strength and pitting resistance rating since these gears are most likely to be subjected to high loads or long service cycles. For the RDW MT in Figure 3.19b, the input gears, the first gears and the 5th gears should be focused on for the same reason. The allowable contact stress s_{ac} and the allowable bending stress s_{at} are related to gear materials, hardness, and heat treatment, and are tabulated in the AGMA standard [15] accordingly. As shown in the equations, the tooth bending strength and pitting resistance formulation largely depends on the various modification factors. The selection of these factors should follow the guidelines in the AGMA standard [11,14,15] and also account for the operation conditions characteristic of the automotive transmissions.

Dynamic factors C_v and K_v: The dynamic factors account for extra gear load in addition to the normal gear load that is generated by the non-perfect conjugate motions in a gear pair. The root cause for this extra load is the gear transmission errors defined as the deviation from the theoretical uniform gear rotations. The values of C_v and K_v are mainly related to the accuracy of the gears and the pitch line speed. In the AGMA standard [11], the dynamic factors, C_v and K_v, are plotted against the pitch line speed for different gear classes. The pitch line speed of transmission gears is approximately in the range 900~2100 ft/min. For transmission gears in AGMA Class 10, the values of C_v and K_v are found to be in the range of $0.84 \sim 0.88$ based on the AGMA standard.

Application factors C_a and K_a: The application factors account for the momentary peak load which is appreciably larger than the nominal or design load. For automotive transmission gears, there are rarely such operational conditions that would create momentary load spikes. If the maximum engine torque is used to calculate the gear forces, as mentioned previously, the values of C_a and K_a can be taken as one.

Size factors C_s and K_s: The size factor accounts for the non-uniformity of material property in gears that have large dimensions. Transmission gears are normally sized and the size factor is set to be one.

Load distribution factors C_m and K_m: The load distribution factor reflects the variations of load distribution over the tooth length caused by manufacturing errors, bearing and shaft misalignments, and deflections. Following the AGMA guidelines, the values of the load distribution factors, C_m and K_m, are selected to be $1.1 \sim 1.2$.

Life factors C_L and K_L: The allowable contact and bending stresses, s_{ac} and s_{at}, are established in the AGMA standard for 10^7 tooth load cycles with 99% reliability. The life factor is used to modify the allowable stresses if the designed life of gears differs from 10^7 tooth load cycles. The load cycles of the final drive ring gear are in the order of 10^8, corresponding to roughly 120,000 miles or 190,000 kilometers. The load cycles of the final drive pinion can then be approximated as $3.0 \sim 5.0\ (10^8)$. For the gears on the input shaft in the FWD MT shown in Figure 3.19a, the load cycles can be calculated based on the percentage of total mileage each gear contributes. This contribution percentage can be estimated as: 1st gear (1%), 2nd gear (3%), 3rd gear (8%), 4th gear (28%), and 5th gear (60%). The load cycles of the gears on the output shaft are the multiplication of the contribution percentage and the final drive pinion load cycles. The load cycles of the gears on the input shaft are then determined by multiplying the respective gear ratio and the load cycles of the gear on the output shaft. For the RWD MT in Figure 3.19b, the load cycles of the gears on the counter shaft for each gear can be determined in similar fashion, i.e. by the multiplication of the respective contribution percentage and the final drive pinion load cycle. Note that the load cycles for the input gear N_i in Figure 3.19b should be calculated as the multiplication of the ratio $\frac{N_o}{N_i}$ and the sum of load cycles of the 1st, 2nd, 3rd, and 5th gears. After the load cycles are determined, the life factors can then be found from the AGMA standard plots to modify the allowable stresses.

Temperature factors C_T and K_T: In general, automotive transmission gears operate at temperatures below 250 °F. The temperature factors are taken as unity.

Reliability factors C_R and K_R: As mentioned previously, the allowable stresses are established for 10^7 tooth load cycles as 99% reliability. This means that statistically there is one failure out of 100 after 10^7 tooth load cycles. This reliability is not sufficient for automotive transmission gears. As recommended by the ASMG standard, the values of C_R and K_R can be taken as 1.25, or 1.5 for higher reliability, corresponding to one failure in a thousand or one failure in ten thousand after 10^7 tooth load cycles.

The factors described above are common to both the pitting resistance and bending resistance. Two additional modification factors – the surface condition factor C_f and the hardness ratio factor C_H – are used for the formulation of pitting resistance. One additional factor, the rim thickness factor K_B, is used for the formulation of bending strength. These factors are explained in the following specifically for transmission gears.

Surface condition factor C_f and **hardness ratio factor** C_H: The surface condition factor reflects the influence of surface finish and residual stress of the tooth surface on pitting resistance. For automotive transmission gears, C_f can be taken as 1.0 because these gears are normally sized and have sufficient accuracy in tooth surface finish. The hardness ratio factor C_H depends on the gear ratio and the hardness of the pinion and the gear. In the AGMA standard [11], this factor is plotted against the gear ratio according to the pinion and gear tooth hardness ratios. For transmission gears, the hardness ratio factor C_H can also be taken as unity since the gear ratio is usually smaller than 4.0 and there is no significant difference in the tooth hardness between the pinion and the gear.

Rim thickness factor K_B: This factor applies to ring gears where gear teeth are manufactured on a thin ring, such as the ring gear of the final drive in the FDW MT in Figure 3.19a. The ring gear of the final drive is bolted to the differential carrier and must be designed with sufficient strength and rigidity to support the gear teeth on it. The rim thickness factor K_B is therefore set as unity for transmission gears.

Example A manual gear box for a RWD vehicle is shown in the figure. The normal diametral pitch, the normal pressure angle and the helical angle of the gears labeled as Ni, Nc, N1c, N1o, N2c, and N2o are given as: 8, 16, and 20° respectively. Gears Ni and Nc are standard helical gears. The numbers of teeth of gears for the first and second speeds are given with the drawing. All gears on the counter shaft are right-handed. Calculate the following for the gear design:

a) The shaft distance between the input shaft and the counter shaft.
b) The gears N10 and N1c have a tooth length of 0.95 inch. The two gears are designed as non-standard gears using the long-short addendum system. It is required that the transverse tooth thickness of gear N1c be increased by 10% compared with the standard tooth thickness. Determine the cutter displacements for the two gears and the contact ratio.
c) Determine the operating pressure angle and the sum of the cutter displacements for gears N2c and N2o.
d) If the engine torque is 200 ft.lb, determine the gear forces on the counter shaft in first gear. Show the directions of gear force components at the mesh positions. Also, find the thrust force applied to the counter shaft bearing.

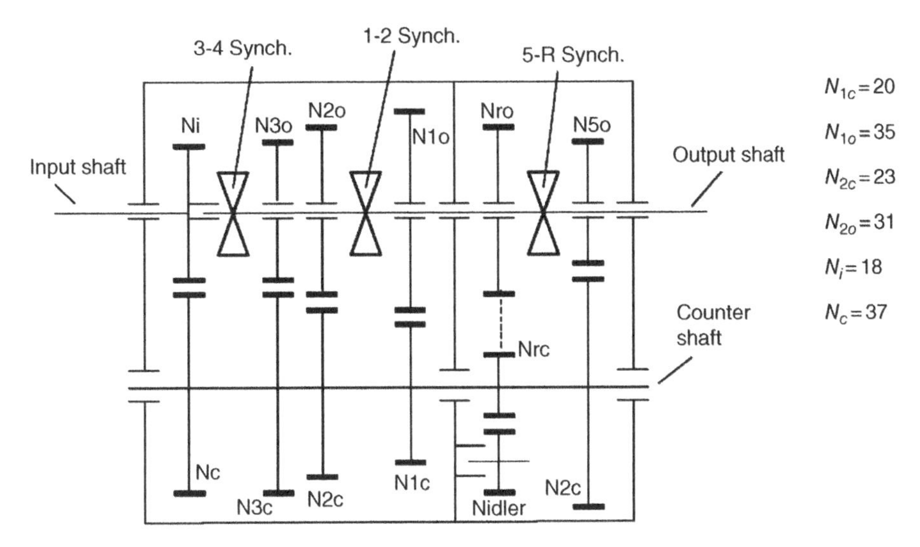

Solution:

a) Shaft distance:

$$P_t = P_n \cos\beta = 8\cos 20^0 = 7.5175$$

$$E = \frac{N_i + N_C}{2P_t} = \frac{18+37}{2(7.5175)} = 3.6581 \text{ (inches)}$$

b) $N_{1c} + N_{1o} = 20 + 35 = 55 \geq 2N_{min}$, so long-short addendum system can be designed.
Standard tooth thickness:

$$t_{std} = \frac{\pi}{2P_t} = 0.2090 \text{ (inch)}$$

Since $t_{N1c} = 1.1 t_{std}$,

$$\frac{\pi}{2P_t} + 2e_p \tan\alpha_t = 1.1(0.2090)$$

$$\alpha_t = \tan^{-1}\frac{\tan\alpha_n}{\cos\beta} = \tan^{-1}\frac{\tan 16^\circ}{\cos 20^\circ} = 16.969^\circ$$

$$e_p = \frac{0.2299 - 0.2090}{2\tan 16.969^\circ} = 0.03432\ (\text{inch})$$

$$e_g = -e_p - 0.03432$$

Pitch radii:

$$r_p = \frac{20}{2(7.5175)} = 1.3302;\ r_g = \frac{35}{2(7.5175)} = 2.3279$$

Pitch radii:

$$r_{bp} = \frac{20\cos 16.969^\circ}{2(7.5175)} = 1.2723;\ r_g = \frac{35\cos 16.969^\circ}{2(7.5175)} = 2.2265$$

Addendum radii:

$$r_{pa} = r_p + \frac{1}{P_n} + e_p = 1.3302 + \frac{1}{8} + 0.03432 = 1.4895$$

$$r_{ga} = r_g + \frac{1}{P_n} + e_g = 2.3279 + \frac{1}{8} - 0.03432 = 2.4186$$

Contact ratio:

$$m_t = \frac{7.5175}{\pi\cos 16.969^\circ}\left[\sqrt{1.4895^2 - 1.2723^2} + \sqrt{2.4186^2 - 2.2265^2} - 3.6581\sin 16.969^\circ\right] = 1.062$$

$$m_l = \frac{P_n L\sin\beta}{\pi} = \frac{8(0.95)\sin 20^\circ}{\pi} = 0.8274$$

$$m = m_t + m_l = 1.062 + 0.82741 = 1.89$$

c) Gears N_{2c} and N_{2o} must be designed as general non-standard gears to accommodate the shaft distance determined in (a).

$$\cos\alpha_t' = \frac{(N_p + N_g)\cos\alpha_t}{2E'P_t} = \frac{(23+31)\cos\sin 16.969^\circ}{2(3.6581)(7.5175)} = 0.9391$$

$$\alpha_t' = \cos^{-1}0.9391 = 20.0990^\circ$$

$$\text{Inv}\left(\alpha_t'\right) = \frac{2\left(e_p + e_g\right)P_t}{N_p + N_g}\tan\alpha_t + \text{Inv}\left(\alpha_t\right)$$

$$\text{Inv}\left(\alpha_t'\right) = \tan\left(\alpha_t'\right) - \alpha_t' = \tan 20.0990 - \frac{20.0990\pi}{180} = 0.01513$$

$$\text{Inv}\left(\alpha_t\right) = \tan\left(\alpha_t\right) - \alpha_t = \tan 16.9695 - \frac{16.9695\pi}{180} = 0.008975$$

$$\left(e_p + e_g\right) = \frac{\left(\text{Inv}\left(\alpha_t'\right) - \text{Inv}\left(\alpha_t\right)\right)}{2P_t\tan\alpha_t}\left(N_p + N_g\right) = \frac{(0.01513 - 0.008975)(54)}{2(7.5175)\tan 16.9695} = 0.07244\ (\text{inch})$$

d) Gear forces:

For gears N_i and N_c : $P_t = 8\cos 20° = 7.5175; r_p = \dfrac{18}{2(7.5175)} = 1.1972$

Contact Force at the mesh of gears N_i and N_c:

$$N = \frac{T_d}{r_p \cos\alpha_n \cos\beta} = \frac{200(12)}{1.1972\cos 16° \cos 20°} = 2219.3\ \text{(lb)}$$

Gear force components on gear N_c:

$$F_r = N\sin\alpha_n = 2219.3\sin 16^0 = 611.7\ \text{(lb)}$$
$$F_t = N\cos\alpha_n \cos\beta = 2219.3\cos 16° \cos 20° = 2004.7\ \text{(lb)}$$
$$F_a = N\cos\alpha_n \sin\beta = 2219.3\cos 16° \sin 20° = 729.6\ \text{(lb)}$$

For gears N_i and N_c : $T_d = 200\left(\dfrac{37}{18}\right) = 411.1\ \text{(ft.lb)}; r_p = \dfrac{20}{2(7.5175)} = 1.3302$

Contact Force at the mesh of gears N_i and N_c:

$$N = \frac{T_d}{r_p \cos\alpha_n \cos\beta} = \frac{411.1(12)}{1.3302\cos 16° \cos 20°} = 4105.8\ \text{(lb)}$$

Gear force components on gear N_{1c}:

$$F_r = N\sin\alpha_n = 4105.8\sin 16^0 = 1131.7\ \text{(lb)}$$
$$F_t = N\cos\alpha_n \cos\beta = 4105.8\cos 16° \cos 20° = 3708.7\ \text{(lb)}$$
$$F_a = N\cos\alpha_n \sin\beta = 4105.8\cos 16° \sin 20° = 1349.7\text{(lb)}$$

Resultant thrust on the counter shaft: $1349.7 - 729.6 = 620.1$ (lb)
Directions of gear forces on the counter shaft:

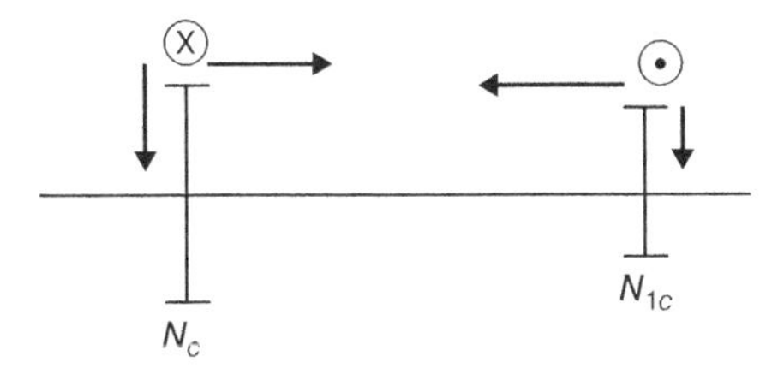

3.8 Planetary Gear Trains

Gears can be designed in series where all gear axes are fixed with respect to each other or in systems where one or more gear axes rotate with respect to one another. Such systems are called planetary gear trains (PGT). There are many PGT types categorized in terms of their structure or configuration. Two PGT types are widely applied in automotive automatic transmissions: simple PGT and Ravigneaux PGT. Simple PGTs can be designed with one planet gear that meshes with the sun gear and the ring gear, as shown in Figure 3.20, or two planet gears in series as shown in Figure 3.21.

In the simple PGT shown in Figure 3.20 there are four elements: sun gear, planet gear, ring gear, and carrier. The number of planet gears does not change the kinematics, but multiple planet gears, usually four, are used in automatic transmissions for load sharing. The sun gear and the ring gear rotate about the same axis. A planet gear participates in two rotations, one about its own axis and the other about the axis of the sun gear. In other words, the planet gear rotates about its axis which rotates about the sun. The angular velocities of the sun gear ω_S, the carrier ω_C, and the ring gear ω_R of the simple PGT in Figure 3.20 are governed by the so-called characteristic equation as follows:

$$\omega_S + \beta\omega_R - (1+\beta)\omega_C = 0 \tag{3.96}$$

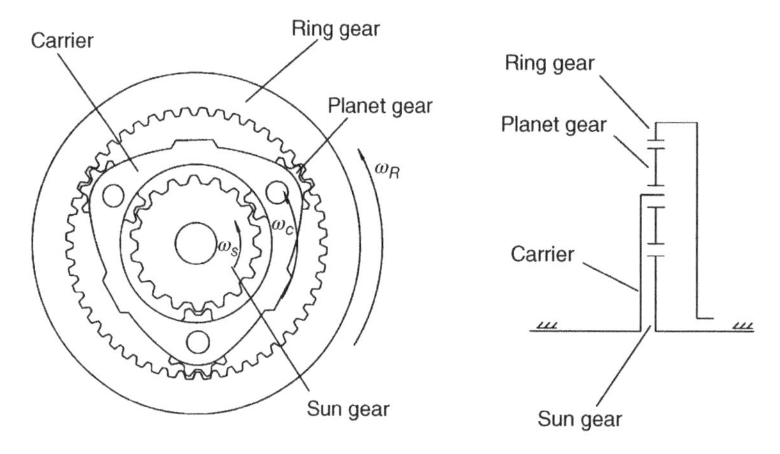

Figure 3.20 Simple planetary gear train.

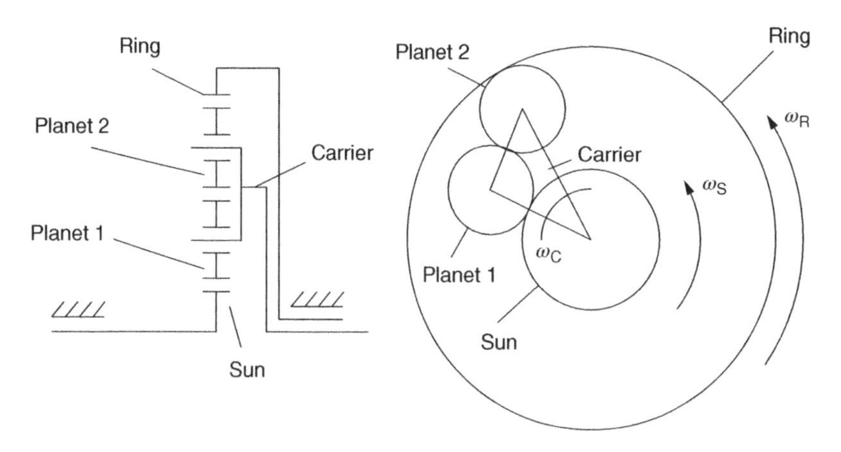

Figure 3.21 Dual-planet simple planetary gear train.

where β is the planetary parameter that is equal to the ratio between the numbers of teeth in the ring gear and in the sun gear ($\beta = \frac{N_R}{N_S}$). As can be observed from Eq. (3.96), the simple planetary train has two degrees of freedom, since only the characteristic equation constrains the three angular velocities. If one of the three angular velocities is given as input, the other two are not defined unless an additional constraint is imposed. In automatic transmissions designed with planetary gear trains, the additional constraints are from hydraulically actuated clutches.

In the dual-planet PGT shown in Figure 3.21, the carrier supports two planet gears in series, and the mesh path is sun gear – planet gear 1 – planet gear 2 – ring gear. Each planet gear group, planet 1 or planet 2, contains multiple planet gears for load sharing in automatic transmission applications. The characteristic equation governing the three angular velocities differs from Eq. (3.96) only by the sign before the planetary parameter β:

$$\omega_S - \beta\omega_R - (1-\beta)\omega_C = 0 \tag{3.97}$$

The structure of a Ravigneaux planetary gear train is shown in Figure 3.22. A Ravigneaux PGT consists of sun gear 1, sun gear 2, long planet gear, short planet gear, carrier, and ring gear. In terms of kinematics, a Ravigneaux PGT can be considered as the combination of a simple planetary PGT and a dual-planet PGT with a shared carrier and a shared ring gear. There are two mesh paths in the Ravigneaux PGT: sun gear 1 – long planet gear – ring gear; and sun gear 2 – short planet gear – long planet gear – ring gear. Here, Eq. (3.96) applies to the first mesh path and Eq. (3.97) applies to the second mesh path. The characteristic equations that govern the angular velocities of a Ravigneaux PGT are:

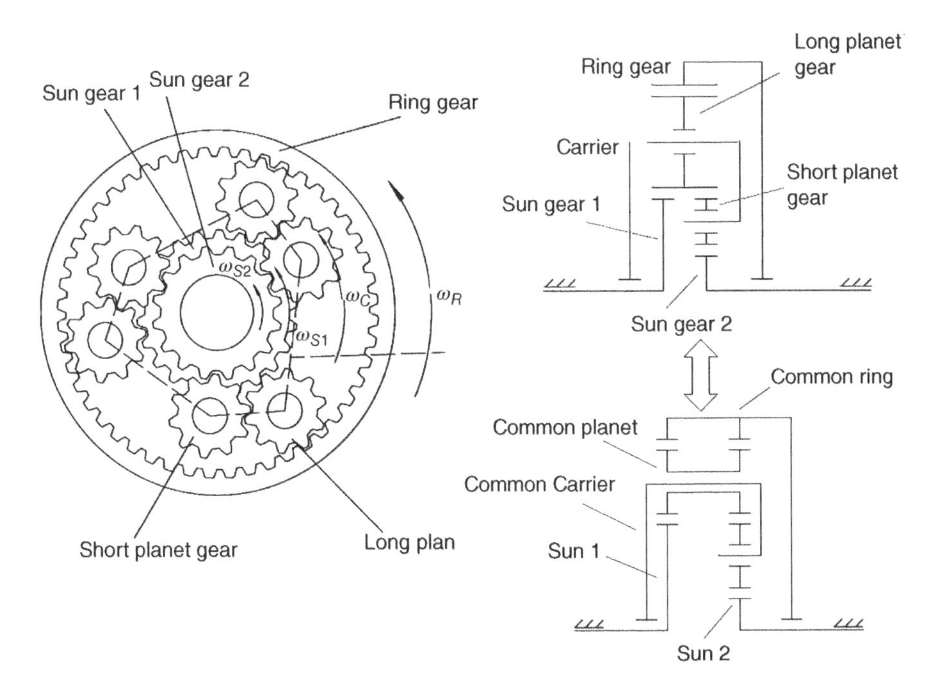

Figure 3.22 Ravigneaux planetary gear train.

$$\omega_{S1} + \beta_1 \omega_R - (1 + \beta_1)\omega_C = 0 \tag{3.98}$$

$$\omega_{S2} - \beta_2 \omega_R - (1 - \beta_2)\omega_C = 0 \tag{3.99}$$

where ω_{S1} and ω_{S2} are the angular velocities of sun gear 1 and sun gear 2 respectively. ω_C and ω_R are the angular velocities of the carrier and the ring gear, β_1 is the ratio between the number of teeth in the ring gear and the number of teeth the sun gear 1 $\left(\frac{N_R}{N_{S1}}\right)$, and β_2 is the ratio between the number of teeth in the ring gear and the number of teeth in the sun gear 2 $\left(\frac{N_R}{N_{S2}}\right)$. It can be observed from Eqs (3.98) and (3.99) that the Ravigneaux PGT also has two degrees of freedom. The four angular velocities, ω_{S1}, ω_{S2}, ω_C, and ω_R, are constrained by two equations. When one angular velocity is given as the input, there must be an additional constraint for the motions to be defined. The Ravigneaux PGT offers a compact design in automatic transmissions since it provides the functionality of two simple PGTs.

When a planetary gear train is loaded in the transmission of power, members in the PGT are subjected to externally or internally applied torque. In transmission applications, external torque can be applied to the sun gear, carrier, or ring gear through the input, reaction, or output, as will be detailed in Chapter 5. The internal torque is applied on the other PGT members. Similar to the angular velocities, the torque magnitudes on PGT members are related to the planetary parameter, and the direction of the internal torque on each member depends on the PGT type.

3.8.1 Simple Planetary Gear Train

The directions of the internal torques on the sun gear, the carrier, and the ring gear of a simple PGT are shown in Figure 3.23a. Note that only the relative directions are shown in the figure for the internal torques. For the simple PGT, the internal torques on the sun gear and on the ring gear have the same direction, but the internal torque on the carrier is in the opposite direction, as shown in Figure 3.23a. The internal torque on the sun gear with a magnitude of T_S is applied to the sun gear via the mesh between the planet gear and the sun gear. The internal torque on the ring gear with a magnitude of T_R is applied to the ring gear via the mesh between the planet gear and the ring gear. The magnitude of the torque on the carrier is the algebraic sum of the torques on the sun gear and on the

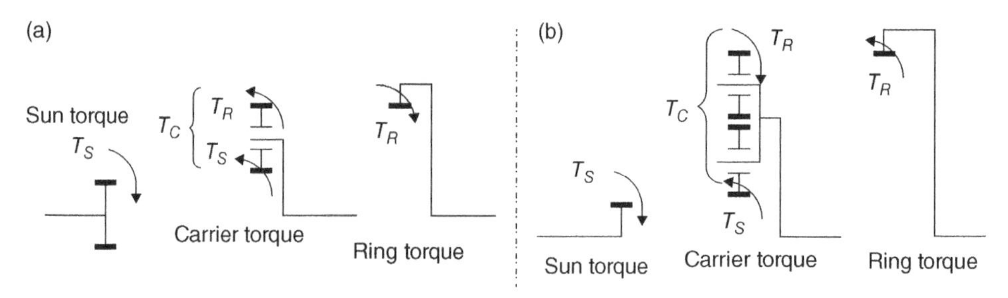

Figure 3.23 Internal torque directions for simple and dual-planet PGTs.

ring gear. For convenience and clarity the carrier and the planet can be considered as one body in drawing the torque diagrams for automatic transmissions that involve multiple planetary gear trains. The torque on the sun gear and the torque on the carrier–planet assembly are action and reaction, and so have the same magnitude but opposite directions. Similarly, the torque on the ring gear and the torque on the carrier-planet assembly are also action and reaction, as shown in Figure 3.23a. The magnitudes of the torque on the sun gear, the torque on the ring gear, and the torque on the carrier are related by the following equations:

$$T_R = \beta T_S \tag{3.100}$$

$$T_C = (1+\beta) T_S \tag{3.101}$$

3.8.2 Dual-Planet Planetary Gear Train

The directions of the internal torques on the sun gear, the carrier, and the ring gear of a dual-planet PGT are shown in Figure 3.23b. For the dual-planet PGT, the internal torques on the sun gear and on the ring gear have the opposite direction. The internal torque on the sun gear with a magnitude of T_S is applied to the sun gear via the mesh between the planet gear 1 and the sun gear. The internal torque on the ring gear, with a magnitude of T_R, is applied to the ring gear via the mesh between the planet gear 2 and the ring gear. The magnitude of the torque on the carrier is the algebraic sum of the torques on the sun gear and the ring gear. The magnitudes of the torques on the sun gear, the ring gear, and the carrier are related by the following:

$$T_R = \beta T_S \tag{3.102}$$

$$T_C = (\beta - 1) T_S \tag{3.103}$$

Since the planetary parameter β is always larger than one, the direction of the internal torque on the carrier is always the same as the direction of the torque on the sun gear in a dual-planet PGT, and the magnitude of the ring gear torque is the sum of the sun gear torque and the carrier torque.

3.8.3 Ravigneaux Planetary Gear Train

The directions of the internal torques on the members of a Ravigneaux PGT are shown in Figure 3.24. Note that the Ravigneaux PGT is treated as the combination of a simple PGT and a dual-planet PGT. The torque magnitudes for a Ravigneaux PGT are related by the following:

$$T_{R1} = \beta_1 T_{S1} \tag{3.104}$$

$$T_{C1} = (1+\beta_1) T_{S1} \tag{3.105}$$

$$T_{R2} = \beta_2 T_{S2} \tag{3.106}$$

$$T_{C2} = (\beta_2 - 1) T_{S2} \tag{3.107}$$

In automatic transmissions, there are multiple PGTs with members structurally interconnected. Equations (3.96–3.99) are used to constrain the angular velocities and accelerations of PGT members and other drive train components in the powertrain system.

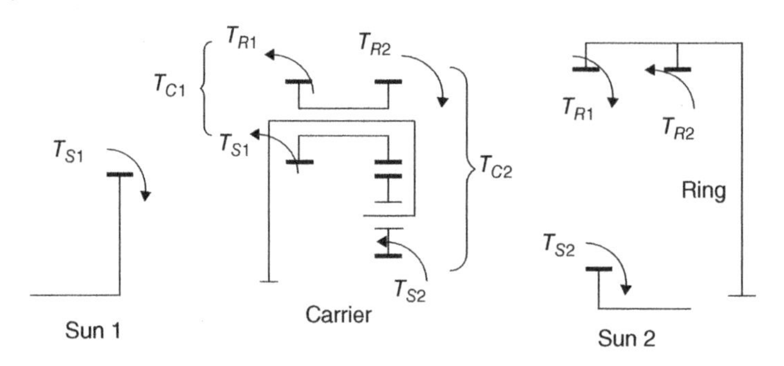

Figure 3.24 Internal torque directions for Ravigneaux PGTs.

Equations (3.100–107) are used to relate the transmission input, output torque, and clutch torques, as detailed in Chapter 5.

References

1 Gear – Wikipedia https://en.wikipedia.org/wiki/Gear.
2 Involute – Wikipedia https://en.wikipedia.org/wiki/Involute.
3 Buckingham, E.: *Analytical Mechanics of Gears,* Dover Publications, 1988, ISBN 13-9780486657127.
4 Litvin, F.L. and Fuentes, A.: *Gear Geometry and Applied Theory, Second Edition, Cambridge University Press,* 2004, ISBN 0-521-81517 7.
5 Radzevich, S.P.: Dudley's Handbook of Practical Gear Design and Manufacture, 3rd Edition, CRC Press, ISBN 13-978-1498753104.
6 AGMA 933-B03: Basic Gear Geometry.
7 AGMA 913-A98: Method for Specifying the Geometry of Spur and Helical Gears.
8 American Gear Manufacturers Association: *Design Manual for Bevel Gears,* ANSI/ AGMA 2005-B88, Ma9 1988, ISBN 1-55589-496-8.
9 AGMA 2000-A88: Gear Classification and Inspection Handbook.
10 AGMA 2002-B88: Tooth Thickness Specification and Measurement.
11 AGMA 2001-D04: Fundamental Rating factors and Calculation Methods for Involute Spur and Helical Gear Teeth.
12 AGMA 918-A93: A summary of Numerical Examples Demonstrating the Procedures for Calculating Geometry Factors for Spur and Helical Gears.
13 AGMA 908-B89: Information Sheet – Geometry Factors for Determining the Pitting Resistance and Bending Strength for Spur, Helical and Herringbone Gear Teeth, Parts 1 & 2.
14 American Gear Manufacturers Association: *Geometry Factors for Determining the Pitting Resistance and Bending Strength of Spur, Helical and Herringbone Gear Teeth,* AGMA 908-B89, April 1989, ISBN 1-55589-525-5.
15 AGMA 6002-B93: Design Guide for Vehicle Spur and Helical Gear Drives.

Problems

1 A six-speed RWD MT is shown in the figure. The following data are given in the gear design:

Input gears: Ni = 26, Nc = 29, normal diametral pitch: 12, normal pressure angle: 20°; helical angle: 25°, tooth length: 0.95 inch.

First speed: N10/N1c = 41/13, normal diametral pitch: 12, normal pressure angle: 20°, helical angle: 25°.

Sixth speed: N60/N6c = 21/34, normal diametral pitch: 12, normal pressure angle: 20°, helical angle: 25°.

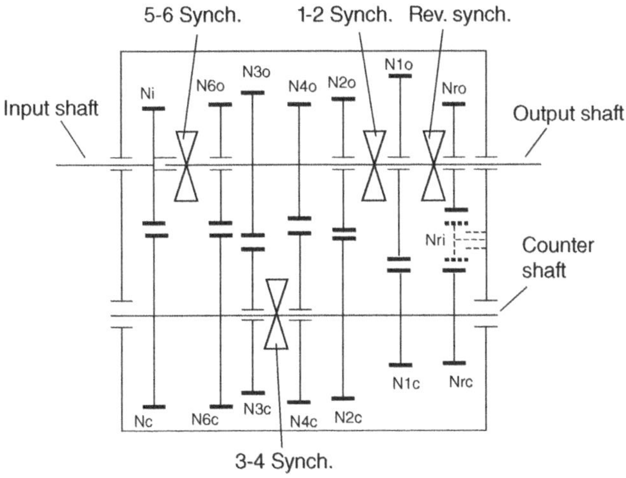

All the gears on the counter shaft are right handed. The transmission efficiency is 0.98 and the final drive efficiency is 0.97.

a) The input gears Ni and Nc are designed as standard gears. Determine the contact ratio.

b) The tooth thickness of gear N1c is designed to be 20% larger than the standard tooth thickness. Determine the tooth thickness of gears N1c and N10.

c) The vehicle is running in 6th gear at a speed of 65 miles on a 1% slope and with an acceleration of 2 ft/s^2. Determine the gear forces on the counter shafts, and the resultant thrust on the bearing on the counter shaft. You must show the direction of each gear force and the thrust.

d) An optional 6th gear ratio of 0.7215 is to be made available for the transmission by redesigning the six gears N60 and N6c only. The redesigned 6th gears must be standard helical gears with normal diametral pitch equal to 12 and normal pressure angle equal to 20°. The number of teeth N6c must be kept the same at 34. Determine the helical angle of the resigned 6th gears.

Vehicle data:

Front axle weight: 1900 lb
Center of gravity height: 14 in.
Air drag coefficient: 0.30
Tire radius: 11.0 in
Mass factor: 1.0
Rear axle weight: 1500 lb
Wheel base: 108 in.
Frontal projected area: 21 sq. ft
Roll resistance coefficient: 0.02
Final drive ratio 3.21

Prob. 2: A six-speed FWD MT is shown in the figure. The gears for the 1st speed, the 6th speed, and the final drive are standard helical gears. The following data are given in gear design:

First speed: gear ratio: 3.3846, number of teeth of gear N_{1i}: 13, normal diametral pitch: 12, normal pressure angle: 20°

Sixth speed: numbers of teeth: N6i/N60 = 36/23, normal diametral pitch: 12, normal pressure angle: 20°, helical angle: 25°, tooth length: 0.8 in.

Final drive: numbers of teeth: 59/18, normal diametral pitch: 10, normal pressure angle: 20°; helical angle: 21°

All gears on the output shaft are left handed. The overall drive line efficiency from transmission input to final drive output is .96 and the final drive efficiency is .97.

a) Determine the contact ratio of the 6th gears.
b) Determine the pitch radii for the gears of the 1st speed.
c) The vehicle is running in 6th gear at a speed of 65 mph on a 1% slope and with an acceleration of 2 ft/s^2. Determine the gear forces on the output shafts and the resultant thrust on the bearing on the output shaft. You must show the direction of each gear force and the thrust.
d) An optional final drive is fitted into the same transmission housing. The numbers of teeth of the optional final drive gears are: 59/16, the normal diametral pitch is 10, the normal pressure angle is 20°, and the helical angle is 21°. The tooth thickness of the pinion (the gear with 16 teeth) is to be designed 40% thicker than the standard tooth; determine the cutter displacements for the two gears of the final drive.

Vehicle data:

Front axle weight: 1900 lb
Rear axle weight: 1500 lb
Center of gravity height: 14 in.
Wheel base: 108 in.
Air drag coefficient: 0.30
Frontal projected area: 21 sq. ft
Tire radius: 11.0 in.
Roll resistance coefficient: 0.02
Mass factor: 1.0

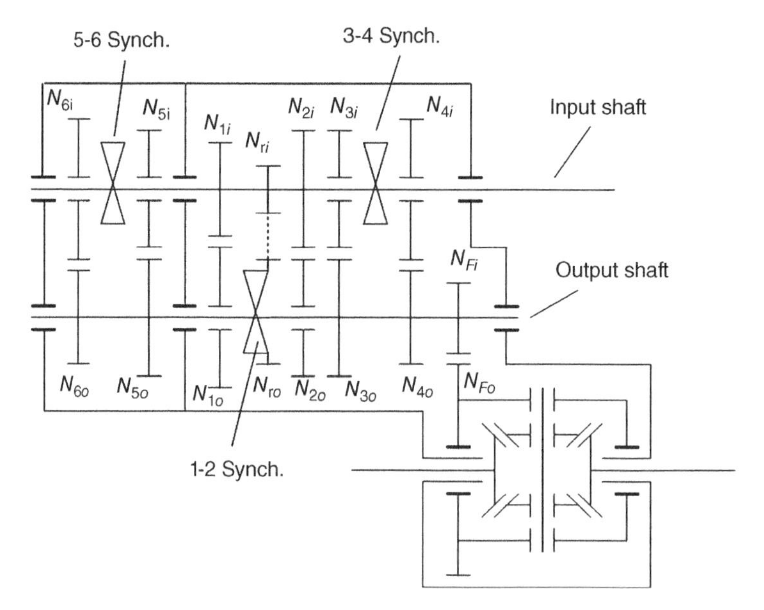

4

Torque Converters

4.1 Introduction

The invention of fluid couples, over a hundred years ago, was credited to H. Foettinger [1]. The fluid couple, with two elements, the impeller as input and the turbine as output, was later improved by adding the reactor between the two elements to become the torque converter. These hydraulic devices were first applied in the driving systems of ship propellers. The development of torque converters for applications in automotive powertrains dates back to the 1920s [1]. In the US automotive industry, fluid couples were already being applied in the 1930s in transmissions for production passenger cars and city buses [2]. The application of torque converters in automatic transmissions of passenger cars started to take off in the mid 1940s. By the late 1940s, automatic transmissions with torque converters were already in mass production [3,4]. The market demand for family cars grew rapidly in the booming economy after WWII, and this called for the development of automatic transmissions for passenger automobiles that would offer comfort and operation easiness. Torque converter, due to its input and output characteristics, proved to be a perfect fit between the engine output and the transmission input. The application of torque converters in the automotive industry was so widespread that more than 70% of passenger automobiles sold in 1960 in the USA were already equipped with automatic transmissions with these devices [5]. It can be stated that torque converter equipped automatic automobiles are indispensable for daily transportation in today's society.

Although torque converters can be applied in a variety of machinery, their application in automotive transmissions is by far the most dominant. Today the vast majority of automatic vehicles, if not all of them, are equipped with one of the three types of automatic transmissions: (1) conventional automatic transmission that is commonly termed just automatic transmission (AT) since it had been applied in the industry almost exclusively for decades before the other two types were applied in significant scale: (2) continuously variable transmissions (CVT), and (3) dual-clutch transmissions (DCT). Torque converters are applied in all ATs and in the vast majority of CVTs. It is not necessary to use torque converters in DCTs since the vehicle launch is realized by controlling the slippage of one of the dual clutches. However, Honda recently developed a DCT that is equipped with a torque converter to improve vehicle launch control and applied it for a popular passenger car [6]. It is likely that the market share of automatic vehicles equipped with torque converters will still grow, as customers in emerging markets are buying more driver-friendly automobiles.

Automotive Power Transmission Systems, First Edition. Yi Zhang and Chris Mi.

The in-depth research and development conducted by the US automotive industry in the 1940s and 1950s established the standards for torque converter design and application for automotive transmissions. Pioneering works on torque converter design, manufacturing, and applications can be found in the SAE publication [4] "Design Practices: Passenger Car Automatic Transmissions". This publication provides a collection of classic technical papers that detailed the operation, analysis, and design optimization of torque converters [3,5,7–11]. The advancement and maturity of torque converter technologies today is thanks to the original technical contributions represented in these papers.

The objective of this chapter is to provide readers a systematic description on the structure, operation principles, and input–output characteristics of torque converters. The chapter starts in Section 4.2 with the basic structure and geometry of torque converters with lock-up capability and a qualitative analysis on the functionality. Then Section 4.3 follows on with the mechanism of fluid circulation and torque multiplication, including topics on fluid velocity diagrams, analysis of the angular momentum of automatic transmission fluid (ATF) circulating inside the torque converter, and torque formulations on the three torque converter elements. Section 4.4 presents the torque capacity formulation and input–output characteristics of torque converters. The section ends with the modeling of the joint operation of the complete vehicle system that consists of engine, torque converter, transmission, and the vehicle itself. The conventions on torque converter terminologies in SAE publications are used throughout this chapter wherever appropriate. For detailed torque converter blade designs, readers are referred to specific SAE publications and guidelines [3,5].

4.2 Torque Converter Structure and Functions

Torque converters used in automatic transmissions today consist of four elements: impeller, reactor, turbine, and pressure plate. The impeller, reactor, and turbine are basic elements that realize the functionality of a torque converter. The pressure plate is designed to lock up the torque converter as a mechanical link between the engine and the transmission input. As shown in Figure 4.1, the impeller, which is also called the "pump", is always rigidly connected to the engine flywheel; the turbine is the output element and is connected to the transmission input; and the reactor or stator is placed between the impeller and the turbine. The reactor is fixed with the outer race of a one-way clutch whose inner race is grounded to the transmission housing and is thus allowed to rotate in only one direction. The torque converter cover is welded to the impeller and forms an enclosure with the impeller, which contains the turbine, the reactor, and the automatic transmission fluid (ATF). The pressure plate is splined to the turbine shaft and is machined with a friction surface. When the pressure plate is pushed against the inside wall of the cover by the ATF under pressure, the friction generated on the contacting surfaces of the cover and the pressure plate will lock up the torque converter, forming a mechanical coupling between the engine and transmission. The pressure plate assembly includes a torsional spring damper which cushions the impacts during locking up.

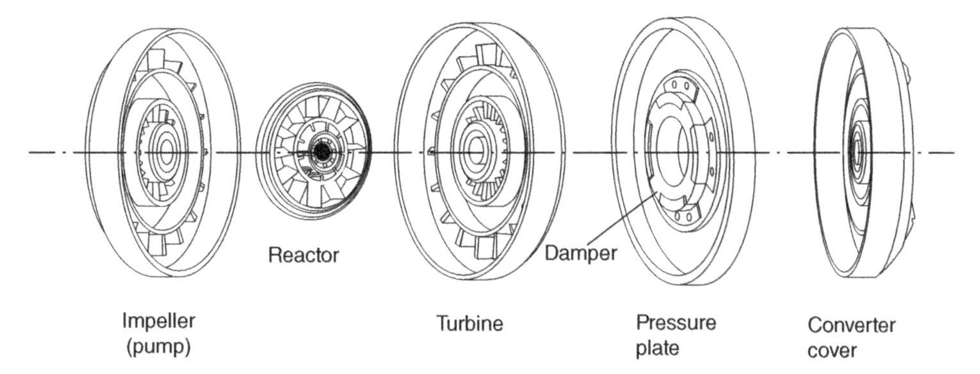

Figure 4.1 Torque Converter Elements.

Torque converters provide several important functions for vehicle powertrain systems:

a) To multiple the engine output torque at lower turbine rotational velocity, which corresponds to low vehicle speed. This function is very useful since it helps launch the vehicle quickly and smoothly while the engine is running in a low torque RPM range. The torque multiplication is realized by the reactor which is not allowed to rotate by the one-way clutch at low turbine speed and thus provides a reaction through the ATF to the turbine. Without the reactor, the impeller and the turbine just form a fluid couple between the engine and the transmission input.
b) To form a fluid couple between the engine and the transmission after the turbine speed gets close to the impeller speed. As the turbine speed increases to a certain point, termed the coupling point, the ATF circulating inside the torque converter will enter the reactor from the turbine in such a way that its impact on the reactor blades starts to turn the reactor in the direction allowed by the one-way clutch. The torque multiplication function no longer exists after the coupling point, and the torque converter becomes a fluid couple. Functioning as a fluid couple, the torque converter dampens the dynamic transients in the powertrain system and enhances transmission shift smoothness.
c) To provide crawl and hill holding capability. This function can be considered as the combined result of functions (a) and (b). When an automatic vehicle is stopped with the engine running at idle, the small engine output torque will be transmitted by the torque converter to the transmission input, generating sufficient driving force to move the vehicle forward at low speed on level ground. If the vehicle is stopped in an uphill position, this driving force prevents the vehicle from slipping backward up to a certain grade percentage even if the brake pedal is not pressed.
d) To provide a mechanical link between the engine and the transmission. When the torque converter functions as a fluid couple, the torque on the turbine is the same as the torque on the impeller, but the turbine speed always lags behind the impeller speed. This represents a power loss due to the friction between the ATF and the converter interior surfaces, internal ATF leakage, and turbulence. This power loss is converted to an ATF temperature rise. After the torque converter is locked by the pressure plate, it just becomes a mechanical link with no more power loss.

4.2.1 Torque Multiplication and Fluid Coupling

The torque multiplication mechanism in a torque converter can be analysed qualitatively as follows. As the impeller turns with the engine, automatic transmission fluid (ATF) enters the torque converter and soon fills it up. The ATF then circulates continuously within the torque converter between the impeller, turbine, and reactor. To simplify the analysis on ATF circulation, let's consider the circulation of a single drop of ATF. This drop enters the impeller at the juncture between the reactor and the impeller, and then travels outward on the surface of an impeller blade due to the centrifugal effect from the impeller rotation. It then enters the turbine at the juncture between the impeller and the turbine, producing an impact on the turbine. This impact from the ATF causes the turbine to rotate. After entering the turbine, the fluid drop travels on the surface of a turbine blade toward the juncture between the turbine and the reactor. When the fluid drop exits the turbine and enters the reactor, it may impact either side of the reactor blade, depending on the direction of the absolute velocity of the fluid drop in its circulation motion. The force from this impact tends to rotate the reactor in the direction conducive to the ATF flow. However, due to the constraint from the one-way clutch, the reactor is only allowed to rotate in one direction, as shown in Figure 4.2. If the fluid drop impacts the blade side such that the rotation of the reactor is not allowed by the one-way clutch, it will be deflected or redirected upon impact, resulting in a reaction to the turbine blade and increasing the torque on the turbine. If the fluid drop impacts the other side of the reactor blade, the impact force will then turn the reactor as allowed by the one-way clutch. The motion of the fluid drop is thus continuous without deflection, resulting in no reaction to the turbine and thus no turbine torque increment. The torque converter then functions as a fluid couple.

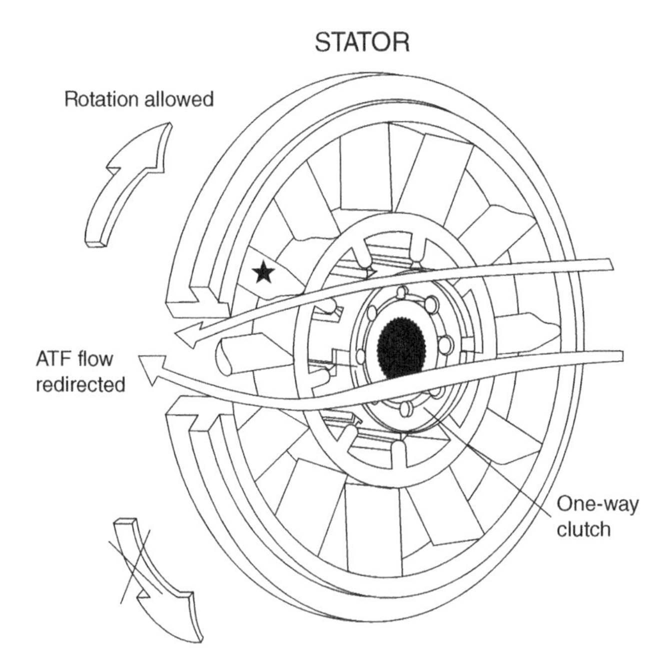

Figure 4.2 Reaction from the reactor.

The absolute velocity of the fluid drop circulating in the torque converter is the vector sum of its relative velocity with respect to the blade and the transfer velocity with the blade. The relative velocity is in the direction tangential to the blade and the transfer velocity is caused by the rotation of a torque converter element and is perpendicular to the radial direction. Therefore, both the magnitude and the direction of the absolute velocity of the fluid drop at the turbine exit, or the reactor entrance, depend on the angular velocity of the turbine. Because of the design of the blade geometry, the ATF impact force when it enters the reactor cannot turn the reactor when the turbine angular velocity is low, resulting in the turbine torque increasing as mentioned previously. The absolute velocity of the fluid drop will change as the turbine angular velocity increases. At some point, the direction of the fluid drop velocity at the turbine exit or the reactor entrance will be such that the fluid drop enters the reactor from the blade side which produces the impact to turn the reactor in the direction allowed by the one-way clutch. This point is defined as the coupling point of the torque converter. If not locked up, the torque converter just behaves as a fluid couple, with equal torque on the impeller and the turbine, and with the turbine speed lagging behind the impeller speed by a small amount.

4.2.2 Torque Converter Locking up

There is an apparent power loss when a torque converter functions as a fluid couple because the turbine speed lags behind the impeller speed. Torque converters with lock-up clutch are designed to salvage that power loss. The unlocked and locked positions of a torque converter are shown in Figure 4.3a and 4.3b respectively. In the locked position shown in Figure 4.3b, applied ATF flows between the turbine shaft and the fixed support that is splined with the reactor, and empties the right side of the pressure plate. The pressure plate is thus pressed by the ATF on its left side against the cover to lock up the torque converter. Note that the AFT enters the pressure plate apply side (left side as shown) from the converter inside through the small gap between the turbine and the impeller and holes drilled on the turbine shell. In the unlocked position shown in Figure 4.3a, released AFT enters the right side of the pressure plate to disengage the clutchFigure. There are other designs of torque converter locking clutch. For example, a multiple disk clutch can be placed between the turbine and the converter cover, with

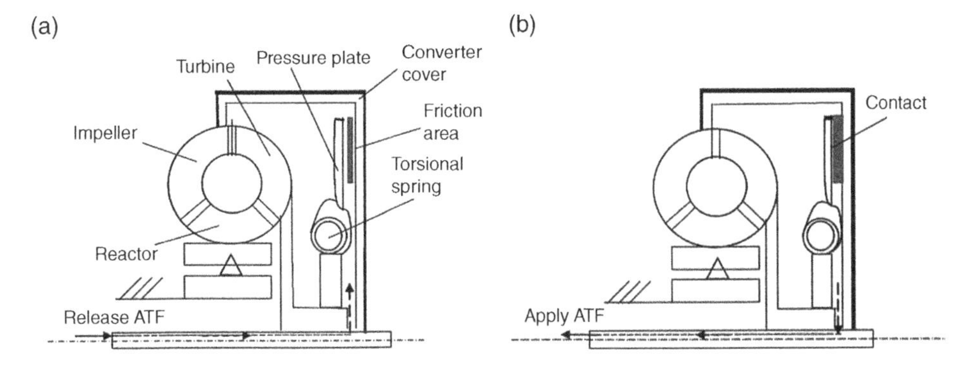

Figure 4.3 Torque converter lock-up mechanism.

friction disks and plates splined to a hub and a drum attached to the turbine and the cover respectively. In this design, the hydraulic circuit for the engagement and release of the lock-up clutch can be separated from the ATF circuit for the torque converter operation.

4.3 ATF Circulation and Torque Formulation

In a torque converter, power is transmitted from the impeller to the turbine by ATF circulation inside the torque converter. Based on Newton's second law, the torque applied on a torque converter element is equal to the change of angular moment with respect to time of the AFT circulating inside the element from the entrance to the exit. The angular momentum of a single drop in the AFT continuum depends on its velocity and position inside the torque converter. Computational fluid dynamics can be applied to study the ATF circulation as a continuum for torque converter dynamic behavior and design optimization [10]. To simplify the problem, the torque formulations in this section will be based on the concept of the mean effective flow, or design path, as defined in SAE publications on torque converters [4]. In this simplified formulation, AFT circulation as a continuum is assumed to be equivalent to the mean effective flow along the design path. For readers' convenience, the text throughout this section will follow SAE conventions in terminologies and definitions on torque converters.

4.3.1 Terminologies and Definitions

The section view (i.e. the view on a plane that contains the axis) of a torque converter is shown in Figure 4.4. This section view is commonly termed the torus section, which is used to define the geometry, dimension and other features of a torque converter as described in the following.

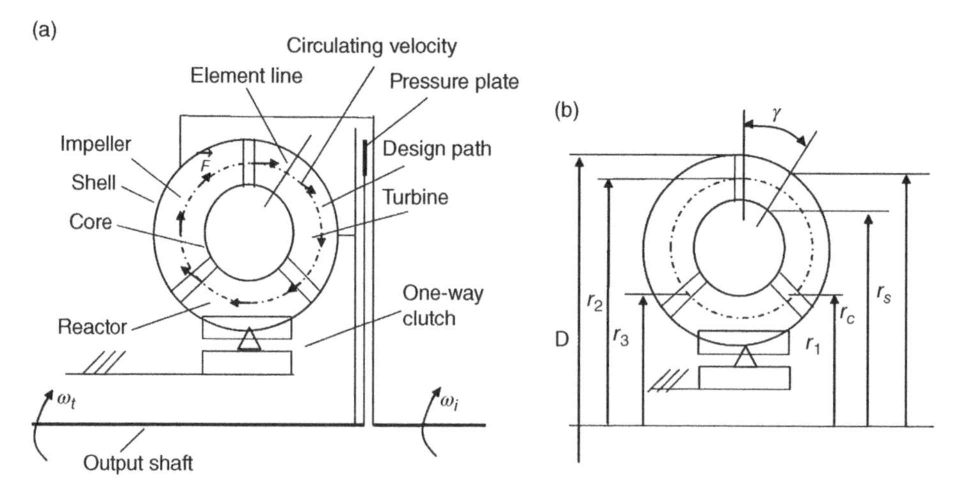

Figure 4.4 Torque converter terminologies and definitions.

Design path: The design path is a circle on the torus section that is assumed to be the mean flow path of ATF circulating as a continuum in the three elements of the torque converter. The ATF circulating velocity is always tangential to the design path, as shown in Figure 4.4a. There are an infinite number of axial sections or torus sections, and the design paths on all torus sections form a circular tube about the converter axis.

Shell: The shell is the outside boundary of the AFT continuum and is physically the inside wall of the shell of a torque converter element. The inside walls of the shells of the three elements jointly form a tube-like surface.

Core: The core is the inside boundary of the ATF continuum and is physically the outside wall of the core of a torque converter element. ATF is contained between the core and the shell in the torque converter. The core is also a tube-like surface.

Element line: The element line is perpendicular to the design path on the torus section and is used for the design of blade geometry [3]. A set of element lines is arranged on the torus section on each element and the position of each element line is defined by the angle γ formed by the element line with respect to the vertical axis. The element line intersects the shell and the core on a torus section at radius r_s and r_c respectively, as shown in Figure 4.4b.

Maximum diameter of flow path: As shown in Figure 4.4b, the maximum diameter of flow path D is defined at the point in the ATF flow circuit that is the farthest from the converter axis. It is actually the outer diameter of the shell. The maximum flow path diameter is the most important design parameter that determines the torque converter torque capacity, as will be shown later.

Entrance and exit radii: These radii are at the junctures between the three elements. The radius at reactor exit and impeller entrance is r_1, the radius at impeller exit and turbine entrance is r_2, and the radius at turbine exit and reactor entrance is r_3. In design practice, $r_1 = r_3$, and r_2 is approximately equal to r_1.

ATF flow area: The flow area is measured between the core and the shell of an element in the direction perpendicular to the AFT circulation velocity. This area is formed by an element line in Figure 4.4b performing a full revolution about the converter axis. The ATF flow area is denoted as A and can be calculated by the following equation in terms of radii r_s and r_c, and angle γ defined by an element line:

$$A = \frac{\pi\left(r_s^2 - r_c^2\right)}{\cos\gamma} \tag{4.1}$$

For optimized torque converter performance, it is beneficial to keep the flow area A approximately a constant along the design path. This is possible by properly designing the shape and dimension of the shell and the core, as demonstrated in the classic paper by V.J. Jandasek [3].

Blade entrance and exit angles: The blade surface of a converter element is highly three dimensional. In each element, the blade surface intersects the circulating tube surface formed by the design path, as shown in Figure 4.5. The curve of this intersection is called the mean blade curve and is used to define the blade geometry. The ATF motion in an element is assumed to be equivalent to the motion of an ATF drop along the mean blade curve.

The motion of an ATF drop along the mean blade curve on an impeller blade is shown in Figure 4.6. At any point P along the mean blade curve, the absolute velocity of the ATF drop is the vector sum of two velocities, namely, the relative velocity whose direction is

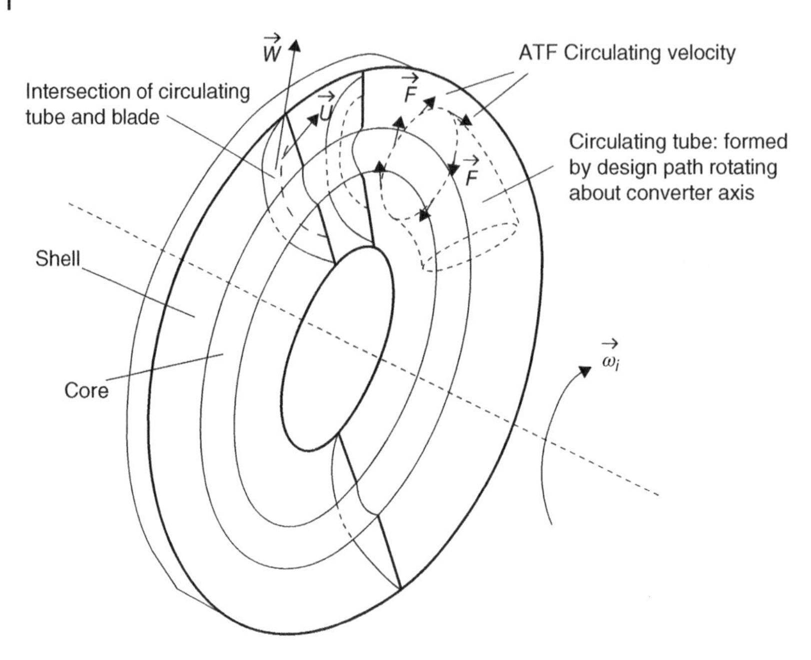

Figure 4.5 Mean blade curve.

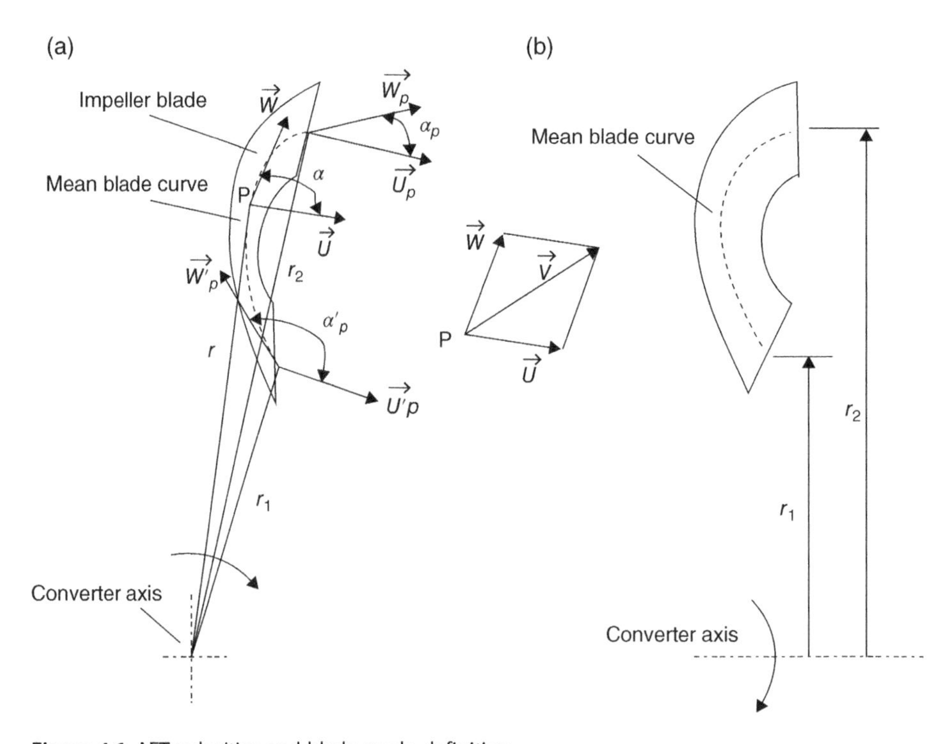

Figure 4.6 AFT velocities and blade angle definition.

along the tangent to the mean blade curve, and the transfer velocity whose direction is perpendicular to radial direction at point P. Following the convention in SAE publications, the absolute velocity, the relative velocity, and the transfer velocity are denoted by V, W, and U respectively, with a subscript indicating element and bold letters or arrows indicating vectors. Denoted by F in Figure 4.5, the circulating velocity is tangent to the design path and is contained in the axial section. A prime is used for a velocity component at the entrance of an element to distinguish it from the exit. The angle between the relative velocity and the transfer velocity is defined as the blade angle. Geometrically, the blade angle is formed between the tangent to the mean blade curve and the normal to the axial plane at point P. The blade angles at the entrance and the exit are the entrance angle and exit angle respectively.

ATF mass circulating rate: It is apparent that the ATF volumetric circulating rate is (AF). The ATF has a constant density ρ since it is considered incompressible. The ATF mass circulating rate, i.e. the ATF mass flow $(AF\rho)$, is therefore also a constant in the converter elements if leakage is not considered. Similar to a centrifugal pump, the mass circulating rate in a torque converter is proportional to the impeller (pump) speed. For a given impeller speed, the mass circulating rate in a converter element can be assumed to be a constant, resulting in a constant circulating velocity F since the circulating area A is a almost constant by design.

Speed ratio: The torque converter speed ratio, denoted as i_s, is defined as the angular velocity of the turbine divided by the angular velocity of the impeller, i.e. $i_s = \frac{\omega_t}{\omega_i}$. As an operating parameter, the speed ratio is used to define a range of variables related to torque converter performance, as will be detailed later.

Torque ratio: The torque converter torque ratio, denoted as i_q, is defined as the torque on the turbine or the output torque divided by the torque on the impeller or the input torque, i.e. $i_q = \frac{T_t}{T_i}$. For a typical torque converter used for automatic transmissions, the torque ratio varies between the stall ratio, which is approximately equal to 2.0, to the coupling torque ratio which is always equal to 1.0.

Efficiency: The torque converter efficiency is the ratio between the output power and the input power. Apparently, the efficiency is equal to the multiplication of the speed ratio and the torque ratio, i.e. $\eta = \frac{\omega_t T_t}{\omega_i T_i} = i_s i_q$.

4.3.2 Velocity Diagrams

Since the AFT circulation is assumed to be along the design path that is always on the axial plane, the transfer velocity is always perpendicular to the circulating velocity. Geometrically, the directions of the circulating velocity and the transfer velocity are respectively along the two mutually perpendicular tangents at point P on the surface of the circulating tube. These two tangents define the tangent plane of the circulating tube surface at point P. Since the mean blade curve is the intersection between the blade surface and the circulating tube surface, the tangent line to the mean blade curve at point P must be on the tangent plane. Therefore, the circulating velocity F, the transfer velocity U, and the relative velocity W are contained within the tangent plane at point P to the surface of the circulating tube shown in Figure 4.5. At any point along the design path, the absolute

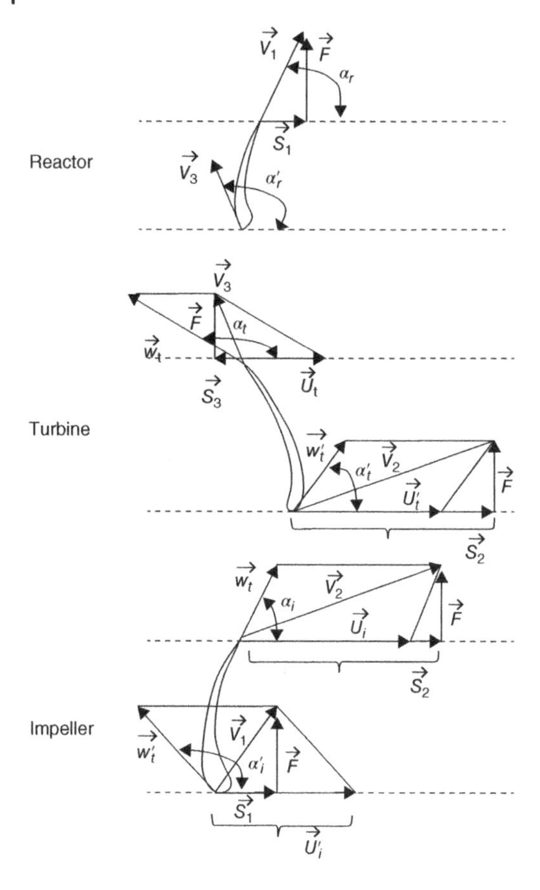

Figure 4.7 Velocity diagrams for impeller, turbine, and reactor.

velocity of an ATF drop is the vector sum of its relative velocity and transfer velocity as follows:

$$V = W + U \tag{4.2}$$

The magnitude of the transfer velocity depends on the element angular velocity and the radial distance to the converter axis. Velocity diagrams for ATF circulation can be constructed for the three converter elements based on Eq. (4.2) and blade geometry under various operation conditions. Figure 4.7 shows the velocity diagrams when the reactor is not turning and the converter functions as a torque multiplier.

While drawing the velocity diagrams shown in Figure 4.7, the angular velocities of the impeller ω_i is considered as given and the circulating speed F is also a given constant. The velocity components are shown on the plane formed by the transfer velocity U and circulating velocity F. The drawing starts at the impeller entrance with radius r_1. The magnitude of the transfer velocity at the impeller entrance, U_i', is equal to $\omega_i r_1$, and the angle between the relative velocity W_i' and the transfer velocity U_i' is the impeller blade entrance angle α_i'. The circulating velocity F is the projection of the absolute velocity V_1 on the tangential direction to the design path and the tangential component S_1 is the projection of the absolute velocity V_1 on the tangential direction of the plane of rotation, with subscript 1 indicating radius r_1.

After entering the impeller, the ATF drop moves along the mean blade curve of the impeller blade surface toward the exit at radius r_2. At the impeller exit, the magnitude of the transfer velocity U_i is higher than that at the entrance and is equal to $\omega_i r_2$. The angle between the relative velocity W_i and the transfer velocity U_i is the impeller exit angle α_i. The velocity diagram is uniquely defined since the magnitude of circulating velocity F is a constant. Based on the velocity diagrams for the impeller at the entrance and exit, there exist the following relations between the impeller velocity components:

$$S_1 = U_i' + \frac{F}{\tan \alpha_i'} = \omega_i r_1 + \frac{F}{\tan \alpha_i'} \tag{4.3}$$

$$S_2 = U_i + \frac{F}{\tan \alpha_i} = \omega_i r_2 + \frac{F}{\tan \alpha_i} \tag{4.4}$$

At the turbine entrance, the ATF absolute velocity V_2 is the same as that at the impeller exit. For a given turbine angular velocity ω_t, the magnitude of the transfer velocity U_t' is calculated as $\omega_t r_2$. The relative velocity W_t' is then determined by the vector parallelogram. At the turbine exit, the magnitude of the transfer velocity U_t is calculated as $\omega_t r_3$. The relative velocity W_t is along the blade tangent that forms the exit angle α_t with the transfer velocity U_t. The velocity diagram at the turbine exit leads to the following relation on the velocity components:

$$S_3 = U_t + \frac{F}{\tan \alpha_t} = \omega_t r_2 + \frac{F}{\tan \alpha_t} \tag{4.5}$$

When the reactor is not turning, the transfer velocity for the ATF moving on the reactor blade is zero. The absolute velocity at the reactor entrance is the same as that at the turbine exit. The velocity diagrams at the reactor entrance and exit can be easily constructed as shown at the top of Figure 4.7.

The velocity diagrams in Figure 4.7 illustrate the velocity components of ATF circulating inside the torque converter and the relationship between them. Several important observations and design recommendations can be made about these diagrams for a better understanding on blade geometry and ATF circulation:

- The geometric blade exit angle is equal to the angle formed between the relative velocity and the transfer velocity, but the same is not true in general for the geometric blade entrance angle. This can be explained by the velocity diagrams at the impeller exit and the turbine entrance in Figure 4.7. The impeller exit angle α_i is the same as the angle formed between the ATF relative velocity W_i and the transfer velocity U_i, whose vector sum determines the AFT absolute velocity V_1. The AFT absolute velocity at the turbine entrance is the same as V_1 due to AFT flow continuity. The relative velocity W_t' is determined by the velocity parallelogram defined by V_1 and the transfer velocity U_t' which depends on the turbine angular velocity ω_t. Therefore, the angle between the relative velocity W_t' and the transfer velocity U_t', denoted as α_t' in Figure 4.7, is generally not equal to the geometric turbine blade entrance angle. As a matter of fact, there exists only one turbine angular velocity at which the geometric blade entrance angle will be equal to the angle formed between the relative velocity W_t' and the transfer velocity U_t'. The same is true for the entrance angles for the reactor and the impeller.
- To minimize ATF flow shock loss at an exit and entrance juncture, the blade entrance angle should be equal to the angle formed between the relative velocity and the transfer velocity. However, this is not possible since torque converter elements turn at different

speeds that affect the ATF velocity direction. According to the research data in SAE publication [4], the angles between the relative velocity and the transfer velocity at the entrances of the impeller and the reactor, α_i' and α_r', as denoted in Figure 4.7, vary in wide ranges from around 30° to 150° while at the turbine entrance, the angle varies in a much narrower range between about 22° to 45°. As recommended in [3,4], the entrance angles of the impeller, turbine, and reactor are optimized at a speed ratio of 0.7 for optimized results in stall torque ratio, efficiency, high coupling speed, and characteristics. Typical blades that provide the optimized trade-off between stall ratio, efficiency, and coupling characteristics are also recommended in SAE publication [4]. For example, a torque converter with a 12 inch design path diameter, the blade angles are designed as: $\alpha_i = 75°$, $\alpha_i' = 105°$; $\alpha_t = 150°$, $\alpha_t' = 32°$; $\alpha_r = 22°$, $\alpha_i' = 90°$.

- The velocity diagrams in Figure 4.7 are constructed for the torque converter operation when the reactor is not turning. As shown at the top of Figure 4.7, the absolute AFT velocity at the reactor entrance V_3 is directed in such a way that the impact of the ATF flow upon the reactor blade cannot turn the reactor because of the constraint from the one-way clutch. As the turbine speed increases, the magnitude of the transfer velocity U_t will also increase and thus the direction of V_3 will change, as shown in the velocity diagram at the turbine exit. The angle between V_3 and the tangential direction on the plane of rotation, α_r' shown in Figure 4.7, varies gradually from a value larger than 90° to a value smaller than 90°. When this happens, the impact of the ATF flow upon the reactor blade will cause the reactor to turn in the direction allowed by the one-way clutch. The torque converter then reaches the coupling point and functions just as fluid couple.

4.3.3 Angular Momentum of ATF Flow and Torque Formulation

For an ATF drop circulating inside the torque converter, its angular momentum about the converter axis is the cross product between its position vector from the axis and its linear momentum. The component of the angular momentum of the ATF drop about the converter axis is equal to (msr), where m is its mass, s is the tangential component of its absolute velocity on the plane of rotation, and r is its radial distance from the converter axis. For simplicity, the AFT is assumed to be circulating inside the torque converter as a continuum along the design path. The change of ATF circulation status over time Δt is shown in Figure 4.8 for the impeller. At any arbitrary time t during torque converter operation, the ATF continuum is fully contained between the core and the shell of the impeller, as shown in Figure 4.8a. At time $(t + \Delta t)$, a small amount of the ATF continuum moves into the impeller at the entrance and the same amount moves out of the impeller at the exit since the ATF is assumed to be incompressible, as shown in Figure 4.8b. The mass of the ATF amount moving into or moving out of the impeller is equal to $(AF\Delta t\rho)$, where ρ is the mass density of the ATF, A is the ATF flow area in Eq. 4.1, and F is the ATF circulating velocity. The angular momentum of the ATF amounts that moves into or out of the impeller are therefore $(AF\Delta t\rho s_1 r_1)$ and $(AF\Delta t\rho s_2 r_2)$ respectively, with r_1 as the radius at the impeller entrance and r_2 as the radius at the impeller exit. As can be observed from Figure 4.8, the only change on the status of the ATF continuum over time Δt occurs at the impeller entrance and exit. According to the principle of angular impulse and momentum, the following equation applies to

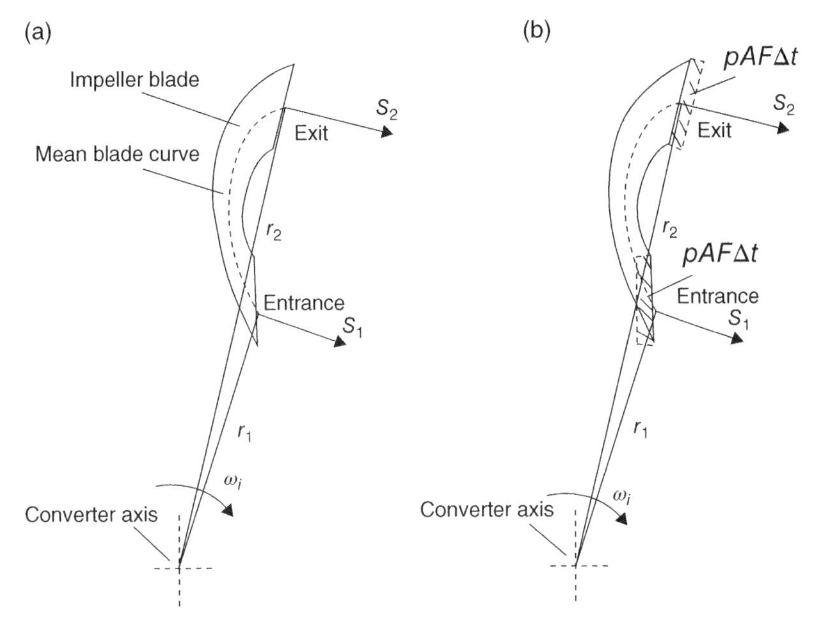

Figure 4.8 Angular momentum change of ATF continuum over time Δt.

the ATF continuum contained between the core and shell of the torque converter impeller:

$$T_i \Delta t = AF\Delta t \rho S_2 r_2 - AF\Delta t \rho S_1 r_1 \tag{4.6}$$

where the left side is the angular impulse of torque T_i that is applied to the ATF continuum over time Δt and the right side is the change of angular momentum of the ATF continuum contained inside the impeller. It is obvious that the torque applied to the impeller by the ATF continuum has the same magnitude as T_i, which can be determined by dividing Eq. (4.6) on both sides by Δt, i.e.

$$T_i = AF\rho S_2 r_2 - AF\rho S_1 r_1 = AF\rho (S_2 r_2 - S_1 r_1) \tag{4.7}$$

The torque equation for the turbine and reactor can be derived by following the same procedure, as presented in the following:

$$T_t = AF\rho S_3 r_3 - AF\rho S_2 r_2 = AF\rho (S_3 r_3 - S_2 r_2) \tag{4.8}$$

$$T_r = AF\rho S_1 r_1 - AF\rho S_3 r_3 = AF\rho (S_1 r_1 - S_3 r_3) \tag{4.9}$$

It is apparent that the sum of the torque on the impeller, turbine, and reactor is equal to zero, i.e. $T_i + T_t + T_r = 0$, since the torque converter as a whole body must be in equilibrium. The torque equations, Eqs (4.7–4.9), can also be expressed as follows by the substitution of S_1, S_2, and S_3 from Eqs (4.3–4.5):

$$T_i = AF\rho (S_2 r_2 - S_1 r_1) = AF\rho \left[\omega_i \left(r_2^2 - r_1^2\right) + F\left(\frac{r_2}{\tan\alpha_i} - \frac{r_1}{\tan\alpha_i'}\right)\right] \tag{4.10}$$

$$T_t = AF\rho(S_3r_3 - S_2r_2) = AF\rho\left[\omega_t r_3 r_2 - \omega_i r_2^2 + F\left(\frac{r_3}{\tan\alpha_t} - \frac{r_2}{\tan\alpha_t'}\right)\right] \tag{4.11}$$

$$T_r = AF\rho(S_1r_1 - S_3r_3) = AF\rho\left[\omega_i r_1^2 - \omega_t r_3 r_2 + F\left(\frac{r_1}{\tan\alpha_i'} - \frac{r_3}{\tan\alpha_t}\right)\right] \tag{4.12}$$

As can be observed from the torque equations above, the torque applied to a converter element depends on many variables, including the angular velocities of converter elements, the ATF circulating velocity, the ATF mass density, the blade geometry, the design path contour, and the converter dimensions.

4.4 Torque Capacity and Input–Output Characteristics

Equations (4.10–4.12) can be further simplified for the formulation of the torque applied to a torque converter element. As an example, let us consider the torque applied to the impeller formulated by Eq. (4.10). For a given torque converter, the following observations and assumptions can be made on the torque converter design parameters and operation variables in Eq. (4.10):

- The radii at the impeller entrance and exit, r_1 and r_2, are proportional to the maximum diameter of the flow path, i.e. $r_1 = \lambda_1 D$ and $r_2 = \lambda_2 D$, where λ_1 and λ_2 are proportionality factors.
- The ATF circulating velocity F is proportional to the mean transfer velocity on the design path and depends on the speed ratio of the torque converter, i.e. $F = \lambda_F \frac{D\omega_i}{2}\lambda(i_s)$, with λ_F as the proportionality factor. Function $\lambda(i_s)$ describes the dependence of the circulating velocity on the speed ratio.
- The AFT flow area A, designed as a near constant, is proportional to the square of maximum diameter of the flow path, i.e. $A = \lambda_A D^2$, with λ_A as the proportionality factor.

With these observations, Eq. (4.10) can then be translated into the following form:

$$T_i = \frac{1}{2}\lambda_A\lambda_F\rho\lambda(i_s)\left[\left(\lambda_2^2 - \lambda_1^2\right) + \frac{1}{2}\lambda_F\lambda(i_s)\left(\frac{\lambda_2}{\tan\alpha_i} - \frac{\lambda_1}{\tan\alpha_i'}\right)\right]D^5\omega_i^2 \tag{4.13}$$

where the ATF mass density ρ is a constant, the proportional factors are also constants, and the entrance angle α_i' and exit angle α_i are determined from the AFT velocity diagrams. As mentioned previously, these angles are not exactly the same as the geometric blade angles and largely depend on the relative angular velocity between the impeller and the turbine which is defined by the speed ratio at a given impeller angular velocity. Therefore, for a given torque converter that turns at a given impeller angular velocity, the torque applied to the impeller can be formulated by:

$$T_i = kD^5\omega_i^2 \tag{4.14}$$

where k represents the whole term before $\left(D^5\omega_i^2\right)$ in Eq. (4.13) and only depends on the speed ratio during the operation of a given torque converter. In the convention of SAE standard publications, the impeller RPM is used in the impeller torque formulation.

By replacing ω_i with $\left(\frac{\pi n_i}{30}\right)$ in the equation above, the torque applied to the impeller is then represented by:

$$T_i = CD^5 n_i^2 \tag{4.15}$$

where C is called the torque capacity coefficient, and is similar to k in Eq. (4.14), depends only on the speed ratio for a given torque converter. It should be emphasized here that the torque capacity coefficient C reflects the effects of the geometry design parameters, such as blade angles, entrance and exit radii, and ATF circulating area, on the performance of a torque converter. Extensive dyno testing is required to obtain the relationship between the torque capacity coefficient and the speed ratio. Based on test data, the torque capacity coefficient is plotted versus the speed ratio for the determination of operation status of a given torque converter [11]. Similar to the formulation of the impeller torque, the turbine torque can also be formulated by an equation that resembles Eq. (4.15), where the impeller RPM n_i is replaced by the turbine RPM n_t, as follows:

$$T_t = C_t D^5 n_t^2 \tag{4.16}$$

where C_t is the capacity coefficient for the turbine torque and, similar to C in Eq. (4.15), depends only on the speed ratio for a given torque converter. It follows directly from Eqs (4.15) and (4.16) that the torque ratio i_q, defined as $\frac{T_t}{T_i}$, also depends on the speed ratio only. It can also be observed from either of these equations that the torque transmitted by a torque converter is proportional to the fifth power of the maximum diameter of the flow path. A 20% increase on the converter size would increase the torque capacity by 2.5 times. This is advantageous for torque converter applications in the automotive industry since a small range of converter dimensions will cover various vehicle power requirements.

4.4.1 Torque Converter Capacity Factor

In torque converter design and applications, the torque capacity factor, or K-factor as commonly termed, is used to present the relationship between the torque and the speed, typically of the impeller, for a torque converter with particular dimension and blade geometry design. The value of the K-factor is equal to the ratio between the impeller speed and the square root of the impeller torque:

$$K = \frac{n_i}{\sqrt{T_i}} \tag{4.17}$$

where n_i is the impeller speed in RPM and T_i is the impeller torque in ft.lb. Comparing Eqs (4.17) and (4.15), it is obvious that the K-factor is related to the torque capacity coefficient C by the following equation:

$$K = \frac{1}{\sqrt{CD^5}} \tag{4.18}$$

For a given converter, D is a constant and thus the K-factor only depends on the speed ratio i_s. If the values of capacity factor C at different speed ratios have been obtained from the test on a given torque converter, then the values of the K-factor can be easily obtained from Eq. (4.18). In practice, the K-factor is often directly obtained through testing. In the dyno test setup, the impeller is connected to an electric motor with controllable torque, and the turbine is connected to a loader with measurable torque load. During the test, the input torque, i.e. the torque applied to the impeller by the motor, is kept as a constant. The test then runs at different torque loads. The speeds of the impeller and the turbine are measured when dynamic equilibrium is achieved on the dyno system. The torques and speeds of the impeller and turbine are then used to calculate the speed ratio, the torque ratio, the efficiency, and the torque capacity factor by Eq. (4.17). Typically, the torque ratio, the torque capacity factor, and the efficiency are plotted versus the speed ratio in the so-called torque converter characteristic plot, as shown in Figure 4.9.

Note that the US customary unit is used in Figure 4.9, where torque is in ft.lb. Similarly, the torque converter characteristics can also be plotted in SI unit, with torque in Nm and the value of K-factor converted accordingly. Several observations can be made on typical torque converter characteristics from Figure 4.9:

- When the speed ratio is zero, i.e. the turbine is not turning, the torque ratio is the highest and is termed the stall ratio. As the speed ratio increases from zero to 0.9, the torque ratio decreases from the stall ratio to 1.0. The torque converter reaches the coupling point when the torque ratio is 1.0 and it then just behaves as a fluid couple if unlocked.
- The curve for K-factor versus speed ratio is upward in shape, with gradual increment with respect to the speed ratio until the coupling point. After the coupling point, the K-factor increases very sharply as the speed ratio approaches 1.0, rendering accurate data reading impractical.

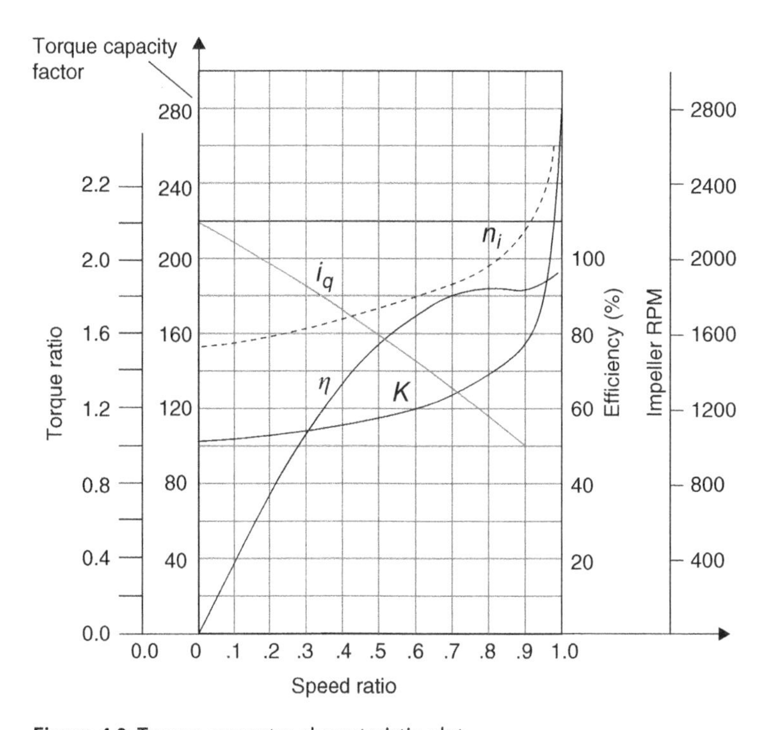

Figure 4.9 Torque converter characteristic plot.

- The torque converter efficiency increases from zero at stall to about 0.9 near the coupling point. It increases further after the coupling point as the speed ratio increases and the torque ratio stays at 1.0. However, there is always a power loss since the turbine always lags behind the impeller if the torque converter is unlocked. To save the power loss, which translates into fuel economy worsening, the torque converter is unlocked only for operation conditions such as launching, crawl and hill holding, and transmission shifting process.
- The stall ratio helps launch the vehicle quickly and smoothly. Typically, torque converters used for passenger vehicle transmissions have a stall ratio in the neighborhood of 2.0.
- As implied by Eq. (4.17), the value of K-factor is equal to the impeller speed scaled by $\frac{1}{\sqrt{T_i}}$. This is validated by the plot in Figure 4.9, which is based on the test setup where a constant torque is applied to the impeller. As shown in Figure 4.9, the curve for torque capacity is parallel to the curve of the impeller speed up to the coupling point. The discrepancy after the coupling point is caused by the sharp increase of K-factor and impeller speed versus the speed ratio.
- In addition, it is worthwhile to point out that the impeller speed (i.e. the engine speed), as shown in Figure 4.9 increases monotonically versus the speed ratio, providing the driver a desirable feel when the vehicle is being accelerated. This feature can be guaranteed by the optimized design of blade entrance and exit angles of the torque converter.

4.4.2 Input–Output Characteristics

As mentioned previously, the plot shown in Figure 4.9 is established based the test data when a constant torque is applied to the impeller. This constant torque, which is used in the test to obtain the input–output characteristics of the torque converter, should be chosen as the maximum or near maximum torque of the engine to which the torque converter is matched. As observed in Eqs (4.17) and (4.18), the torque capacity factor is theoretically unrelated to the input torque on the impeller and is only the characteristics of the torque converter dimension and geometry design. Therefore, the relationship between the torque ratio, the torque capacity factor, and the efficiency versus the speed ratio shown in Figure 4.9 should be applicable for other operation conditions of the torque converter within allowable discrepancy in engineering practice. The small discrepancy is mainly due to the shock loss of ATF flow caused by turbulence at blade entrance and exit, friction between AFT flow and surfaces of blades, core, and shell, and AFT leakages inside the torque converter.

The operation status of a torque converter is defined by four parameters: the angular velocities of the impeller and turbine, ω_i and ω_t, and the torques on the impeller and turbine, T_i and T_t. These four operation parameters are linked to each other by the characteristics of the torque converter through the following equations:

$$i_s = \frac{\omega_t}{\omega_i} = \frac{n_t}{n_i} \tag{4.19}$$

$$i_q = f_1(i_s) \tag{4.20}$$

$$K = f_2(i_s) \tag{4.21}$$

$$T_i = \frac{n_i^2}{K^2} \tag{4.22}$$

$$T_t = i_t T_i \tag{4.23}$$

where the torque ratio i_q and the torque capacity factor K are functions of the speed ratio i_s, in the form of a data file or a plot, and are characteristic of a given torque converter, as illustrated in Figure 4.9. Obviously, the speed ratio i_s is the key to the determination of the torque converter operation status.

4.4.3 Joint Operation of Torque Converter and Engine

When a torque converter is matched to an engine, it is important to know how the turbine torque, i.e. the input torque to the transmission, behaves with respect to the speed ratio. For a given engine, the output torque is a function of its RPM with a specific throttle opening. The engine RPM is the same as the impeller speed n_i. If the impeller–flywheel inertia is not considered, then the impeller torque T_i is the same as the engine output torque T_e, and the turbine torque is then the engine torque multiplied by the torque ratio, that is, $T_t = i_q T_e$. As a function of the speed ratio i_s, the torque ratio is defined by the joint operation status of the torque converter and the engine. The joint operation between the engine and the torque converter can be quantitatively analysed by model simulation to be discussed later in this section. A simple graphic method, proposed in SAE publications [4], can be used to determine the speed ratio and torque capacity factor for the converter–engine joint operation, as demonstrated in Figure 4.10.

The engine torque plotted in Figure 4.10 is at a fixed throttle opening. For example, the wide open throttle (WOT) engine torque can be plotted for acceleration performance analysis. The engine torque capacity factor, which is denoted as K_e and is equal to $\frac{n_e}{\sqrt{T_e}}$, is then plotted accordingly. The K_e factor is characteristic of the engine at a given

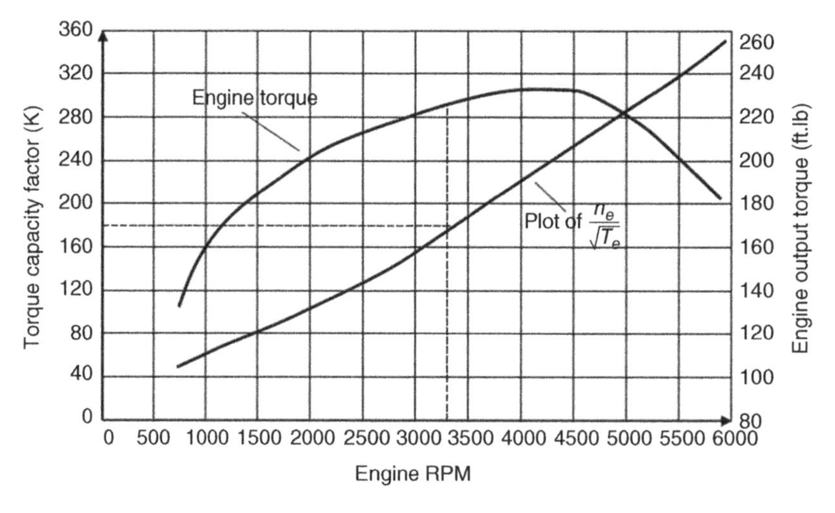

Figure 4.10 Joint converter–engine operation status.

throttle opening. At any given speed ratio i_s, the converter torque capacity factor K and the torque ratio i_q can be found from the characteristic plot of the torque converter matched to the engine, such as the one shown in Figure 4.9. When the converter–engine system reaches dynamic balance, the converter torque capacity factor is equal to the engine torque capacity factor, i.e. $K = K_e$. A horizontal line is drawn at the value of the converter torque capacity factor K in Figure 4.10. The intersection of this horizontal line and the plot of the engine torque capacity factor K_e defines the converter–engine joint operation status. The vertical line drawn at the intersection will then intersect the engine torque plot and the horizontal axis, leading to the determination of the engine torque T_e or impeller torque T_i, and the corresponding engine RPM. The turbine torque and speed are then determined as $i_q T_e$ and $i_s n_e$ respectively. Repeating the steps by varying the speed ratio in the interval [0, 1] monotonically, the turbine torque T_t and turbine speed n_t can be found for the speed ratio from the stall status to the coupling point, corresponding to the engine output at a specified throttle opening. Note that the graphical method described here can be used to assess how well the torque converter is matched to the engine in terms of intended objectives. For example, the converter–engine joint operation status corresponding to the stall ratio near the engine maximum torque would provide the best acceleration performance at vehicle launch.

If the converter torque capacity factor and the torque ratio are given as functions of the speed ratio, and the engine torque at a specified throttle opening is given as a function of its RPM, then the joint converter–engine operation status from stall to coupling, with the speed ratio i_s varying from 0 to 1, can be determined numerically by the following set of equations:

$$K_e(n_e) = \frac{n_e}{\sqrt{T_e}} \tag{4.24}$$

$$K = f_2(i_s) = K_e(n_e) \tag{4.25}$$

$$T_i = T_e = T_e(n_e) \tag{4.26}$$

$$T_t = i_q T_i = f_1(i_s) T_i \tag{4.27}$$

$$n_t = i_s n_e \tag{4.28}$$

where Eq. (4.24) defines the engine torque capacity factor K_e at a given throttle opening as a function of the engine speed n_e. The engine speed is determined by Eq. (4.25) by iteration for a speed ratio in the interval [0, 1], based on the equality of K and K_e. The torque on the impeller and turbine, and the turbine speed, are then determined by Eqs (2.26), (2.27) and (2.28) respectively.

4.4.4 Joint Operation of Torque Converter and Vehicle Powertrain

As a coupling that connects the engine output with the transmission input, the torque converter interacts dynamically with the vehicle powertrain system. The mass moment of inertia of the impeller–flywheel assembly is a part of the system and should be considered in the system dynamics. Figure 4.11 illustrates the dynamic model for vehicle powertrain systems equipped with a torque converter. As shown in the figure, the vehicle mass has been replaced by its equivalent mass moment of inertia (I_v) and the road load by its equivalent torque (T_{load}) on the output of the final drive. The overall powertrain ratio is the multiplication of the transmission ratio and the final drive ratio ($i_t i_a$).

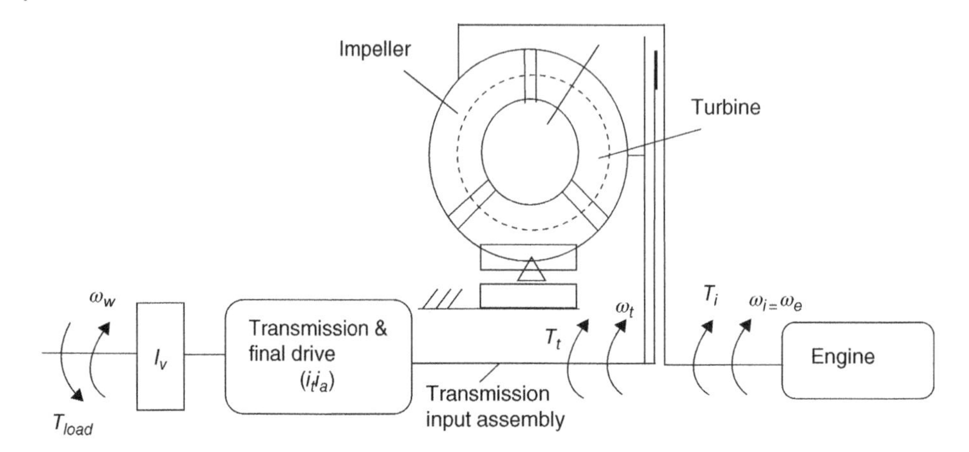

Figure 4.11 Torque converter equipped powertrain system.

There are clearly two degrees of freedom for the dynamics of the lumped mass system shown in Figure 4.11: one is the angular velocity of the impeller–flywheel assembly, ω_e or ω_i, and the other is the angular velocity of the turbine, ω_t, which is related to the vehicle speed V and tire radius r as $\omega_t = i_t i_a \frac{V}{r}$, if tire slippage is not considered. Based on Newton's second law, two equations of motion can be written for the system:

$$T_e - T_i = I_i \dot{\omega}_i \tag{4.29}$$

$$\eta i_t i_a T_t - \frac{T_{load}}{i_t i_a} = \frac{1}{i_t i_a} I_v \dot{\omega}_t \tag{4.30}$$

where I_i is the mass moment of inertia of the impeller–flywheel assembly, η is the combined efficiency of the transmission and final drive, and T_e is the engine output torque. Note that the engine torque is not the same as the impeller torque T_i since the impeller–flywheel assembly has a mass moment of inertia that can be accelerated quickly. The equivalent road load torque is calculated from the road load formulated by Eq. (1.14) in Chapter 1 by the following equation:

$$T_{load} = r\left(fW + W \sin\theta + 0.00118 C_D A V^2\right) \tag{4.31}$$

where the vehicle speed is related to the turbine angular velocity as: $V = \frac{\omega_t}{i_t i_a} r$. The system of two ordinary differential equations formed by Eqs (4.29) and (4.30) governs the motion of the whole vehicle. An initial condition must be provided to solve the equation system for the dynamic status of the vehicle system. When the vehicle is launched from rest at a specified engine throttle opening, the vehicle speed or the turbine angular velocity and the speed ratio are both zero at time zero, and the impeller angular velocity at time zero, n_e^0, can be found using the graphical method described previously or using Eqs (4.24) and (4.25). The initial condition for the equation system is then defined as:

$$t = 0; \omega_i = \frac{\pi n_e^0}{30}; \omega_t = i_s \frac{\pi n_e^0}{30} \tag{4.32}$$

At a specified engine throttle opening, the engine output torque T_e is a function of its angular velocity. The impeller torque and the turbine torque are determined by the torque converter input–output characteristics in terms of the speed ratio. Therefore, combined with the initial condition defined by Eq. (4.32), the equation system formed by Eqs (4.29) and (4.30) can be solved for the velocity–time relationship of the vehicle during launch. If wide open engine throttle opening is specified, the solution will lead to the capacity performance in terms of vehicle acceleration.

Example A vehicle of the following data is equipped with a four-speed AT. The transmission ratios are: 2.84 (1st), 1.60 (2nd), 1.0 (3rd), 0.7 (4th). The engine WOT torque output plot and the torque converter characteristic plot are shown in Figure 4.12. The vehicle is being driven at WOT on level ground in first gear with the torque converter unlocked.

a) Plot the engine torque capacity factor at WOT versus the engine RPM.
b) Suppose the vehicle is launched with WOT and the vehicle is braked until the engine and the torque converter reach dynamic balance in the joint operation, determine the vehicle acceleration at launch.
c) Determine the vehicle speed and acceleration when the torque converter speed ratio is 0.6.
d) What are the vehicle speed and acceleration when the torque converter becomes a fluid couple?

Vehicle data:

Vehicle weight: 3400 lb
Center of gravity height: 16 in.
Air drag coefficient: 0.30
Tire radius: 12.0 in.
Powertrain efficiency: 0.86
Wheel base: 108 in.
Frontal projected area: 22 sq.ft
Roll resistance coefficient: 0.02

Solution:

a) A set of points is selected from the engine WOT torque plot. At each point, the engine torque capacity factor K_e is calculated by $\frac{n_e}{\sqrt{T_e}}$ for corresponding engine torque and RPM values. This leads to a set of points that correlate the engine torque capacity factor to the engine RPM. The engine torque capacity factor is then plotted against engine RPM by connecting these points as shown.

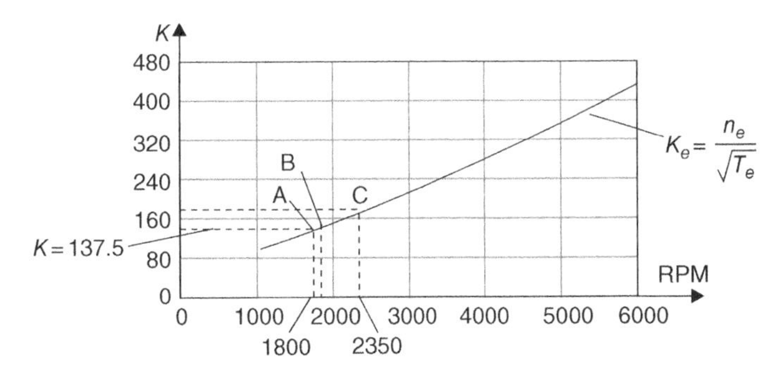

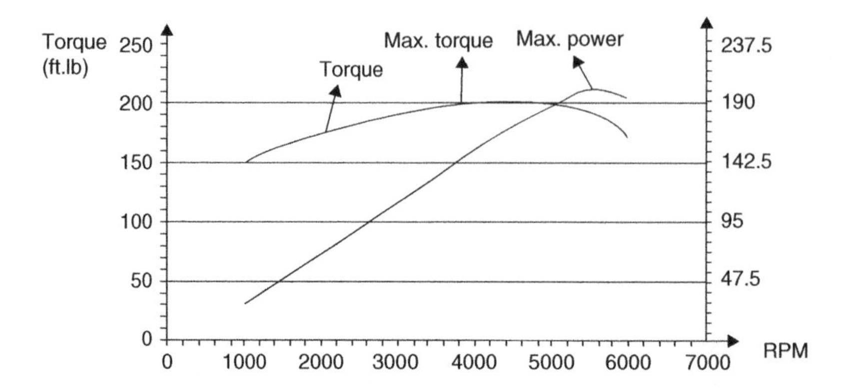

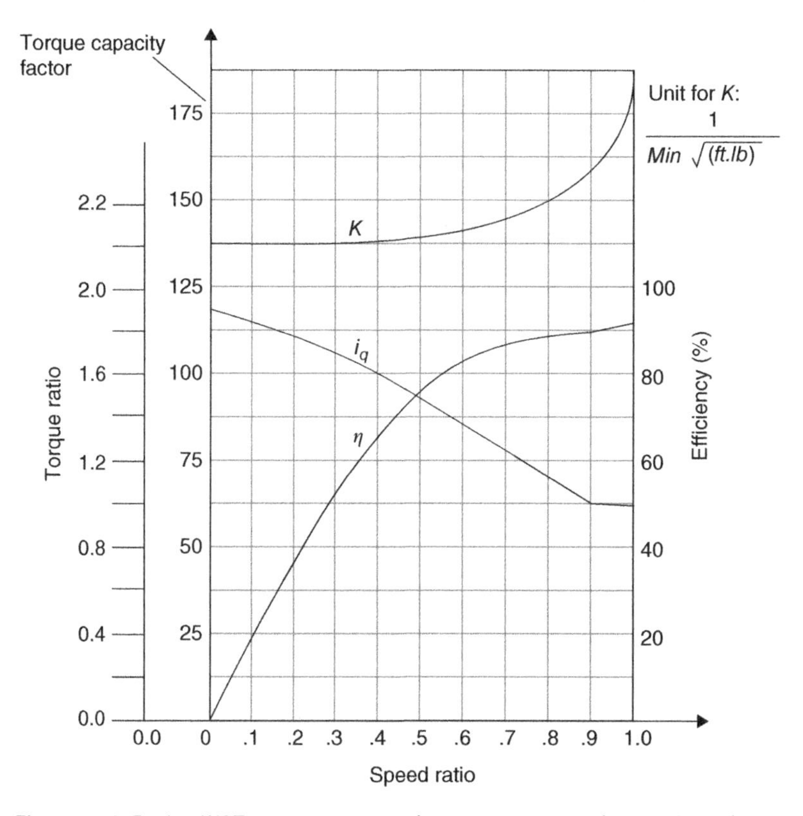

Figure 4.12 Engine WOT output torque and torque converter characteristic plot.

b) At vehicle launch, $n_t = 0$ and $i_s = 0$. Torque ratio $i_q = 1.9$ and the torque capacity factor $K = 137.5$ from the converter characteristic plot. The joint status between the engine and the converter is defined at point A in the graph, corresponding to an engine RPM of 1800.

$$T_i = T_e = \frac{n_e^2}{K_e^2} = \frac{1800^2}{137.5^2} = 173.4\,(\text{ft.lb})$$

Note that you can also find the engine torque from the WOT engine torque plot. At 1800 RPM, the engine WOT torque is 173.4 ft.lb.

$$P = \frac{\eta i_a i_t i_q T_i}{r} = \frac{.86(2.84)(2.840(1.9)(173.4)}{12/12} = 2258.5\,(\text{lb})$$

$$a = \frac{P-R}{W/g} = \frac{2258.5 - 0.02(3400)}{3400/32.2} = 20.75\left(\frac{\text{ft}}{\text{s}^2}\right)$$

c) When $i_s = 0.6$, $i_q = 1.39$, $K = 141$. The joint status is defined at point B with an engine RPM of about 1900.

$$\omega_t = i_s\omega_i = 0.6\left(\frac{1900\pi}{30}\right) = 119.4\left(\frac{1}{\text{s}}\right)$$

$$V = \frac{\omega_t r}{i_a i_t} = \frac{119.4(1.0)}{(2.84)(2.84)} = 14.8\left(\frac{ft}{s}\right) = 10.14\,(\text{mph})$$

$$T_i = \frac{1900^2}{141^2} = 181.6\,(\text{ft.lb})$$

$$P = \frac{\eta i_a i_t i_q T_i}{r} = \frac{.86(2.84)(2.840(1.39)(181.6)}{12/12} = 1750.7\,(\text{lb})$$

$$a = \frac{P-R}{W/g} = \frac{1750.7 - 0.02(3400) - 0.00118(.3)(22)(14.8^2)}{3400/32.2} = 15.91\left(\frac{\text{ft}}{\text{s}^2}\right)$$

d) At the coupling point. $i_s = 0.95$, $i_q = 1.0$, and $K = 169$. The joint status is at point C, at which the engine speed is 2350 RPM.

$$\omega_t = i_s\omega_i = 0.95\left(\frac{2350\pi}{30}\right) = 233.79\left(\frac{1}{\text{s}}\right)$$

$$V = \frac{233.79(1.0)}{(2.84)(2.84)} = 28.98\left(\frac{ft}{s}\right) = 19.85\,(\text{mph})$$

$$T_i = \frac{2350^2}{168^2} = 195.7\,(\text{ft.lb})$$

$$P = \frac{.86(2.84)(2.840(1.0)(195.7)}{12/12} = 1357.46(\text{lb})$$

$$a = \frac{1357.46 - 0.02(3400) - 0.00118(.3)(22)(28.98^2)}{3400/32.2} = 12.1\left(\frac{\text{ft}}{\text{s}^2}\right)$$

References

1 Gott, P.G., *Changing Gears: The Development of the Automotive*, Society of Automotive Engineers, 1991, ISBN 1-56091-099-2.

2 Jandasek, J., *Turbine Clutch Smooths Car Operation – Reduces Stress and Fatigue Failures*, Automotive Industries, September 12. 1931, pp. 396–400.

3 Jandasek, J., *Design of Single-Stage, Three-Element Torque Converter*, SAE Special Publication SP-186, Jan. 1961.
4 SAE Transmission/Axle/Driveline Forum Committee, *Design Practices: Passenger Car Automatic Transmissions*, Third Edition, AE-18, SAE Publication, 1994, ISBN 1-56091-506-4.
5 Upton, E.W., *Application of Hydrodynamic Drive Units to Passenger Car Automatic Transmissions*, Volume 1, 1962.
6 Machida, S., Yagi, N., Miyata, K., Sadakiyo, M., Okaji, T., and Yamane, T., *Development of 8-speed DCT with Torque Converter for Midsize Vehicles*, Article of Honda R&D Technical Review, Vol. 26, No. 2., 2014.
7 By R.R. and Mahoney, J.E., *Technology Needs for the Automotive Torque Converter – Part 1: Internal Flow, Blade Design and Performance*, SAE Paper No. 880482.
8 Blomquist, A.P. and Mikel, S.A., *The Chrysler Torque Converter Lock-up Clutch*, SAE Paper No. 780100.
9 Hiramatsu, T., Akagi, T., and Yoneda, H., *Control Technology of Minimal Slip-Type Torque Converter Clutch*, SAE Paper No. 850460.
10 Tsangarides, M.C. and Tobler W.E., *Dynamic Behaviour of a Torque Converter with Centrifugal Bypass Clutch*, SAE Paper No. 850461.
11 Numazawa, A., Ushijima, F. and Fukumura, K., *An Experimental Analysis of Fluid Flow in a Torque Converter*, SAE Paper No. 830571.

Problem

A vehicle of the following data is equipped with a four-speed AT. The ratios are: 2.84 (1st), 1.60 (2nd), 1.0 (3rd), 0.7 (4th). The final drive ratio is 2.84. The torque converter characteristics and the engine WOT output torque are the same as in the example problem solved previously in this chapter. The vehicle is being driven on level ground in the first gear with the torque converter unlocked.

a) Present the differential equation system with the initial condition for the WOT performance simulation model of the vehicle in first gear. Assume the vehicle is launched at the same condition as in the example.
b) Determine the engine angular acceleration and vehicle acceleration when the vehicle is just launched from standstill.
c) Starting from time zero and using a step size of $t = 0.2$, solve the differential equation system by hand for one step, i.e. find the engine RPM and vehicle speed 0.2 second after launch.
d) Establish the computer model to implement the dynamic model shown in Figure 4.11, using the formulation consisting of Eqs (4.29) and (4.30) and other related equations in this chapter. Use your model to simulate vehicle performance dynamics during launch from rest to the time when the torque converter becomes a fluid couple. Plot the engine speed, engine torque, vehicle speed and acceleration, and the speed ratio against time during the vehicle launch process.

Vehicle data:

Front axle weight: 1290 lb	Rear axle weight: 1240 lb
Center of gravity height: 18 inch	Wheel base: 100 in.
Air drag coefficient: 0.31	Frontal projected area: 20 sq.ft
Tire radius: 10.0 in.	Roll resistance coefficient: 0.02
Powertrain efficiency: 0.92	Mass moment of inertia of engine–impeller: 0.3 lb.ft^2

5

Automatic Transmissions

Design, Analysis, and Dynamics

5.1 Introduction

Automatic transmissions (AT) that use planetary gear trains (PGT) and clutches for power transmission and ratio changes are the dominating transmission type in the automotive industry. This type of transmission dates back to the late 1930s [1,2], and they are conventionally termed automatic transmissions since they were the only production type for many decades before other types of automatic transmission, such as CVT and DCT, were applied in scale in the automotive industry. Early automatic transmissions [3] for rear wheel drive (RWD) vehicles provided three forward and one reverse gears, and were designed on an architecture that consisted of a torque converter, planetary gear train sets, and hydraulically controlled clutches, as illustrated in Figure 5.1. This design architecture or its variations stayed almost unchanged for more than three decades. Before the 1970s, passenger vehicles were almost exclusively RWD, and the vast majority of automatic transmissions had three forward speeds. Only a few high end models were equipped with four-speed automatic transmissions, with the fourth gear as a direct drive [4]. The oil crisis in the 1970s called for the development of more fuel efficient passenger vehicles that are lighter and front wheel driven (FWD). An early type of FWD AT that provided three forward and one reverse gears is illustrated in Figure 5.2. As shown in the figure, in an FWD transmission, the final drive and differential assembly is integrated with the transmission, but the ratio change portion of the transmission is similar to that in RWD transmissions illustrated in Figure 5.1.

As discussed in Chapter 1, the number of speeds in a transmission affects directly how well it can be matched to the engine in terms of performance and fuel economy. In general, automatic transmissions with more speeds provide better vehicle performance, improved fuel economy and enhanced drivability. To achieve the best trade-off between performance and fuel economy for passenger vehicles, the automotive industry started to develop four-speed automatic transmissions with the top gear as an overdrive in the middle of 1970s. Since then, both four-speed and three-speed automatic transmissions have been used across the product lines, and by 1990, three-speed automatics had been phased out from the industry. Four-speed ATs were the mainstream for the industry until the turn of the century, with five-speed types also taking up a significant market share. Today, six-speed ATs are the standard equipment across the industry and transmissions with eight or more speeds are already applied in high end passenger cars and SUVs. It can be predicted that transmissions with around ten speeds will be developed and applied in

Automotive Power Transmission Systems, First Edition. Yi Zhang and Chris Mi.

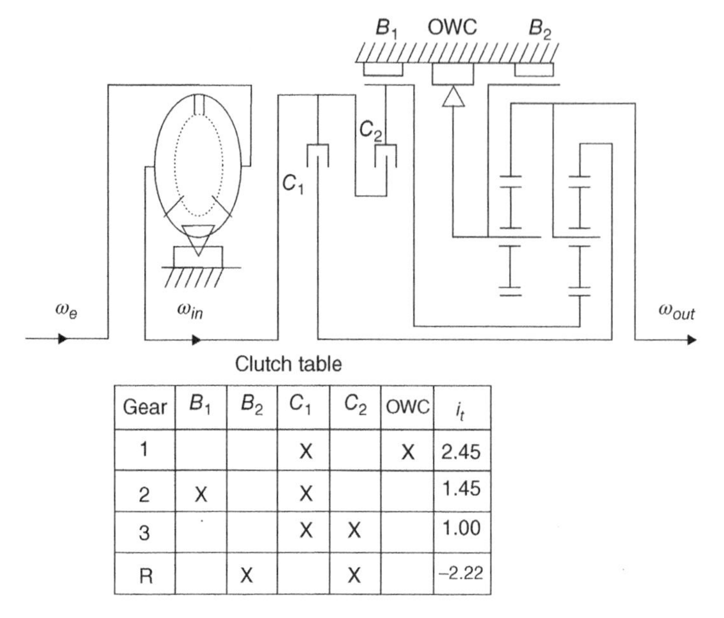

Gear	B_1	B_2	C_1	C_2	OWC	i_t
1			X		X	2.45
2	X		X			1.45
3			X	X		1.00
R		X		X		−2.22

Figure 5.1 Structure of an early three-speed automatic transmission.

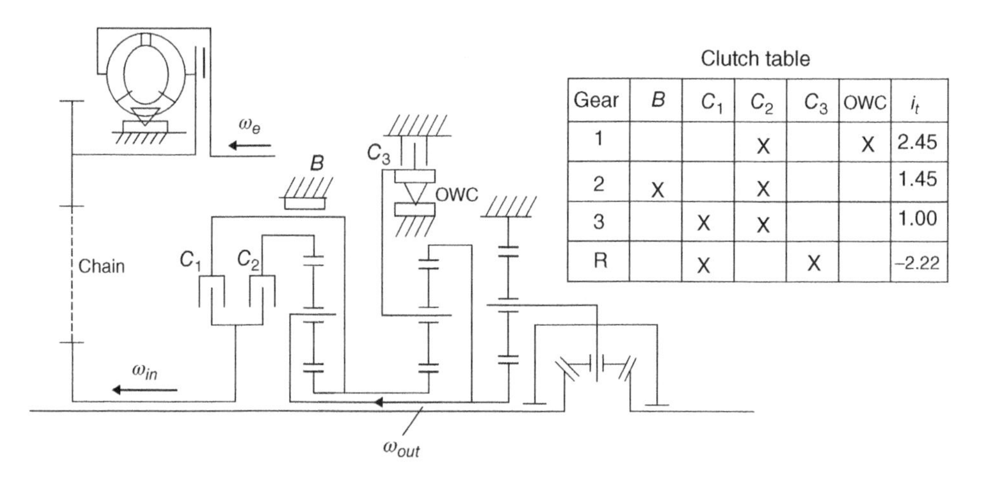

Gear	B	C_1	C_2	C_3	OWC	i_t
1			X		X	2.45
2	X		X			1.45
3		X	X			1.00
R		X		X		−2.22

Figure 5.2 Structure of an early three-speed FWD automatic transmission.

the industry in the next few years. By then, the benefits of more speeds shall diminish and the pursuit for more speeds will probably come to a halt.

This chapter will cover the design, kinematics, and dynamics of automatic transmissions. Section 5.2 will present the architectural configurations of automatic transmissions with various speeds, from the three-speed of the early types to current generation of transmissions with eight or nine speeds. Standard stick diagrams drawn in conventional symbols are used in the chapter to illustrate transmission structures and layouts. Although automatic transmissions are primarily based on planetary gear train structure, this section will include examples of lay-shaft transmission designs for a complete overview.

Section 5.3 will cover the kinematics of automatic transmissions based on the characteristics of planetary gear trains discussed in Chapter 3. The section focuses on how multiple gear ratios are achieved by a combination of clutches and planetary gear trains. A systematic method will be presented in the section for the design and analysis on the gear ratios of automatic transmissions. Case studies on production transmissions will be included to demonstrate the principles for design and analysis. The chapter then presents the analysis on the dynamics of automatic transmissions in Section 5.4. As a continuation of the case studies, this section starts with the free-body diagrams of the assemblies that form the transmission system, and then derives the equations of motion for each of the assemblies. These equations are then transformed by eliminating the internal torques on planetary gear train members to form the state variable equation system. Static clutch torques in each gear are then determined as a special case of the state variable equation system.

In Section 5.5, a qualitative analysis on shift processes will be provided to demonstrate how transmission control and state variables affect shift quality. The concept of torque phase and inertia phase during transmission shifts will be introduced and transmission shift dynamics during the two shift phases will be discussed in this section. After the qualitative analysis in Section 5.5, Section 5.6 will present general vehicle powertrain dynamics in a systematic approach, using an eight-speed automatic transmission as the example. System formulation and modeling will be presented in this section for the simulation, analysis and control of vehicles equipped with automatic transmissions. Finally, Section 5.7 will discuss model simulations and applications for vehicle powertrain systems under various operation conditions, such as fixed gear operation, shifting processes, and operations on specified drive ranges.

5.2 Structure of Automatic Transmissions

Automatic transmissions based on planetary gear train designs have evolved over the past seven decades from the first generation with three speeds to the current generation with eight or more speeds. In terms of mechanical structure, however, these transmissions are still built from the same components, which include a torque converter, planetary gear trains, and clutches. The differences between a first generation three-speed AT and a current generation eight-speed AT mainly lie in the number of planetary gear trains and their layouts, as well as the combination of clutches that realize multiple gear ratios. Stick diagrams of production transmissions are provided in this section as examples to illustrate the PGT layout and clutch combination for transmissions from three speeds to nine speeds.

A three-speed RWD automatic transmission of the early type and the clutch table are shown in Figure 5.1. This transmission was developed by Chrysler in the mid 1950s based on the Simpson-type planetary gear set and had been applied across the Chrysler product line. This transmission was representative of similar products in the automotive industry until the 1970s. As illustrated in Figure 5.1, the Simpson-type planetary gear set consists of two simple planetary gear trains, with the two sun gears structurally connected and the ring gear of the front train structurally connected to the carrier of the rear train. The transmission output is from the ring-carrier assembly. The transmission uses two multiple disk hydraulic clutches, C_1 and C_2, two band clutches, B_1 and B_2, and a one-way clutch to realize three forward speeds and one reverse. The two band clutches are reaction clutches that ground the sun gear assembly and the carrier of the front PGT respectively when applied. As can be noticed from the clutch table, the off-going clutch for the 1–2 upshift is the one-way clutch. This makes the 1–2 shift control less challenging and is an important factor in achieving shift smoothness in the absence of advanced control technology at the time when the transmission was developed. The other upshift, i.e. the 2–3 shift, is clutch to clutch, which is generally more difficult to control, but in this case, the shift would occur at higher vehicle speed and is less sensible to the driver. Here it is worthwhile to introduce some terminologies commonly used in the transmission area: **clutch to one-way clutch shift, clutch to clutch shift, reaction clutch** and **coupling clutch**. In a clutch to one-way clutch shift, the off-going clutch is a one-way clutch, and in a clutch to clutch shift, the off-going clutch is a regular hydraulically actuated clutch. A reaction clutch grounds a component and a coupling clutch couples two components when applied. A band clutch can only be applied as reaction clutch due to its structure.

An early type three-speed FWD automatic transmission with its clutch table is shown in Figure 5.2. This transmission was developed by General Motors for FWD passenger cars in the 1970s and was applied across the GM product line for almost two decades. It uses two planetary gear trains for ratio change and a third PGT for the final drive. The transmission uses three multiple-disk clutches and one band clutch. The one-way clutch serves as the reaction clutch in first gear and makes the 1–2 upshift as a clutch to one-way clutch shift. Clutch C_3 serves as the reaction clutch for the reverse gear and also for the first gear when the driver selects the manual range for engine braking in downhill operations. To accommodate the assembly condition for transversely mounted engine, a chain is used to transfer power from the converter turbine to the transmission input.

It was in the 1980s that the four-speed automatic transmissions started to enter the market in large scale, due to the ever more stringent demand for fuel economy. A four-speed FWD automatic transmission developed by Ford is illustrated in Figure 5.3. This transmission and its variations were applied in Ford passenger vehicles for more than two decades until the introduction of the current generation of six-speed automatics. Unlike the Simpson-type, the planetary gear train set features two structural connections: front ring and rear carrier assembly, and front carrier and rear ring assembly. The latter is the output assembly, with a chain connecting it to the final drive input. The ATF pump is driven by the engine via a shaft passing through the center line of the transmission main structure. As shown in the clutch table, one-way clutches are used to facilitate 1–2 and 3–4 upshifts. Clutch C_3 is only applied in manual ranges in mountainous areas or in snowy conditions.

As discussed in Chapter 3, the Ravigneaux planetary gear train is structurally the combination of a dual-planet PGT and a simple PGT with a shared ring gear and a shared carrier, as shown in Figure 3.22. Transmission designs based on Ravigneaux PGTs are more compact in comparison with those based on Simpson-type PGTs. Figure 5.4 shows a four-speed RWD transmission based on a Ravigneaux planetary gear train. This transmission was developed by Ford in the 1980s and was used in Ford pickup trucks for almost two decades. The transmission uses a total of eight clutches: four multiple disk

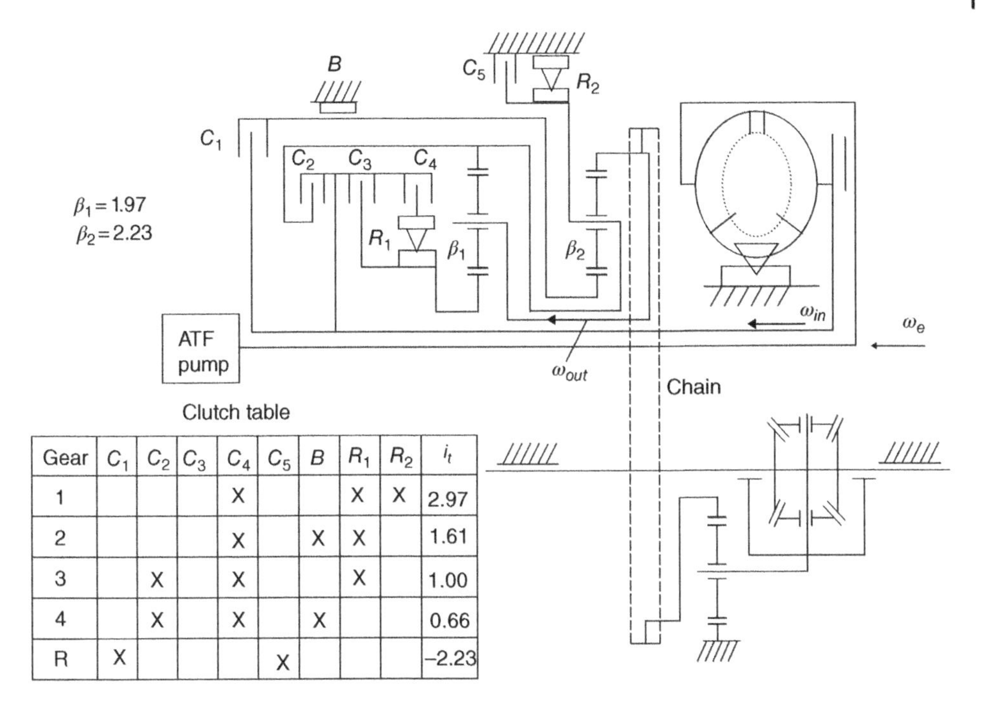

Gear	C_1	C_2	C_3	C_4	C_5	B	R_1	R_2	i_t
1				X			X	X	2.97
2				X		X	X		1.61
3		X		X			X		1.00
4		X		X		X			0.66
R	X				X				−2.23

Figure 5.3 Structure of an early four-speed FWD automatic transmission.

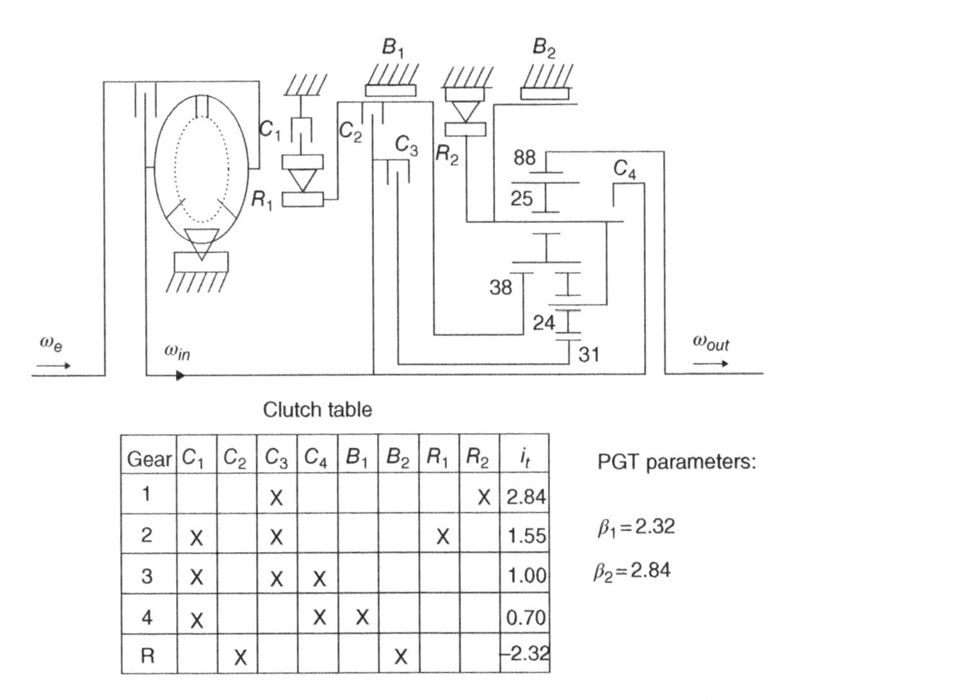

Gear	C_1	C_2	C_3	C_4	B_1	B_2	R_1	R_2	i_t
1			X					X	2.84
2	X		X				X		1.55
3	X		X	X					1.00
4	X			X	X				0.70
R		X				X			−2.32

Figure 5.4 A Ford four-speed RWD Ravigneaux PGT automatic transmission.

clutches, two band clutches, and two one-way clutches. As shown in Figure 5.4, the 1–2 and 2–3 upshifts are clutch to one-way clutches. In manual ranges, band clutch B_1 is applied in first gear and the band clutch B_2 is applied in second gear.

Many different layouts for four-speed ATs existed for production transmissions between 1980s and the beginning of the 21st century. Among the many AT structural designs, the FWD four-speed AT named as Hydra-matic 4 T80-E and developed by General Motors in the early 1990s, stands out in terms of its seamless shift smoothness. As shown in Figure 5.5, the transmission uses two simple planetary gear trains, with the front ring gear and the rear carrier structurally connected as the output assembly. The front carrier and the rear ring gear are not structurally connected but coupled through four clutches, C_4, C_5, R_2, and R_3, in different gears. A chain with a ratio of 1 connects the turbine with the transmission input. With a total of 10 clutches: five multiple disk clutches, two band clutches and three one-way clutches; this is the transmission that uses the highest number of clutches. Clutch C_5 is a coast clutch that is only applied to enable engine braking in manual ranges. As can be seen in the clutch table, all upshifts, 1–2, 2–3, and 3–4, are clutch to one-way clutches. This simplifies the upshift control and results in excellent shift quality. Clearly, there is a price to be paid for the enhanced shift smoothness since the ten clutches used in the transmission definitely increase hardware costs. This transmission and its variations were used across the product line of General Motors until around 2005, by which time the current generation of six-speed FWD automatic transmissions had started to be the mainstream transmissions.

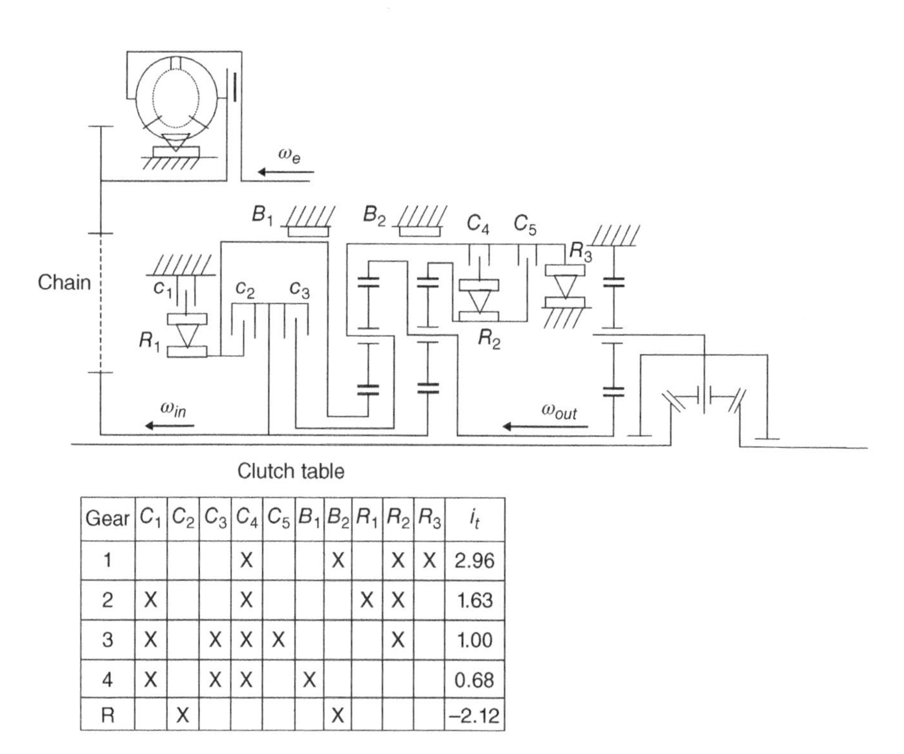

Clutch table

Gear	C_1	C_2	C_3	C_4	C_5	B_1	B_2	R_1	R_2	R_3	i_t
1				X			X		X	X	2.96
2	X			X				X	X		1.63
3	X		X	X	X				X		1.00
4	X		X	X		X					0.68
R		X					X				−2.12

Figure 5.5 A four-speed FWD AT with all clutch to one-way clutch upshifts.

Unlike General Motor's four-speed AT discussed earlier, the four-speed FWD AT developed by Chrysler in the late 1980s uses a lower number of clutches. As shown in Figure 5.6, the Chrysler transmission uses only five clutches to realize four forward speeds and one reverse. All shifts are clutch to clutch since no one-way clutch is used in this transmission. In addition, this is the first transmission that uses only multiple disk clutches. In the absence of one-way clutches during shifts, this transmission was the first that relied solely on advanced electronic control technology to realize shift smoothness. Note that the design of this transmission reflects the trend in the current generation of six-speed and eight-speed ATs, that use only multiple disk clutches in the minimum number possible, to reduce hardware cost and rely on control technology to guarantee shift quality.

In addition to planetary gear train designs, four-speed automatic transmissions had also been developed based on lay-shaft gear designs. Lay-shaft transmissions differ from planetary gear train transmissions mainly in structural layouts, the control systems for both types are similar in nature. Honda and the Saturn division of General Motors are the two major brand names that had developed and used lay-shaft automatics for various passenger cars and SUVs. Figure 5.7 shows the layout of a Honda four-speed lay-shaft AT developed in the 1980s. The transmission is equipped with a torque converter, and gear shifts are realized by hydraulically actuated multiple disk clutches. As shown in the stick diagram and the clutch table, a hydraulically actuated shifter, called the "servo" in the stick diagram, engages forward gears in the D range and the reverse gear in the R range. A one-way clutch is attached to the first gear on the counter shaft which

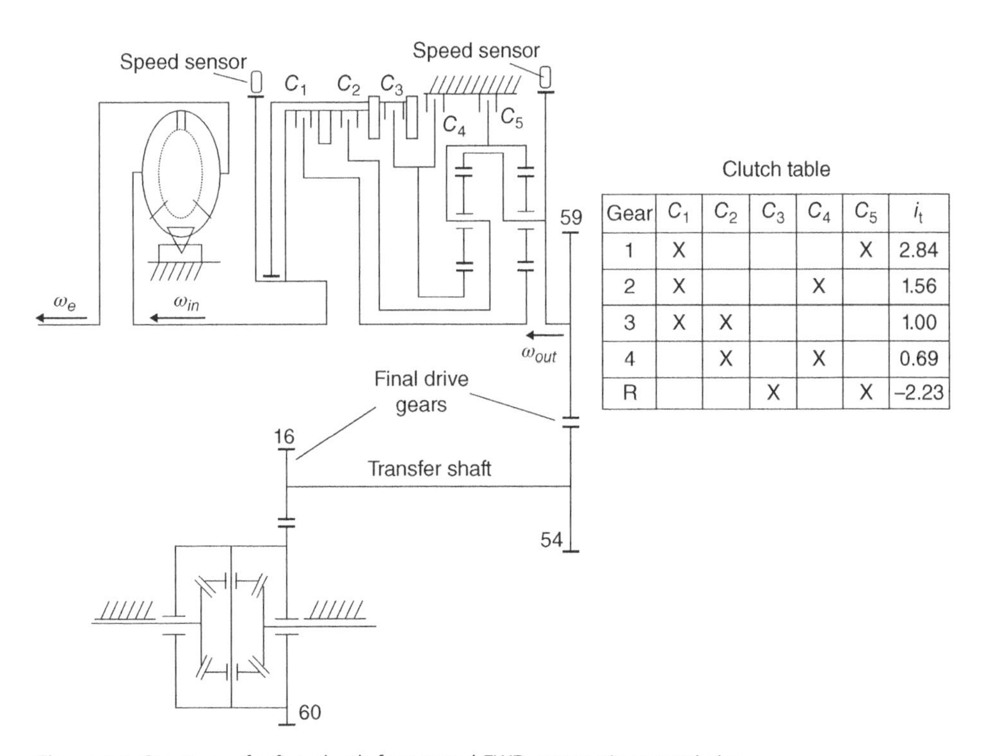

Gear	C_1	C_2	C_3	C_4	C_5	i_t
1	X				X	2.84
2	X			X		1.56
3	X	X				1.00
4		X		X		0.69
R			X		X	−2.23

Figure 5.6 Structure of a five-clutch four-speed FWD automatic transmission.

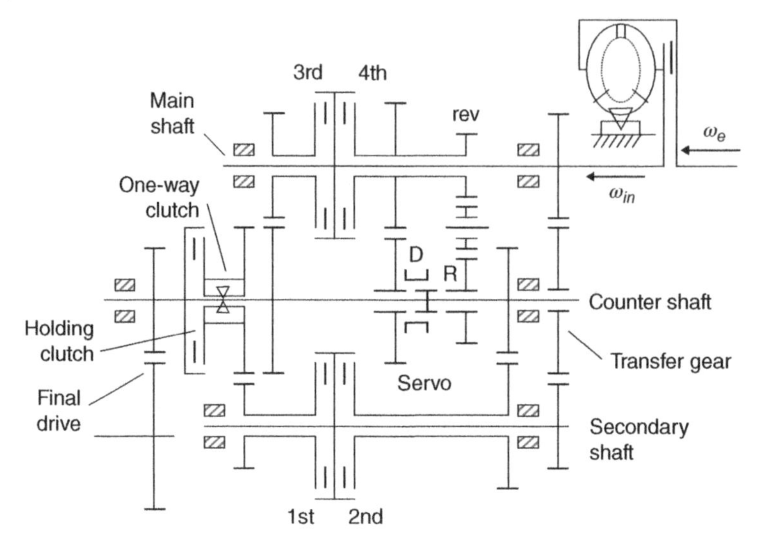

Clutch table

Gear	1st	2nd	3rd	4th	Holding	One-way	Servo
1	X					X	D
2	X	X					D
3	X		X				D
4	X			X			D
1H	X				X	X	D
R				X			R

Figure 5.7 Honda lay-shaft four-speed FWD automatic transmission.

enables the 1–2 upshift to be a clutch to one-way clutch. The 2–3 and 3–4 upshifts are clutch to clutch. The first gear holding clutch is applied in manual ranges to enable engine braking when driving conditions deems it necessary. In comparison with planetary gear train automatic transmissions, lay-shaft automatics have the advantage that many of the mechanical components are similar to those in manual transmissions that have similar lay-shaft designs. It was more cost effective to develop a lay-shaft automatic transmission based on existing manual transmission technology and technical resources than to develop a planetary gear train automatic transmission from scratch. However, lay-shaft automatic transmissions are not as compact as their planetary gear train counterparts. In a lay-shaft transmission, as shown in Figure 5.7, gear shafts are laid out parallel instead of coaxial, and multiple disk clutches are assembled separately on the lay-shafts instead of nested in planetary gear train automatics. This makes lay-shaft automatic transmissions bigger than planetary gear train counterparts. For transmissions with four, five, or even six forward speeds, it is still technically feasible to pack all the components, including, clutches, gears, and shafts into the limited volume allowed by a transversely mounted engine. Due to the intrinsic structural limitations, it would be very difficult to design an eight-speed lay-shaft automatic transmission that could be fitted to an FWD passenger car.

Improved versions were developed by Honda based the lay-shaft automatic transmission shown in Figure 5.7. The one-way clutch and the first gear holding clutch were eliminated in a later version, as shown in Figure 5.8. This version is smaller in overall dimensions due to the elimination of redundant components, and gear shifts are all clutch to clutch. Lay-shaft transmissions was used across the Honda product line and were eventually replaced by other types of automatic transmissions with six or more speeds.

Automatic transmissions with four speeds had been the mainstream for FWD passenger cars or SUVs with 4WD extended from FWD layouts until around 2005. Five-speed transmissions were also developed and used for various vehicles during the transition. Figure 5.9 shows a five-speed RWD transmission developed by Ford. In this transmission, a third simple PGT is added in front a Simpson PGT set. Clutch C_1 is only applied in manual ranges for engine braking. The fourth gear is a direct drive and the fifth gear provides an overdrive ratio of 0.69.

A six-speed FWD automatic transmission of the current generation is shown in Figure 5.10. This transmission was jointly developed by Ford and General Motors and is currently applied across the product lines of both companies in different versions. The transmission uses three simple planetary gear trains in parallel and six clutches: five multiple disk clutches and one one-way clutch. The transmission output assembly consists of the first ring gear and the third carrier, and is connected to the differential carrier through two gear pairs via a transfer shaft. As shown in the clutch table, the one-way clutch is only applied in first gear so that the 1–2 upshift is a clutch to one-way clutch. Technically, the one-way clutch F is redundant in kinematics and can be eliminated since it can be replaced by clutch D in first gear. The transmission would then have only five multiple disk clutches with all shifts clutch to clutch. Structurally, this transmission features an input shaft that passes through the whole transmission, with two input clutches, A and B, nested at the shaft end. Note that there is no band clutch in this transmission.

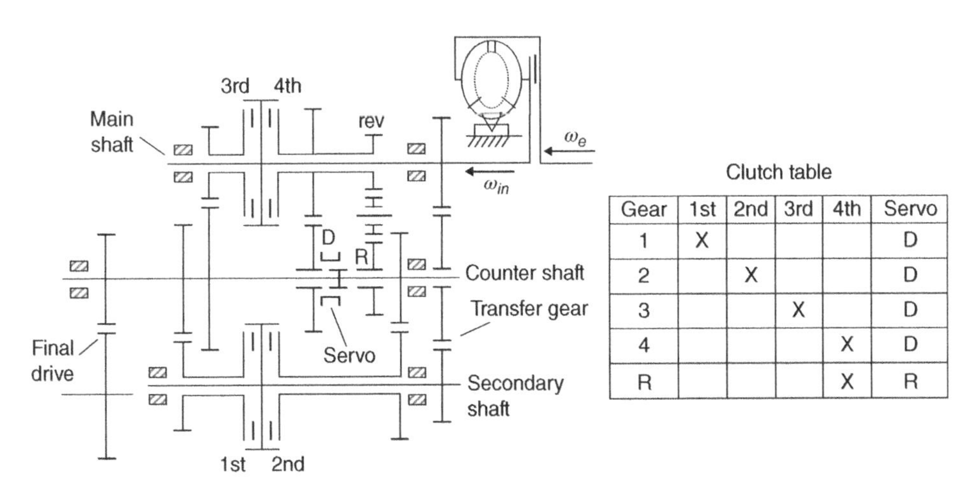

Gear	1st	2nd	3rd	4th	Servo
1	X				D
2		X			D
3			X		D
4				X	D
R				X	R

Figure 5.8 Honda lay-shaft four-speed FWD automatic transmission without one-way clutch.

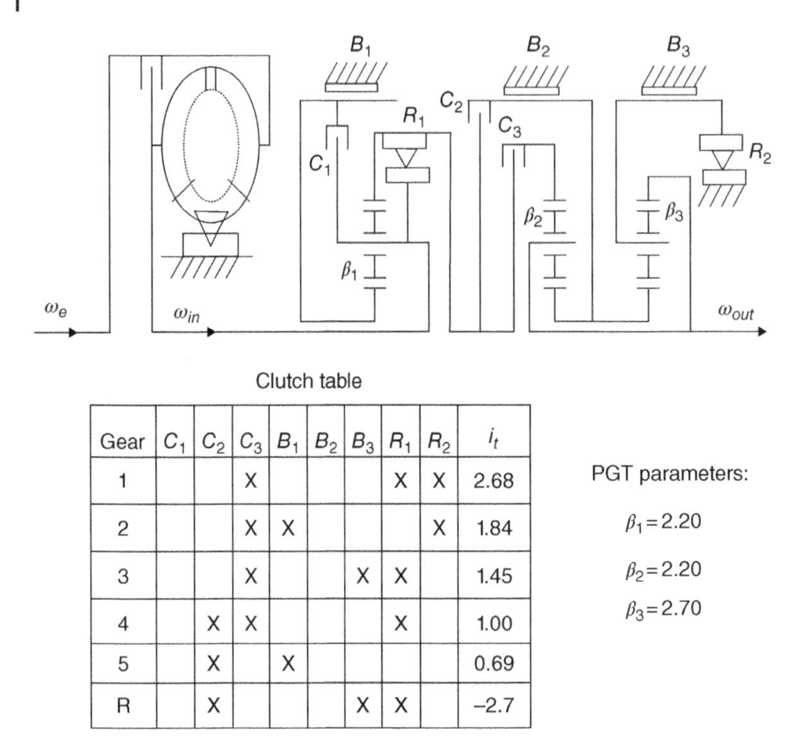

Clutch table

Gear	C_1	C_2	C_3	B_1	B_2	B_3	R_1	R_2	i_t
1			X				X	X	2.68
2			X	X				X	1.84
3			X			X	X		1.45
4		X	X				X		1.00
5		X		X					0.69
R		X				X	X		–2.7

PGT parameters:

$\beta_1 = 2.20$

$\beta_2 = 2.20$

$\beta_3 = 2.70$

Figure 5.9 A five-speed RWD automatic transmission.

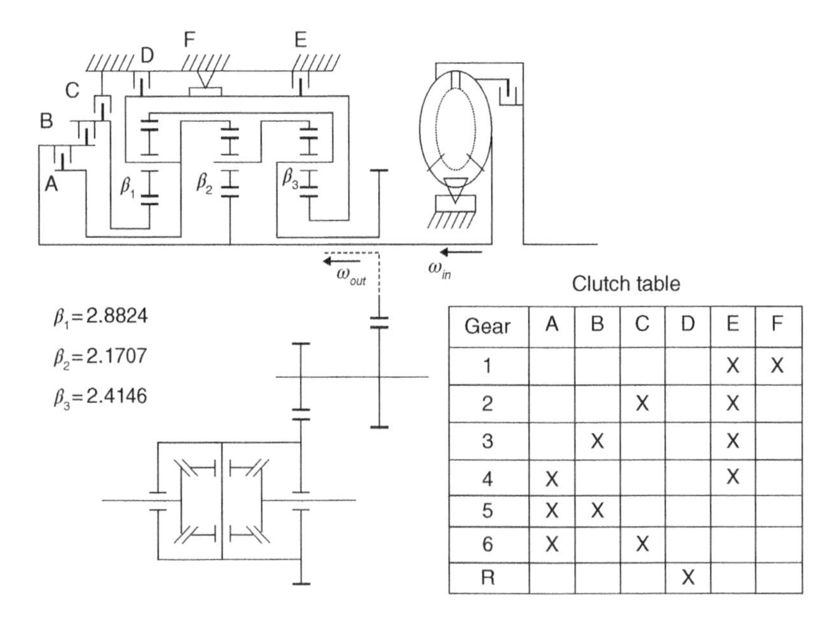

$\beta_1 = 2.8824$

$\beta_2 = 2.1707$

$\beta_3 = 2.4146$

Clutch table

Gear	A	B	C	D	E	F
1					X	X
2			X		X	
3		X			X	
4	X				X	
5	X	X				
6	X		X			
R				X		

Figure 5.10 A Ford six-speed FWD automatic transmission.

Band clutches require actuation pistons and anchors that cannot be symmetrically mounted on the cylindrical transmission structure and are generally not as compact as multiple disk clutches. In addition, band clutches are not as responsive as multiple disk clutches in engagement or disengagement during transmission shifts. For these reasons, band clutches are no longer applied in the current generation of six-speed or eight-speed automatic transmissions and are now obsolete even though they have been applied in the automotive industry for almost 70 years since the first automatic transmission came into production.

The six-speed RWD automatic transmission developed by Ford is based on a planetary gear train setup different from the FWD version shown in Figure 5.10. This transmission is currently used in Ford RWD vehicles, including the popular F-series pickups. The six-speed RWD automatic is based on a setup that consists of a Ravigneaux planetary gear train and a simple planetary gear train, as shown in Figure 5.11. The transmission architecture can be considered as an extension of the four-speed AT structure shown in Figure 5.4, with a simple PGT added in front the Ravigneaux PGT. The sun gear of the front simple PGT is fixed structurally, and the ring gear is connected to the transmission input. All shifts are clutch to clutch since no one-way clutch is used in the transmission. This transmission uses five multiple disk clutches to realize six forward speeds and one reverse speed. In comparison, the previous generation four-speed RWD automatics, shown in Figure 5.4, uses eight clutches to realize only four forward speed and one reverse speed. The improvements are significant in terms of not only transmission performance specifications but also component costs since the number of clutches is reduced from eight to five. These improvements are made possible thanks to the

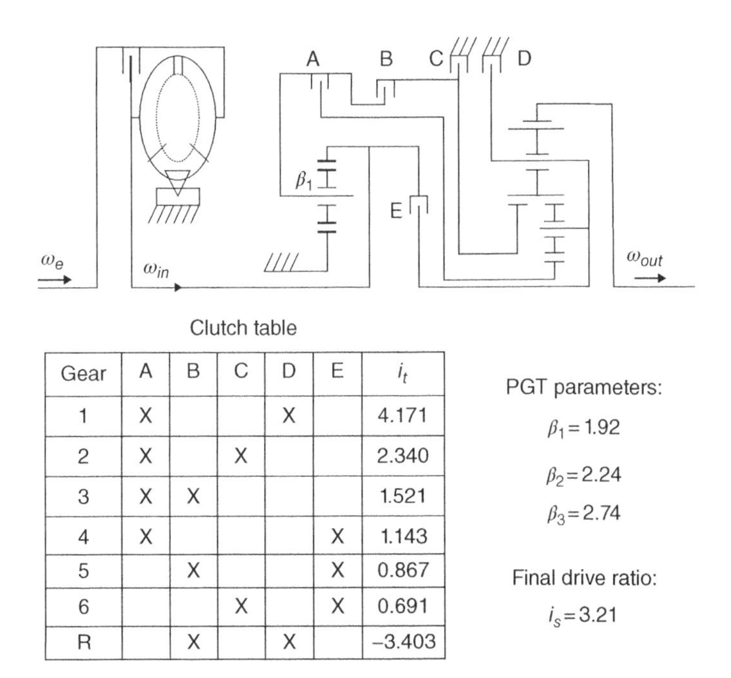

Clutch table

Gear	A	B	C	D	E	i_t
1	X			X		4.171
2	X		X			2.340
3	X	X				1.521
4	X				X	1.143
5		X			X	0.867
6			X		X	0.691
R		X		X		−3.403

Figure 5.11 Ford six-speed RWD Ravigneaux PGT automatic transmission.

advanced technology in the transmission control hardware and software, as will be discussed in the next chapter.

The ever more stringent demand on fuel economy led to the development of automatic transmissions with eight or more speeds. These transmissions have a wide ratio spread defined by the first gear ratio and the ratio of the highest gear, and enable engine matching close to the ideal condition due to the availability of multiple gear ratios. In addition, transmission ratios between neighboring gears are closer to each other in comparison with transmissions with six or fewer speeds, reducing intrinsically the harshness caused by shift dynamics. The design of transmissions with eight or more speeds is technically challenging because of the constraints on the overall transmission dimension and the number of clutches. In general, these transmissions need a planetary gear train setup that is difficult to achieve based on the extension from existing transmissions with fewer speeds. In the transmissions with eight or more speeds currently in production, three types of planetary gears trains – simple PGT with single planet, dual-planet simple PGT, and Ravigneaux PGT – are used in various layouts to provide the ratio multiplicity. By using a dual-planet PGT in a layout, more transmission ratio flexibility may appear because of the sign variations of the characteristic equation. The elements of these planetary gear trains are connected structurally and through clutches in a variety of layouts to form multiple power flows from the input to the output. In kinematics, there are six types of structural connections: sun-sun, sun-ring, ring-carrier, ring-ring, carrier-carrier, and sun-carrier. The planet gear only meshes with the ring gear and the sun gear, without being connected to any other element either by a clutch or structurally. Sun gear, carrier and ring gear, or an assembly containing one of them can be connected to the input by a coupling clutch, and can be grounded to the housing by a reaction clutch. Output is usually taken from a ring gear a carrier, or an assembly containing one of them. Automatic transmissions with eight or nine speeds shown below will demonstrate the basic guidelines on the mechanical structure of these transmissions described herein.

The first eight-speed RWD automatic transmission, shown in Figure 5.12, was developed and used for Lexus vehicle models of the Toyota Motor Company. This transmission can be fitted to any RWD or 4WD vehicle. As shown in the stick diagram, the transmission is based on a setup that consists of a simple dual-planet PGT and a Ravigneaux PGT. The ratio spread of the transmission is 6.71 and the intermediate ratios are closely populated between the first and the eight speeds. In the dual-planet PGT, the sun is structurally fixed to the housing and the carrier is structurally connected to the input. The ring gear of the Ravigneaux PGT is the output element also structurally. It is meaningful to notice that none of the planetary gear train members are structurally interconnected. This maximizes the number of PGT elements that can be connected to the input or grounded and thus the number of power flows. There are seven clutches, including the one-way clutch G that enables 1–2 upshift as a clutch to one-way clutch. The one-clutch G is redundant in kinematics and if eliminated, the transmission shifts would just be all clutch to clutch. As shown in the clutch table, two clutches are applied and five are open in each gear. The transmission realizes eight speeds in a compact design, using a planetary gear train setup that is not much more complicated than what is in a six-speed automatics, such as the one shown in Figure 5.11, and should be credited as a significant engineering achievement. However, it uses seven clutches, even though one of them is a one-way clutch that can be eliminated in later versions, and five clutches (or four without considering the one-way clutch) stay open in each gear. These open

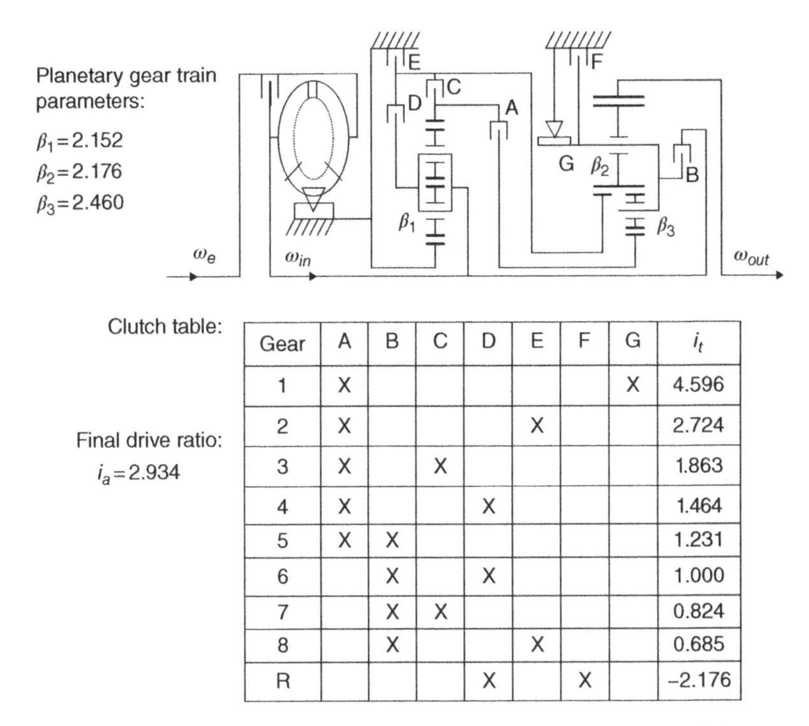

Gear	A	B	C	D	E	F	G	i_t
1	X						X	4.596
2	X				X			2.724
3	X		X					1.863
4	X			X				1.464
5	X	X						1.231
6		X		X				1.000
7		X	X					0.824
8		X			X			0.685
R				X		X		-2.176

Figure 5.12 Lexus eight-speed RWD Ravigneaux PGT automatic transmission.

clutches create drag losses that work against the fuel economy improvement gained from the availability of the eight transmission ratios.

Hyundai was a latecomer to the development of automatic transmissions. However, it has made a great leap forward in transmission technologies over the past decade. The company has developed its own eight-speed automatic transmission and successfully applied it in RWD models of its luxury Genesis brand. Figure 5.13 shows the structure of the eight-speed automatics with the clutch table. The transmission is built on a setup that consists of three planetary gear trains, a simple PGT in the front, a dual-planet PGT in the middle, and a Ravigneaux PGT in the rear. The PGT setup differs from the Lexus eight-speed AT (shown in Figure 5.12) in the added front simple PGT. The transmission also has seven multiple disk clutches, including one-way clutch G that facilitates the 1–2 upshift. Similar to the Lexus eight-speed AT, two clutches are applied and five stay open in each gear. As shown in the clutch table, the fifth gear is a direct drive and the next three higher gears are all overdrives. By the combination of planetary gear train parameters shown with the stick diagram, the transmission achieved a first gear ratio of 3.665 and a eighth gear ratio of 0.556, with a ratio spread close to 7.0. It can be noticed from the clutch table that the transmission ratios are lower in comparison to those of the Lexus eight-speed AT, a relatively high final drive ratio of 3.909 is therefore used for compensation of the overall powertrain ratio. The first four gear ratios are spread more widely than the higher gear ratios, as shown in the list by the ratio column. Note that different transmission ratios can be achieved by changing the values of the PGT parameters, and this is true generally for all other transmissions. Based on the ratio formulae

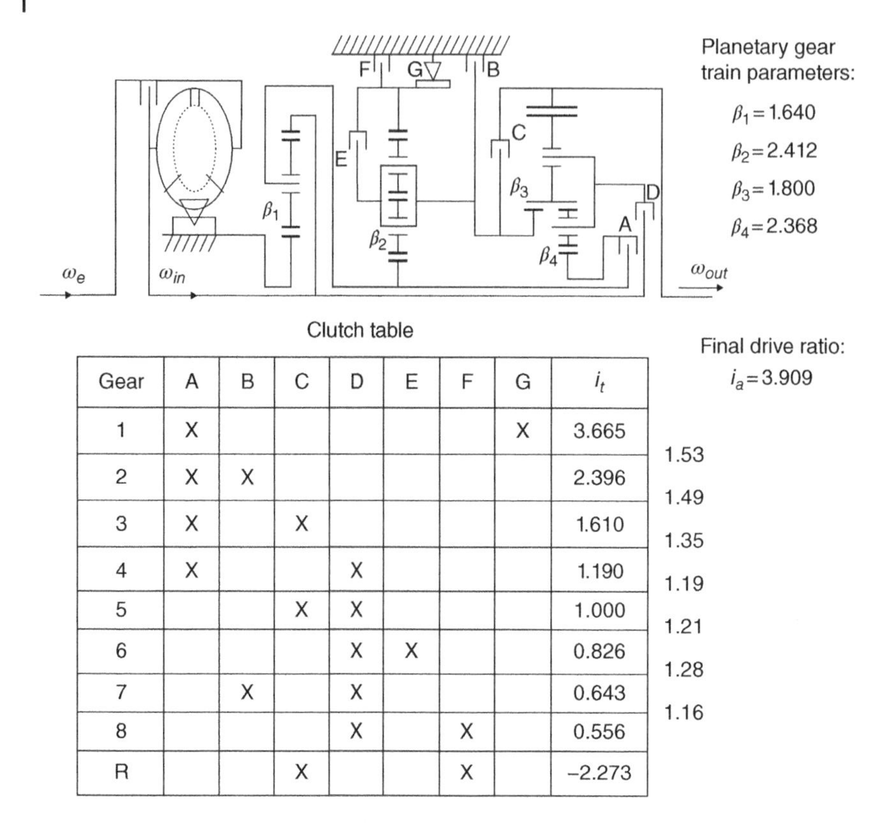

Gear	A	B	C	D	E	F	G	i_t
1	X						X	3.665
2	X	X						2.396
3	X		X					1.610
4	X			X				1.190
5			X	X				1.000
6				X	X			0.826
7		X		X				0.643
8				X		X		0.556
R			X			X		−2.273

Figure 5.13 Hyundai eight-speed RWD Ravigneaux PGT automatic transmission.

for all gears, the four PGT parameters can be selected to optimize the combination of gear ratios that fits the requirement of a specific application. Due to the added front PGT, the Hyundai eight-speed AT is somewhat larger in size in comparison to the Lexus eight-speed automatics. The attributes in transmission efficiency of the two transmissions should be similar because both have six multiple disk clutches and one one-way clutch, with two engaged and five open in each gear.

The ZF eight-speed RWD AT, shown in Figure 5.14, features perhaps the most compact structural design for the current generation of eight-speed automatics. This transmission is built on a planetary gear train setup that consists of four simple PGTs. Only five multiple disk clutches are used for the eight forward speeds and one reverse gear. The most remarkable aspect of this transmission is that in each gear there are only two open clutches. This minimizes the parasitic clutch drag loss and is an apparent advantage over the designs of the Lexus or Hyundai eight-speed automatics. The sixth gear is a direct drive and the seventh and eighth gears are overdrives. The ratio spread is more than 7.0, resulting in close to optimized engine matching for acceleration and fuel economy performances. Note that the reverse gear in this transmission is achieved by a specific combination of PGT motions, instead of the common reverse gear realization where a sun gear is connected to the input, the carrier is grounded and the ring gear is connected to the output. The ZF eight-speed RWD AT has been well received by

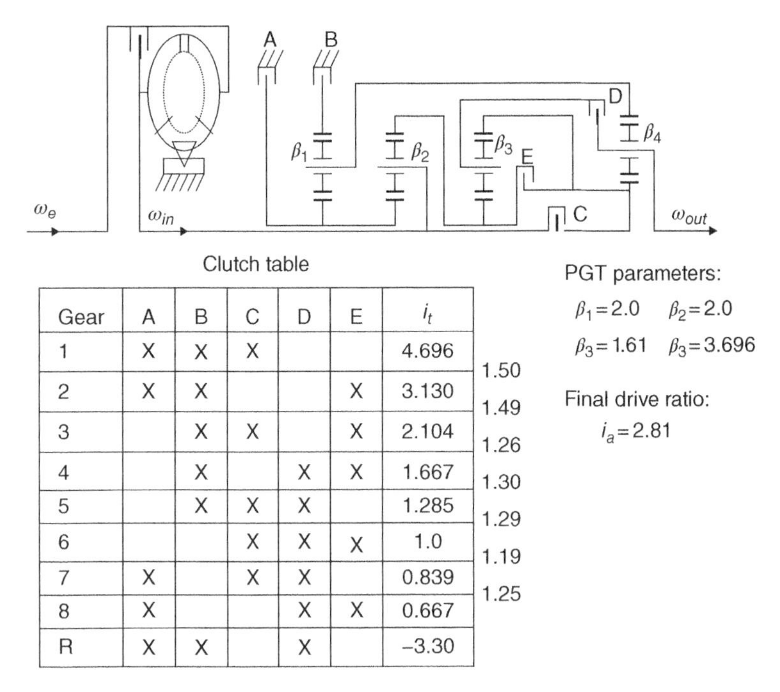

Gear	A	B	C	D	E	i_t
1	X	X	X			4.696
2	X	X			X	3.130
3		X	X		X	2.104
4		X		X	X	1.667
5		X	X	X		1.285
6			X	X	X	1.0
7	X		X	X		0.839
8	X			X	X	0.667
R	X	X		X		−3.30

Figure 5.14 ZF eight-speed RWD automatic transmission.

the automotive industry and has been applied for not only luxury brands, such as BMW and Mercedes Benz, but also less expensive brands such as Chrysler and Jeep. The next section of this chapter will provide a detailed study on the structure, kinematics, and dynamics of this transmission as an example for all other transmissions.

Automatic transmissions with eight speeds were first developed for luxury vehicle models which usually prefer rear wheel drive. The structural design of FWD ATs with multiple gear ratios is generally more challenging since the packaging conditions for transversely mounted engines are more stringent in comparison with longitudinal engine layouts. However, development of FWD automatic transmissions with eight or more speeds does not lag far behind and is proceeding rapidly in the automotive industry. The ZF nine-speed FWD AT, shown in Figure 5.15, is the very first FWD automatic transmission that has nine forward speeds. This transmission is built on a planetary gear train setup that is remarkably different from all existing transmission designs. As shown in the stick diagram, there are four simple planetary gear trains, with planetary parameter β labeled with subscripts 1, 2, 3, and 4 from left to right respectively. There are two structural connections between PGT 1 and PGT 2: sun gear 1 with sun gear 2 and ring gear 1 with carrier 2. What distinguishes this transmission are the two structural connections between PGT 3 and PGT 4: carrier 3 with carrier 4 and sun 3 with ring 4. The transmission uses four multiple disk clutches, B, C, D, and E, and two shifters, A and F, to achieve nine forward speeds and one reverse, as shown in the clutch table. The two shifters, which are actually sliding sleeves with internal splines similar to those in manual transmissions discussed in Chapter 2, are hydraulically actuated to move axially to engage or

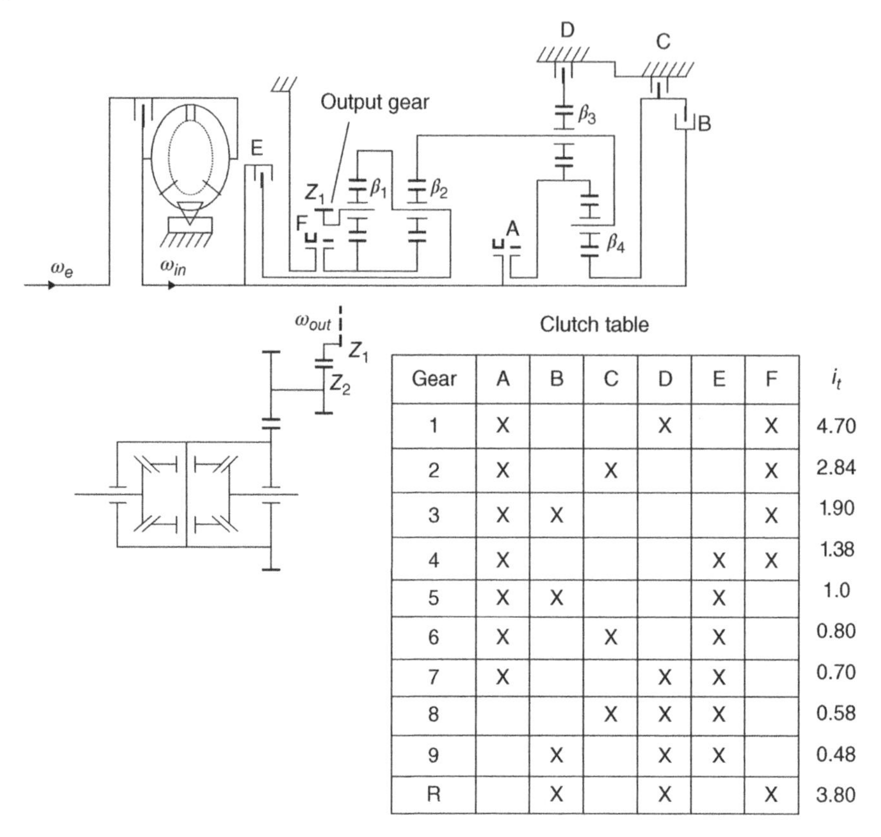

Gear	A	B	C	D	E	F	i_t
1	X			X		X	4.70
2	X		X			X	2.84
3	X	X				X	1.90
4	X				X	X	1.38
5	X	X			X		1.0
6	X		X		X		0.80
7	X			X	X		0.70
8			X	X	X		0.58
9		X		X	X		0.48
R		X		X		X	3.80

Figure 5.15 ZF nine-speed FWD automatic transmission.

disengage the dog teeth on the sun 1–sun 2 assembly and sun 3–ring 4 assembly respectively. Shifts involving the engagement or disengagement of either shifter A or shifter F must be well synchronized to avoid impact or grinding between shift sleeve and dog teeth. This can be technically challenging, especially in downshifts to the fourth or seventh gears where either shifter F or shifter A is the oncoming engagement element. The transmission has four overdrives, with the fifth gear as a direct drive. The nine forward gear ratios are spread between the first gear ratio of 4.70 and the ninth gear ratio of 0.48, resulting a high ratio spread of 9.8. The transmission output is from the carrier of PGT 1 and is transmitted to the differential carrier through two pairs of transfer gears. Because of the compact planetary gear train setup and the two space saving shifters, the transmission realizes nine forward gears and one reverse gear in a dimension comparable to that of a six-speed FWD automatic. The Jeep Cherokee SUV is the first vehicle model that is equipped with this transmission in both FWD and RWD versions. In the RWD version, power is transferred by a spiral bevel gear pair with the pinion attached to the differential carrier to the rear axle via the drive shaft. It has been reported that the Acura MDX will also be equipped with the ZF nine-speed automatic in the 2017 model year.

5.3 Ratio Analysis and Synthesis

Gear ratios in automatic transmissions based on planetary gear train designs are achieved by connecting and grounding different PGT elements to the input and to the housing respectively through a particular clutch combination in each speed. The kinematics of planetary gear trains, covered in Chapter 3 (Section 3.8), is the key to the analysis and synthesis of transmission gear ratios. In general, three types of planetary gear trains are used in the transmission structure: simple PGT, dual-planer PGT, and Ravigneaux PGT. The characteristic equation for each of these three types of PGT was derived in Section 3.8 and is represented in the following for the systematic analysis and synthesis of transmission gear ratios.
Simple planetary gear train:

$$\omega_S + \beta\omega_R - (1+\beta)\omega_C = 0 \tag{5.1}$$

Dual-planet planetary gear train:

$$\omega_S - \beta\omega_R - (1-\beta)\omega_C = 0 \tag{5.2}$$

Ravigneaux planetary gear train:

$$\omega_{S1} + \beta_1\omega_R - (1+\beta_1)\omega_C = 0 \tag{5.3}$$

$$\omega_{S2} - \beta_2\omega_R - (1-\beta_2)\omega_C = 0 \tag{5.4}$$

As detailed in Section 3.8, a Ravigneaux PGT is structurally a combination of a simple PGT and a dual-planet PGT with shared carrier and ring gear. There are two mesh paths in a Ravigneaux PGT: Eq. (5.3) is the characteristic equation for the first mesh path "Sun S_1 – Long Planet – Ring" and Eq. (5.4) is the characteristic equation for the other mesh path "Sun S_2 – Short Planet – Long Planet – Ring". The PGT characteristic equations listed above and the constraint equations caused by clutch engagements will be used systematically for the analysis and synthesis of transmission ratios. In this section, three production transmissions are chosen as examples for the ratio analysis and synthesis:– the Ford FWD six-speed AT, designed with simple PGTs (Figure 5.10); the Ford six-speed RWD AT, designed with a simple PGT and a Ravigneaux PGT (Figure 5.11); and the ZF eight-speed RWD AT, designed with four simple PGTs (Figure 5.14). These transmissions feature structural designs typical of the current generation of multiple ratio automatics. The concepts and methods for ratio analysis and synthesis in the three examples can be readily extended to all other automatic transmissions with different structural layouts.

5.3.1 Ford FWD Six-Speed AT

As illustrated in the stick diagram (Figure 5.10), the Ford FWD six-speed automatics uses three simple planetary gear trains in parallel in the structure layout. The planetary gear train parameters of respectively are labeled as β_1, β_2, and β_3, with the subscripts indicating PGT 1, PGT 2, and PGT 3. The three characteristic equations in the following are intrinsic of the three PGTs in the transmission:

$$\omega_{S1} + \beta_1\omega_{R1} - (1+\beta_1)\omega_{C1} = 0 \tag{5.5}$$

$$\omega_{S2} + \beta_2\omega_{R2} - (1+\beta_2)\omega_{C2} = 0 \tag{5.6}$$

$$\omega_{S3} + \beta_3\omega_{R3} - (1 + \beta_3)\omega_{C3} = 0 \tag{5.7}$$

In the equations above, there are nine angular velocities, three for each planetary gear train. These are two types of constraints on these nine angular velocities: structural constraints and clutch constraints. Structural constraints are from the interconnections between planetary gear train elements or members in the transmission design. As can be found from the stick diagram (Figure 5.10), there are four structural connections: Carrier $C_1 \Leftrightarrow$ Ring R_2; Carrier $C_2 \Leftrightarrow$ Ring R_3; Sun $S_2 \Leftrightarrow$ Input; Ring $R_1 \Leftrightarrow$ Carrier $C_3 \Leftrightarrow$ Output. These four structural connections yield four constraint equations upon the related angular velocities, as follows:

$$\omega_{C1} = \omega_{R2};\ \omega_{C2} = \omega_{R3};\ \omega_{S2} = \omega_{in};\ \omega_{R1} = \omega_{C3} = \omega_{out}$$

These four constraints are then superimposed upon Eqs (5.5–5.7), resulting in a new set of three linear equations with only six independent angular velocities, including ω_{in} and ω_{out}:

$$\omega_{S1} + \beta_1\omega_{out} - (1 + \beta_1)\omega_{C1} = 0 \tag{5.8}$$

$$\omega_{in} + \beta_2\omega_{C1} - (1 + \beta_2)\omega_{C2} = 0 \tag{5.9}$$

$$\omega_{S3} + \beta_3\omega_{C2} - (1 + \beta_3)\omega_{out} = 0 \tag{5.10}$$

After the structural connections are considered, there are six independent angular velocities that are governed by the three equations above, which are characteristic of the transmission and are satisfied for all speeds. Among these six angular velocities, or six motions, one must be provided as the input, and the remaining five angular velocities, including the output, must be uniquely defined in kinematics. It is apparent that there are two equations or two constraints short for the unique determination of these five angular velocities and this shortage in constraints is filled up by the clutch engagement. As shown in the clutch table in Figure 5.10, two clutches are applied in each gear. Therefore, two additional constraints are provided by the clutch engagement in each gear and are combined with Eqs (5.8–5.10) for the unique solution of the angular velocities in terms of the input angular velocity.

5.3.1.1 Ratio Analysis

Ratio analysis concerns the determination of all gear ratios for a given transmission design and clutch table and is relatively straightforward. The first step is to write down the characteristic equations for the planetary gear trains in the transmission structure. The second step is to superimpose the structural constraints in the transmission upon the set of characteristic equations. These two steps have been completed above for the Ford FWD six-speed AT in the example. Now, the constraints from clutch engagement in each gear are combined with Eqs (5.8–5.10) for the determination of all gear ratios for the six forward gears and one reverse gear.

First gear: In first gear, clutches E and F are applied. Sun gear S_3 and carrier C_1 are both grounded to the housing since both clutch E and clutch F are reaction clutches. With $\omega_{S3} = 0$ and $\omega_{C1} = 0$, Eqs (5.8–5.10) form a linear equation system in terms of three unknown angular velocities ω_{S1}, ω_{C1}, and ω_{out} with ω_{in} given as:

$$\begin{aligned} \omega_{S1} + \beta_1\omega_{out} &= 0 \\ \omega_{in} - (1 + \beta_2)\omega_{C2} &= 0 \\ \beta_3\omega_{C2} - (1 + \beta_3)\omega_{out} &= 0 \end{aligned}$$

This equation system can be easily solved for angular velocities ω_{S1}, ω_{C2}, and ω_{out} in terms of ω_{in}, as well as the first gear ratio i_1 as follows:

$$\omega_{out} = \frac{\beta_3}{(1+\beta_2)(1+\beta_3)}\omega_{in}$$

$$\omega_{S1} = \frac{-\beta_1\beta_3}{(1+\beta_2)(1+\beta_3)}\omega_{in}$$

$$\omega_{C2} = \frac{1}{(1+\beta_2)}\omega_{in}$$

$$i_1 = \frac{\omega_{in}}{\omega_{out}} = \frac{(1+\beta_2)(1+\beta_3)}{\beta_3}$$

Second gear: In second gear, clutches E and C are applied. Sun gear S_3 and sun gear S_1 are both grounded to the housing since both clutch E and clutch C are reaction clutches. With $\omega_{S3} = 0$ and $\omega_{S1} = 0$, Eqs (5.8–5.10) form a linear equation system in terms of three unknown angular velocities ω_{C1}, ω_{C2}, and ω_{out} with ω_{in} given as:

$$\beta_1\omega_{out} - (1+\beta_1)\omega_{C1} = 0$$

$$\omega_{in} + \beta_2\omega_{C1} - (1+\beta_2)\omega_{C2} = 0$$

$$\beta_3\omega_{C2} - (1+\beta_3)\omega_{out} = 0$$

This equation system leads to the solution for the second gear ratio and angular velocities ω_{C1} and ω_{C2} in terms of ω_{in}:

$$i_2 = \frac{\omega_{in}}{\omega_{out}} = \frac{(1+\beta_2)(1+\beta_3)}{\beta_3} - \frac{\beta_1\beta_2}{1+\beta_1}$$

$$\omega_{C1} = \frac{\beta_1}{(1+\beta_1)}\frac{\omega_{in}}{i_2}$$

$$\omega_{C2} = \frac{(1+\beta_3)}{\beta_3}\frac{\omega_{in}}{i_2}$$

Third gear: In third gear, clutches E and B are applied. Sun gear S_3 and sun gear S_1 are grounded to the housing or connected to the input. With $\omega_{S3} = 0$ and $\omega_{S1} = \omega_{in}$, Eqs (5.8–5.10) form a linear equation system in terms of three unknown angular velocities ω_{C1}, ω_{C2}, and ω_{out} with ω_{in} given as:

$$\omega_{in} + \beta_1\omega_{out} - (1+\beta_1)\omega_{C1} = 0$$

$$\omega_{in} + \beta_2\omega_{C1} - (1+\beta_2)\omega_{C2} = 0$$

$$\omega_{C2} - (1+\beta_3)\omega_{out} = 0$$

This equation system leads to the solution for the third gear ratio and angular velocities ω_{C1} and ω_{C2} in terms of ω_{in}:

$$i_3 = \frac{\omega_{in}}{\omega_{out}} = \frac{(1+\beta_1)(1+\beta_2)(1+\beta_3) - \beta_1\beta_2\beta_3}{\beta_2\beta_3 + \beta_3(1+\beta_1)}$$

$$\omega_{C1} = \frac{(1+\beta_2)(1+\beta_3)}{\beta_2\beta_3}\frac{\omega_{in}}{i_3} - \frac{\omega_{in}}{\beta_2}$$

$$\omega_{C2} = \frac{(1+\beta_3)}{\beta_3}\frac{\omega_{in}}{i_3}$$

Fourth gear: In fourth gear, clutches E and A are applied. Sun gear S_3 is grounded to the housing and carrier C_1 is connected to the input. With $\omega_{S3} = 0$ and $\omega_{C1} = \omega_{in}$, Eqs (5.8–5.10) form a linear equation system in terms of three unknown angular velocities ω_{S1}, ω_{C2}, and ω_{out} with ω_{in} given as:

$$\omega_{S1} + \beta_1\omega_{out} - (1+\beta_1)\omega_{in} = 0$$
$$\omega_{in} + \beta_2\omega_{in} - (1+\beta_2)\omega_{C2} = 0$$
$$\beta_3\omega_{C2} - (1+\beta_3)\omega_{out} = 0$$

This equation system leads to the fourth gear ratio and angular velocities ω_{S1} and ω_{C2} in terms of ω_{in}:

$$i_4 = \frac{\omega_{in}}{\omega_{out}} = \frac{(1+\beta_3)}{\beta_3}$$
$$\omega_{C2} = \omega_{in}$$
$$\omega_{S1} = (1+\beta_1)\omega_{in} - \beta_1\frac{\omega_{in}}{i_4}$$

Fifth gear: In fifth gear, clutches B and A are applied, coupling sun gear S_1 and carrier C_1 to the input at the same time, and thus making PGT1 turn in unity. The three planetary gear trains turn as a whole body due to the structural connections, resulting in a direct drive with a gear ratio of one. This can also be demonstrated by the transmission characteristic equations. With $\omega_{S1} = \omega_{C1} = \omega_{in}$, it is from Eqs (5.8–5.10) that all angular velocities must be the same as the input angular velocity ω_{in}.

Sixth gear: In sixth gear, clutches C and A are applied, grounding sun gear S_1 to the housing and coupling carrier C_1 to the input. With $\omega_{S1} = 0$ and $\omega_{C1} = \omega_{in}$, the transmission characteristic equations (5.8–5.10) form a system of linear equations in terms of ω_{C2}, ω_{S3}, and ω_{out} with the input angular velocity ω_{in} as given:

$$\beta_1\omega_{out} - (1+\beta_1)\omega_{in} = 0$$
$$\omega_{in} + \beta_2\omega_{in} - (1+\beta_2)\omega_{C2} = 0$$
$$\omega_{S3} + \beta_3\omega_{C2} - (1+\beta_3)\omega_{out} = 0$$

Since sun S_1 is grounded, carrier C_1 is connected to the input and ring R_1 is the output, so the sixth gear of the transmission is an overdrive. The overdrive ratio and angular velocities ω_{C2} and ω_{S3} are determined by:

$$i_6 = \frac{\omega_{in}}{\omega_{out}} = \frac{\beta_1}{1+\beta_1}$$
$$\omega_{C2} = \omega_{in}$$
$$\omega_{S3} = -\beta_3\omega_{in} + (1+\beta_3)\frac{\omega_{in}}{i_6}$$

Reverse gear: In reverse gear, clutches B and D are applied, connecting sun gear S_1 to the input and grounding carrier C_1 to the housing. This is a typical reverse gear design used in the majority of automatic transmissions. With $\omega_{S1} = \omega_{in}$ and $\omega_{C1} = 0$, the transmission characteristic equations (5.8–5.10) form a system of linear equations in terms of ω_{C2}, ω_{S3}, and ω_{out} with the input angular velocity ω_{in} given as:

$$\omega_{in} + \beta_1 \omega_{out} = 0$$
$$\omega_{in} - (1+\beta_2)\omega_{C2} = 0$$
$$\omega_{S3} + \beta_3 \omega_{C2} - (1+\beta_3)\omega_{out} = 0$$

These equations lead to the solution of the reverse gear ratio and angular velocities ω_{C2} and ω_{S3} as:

$$i_R = \frac{\omega_{in}}{\omega_{out}} = -\beta_1$$
$$\omega_{C2} = \frac{1}{(1+\beta_2)}\omega_{in}$$
$$\omega_{S3} = (1+\beta_3)\frac{\omega_{in}}{i_R} - \frac{\beta_3}{(1+\beta_2)}\omega_{in}$$

In summary, plugging $\beta_1 = 2.8824$, $\beta_2 = 2.1707$, and $\beta_3 = 2.4146$ into the ratio formulae just derived, the transmission ratios are: $i_1 = 4.484$, $i_2 = 2.872$, $i_3 = 1.842$, $i_4 = 1.414$, $i_5 = 1.000$, $i_6 = 0.742$, $i_R = -2.882$. The minus sign for the reverse gear ratio means that the output angular velocity is opposite to that of forward gears.

5.3.1.2 Ratio Synthesis

The objective of ratio synthesis for automatic transmissions is to obtain the maximum number of transmission ratios with optimized ratio spread in the most compact structure with a clutch engagement sequence conducive to shift control. Ratio synthesis is implemented in three aspects: **Structural synthesis**, **ratio formulation**, and **clutch sequencing**. In the following, the Ford FWD six-speed AT, as analysed, is used as an example for transmission ratio synthesis. The techniques and methods used in the example can be extended to all other transmissions with multiple ratios.

Structural synthesis: Automatic transmissions are designed in various configurations based on a series of planetary gear trains. As mentioned previously, three PGTs types, simple PGT, dual-plane PGT, and Ravigneaux PGT, are commonly used in automatic transmission structures. The structural synthesis is the first step in ratio synthesis and focuses mainly on the number of PGTs, the structural connections between PGT elements, the number of clutches, and the element or assembly to which each clutch is attached. Although structural synthesis is a trial and error practice by nature, certain guidelines can be followed to make the practice less time consuming and more productive. In general, transmissions with six speeds can be designed with three PGTs, and the design of transmissions with eight or more speeds require three or four PGTs. The use of dual-planet PGTs in the transmission structure may provide more flexibility in the ratio formulae that generate more useful ratios.

The key to structural synthesis is the structural interconnections between PGT elements or members. In general, there are six possible structural connections between two members in two individual PGTs:

Ring Gear ⇔ Carrier; Sun Gear ⇔ Ring Gear; Sun Gear ⇔ Sun Gear; Carrier ⇔ Carrier;
Ring Gear ⇔ Ring Gear; Sun Gear ⇔ Carrier

Note that planet gears are only internal PGT members that cannot be used for either input or output or reaction. In addition to the structural connections between PGT members, one PGT member or an assembly containing a PGT member must be structurally connected to the transmission output to avoid the coupling of a PGT member or assembly through a clutch to the vehicle inertia. Transmission input can also be structurally connected to a PGT member or assembly. But the connection between PGT members must not cause interference in the structural layout and must facilitate the installment of the related reaction and coupling clutches. To visualize the structural connections, the planetary gear trains to be used in the transmission can be drawn first in the stick diagram independently, i.e. without any structural connections. Possible structural connections can then be drawn in the stick diagram between PGT members or between a PGT member and the input or the output. The following tips should be useful in the process of structural synthesis:

- The total number of angular velocities or motions, N_ω, is equal to $3n$, with n as the number of PGTs. Since these N_ω angular velocities must satisfy the n characteristic equations in kinematics, the number of independent angular velocities, N_I, before any structural constraints are imposed, is equal to $(N_\omega - n)$.
- A structural connection between a PGT member or assembly and the input is counted as a structural constraint without reducing the number of the independent angular velocities, but the structural connection between a PGT member or assembly and the output is not counted as a structural constraint.
- Each structural interconnection between two PGTs is counted as a structural constraint and reduces the number of the independent angular velocities by one.
- The total number of structural constraints, N_{SC}, is equal to the sum of the number of structural interconnections between two PGTs, N_{PP}, and the number of structural connections between a PGT member or assembly with the input, N_{PI}, i.e. $N_{SC} = N_{PP} + N_{PI}$.
- The number of additional constraints from clutch engagement, or the number of clutches, N_C, to be applied in each gear, is equal to the total number of independent angular velocities, N_I, minus the total number of structural constraints, i.e. $N_C = N_I - N_{SC}$.
- The number of angular velocities available for ratio synthesis, N_{RS}, is equal to the difference between the total number of angular velocities, N_ω, and the number of structural interconnections between two PGTs, N_{PP}, minus one. That is, $N_{RS} = (N_\omega - N_{PP}) - 1$. This is because one of the angular velocities has to be the output angular velocity structurally.
- The number of possible transmission ratios, N_R, is then equal to the combinations of the number of clutches, N_C, out of the number of angular velocities available for ratio synthesis, N_{RS}, i.e. $N_R = C_{N_{RS}}^{N_C}$.
- Some of the possible power flows are useless for gear ratios and are eliminated in the rest of the ratio synthesis process.

The number of possible structural connections depends on the number of gears to be designed for the transmission. If an automatic transmission with six forward speeds and one reverse, such as the Ford FWD six-speed AT, is to be designed with three PGTs, the structural connections can be established in the following logic. Firstly, there are nine angular velocities if the three PGTs are independent of each other, i.e. if no structural connection exists, and these nine angular velocities are only constrained by the three characteristic equations, one for each PGT. Each structural connection between two PGTs will reduce the number of angular velocities by one. For example, if three such structural connections are imposed upon the three PGTs, as is the case in the Ford FWD six-speed AT, the number of angular velocities would then be reduced to six. The other structural connection between the input and sun gear S_2 does not reduce the number of angular velocities. Secondly, these six angular velocities must still satisfy the three characteristic equations with one as the given input, so two additional constraints, from two clutches separately, must be placed in each gear upon the other five angular velocities for their unique determination. Since one of the six angular velocities must be chosen as the output structurally, the number of angular velocities available for ratio synthesis is five. Therefore the number of possible power flow paths or gear ratios is equal to the number of combinations of two out of five, i.e. $C_5^2 = 10$. Since the number of possible gear ratios is larger than the required number of gear ratios, that is seven for all forward and reverse speeds, for the AT to be designed, the structural synthesis can be continued on the three PGTs with four structural connections. There may be multiple configurations based on the three PGTs with four structural connections, some of them may not be applicable for transmission designs because of the structural complexity, and others are useless gear ratio values or unacceptable clutch sequence. These useless configurations can often be eliminated by simple observation. The Ford FWD six-speed AT shown in Figure 5.10 is based on one of the useful configurations and should be the one that offers the best ratio spread, structural compactness, and clutch sequence. It is interesting to note that three of the ten possible power flows, corresponding to clutch combinations AD, BC, and CD, are useless since they result in transmission lock-up. All the other seven clutch combinations are used for the six forward speeds and one reverse.

Consider now that four structural interconnections between two PGTs are imposed on the three planetary gear trains. This would reduce the number of angular velocities from nine to five. The remaining five angular velocities must still satisfy the three characteristic equations, so two clutches must be applied to provide the two additional constraints. One of five angular velocities is used as the output structurally and the number of angular velocities available for ratio synthesis is equal to four. Therefore, the number of possible power flow paths or gear ratios is equal to the number of combinations of 2 out of 4, i.e. $C_4^2 = 6$. This will not work since the number of possible gear ratios is less than the number of gear ratios required in the six-speed transmission. Ratio synthesis cannot be continued on the three PGTs with four such structural connections.

These guidelines can also be applied for the structural synthesis of automatic transmissions with more than six forward speeds and should make the trial and error process more effective.

5.3.1.3 Ratio Formulation

Once a configuration is chosen from the possible combinations in the structural synthesis discussed, the ratio formulae can then be derived in terms of the PGT parameters based on the characteristic equations and the clutch constraints, as demonstrated in the ratio analysis, in which the Ford FWD six-speed AT was used as the example. If a simple PGT in the structural configuration is replaced by a dual-planet PGT, the corresponding PGT parameter then becomes negative in the ratio formulae. Using the ratio formulae derived for all speeds, the values of these PGT parameters can be optimized in a certain range in terms of ratio values and ratio spread. Once these PGT parameters are selected, the ratio values for all forward speeds are then calculated and arranged in a sequence from the highest to the lowest, i.e. from the first gear to the highest gear. The corresponding clutch combinations are also arranged in a sequence of the same order.

5.3.1.4 Clutch Sequencing

The principle in clutch sequencing is that all transmission shifts shall only involve two clutches, one oncoming and one off-going. This requires the clutch combination sequence in a specific pattern, as shown in the clutch table for the Ford FWD six-speed AT (Figure 5.10). The sequence of clutch combinations is mainly affected by the transmission structure, i.e. the result of structural syntheses, but is also affected to a lesser degree by the selection of planetary gear train parameters. It is possible that some clutch combinations fall out at the right sequence for some initial selections of planetary gear train parameters. By modifying the selected PGT parameters, it is also possible to vary the ratio values so that the clutch combinations will be in the right sequence that makes all shifts with only two clutches involved.

5.3.2 Ford six-speed RWD Ravigneaux AT

Since a Ravigneaux PGT is structurally a combination of a simple PGT and a dual-planet PGT with shared carrier and ring gear, automatic transmissions based on Ravigneaux PGTs are generally advantageous in terms of compactness. The procedure for the ratio analysis and synthesis of this type of automatic transmission is similar in nature to that for ATs with only simple PGTs. As shown in Figure 5.11, the Ford six-speed RWD AT is designed structurally with a combination of a simple PGT and a Ravigneaux PGT. Three characteristic equations can be written for these two PGTs:

$$\omega_{S1} + \beta_1 \omega_{R1} - (1 + \beta_1)\omega_{C1} = 0 \tag{5.11}$$

$$\omega_{S2} + \beta_2 \omega_{R2} - (1 + \beta_2)\omega_{C2} = 0 \tag{5.12}$$

$$\omega_{S3} - \beta_3 \omega_{R2} - (1 - \beta_3)\omega_{C2} = 0 \tag{5.13}$$

There are seven angular velocities, three for the simple PGT and four for the Ravigneaux PGT. The first equation, Eq. (5.11), is written for the simple PGT with β_1 as the PGT parameter, and the other two, Eqs (5.12) and (5.13), are written for the two mesh paths of the Ravigneaux PGT with PGT parameters β_2 and β_3 respectively. As can be seen from the stick diagram (Figure 5.11), sun gear S_1 is structurally fixed and ring gear R_1 is structurally connected to the input, providing two structural constraints in the transmission layout. The output member of the transmission is structurally the ring gear of the Ravigneaux PGT. Therefore, Eqs (5.11–5.13), become a new set of three linear equations with six independent angular velocities, including ω_{in} and ω_{out}:

$$\beta_1 \omega_{in} - (1 + \beta_1)\omega_{C1} = 0 \tag{5.14}$$

$$\omega_{S2} + \beta_2 \omega_{out} - (1 + \beta_2)\omega_{C2} = 0 \tag{5.15}$$

$$\omega_{S3} - \beta_3 \omega_{out} - (1 - \beta_3)\omega_{C2} = 0 \tag{5.16}$$

These three equations are characteristic of the transmission and must be satisfied in all the six forward gears and the one reverse gear. It is apparent that with the input angular velocity given, two additional conditions are needed to uniquely define the other five angular velocities. As shown in the clutch table in Figure 5.11, two clutches are applied in each gear. Therefore, the two additional constraints provided by the clutch engagement in each gear are combined with Eqs (5.14–5.16) for the unique solution of the angular velocities in terms of the input angular velocity.

5.3.2.1 Ratio Formulation

The ratio formulation for the Ford RWD six-speed AT is similar in nature to that for the Ford FWD six-speed AT. For example, clutch B and E are applied in the fifth gear, respectively coupling sun S_2 with carrier C_1 and connecting carrier C_2 to the input. The three equations, Eqs (5.14–5.16), then become a system of linear equations in terms of three unknowns, ω_{C1}, ω_{S3}, and ω_{out}, with ω_{in} given as:

$$\beta_1\omega_{in} - (1+\beta_1)\omega_{C1} = 0$$
$$\omega_{C1} + \beta_2\omega_{out} - (1+\beta_2)\omega_{in} = 0$$
$$\omega_{S3} - \beta_3\omega_{out} - (1-\beta_3)\omega_{in} = 0$$

This equation system directly leads to the fifth gear ratio formula in terms of the PGT parameters and the solutions of angular velocities, ω_{C1} and ω_{S3}:

$$i_5 = \frac{\omega_{in}}{\omega_{out}} = \frac{(1+\beta_2)\beta_2}{(1+\beta_2)(1+\beta_1)-\beta_1}$$

$$\omega_{C1} = (1+\beta_2)\omega_{in} - \frac{\beta_2}{i_5}\omega_{in}$$

$$\omega_{S3} = \frac{\beta_3\omega_{in}}{i_5} + (1-\beta_3)\omega_{in}$$

Similarly, the transmission ratio formulae for all speeds can be derived by combining the characteristic equations, Eqs (5.14–5.16), with the constraints from the respective clutch engagement shown in Figure 5. 11:

$$i_1 = \frac{(1+\beta_1)\beta_3}{\beta_1};\ i_2 = \frac{(1+\beta_1)(\beta_2+\beta_3)}{\beta_1(1+\beta_2)};\ i_3 = \frac{(1+\beta_1)}{\beta_1};\ i_4 = \frac{(1+\beta_1)\beta_3}{\beta_3(1+\beta_1)-1}$$
$$i_5 = \frac{(1+\beta_2)\beta_2}{(1+\beta_1)(1+\beta_2)-\beta_1};\ i_6 = \frac{\beta_2}{(1+\beta_2)};\ i_R = -\frac{(1+\beta_1)\beta_2}{\beta_1}$$

The three PGT parameters, $\beta_1 = 1.92$, $\beta_2 = 2.24$, and $\beta_3 = 2.74$, as shown in Figure 5.11, result in the six forward ratios and the reverse ratio listed in the last column of the clutch table. These three PGT parameters also result in the clutch sequence as shown. Note that, in the Ford RWD six-speed AT, there is no direct drive and the first PGT serve as a reduction unit for the first four speeds and the reverse speed.

For the sake of structural synthesis, it is interesting to compare the structures of the Ford RWD six-speed AT shown in Figure 5.11 and the Lexus RWD eight-speed AT shown in Figure 5.12. There is no significant difference in the planetary gear train layout between the two transmissions. Both use a Ravigneaux PGT, and in front of it there is a simple PGT in the Ford design and a dual-planet PGT in the Lexus design. The Ford design uses the simple PGT as a reduction unit with the carrier as the unit output for the first four speed and reverse speed, while the Lexus design uses the dual-planet PGT as a reduction unit with the ring gear as the unit output for the first five forward speeds. In a configuration that resembles the Ford design in PGT setup, the Lexus AT realizes eight forward speeds by adding one more multiple disk clutch in the transmission structure. The one-way clutch in the Lexus AT is redundant in kinematics and is only used in first gear to enable clutch to one-way clutch 1–2 upshift.

5.3.3 ZF RWD Eight-Speed AT

The ZF RWD eight-speed AT is designed with four simple planetary gear trains and five multiple disk clutches in the structure shown in Figure 5.14. The design has many advantages in comparison with other six-speed or eight-speed AT designs. Firstly, the ZF eight-speed AT uses only five clutches, while other eight-speed ATs, such as the Lexus eight-speed AT and the Hyundai eight-speed AT shown in Figures 5.12 and 5.13, use six clutches even without counting the one-way clutch. Secondly, among the five clutches in ZF eight-speed AT, three are applied and only two are open in all speeds, minimizing the parasitic clutch drag loss. Thirdly, due to the smaller number of clutches and other components, the ZF eight-speed AT excels in hardware cost and compactness. There are many possible structural layouts based on a setup consisting four planetary gear trains, the ZF eight-speed AT design should be one of the best and might be a unique layout that features the advantages mentioned.

For the four simple planetary gear trains in the ZF RWD eight-speed AT, there are four structural interconnections, as shown in Figure 5.14. In addition, the second carrier C_2 is structurally connected to the input shaft and the output of the transmission is structurally from carrier C_4. There are five constraints on the angular velocities of related PGT members from structural connections, including the constraint from the input structural connection. The structural connections and the corresponding constraints are as follows:

$$\left.\begin{array}{l} Input \Leftrightarrow C_2 \\ S_1 \Leftrightarrow S_2 \\ R_2 \Leftrightarrow S_3 \\ R_3 \Leftrightarrow S_4 \\ C_1 \Leftrightarrow R_4 \\ C_4 \Leftrightarrow Output \end{array}\right\} \Rightarrow \left\{\begin{array}{l} \omega_{C2} = \omega_{in} \\ \omega_{S1} = \omega_{S2} \\ \omega_{R2} = \omega_{S3} \\ \omega_{R3} = \omega_{S4} \\ \omega_{C1} = \omega_{R4} \\ \omega_{C4} = \omega_{out} \end{array}\right\}$$

There are 12 angular velocities and four characteristic equations for the four PGTs in the transmission. The transmission characteristic equations are formed by superimposing the constraints listed above upon the four PGT characteristic equations:

$$\omega_{S1} + \beta_1\omega_{R1} - (1+\beta_1)\omega_{C1} = 0 \tag{5.17}$$

$$\omega_{S1} + \beta_2\omega_{R2} - (1+\beta_2)\omega_{in} = 0 \tag{5.18}$$

$$\omega_{R2} + \beta_3\omega_{R3} - (1+\beta_3)\omega_{C3} = 0 \tag{5.19}$$

$$\omega_{R3} + \beta_3\omega_{C1} - (1+\beta_3)\omega_{out} = 0 \tag{5.20}$$

There are eight independent angular velocities, including the input angular velocity ω_{in}, that are governed by these four transmission characteristic equations. With the input angular velocity provided to the transmission, the remaining seven angular velocities must be uniquely determinant in kinematics for each gear. Therefore, three of the five clutches must be applied to provide the three additional constraints required. The number of clutch combinations is then equal to the combination of three in five, i.e. $C_5^3 = 10$. One of these ten clutch combinations, AEC, is useless since it locks up the transmission. The remaining nine clutch combinations provide eight forward speed and one reverse ratios, as listed sequentially in the clutch table in Figure 5.14. The constraints on the angular velocities in each gear are listed in the following table in accordance with the clutch engagement:

Clutch constraints

Gear	Clutches applied	Constraints on angular velocities	Ratio
1st	A, B, C	$\omega_{S1} = 0$; $\omega_{R1} = 0$; $\omega_{R3} = \omega_{in}$	i_1
2nd	A, B, E	$\omega_{S1} = 0$; $\omega_{R1} = 0$; $\omega_{R3} = \omega_{R2}$	i_2
3rd	B, C, E	$\omega_{R1} = 0$; $\omega_{R3} = \omega_{in}$; $\omega_{R3} = \omega_{R2}$	i_3
4th	B, D, E	$\omega_{R1} = 0$; $\omega_{C3} = \omega_{out}$; $\omega_{R3} = \omega_{R2}$	i_4
5th	B, C, D	$\omega_{R1} = 0$; $\omega_{R3} = \omega_{in}$; $\omega_{C3} = \omega_{out}$	i_5
6th	C, D, E	$\omega_{R3} = \omega_{in}$; $\omega_{C3} = \omega_{out}$; $\omega_{R3} = \omega_{R2}$	i_6
7th	A, C, D	$\omega_{S1} = 0$; $\omega_{R3} = \omega_{in}$; $\omega_{R3} = \omega_{out}$	i_7
8th	A, D, E	$\omega_{S1} = 0$; $\omega_{C3} = \omega_{out}$; $\omega_{R3} = \omega_{R2}$	i_8
Rev	A, B, D	$\omega_{S1} = 0$; $\omega_{R1} = 0$; $\omega_{C3} = \omega_{out}$	i_R

All sequential upshifts or downshifts in the ZF eight-speed AT are the so-called direct shift, or clutch to clutch shift, with one oncoming and another off-going. In addition to sequential shifts, the transmissions can also make direct non-sequential shifts, should vehicle operations deem them necessary in terms of fuel economy or performance. With upshifts and downshifts counted separately, there are a total of 36 direct shifts, sequential and non-sequential combined, as shown in Figure 5.16. Note that other non-sequential shifts are possible in kinematics, but it would then involve three clutches during shift processes, making shift control much more technically challenging. The 36 available directs should be sufficient for the selections of gears that best fit the driver's intention and vehicle operation conditions.

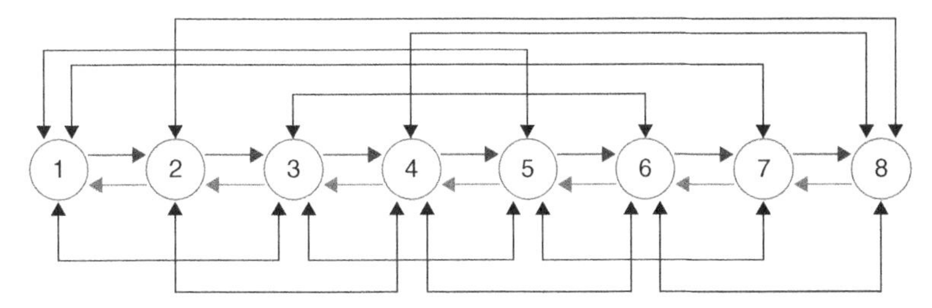

Figure 5.16 Direct shifts in ZF RWD eight-speed automatic transmission.

5.3.3.1 Ratio Formulation

Similar to the two previous examples, the ratio formulation for the ZF Ford RWD eight-speed AT can be derived using the combination of the characteristic equations from Eqs (5.17–5.20) and the clutch constraints on the related angular velocities in each gear. For example, the fifth gear ratio is determined in terms of the planetary parameters by plugging in the clutch constraints of the fifth gear, namely, $\omega_{R1} = 0$; $\omega_{C3} = \omega_{out}$; and $\omega_{R3} = \omega_{in}$, in Eqs (5.17–5.20):

$$\begin{aligned}
\omega_{S1} - (1+\beta_1)\omega_{C1} &= 0 \\
\omega_{S1} + \beta_2\omega_{R2} - (1+\beta_2)\omega_{in} &= 0 \\
\omega_{R2} + \beta_3\omega_{in} - (1+\beta_3)\omega_{out} &= 0 \\
\omega_{in} + \beta_3\omega_{C1} - (1+\beta_3)\omega_{out} &= 0
\end{aligned}$$

With ω_{in} given, the other four angular velocities, ω_{S1}, ω_{C1}, ω_{R2}, and ω_{out} are solved from the linear equation system above and the fifth gear ratio is formulated in terms of the four planetary parameters β_1, β_2, β_3, and β_4 by using:

$$i_5 = \frac{\omega_{in}}{\omega_{out}} = \frac{(1+\beta_1)(1+\beta_4) + \beta_2\beta_3(1+\beta_3)}{(1+\beta_1) + \beta_4(1+\beta_2) + \beta_2\beta_3\beta_4}$$

All other gear ratios can be formulated in a similar fashion in terms of the planetary parameters:

$$\begin{aligned}
& i_1 = 1+\beta_4;\, i_2 = \frac{\beta_2(1+\beta_4)}{(1+\beta_2)};\, i_3 = \frac{(1+\beta_1)(1+\beta_4)}{1+\beta_1+\beta_4};\, i_4 = \frac{(1+\beta_1+\beta_2)}{(1+\beta_2)} \\
& i_5 = \frac{(1+\beta_1)(1+\beta_4) + \beta_2\beta_3(1+\beta_3)}{(1+\beta_1) + \beta_4(1+\beta_2) + \beta_2\beta_3\beta_4};\, i_6 = 1;\, i_7 = \frac{\beta_2(1+\beta_3)}{(1+\beta_2+\beta_2\beta_3)};\, i_8 = \frac{\beta_2}{(1+\beta_2)} \\
& i_R = \frac{\beta_2(1-\beta_3\beta_4)}{1+\beta_2}
\end{aligned}$$

It can be observed from these formulations that the sixth gear is a direct drive, and the 7th and 8th gears are overdrives. The reverse gear ratio is negative since the term in the parenthesis is negative. With the four planetary parameters given as $\beta_1 = 2.0$, $\beta_2 = 2.0$, $\beta_3 = 1.61$, and $\beta_4 = 3.696$, the eight forward gear ratios and the one reverse ratio are calculated by the formulae above and listed in the clutch table as shown in Figure 5.14.

5.4 Transmission Dynamics

As a key component in the vehicle drive line, transmission dynamics is crucial to powertrain system responses during gear shifts. Transmission shift control is based on the understanding of transmission dynamics and quantification of transient variables during shift operations. This section presents a systematic approach to the modeling and analysis of transmission dynamics under fixed gear operations and shifting processes. The focus of the section will be on the derivation of the state variable equations of the vehicle systems and how these equations are solved for various vehicle operating statuses. Starting from the free body diagrams of powertrain components, the section shows how the equations of motion in the component level are integrated to form the vehicle state variable equation system. Under static conditions, this equation system will determine the torque of the involved clutches and the transmission output in terms of the input torque. During transmission shifts, this equation system is used to quantitatively analyse transient variables, such as speed and torque of the involved clutches, transmission output torque variations, shift time, and other variables that affect transmission shifting performance. Note that this section only concerns the rotational dynamics of the powertrain system so that the equations of motion involve only torque and angular velocity. In addition, transmission components are modeled as lumped masses with mass moments of inertia. The approach in the section is demonstrated with three case studies using the same three example transmissions as in Section 5.3.

As previously mentioned, automatic transmissions are designed with three types of planetary gear trains: simple PGT, dual-planet PGT, and Ravigneaux PGT. As discussed in Section 3.8, the relationship between the magnitudes of the torque on the sun, carrier, and ring is characteristic of a planetary gear train, which is repeated here for the three types of PGTs.

Simple planetary gear train:

$$T_R = \beta T_S \tag{5.21}$$

$$T_C = (1+\beta)T_S \tag{5.22}$$

Here the torque on the sun gear and the torque on the ring gear are in the same direction. Clearly the torque on the carrier is the algebraic sum of the torque on the sun gear and the torque on the ring gear for equilibrium.

Dual-planet planetary gear train:

$$T_R = \beta T_S \tag{5.23}$$

$$T_C = (\beta-1)T_S \tag{5.24}$$

Here the torque on the ring gear is opposite in direction to the torque on the sun gear, and the torque on the carrier is opposite in direction to the torque on the ring gear since β is larger than 1.

Ravigneaux planetary gear train:

$$T_{R1} = \beta_1 T_{S1} \tag{5.25}$$

$$T_{C1} = (1+\beta_1)T_{S1} \tag{5.26}$$

$$T_{R2} = \beta_2 T_{S2} \tag{5.27}$$

$$T_{C2} = (\beta_2-1)T_{S2} \tag{5.28}$$

Here the Ravigneaux PGT is decomposed into two PGTs, the first two equations are for the simple PGT and the second two equations are for the dual planet PGT, as shown in Section 3.8.

5.4.1 Ford FWD Six-Speed AT

Free body diagrams: The ratio change portion of the Ford FWD six-speed automatic transmission illustrated Figure 5.10 contains six subassemblies: input shaft assembly, sun gear S_1 assembly, output assembly, carrier C_1 – ring gear R_2 assembly, carrier

C_2 – ring gear R_3 assembly, and sun gear S_3 assembly, as shown in Figure 5.17. The gear ratio from the gear pairs after the output assembly is the constant final drive ratio.

The input torque T_{in} and the output torque T_{out}, as well as the reaction torques, T_C, T_D, T_E, and T_F, which are applied by the four reaction clutches to respective transmission subassemblies, are shown in the free body diagram of the transmission ratio change portion, as demonstrated in Figure 5.17. The directions of the angular velocity and the torque on the input and output are known for a given transmission. The directions of reaction torques are assumed to be the same as the input torque in the free body diagram. The free body diagram for each of the six subassemblies are shown separately in Figure 5.18.

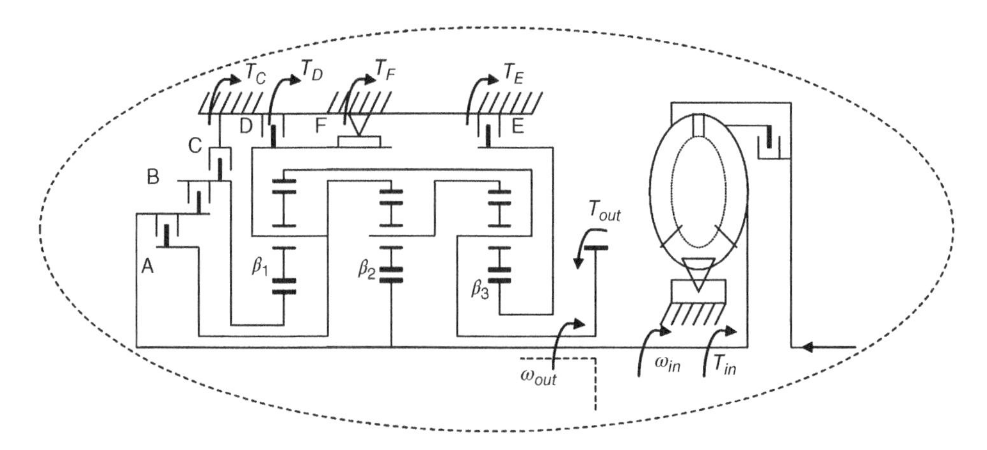

Figure 5.17 Ratio change portion of Ford FWD six-speed AT.

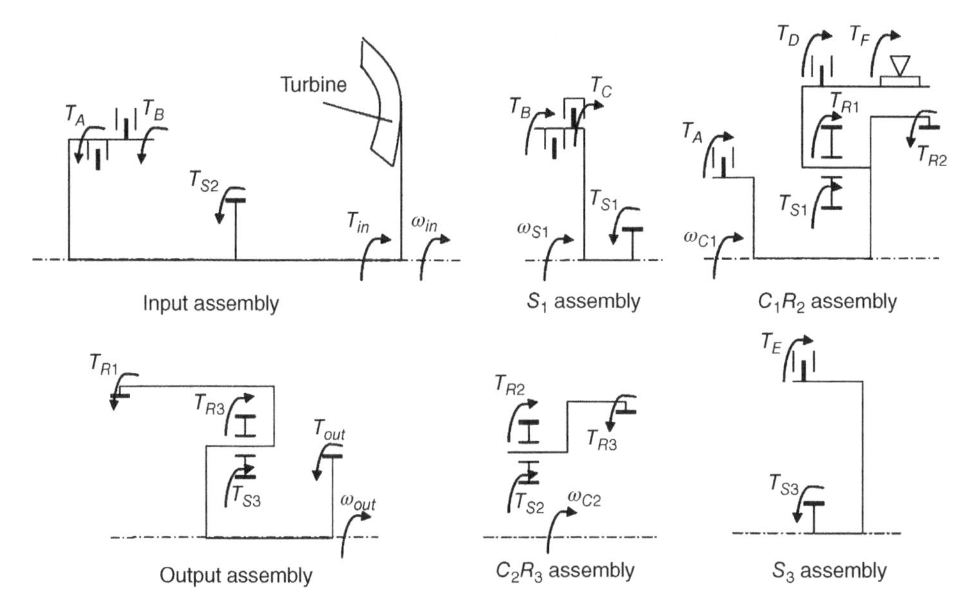

Figure 5.18 Free body diagram of the subassemblies of Ford FWD six-speed AT.

While drawing the free body diagrams (FBD) for the subassemblies, the directions of the internal torques, including the torques applied by the coupling clutches and gear torques, are not known and must be assumed. Once the direction of an internal torque is assumed in an FBD, it must be consistent for all other FBDs. For example, torque T_A applied by clutch A to the input assembly is an internal torque and is assumed to be counter clockwise in the FBD of the input assembly. When applied, clutch A couples the C_1R_2 assembly with the input assembly. Therefore, the torque applied by clutch A to the C_1R_2 assembly is opposite in direction to the torque applied by clutch A to the input assembly. The direction of torque T_A is thus clockwise in the FBD of the C_1R_2 assembly. The directions of gear torques with unknown directions are treated similarly. For example, T_{S1} is the torque applied to sun gear S_1 of the S_1 assembly and is assumed to be counter clockwise in the FBD. Sun gear S_1 meshes with the planet gear in the C_1R_2 assembly. The torque applied on the planet gear in the C_1R_2 assembly by the sun gear in the S_1 assembly has the same magnitude but is opposite in direction to the torque applied by the planet gear in the C_1R_2 assembly to the sun gear in the S_1 assembly. The direction of T_{S1} is thus clockwise in the FBD of the C_1R_2 assembly. A planet gear is supported on a carrier and is considered as a part of the assembly that contains the carrier in the FBD. While drawing the FBD of an assembly containing a carrier, the ring gear torque and the sun gear torque are shown rather than the carrier torque directly. The torque on the carrier is of course the algebraic sum of the ring gear torque and the sun gear torque. In addition, it is worthwhile to note that a simple planet gear train, the ring gear torque, and the sun gear torque are in the same direction, as can be observed in the free body diagrams in Figure 5.18.

Equations of motion: After the FBDs for all subassemblies are correctly drawn, the equations of motion can then be written easily based on the Newton's second law. These equations are:

$$T_{in} - T_{S2} - T_A - T_B = I_{in}\dot{\omega}_{in} \tag{5.29}$$

$$T_C + T_B - T_{S1} = I_{S1}\dot{\omega}_{S1} \tag{5.30}$$

$$T_A + T_D + T_F + (T_{S1} + T_{R1}) - T_{R2} = I_{C1R2}\dot{\omega}_{C1} \tag{5.31}$$

$$(T_{S3} + T_{R3}) - T_{R1} - T_{out} = I_{out}\dot{\omega}_{out} \tag{5.32}$$

$$(T_{S2} + T_{R2}) - T_{R3} = I_{C2R3}\dot{\omega}_{C2} \tag{5.33}$$

$$T_E - T_{S3} = I_{S3}\dot{\omega}_{S3} \tag{5.34}$$

where the I terms before the angular accelerations are the mass moments of inertia of the respective assemblies. In the study of transmission dynamics, the output torque and clutch torques are the key variables affecting shifting performances. These equations can be streamlined by eliminating the gear torques by various substitutions. Using Eqs (5.29), (5.30), and (5.34), the sun gear torques can be directly represented as:

$$T_{S2} = -I_{in}\dot{\omega}_{in} + T_{in} - T_A - T_B \tag{5.35}$$

$$T_{S1} = -I_{S1}\dot{\omega}_{S1} + T_C + T_B \tag{5.36}$$

$$T_{S3} = -I_{S3}\dot{\omega}_{S3} + T_E \tag{5.37}$$

Plugging T_{S1}, T_{S2}, and T_{S3} expressed by these three equations into Eqs (5.31), (5.32) and (5.33) and noting that $T_R = \beta T_S$, the system equation of transmission dynamics can then be represented with only the torques of interest, namely the output torque and clutch torques remaining in the equation:

$$T_A + T_D + T_F + (1+\beta_1)(-I_{S1}\dot{\omega}_{S1} + T_C + T_B) - \beta_2(-I_{in}\dot{\omega}_{in} + T_{in} - T_A - T_B) = I_{C1R2}\dot{\omega}_{C1} \tag{5.38}$$

$$(1+\beta_3)(-I_{S3}\dot{\omega}_{S3} + T_E) - \beta_1(-I_{S1}\,\omega_i + T_C + T_B) - T_{out} = I_{out}\dot{\omega}_{out} \tag{5.39}$$

$$(1+\beta_2)(-I_{in}\dot{\omega}_{in} + T_{in} - T_A - T_B) - \beta_3(-I_{S3}\dot{\omega}_{S3} + T_E) = I_{C2R3}\dot{\omega}_{C2} \tag{5.40}$$

Note that Eqs (5.35–5.37) are used to calculate the dynamic loads on the gears once other variables are solved from system dynamics. These equations can be further rearranged as follows and will be integrated into the vehicle powertrain system for the modeling and analysis of the vehicle system dynamics, as will be discussed later.

$$\begin{aligned}&\beta_2 I_{in}\dot{\omega}_{in} - (1+\beta_1)I_{S1}\dot{\omega}_{S1} - I_{C1R2}\dot{\omega}_{C1} - \beta_2 T_{in} + (1+\beta_2)T_A + (1+\beta_1+\beta_2)T_B +\\ &\qquad (1+\beta_1)T_C + T_D + T_F = 0\end{aligned} \tag{5.41}$$

$$\beta_1 I_{S1}\dot{\omega}_{S1} - (1+\beta_3)I_{S3}\dot{\omega}_{S3} - I_{out}\dot{\omega}_{out} - \beta_1 T_B - \beta_1 T_C + (1+\beta_3)T_E - T_{out} = 0 \tag{5.42}$$

$$\begin{aligned}&-(1+\beta_2)I_{in}\dot{\omega}_{in} + \beta_3 I_{S3}\dot{\omega}_{S3} - I_{C2R3}\dot{\omega}_{C2} + (1+\beta_2)T_{in}\\ &\quad -(1+\beta_2)T_A - (1+\beta_2)T_B - \beta_3 T_E = 0\end{aligned} \tag{5.43}$$

Static torque magnitudes: In static terms, the angular accelerations in Eqs (5.41–5.43) are zero and all inertia terms drop out. So the input torque, the output torque, and the clutch torques are determined by the following three equations:

$$-\beta_2 T_{in} + (1+\beta_2)T_A + (1+\beta_1+\beta_2)T_B + (1+\beta_1)T_C + T_D + T_F = 0 \tag{5.44}$$

$$-\beta_1 T_B - \beta_1 T_C + (1+\beta_3)T_E - T_{out} = 0 \tag{5.45}$$

$$(1+\beta_2)T_{in} - (1+\beta_2)T_A - (1+\beta_2)T_B - \beta_3 T_E = 0 \tag{5.46}$$

For the Ford FWD six-speed AT with $\beta_1 = 2.8824$, $\beta_2 = 2.1707$, and $\beta_3 = 2.4146$, the three equations above are of the following form:

$$-2.1707T_{in} + 3.1707T_A + 6.0531T_B + 3.8824T_C + T_D + T_F = 0$$

$$-2.8824T_B - 2.8824T_C + 3.4146T_E - T_{out} = 0$$

$$3.1707T_{in} - 3.1707T_A - 3.1707T_B - 2.4145T_E = 0$$

In each of the six forward gears and the one reverse gear, only two clutches are applied and the other four clutches are open. Considering the input torque as given, the two clutch torques and the output torque are uniquely determined from the linear equation system formed by the three equations above. Note that it is possible to determine the clutch torques on static conditions by just using the Eqs (5.21) and (5.22) without using Eqs (5.41–5.43). For example, in first gear, $T_A = T_B = T_C = T_D = 0$ since clutches A, B, C, D are open. It can be observed from the input assembly FBD in Figure 5.18 that $T_{S2} = T_{in}$. From the C_1R_2 assembly FBD, we can see that $T_F = T_{R2} = \beta_2 T_{in}$. It can also be seen that since $T_E = T_{S3}$ from the S_3 assembly, then from the C_2R_3 assembly, $T_{S3} = {}^{T_{R3}}/_{\beta_3} = \dfrac{(1+\beta_2)T_{S2}}{\beta_3} = \dfrac{(1+\beta_2)T_{in}}{\beta_3}$. For transmissions with more sophisticated structures, such as transmissions using Ravigneaux PGTs, it is more convenient to derive the equations of motion based on the FBDs and solve these equations after dropping the inertia terms for the clutch torques of interest. It is a common practice to express the output torque and the clutch torques as multiples of the input torque. The clutch torques and output torque for the Ford FWD six-speed AT are tabulated here, with the input torque unitized. Note that if the clutch torque is negative, it means that the real direction of the torque is just opposite to what is assumed in the free body diagram. A box left blank in the following table indicates that the referred clutch is open.

Gear	Applied Clutches	T_A	T_B	T_C	T_D	T_E	T_F	T_{out}
1st	E, F					1.3131	2.1707	4.4838
2nd	E, C			0.5591		1.3131		2.8722
3rd	E, B		0.3586			0.8422		1.8422
4th	E, A	0.6846				0.4141		1.4141
5th	A, B	1.3464	−0.3464					1.00
6th	A, C	1.00		−0.2576				0.7424
Rev.	B, D		1.00		−3.8824			−2.8824

General state variable equation system: The motions, i.e. the angular accelerations, of the transmission subsystems, are governed by the equations of motion represented by Eqs (5.41–5.43). In addition, the angular accelerations must also satisfy the transmission characteristic equations represented by Eqs (5.8–5.10). The state variable equation system for the transmission is formed by combining the equations of motion and the transmission characteristic equations with the angular accelerations differentiated with respect to time:

$$\dot{\omega}_{S1} + \beta_1 \dot{\omega}_{out} - (1+\beta_1)\dot{\omega}_{C1} = 0 \tag{5.47}$$

$$\dot{\omega}_{in} + \beta_2 \dot{\omega}_{C1} - (1+\beta_2)\dot{\omega}_{C2} = 0 \tag{5.48}$$

$$\dot{\omega}_{S3} + \beta_3 \dot{\omega}_{C2} - (1+\beta_3)\dot{\omega}_{out} = 0 \tag{5.49}$$

$$\begin{aligned} &\beta_2 I_{in}\dot{\omega}_{in} - (1+\beta_1)I_{S1}\dot{\omega}_{S1} - I_{C1R2}\dot{\omega}_{C1} - \beta_2 T_{in} + (1+\beta_2)T_A + (1+\beta_1+\beta_2)T_B + \\ &(1+\beta_1)T_C + T_D + T_F = 0 \end{aligned} \tag{5.50}$$

$$\beta_1 I_{S1}\dot{\omega}_{S1} - (1+\beta_3)I_{S3}\dot{\omega}_{S3} - I_{out}\dot{\omega}_{out} - \beta_1 T_B - \beta_1 T_C + (1+\beta_3)T_E - T_{out} = 0 \tag{5.51}$$

$$\begin{aligned} &-(1+\beta_2)I_{in}\dot{\omega}_{in} + \beta_3 I_{S3}\dot{\omega}_{S3} - I_{C2R3}\dot{\omega}_{C2} + (1+\beta_2)T_{in} - (1+\beta_2)T_A \\ &-(1+\beta_2)T_B - \beta_3 T_E = 0 \end{aligned} \tag{5.52}$$

Specific state variable equation system: The general state variable equations Eqs (5.47–5.52) apply for all gears and for all vehicle operation conditions. However, note that the initial condition and clutch engagement constraints must be superimposed upon the general state variable equation system to form the specific state variable equation system for the unique solution of the state variables. For example, when the vehicle equipped with the Ford FWD six-speed AT runs at the first gear, two clutch constraints, $\dot{\omega}_{C1} = 0$ and $\dot{\omega}_{S3} = 0$, will be superimposed upon the general state variable equation system represented by Eqs (5.47–5.52). In addition, since clutches A, B, C, and D are open in first gear, $T_A = T_B = T_C = T_D = 0$. There are seven state variables remaining in the specific state variable equation system: $\dot{\omega}_{in}$, $\dot{\omega}_{S1}$, $\dot{\omega}_{C2}$, $\dot{\omega}_{out}$, T_{in}, T_E, T_F, T_{out}. As will be discussed in Section 5.7, the combination of the specific state variable equation system and the vehicle equations of motion will lead to the unique solution of all state variables, with the input torque T_{in} as a given variable.

5.4.2 Ford RWD Six-Speed AT

Free body diagrams: The Ford RWD six-speed automatic transmission illustrated in Figure 5.11 contains six subassemblies: input shaft assembly, carrier C_1 assembly, sun gear S_2 assembly, sun gear S_3 assembly, carrier C_2 assembly, and ring gear R_3 or output assembly. The free body diagram for the whole transmission is shown in Figure 5.19 below, with the reaction torques assumed to be in the same direction as the input torque.

The free body diagrams for the Ford RWD six-speed AT are drawn in a similar way as in the previous example involving the Ford FWD six-speed AT. While drawing the FBDs for the transmission subassemblies, the Ravigneaux PGT is treated as a combination of two separate PGTs, one a simple PGT and the other a dual-planet PGT, as discussed in Section 3.8. The six subassembly FBDs are shown in Figure 5.20.

As mentioned previously, the directions of the sun gear torque and the ring gear torque are the same in a simple PGT and are the opposite in a dual-planet PGT. This is clearly demonstrated in the FBD for the C_2 assembly in Figure 5.20. The carrier C_2 assembly is commonly shared by the simple PGT $S_2 - C_2 - R_2$ and the dual-planet PGT $S_3 - C_2 - R_3$. The ring gear in the Ravigneaux PGT is also commonly shared by the two separate PGTs and is subject to two ring gear torques, T_{R2} from the simple PGT and T_{R3} from the dual-planet PGT. In the FBD of the C_2 assembly, it is apparent that T_{R2} and T_{S2} are in the same direction, and T_{R3} and T_{S3} are in the opposite directions. It is pointed out here again that the magnitudes of sun gear torque and ring gear torque are related as $T_R = \beta T_S$.

Equations of motion and static torque magnitudes: Based on the FBDs in Figure 5.20, the equations of all six subassemblies are derived as:

$$T_{in} - T_E - T_{R1} = I_{in}\dot{\omega}_{in} \tag{5.53}$$

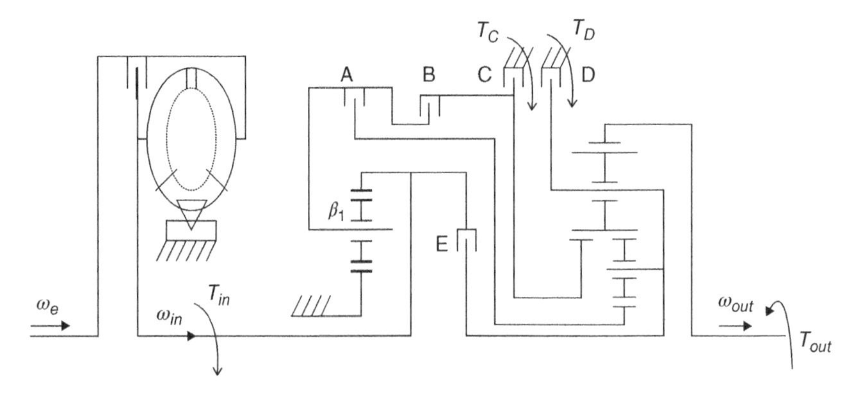

Figure 5.19 FBD of Ford RWD six-speed AT.

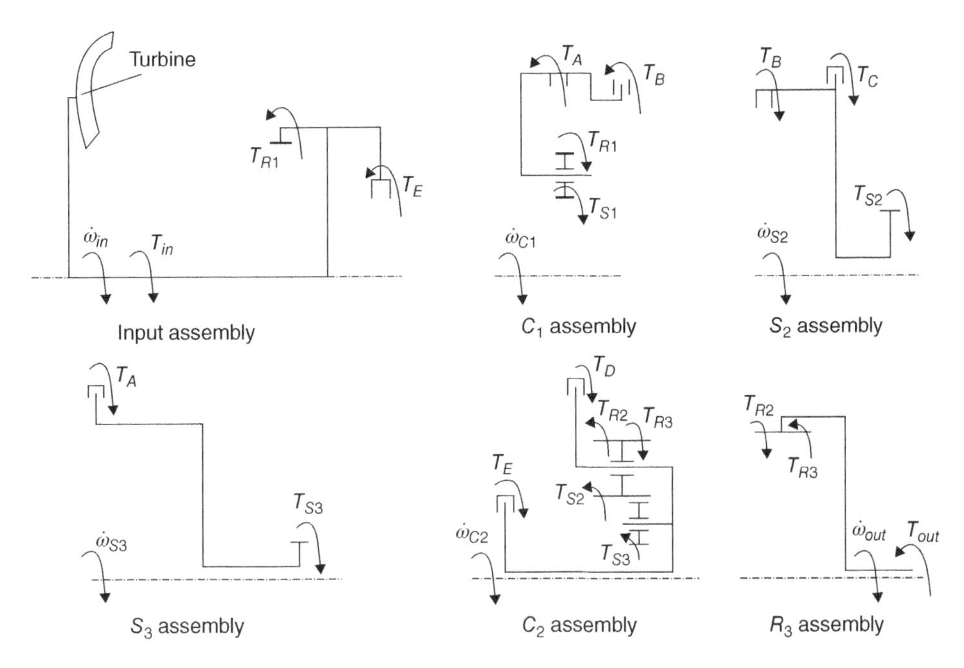

Figure 5.20 FBDs for the subassemblies of Ford RWD six-speed AT.

$$T_{S1} + T_{R1} - T_A - T_B = I_{C1}\dot{\omega}_{C1} \tag{5.54}$$

$$T_{S2} + T_C + T_B = I_{S2}\dot{\omega}_{S2} \tag{5.55}$$

$$T_{S3} + T_A = I_{S3}\dot{\omega}_{S3} \tag{5.56}$$

$$T_E + T_D + T_{R3} - T_{S3} - T_{S2} - T_{R2} = I_{C2}\dot{\omega}_{C2} \tag{5.57}$$

$$T_{R2} - T_{R3} - T_{out} = I_{R3}\dot{\omega}_{out} \tag{5.58}$$

Similar to the previous example, the six equations can be simplified by eliminating the internal gear torques by variable substitution. This can be done by first solving T_{R1}, T_{S2}, and T_{S3} from Eqs (5.53), (5.55) and (5.56) respectively and then plugging them into Eqs (5.54), (5.57) and (5.58). The remaining three equations can then be combined the transmission characteristic equations represented by Eqs (5.14–5.16) to form the general state variable equation system.

The magnitudes of static clutch torque and output torque are determined following the same steps as in the previous example. Firstly, all of the inertia terms are dropped in the general state variable equation system. Then torque magnitudes of open clutches in each gear are dropped from the remaining equations respectively. The torque magnitude of the applied clutches and the output torque are then solved in terms of a unity input torque. The static clutch torque and output torque for the Ford RWD six-speed AT are as tabulated:

Gear	Applied clutches	T_A	r_B	T_c	T_D	T_E	T_{out}
1st	A, D	1.521			2.646		4.168
2nd	A, C	1.521		0.817			2.338
3rd	A, B	0.989	0.531				1.520
4th	A, E	0.417				0.726	1.143
5th	B, E		−0.337			1.255	0.868
6th	C, E			−0.309		1.00	0.691
Rev.	B, D		1.521		−4.928		−3.928

5.4.3 ZF RWD Eight-Speed AT

As shown in Figure 5.14, the ZF RWD eight-speed AT contains eight subassemblies: input assembly, S_1S_2 assembly, R_1 assembly, C_4 or output assembly, C_1R_4 assembly, R_2S_3 assembly, C_3 assembly, and R_3S_4 assembly. The free body diagram of the whole transmission is shown in Figure 5.21, with the reaction torque from clutches A and B assumed to be in the same direction as the input torque. The input torque T_{in} is the sum of turbine torque T_T and the converter lock-up clutch torque T_L, i.e. $T_{in} = T_T + T_L$.

The FBDs for the eight subassemblies are drawn in Figure 5.22. Since all four planetary gear trains in the transmission are of the simple type, the sun gear torque and the ring gear torque shown on the planet gear are in the same direction Figure.

Equations of motion: The equations of motion for all of the eight subassemblies are as follows based on Newton's second law corresponding to the respective FBD:

$$T_{in} - (T_{S2} + T_{R2}) - T_C = I_{in}\dot{\omega}_{in} \tag{5.59}$$

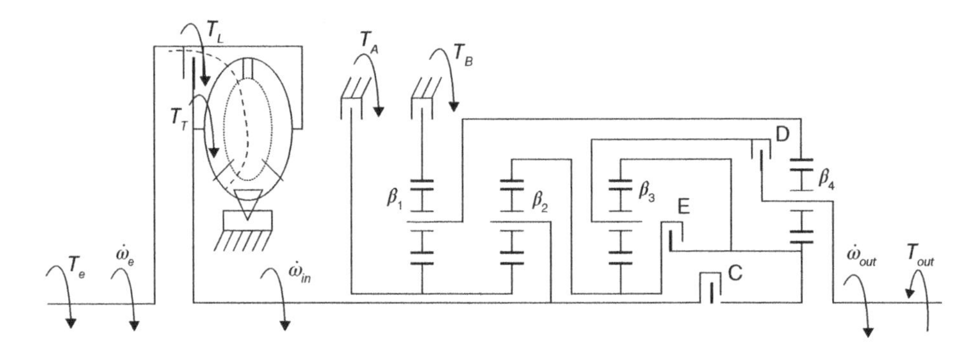

Figure 5.21 FBD for ZF RWD eight-speed AT.

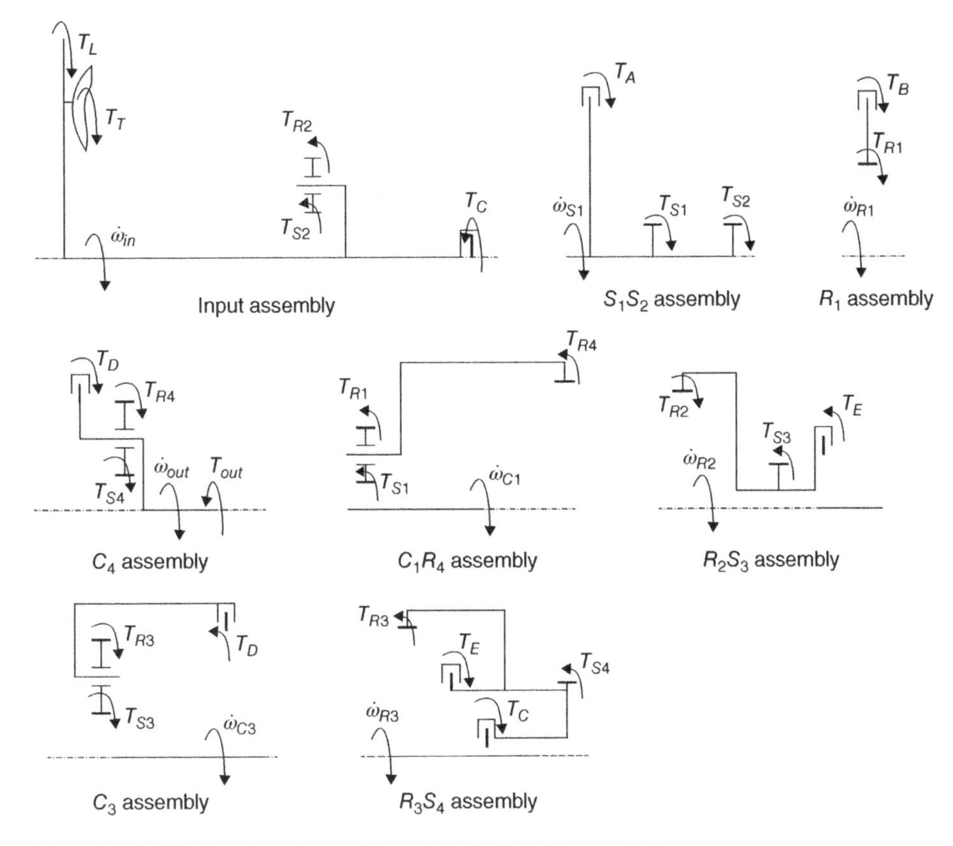

Figure 5.22 FBDs for the subassemblies of ZF RWD eight-speed AT.

$$T_A + T_{S1} + T_{S2} - = I_{S1S2}\dot{\omega}_{S1} \tag{5.60}$$

$$T_{R1} + T_B = I_{R1}\dot{\omega}_{R1} \tag{5.61}$$

$$(T_{S4} + T_{R4}) + T_D - T_{out} = I_{C4}\dot{\omega}_{out} \tag{5.62}$$

$$-(T_{S1} + T_{R1}) - T_{R4} = I_{C1R4}\dot{\omega}_{C1} \tag{5.63}$$

$$T_{R2} - T_{S3} - T_E = I_{R2S3}\dot{\omega}_{R2} \tag{5.64}$$

$$(T_{S3} + T_{R3}) - T_D = I_{C3}\dot{\omega}_{C3} \tag{5.65}$$

$$T_E + T_C - T_{R3} - T_{S4} - = I_{R3S4}\dot{\omega}_{R3} \tag{5.66}$$

where the ring gear torque and the sun gear torque are related as $T_{Ri} = \beta_i T_{Si}$ with $i = 1, 2, 3, 4$. The internal torques on the four sun gears can be solved from Eqs (5.59), (5.61), (5.62), and (5.65) in terms of other variables and substituted into the other four equations: Eqs (5.60), (5.63), (5.64), and (5.66). The remaining four independent equations of motion can then be represented in the following arrangement with the internal gear torques eliminated:

$$-\frac{1}{(1+\beta_2)}I_{in}\,\dot{\omega}_{in} - I_{S1S2}\dot{\omega}_{S1} + \frac{1}{\beta_1}I_{R1}\dot{\omega}_{R1} + T_A - \frac{1}{\beta_1}T_B - \frac{1}{(1+\beta_2)}T_C + \frac{1}{(1+\beta_2)}T_{in} = 0 \tag{5.67}$$

$$\frac{(1+\beta_1)}{\beta_1}I_{R1}\dot{\omega}_{R1}+I_{C1R4}\dot{\omega}_{C1}+\frac{\beta_4}{(1+\beta_4)}I_{C4}\dot{\omega}_{out}-\frac{(1+\beta_1)}{\beta_1}T_B-\frac{\beta_4}{(1+\beta_4)}T_D$$
$$+\frac{\beta_4}{(1+\beta_4)}T_{out}=0 \tag{5.68}$$

$$-\frac{(1+\beta_2)}{\beta_2}I_{in}\dot{\omega}_{in}-I_{R2S3}\dot{\omega}_{R2}-\frac{1}{(1+\beta_3)}I_{C3}\dot{\omega}_{C3}-\frac{(\beta_2)}{(1+\beta_2)}T_C$$
$$-\frac{1}{(1+\beta_3)}T_D-T_E+\frac{\beta_2}{(1+\beta_2)}T_{in}=0 \tag{5.69}$$

$$-I_{R3S4}\dot{\omega}_{R3}-\frac{\beta_3}{(1+\beta_3)}I_{C3}\dot{\omega}_{C3}-\frac{1}{(1+\beta_4)}I_{C4}\dot{\omega}_{out}+T_C$$
$$+\left[\frac{1}{(1+\beta_4)}-\frac{\beta_3}{(1+\beta_3)}\right]T_D+T_E-\frac{1}{(1+\beta_4)}T_{out}=0 \tag{5.70}$$

Static torque magnitudes: In static terms, the inertia terms drop out and Eqs (5.67–5.70) take the following form, as a linear equation system with the clutch torques and the output torque as unknowns:

$$T_A-\frac{1}{\beta_1}T_B-\frac{1}{(1+\beta_2)}T_C+\frac{1}{(1+\beta_2)}T_{in}=0$$
$$-\frac{(1+\beta_1)}{\beta_1}T_B-\frac{\beta_4}{(1+\beta_4)}T_D+\frac{\beta_4}{(1+\beta_4)}T_{out}=0$$
$$-\frac{(\beta_2)}{(1+\beta_2)}T_C-\frac{1}{(1+\beta_3)}T_D-T_E+\frac{\beta_2}{(1+\beta_2)}T_{in}=0$$
$$T_C+\left[\frac{1}{(1+\beta_4)}-\frac{\beta_3}{(1+\beta_3)}\right]T_D+T_E-\frac{1}{(1+\beta_4)}T_{out}=0$$

These equations apply for all eight forward speeds and reverse speed. With the input torque given, there are six torques to be determined, including five clutch torques and the output torque. In the ZF RWD eight-speed AT, three clutches are applied in each gear. The clutch torques of the two open clutches are equal to zero. Therefore, the three clutch torques and the output torque are uniquely determined by the four equations above with the respective clutch constraints in each gear. For example, in the fifth gear, $T_A = T_E = 0$ since clutches A and E are open. The four equations then take the following form for the ZF RWD eight-speed AT with $\beta_1 = 2.0$, $\beta_2 = 2.0$, $\beta_3 = 1.61$, and $\beta_4 = 3.696$.

$$T_A-0.5T_B-0.3333T_C=-0.3333T_{in}$$
$$-1.5T_B-0.787T_D+0.787T_{out}=0$$
$$-0.6667T_C-0.3831T_D-T_E=-0.6667T_{in}$$
$$T_C-0.4039T_D+T_E-0.2129T_{out}=0$$

The three clutch torques and the output torque are then directly solved from these four equations in terms of the input torque: $T_B = 0.282T_{in}$, $T_C = 0.573T_{in}$, $T_D = 0.743T_{in}$, and $T_{out} = 1.282T_{in}$. All other static clutch torques and the output torque are determined similarly and tabulated as shown in terms of unitary input torque.

	Engaged clutches	T_A	T_B	T_C	T_D	T_E	T_{out}
1st	A, B, C	1.232	2.464	1.0			4.697
2nd	A, B, E	0.488	1.643			0.667	3.131
3rd	B, C, E		1.104	-0.656		1.104	2.104
4th	B, D, E		0.667		0.396	0.515	1.667
5th	B, C, D		0.287	0.573	0.743		1.282
6th	C, D, E			1.0	1.0	-0.383	1.0
7th	A, C, D	-0.161		0.518	0.839		0.839
8th	A, D, E	-0.333			0.667	0.411	0.667
Rev	A, B, D	-1.655	-2.644		1.740		-3.30

General state variable equation system: The general state variable equation system is formed by combining the transmission characteristic equations systems represented by Eqs (5.17–5.20) differentiated with respect to time, and the equations of motion represented by Eqs (5.67–5.70), as shown, with the given β values.

$$\dot{\omega}_{S1} + 2.0\dot{\omega}_{R1} - 3.0\dot{\omega}_{C1} = 0 \tag{5.71}$$

$$\dot{\omega}_{S1} + 2.0\dot{\omega}_{R2} - 3.0\dot{\omega}_{in} = 0 \tag{5.72}$$

$$\dot{\omega}_{R2} + 1.61\dot{\omega}_{R3} - 2.61\dot{\omega}_{C3} = 0 \tag{5.73}$$

$$\dot{\omega}_{R3} + 3.696\dot{\omega}_{C1} - 4.696\dot{\omega}_{out} = 0 \tag{5.74}$$

$$-0.3333I_{in}\dot{\omega}_{in} - I_{S1S2}\dot{\omega}_{S1} + 0.5I_{R1}\dot{\omega}_{R1} + 0.3333T_{in} + T_A - 0.5T_B - 0.3333T_C = 0 \tag{5.75}$$

$$1.5I_{R1}\dot{\omega}_{R1} + I_{C1R4}\dot{\omega}_{C1} + 0.787I_{C4}\dot{\omega}_{out} - 1.5T_B - 0.787T_D + 0.787T_{out} = 0 \tag{5.76}$$

$$-0.6667I_{in}\dot{\omega}_{in} - I_{R2S3}\dot{\omega}_{R2} - 0.3831I_{C3}\dot{\omega}_{C3} + 0.6667T_{in} - 0.6667T_C - 0.3831T_D - T_E = 0 \tag{5.77}$$

$$-I_{R3S4}\dot{\omega}_{R3} - 0.6168I_{C3}\dot{\omega}_{C3} - 0.2129I_{C4}\dot{\omega}_{out} + T_C - 0.4039T_D + T_E - 0.2129T_{out} = 0 \tag{5.78}$$

As discussed in the previous example, the general state variable equation system is characteristic of the transmission and applies for all vehicle operations, including fixed gear operations and gear shifting processes. The clutch constraints in each operation are respectively superimposed on the general state variable equation system to form the specific state variable equation system for the unique determination of transmission system dynamic status, as will be detailed in Section 5.6.

5.5 Qualitative Analysis on Transmission Shifting Dynamics

Qualitative analysis on transmission shift dynamics provides an understanding of the trends of transient behavior of dynamic variables during shifts and their effects on shift quality [5]. In this section, the Ford FWD six-speed AT with the planetary parameters shown in Figure 5.10 will be used as the example for the qualitative analysis of shift dynamics. The general state variable equation system has been represented in Eqs (5.47–5.52). As mentioned previously, clutch constraints in each gear are superimposed upon these equations to define the transmission operation status. Without losing generality, the 1–2 shift process is used in this section as a case study for transmission shift dynamics during upshifts. For the convenience of analysis, the specific state variable equations will be derived in the following, respectively for the first gear operation, torque phase, and inertia phase in the 1–2 shift process.

First gear operation: In first gear, of the Ford FWD six-speed AT, clutches E and F are applied, resulting in two clutch constraints: $\dot{\omega}_{C1} = 0$ and $\dot{\omega}_{S3} = 0$. Superimposing these two constraints in Eqs (5.47–5.49), we can solve other angular accelerations in terms of the input angular acceleration, i.e., $\dot{\omega}_{C2} = 0.3154\dot{\omega}_{in}$, $\dot{\omega}_{S1} = -6428\dot{\omega}_{in}$, $\dot{\omega}_{out} = 0.2230\dot{\omega}_{in}$. These angular accelerations are then plugged into Eqs (5.50–5.52) for the solution of the clutch torques, T_E and T_F, and the output torque T_{out}, as follows:

$$T_F = 2.1707T_{in} - (2.1707I_{in} + 2.4956I_{S1})\dot{\omega}_{in} \tag{5.79}$$

$$T_E = 1.3131T_{in} - (1.3131I_{in} + 0.1306I_{C2R3})\dot{\omega}_{in} \tag{5.80}$$

$$T_{out} = 4.4837T_{in} - (4.4837I_{in} + 0.4459I_{C2R3} + 1.8528I_{S1} + 0.2230I_{out}))\dot{\omega}_{in} \tag{5.81}$$

These equations represent the clutch torque and the output torque in terms of the input torque and the input angular acceleration. The transmission output torque is directly related to the vehicle acceleration by the equation of motion of the vehicle:

$$\frac{W}{g}\frac{dV}{dt} = \frac{W}{g}\frac{\dot{\omega}_{out}r}{i_a} = \frac{\eta i_a T_{out}}{r} - R \tag{5.82}$$

where i_a is the final drive ratio, η is the final drive efficiency, r is the tire rolling radius, W is the vehicle weight, and R is the road load formulated by Eq. (1.14). It is apparent from Eq. (5.82) that the vehicle acceleration varies proportionally with the transmission output torque. If the transmission input torque T_{in} is given, the status of the vehicle dynamics and all of the state variables are uniquely defined by the combination of Eqs (5.79–5.82).

1–2 shift process: It must be understood that any transmission gear shift is a dynamic process during which the gear ratio is changed gradually. Let's consider the 1–2 shift process in the Ford FWD six-speed AT. This involves the releasing of clutch F and engaging of clutch C. As soon as the transmission enters the 1–2 shift process, the ongoing clutch C will be pressurized to apply a reaction torque to the transmission, and meanwhile, the torque in clutch F will decrease as clutch C is being pressurized. However, the off-going clutch F will stay applied until a certain point of time during the shift process. As long as the off-going clutch in this 1–2 upshift, or any upshift in general, stays closed, the first gear ratio, i.e. the current gear ratio, will remain unchanged, even though the oncoming clutch is being applied. This is because the clutch constraints on the transmission kinematics are still the same as in first gear, or current gear in any upshift, until the off-going clutch starts to slip. As the torque in the oncoming cutch increases and the torque in the off-going clutch decreases, the off-going clutch will start to slip at some point in the process. As soon as the off-going clutch starts to slip, the constraint placed by the off-going clutch on transmission kinematics is released and the transmission ratio starts to change toward the target gear ratio until the completion of the upshift. Based on this analysis, the 1–2 shift process, or any shift process, consists of two phases, torque phase and inertia phase, defined in general as follows.

Torque phase: The period from the time at which the oncoming clutch is pressured to the time at which the off-going clutch starts to slip is defined as the torque phase. During the torque phase, the transmission ratio remains to be the same as the current gear. Note that, during the torque phase, torque exists in both the off-gong clutch and the oncoming clutch. The torque in the oncoming clutch is a control variable that depends on the clutch pressure, and the torque in the off-going clutch is a state variable that depends on transmission dynamics.

Inertia phase: The period from the time at which the off-going clutch starts to slip to the completion of the shift process is defined as the inertia phase. It is during the inertia phase that the current transmission ratio is gradually changed to the target transmission ratio. If the off-going clutch in an upshift is a one-way clutch, such as the 1–2 shift in the Ford FWD six-speed AT, the off-going clutch torque is zero during the inertia phase. This is because once the one-way clutch starts to slip, it only exerts a negligible drag torque. If the off-going clutch is a regular clutch, the torque in the off-going clutch may exist after the inertia phase starts. Following the industry convention, a shift is termed a "clutch to clutch shift" if it involves two regular clutches, one off-going and one oncoming. If a shift involves a regular clutch and a one-way clutch, it is termed a "clutch to one-way clutch shift". For clutch to clutch shifts, both the off-going clutch torque and the oncoming clutch torque are control variables during the inertia phase.

Torque phase of 1–2 shift process: As mentioned in the definition of torque phase, the clutch constraints in the torque phase in the 1–2 upshift are the same as in first gear. However, clutch C is now being applied and clutch torque T_C in Eq. (5.50) is no longer equal to zero but rather a control variable. The relations for the angular accelerations are still the same as in first gear operation since the clutch constraints are the same. Eqs (5.50–5.52) lead to the solution of clutch torque, T_E and T_F, and the output torque T_{out}, as follows:

$$T_F = 2.1707T_{in} - (2.1707I_{in} + 2.4956I_{S1})\dot{\omega}_{in} - 3.8824T_C \tag{5.83}$$

$$T_E = 1.3131T_{in} - (1.3131I_{in} + 0.1306I_{C2R3})\dot{\omega}_{in} \tag{5.84}$$

$$T_{out} = 4.4837T_{in} - 2.8824T_C - (4.4837I_{in} + 0.4459I_{C2R3} + 1.8528I_{S1} + 0.2230I_{out})\dot{\omega}_{in} \tag{5.85}$$

During the torque phase, the gear ratio remains the same as the first gear and $\dot{\omega}_{in}$ is small in magnitude since it is coupled to the vehicle acceleration. The inertia terms in the Eqs (5.83–5.85) can be dropped without affecting the results in a qualitative analysis. As observed in Eq. (5.85), the torque applied by the oncoming clutch T_C reduces the output torque by 2.8824 times its magnitude. Meanwhile, the torque in the off-going clutch T_F also decreases as the oncoming clutch torque T_C increases and will become equal to zero when $T_C \approx 0.5591T_{in}$. Because the off-going clutch in the 1–2 shift is a one-way clutch, it will start to slip once its torque becomes equal to zero. The shift then enters the inertia phase. Furthermore, the torque in the one-way clutch will remain at zero until the completion of the 1–2 shift.

If the off-going clutch in the 1–2 upshift is not a one-way clutch but rather a regular clutch, such as clutch D shown in the stick diagram of the Ford FWD six-speed AT, then the slippage of the off-going clutch is not self-actuated but depends on the ATF pressure in the piston chamber of clutch D. This pressure defines the torque capacity of the off-going clutch D as indicated by Eq. (2.8). In this case, the torque in the off-going clutch,

T_D, is determined by the transmission system dynamics up to the point of slippage. The threshold of clutch slippage is at the point when the off-going clutch torque determined by the transmission system dynamics as a state variable is equal to the clutch torque capacity. As soon as the off-going clutch D slips, the 1–2 shift enters the inertia phase and the transmission ratio starts to change. However, the torque in clutch D is not equal to zero after the 1–2 shift enters the inertia phase but rather a control variable that depends on the clutch pressure. This residual torque must be brought to zero as quickly as possible to optimize the shift response and smoothness. If the ATF pressure in clutch D is not brought down quickly after the 1–2 shift is initiated, it can happen that even when the clutch torque T_D determined by transmission system dynamics is zero already, clutch D is still not slipping. This causes the phenomenon termed "clutch tie-up", which is the worst-case scenario and must definitely be avoided during any shift. To avoid clutch tie-up, the off-going clutch torque capacity must be controlled to drop quickly so that it will be smaller than the off-going clutch torque determined as a state variable before it becomes equal to zero.

Summarizing the analysis above, it can be concluded in general that the torque phase in upshifts begins as soon as the oncoming clutch is pressurized and ends at the point when the off-going starts to slip. The transmission ratio remains to as the current gear ratio and the input angular acceleration $\dot{\omega}_{in}$ is almost unchanged during the torque phase. The torque in the off-going clutch is a state variable that is determined by transmission system dynamics. Generally, a clutch to one-way clutch shift is more favorable than a clutch to clutch shift for shift quality control. The transmission output torque T_{out} drops in proportion to the oncoming clutch torque that is a control variable depending on the clutch pressure. The behaviors of key variables during the torque phase in the 1–2 upshift, including transmission input angular velocity ω_{in}, off-going clutch torque T_F or T_D, oncoming clutch torque T_C, and output torque T_{out}, are demonstrated in Figure 5.22. Note here that the qualitative analysis on the 1–2 shift of the Ford FWD six-speed AT is generally applicable to all transmission upshifts.

Inertia phase of 1–2 shift process: The constraint on angular acceleration $\dot{\omega}_{C1}$ is released once the 1–2 shift enters the inertia phase, i.e. $\dot{\omega}_{C1} \neq 0$. With clutch E applied during the inertia phase of the 1–2 shift, $\dot{\omega}_{S3}$ drops out in Eqs (5.47–5.49). This leads to the solution of angular accelerations, $\dot{\omega}_{S1}$, $\dot{\omega}_{C1}$, and $\dot{\omega}_{C2}$, in terms of $\dot{\omega}_{in}$ and $\dot{\omega}_{out}$ as:

$$\dot{\omega}_{S1} = -1.7886\dot{\omega}_{in} + 5.1367\dot{\omega}_{out} \tag{5.86}$$

$$\dot{\omega}_{C1} = -0.4607\dot{\omega}_{in} + 2.0655\dot{\omega}_{out} \tag{5.87}$$

$$\dot{\omega}_{C2} = 1.4141\dot{\omega}_{out} \tag{5.88}$$

These relations on the angular accelerations and the clutch conditions with $T_A = T_B = T_D = 0$ lead to the simplification of Eqs (5.50–5.52) in the following form:

$$(2.1707I_{in} + 6.9441I_{S1} + 0.4607I_{C1R2})\dot{\omega}_{in} - (19.9427I_{S1} + 2.0655I_{C1R2})\dot{\omega}_{out} - 2.1707T_{in} + 3.8834T_C + T_F = 0 \tag{5.89}$$

$$-5.1555I_{S1}\dot{\omega}_{in} + (14.8060I_{S1} - I_{out})\dot{\omega}_{out} - 2.8824T_C + 3.4146T_E - T_{out} = 0 \tag{5.90}$$

$$-3.1707I_{in}\dot{\omega}_{in} - 1.4141I_{C2R3}\dot{\omega}_{out} + 3.1707T_{in} - 2.4146T_E = 0 \tag{5.91}$$

In these three equations, T_C is the torque of the oncoming clutch, which is a control variable. T_F is the one-way clutch torque and is equal to zero during the inertia phase.

If the off-going clutch is not a one-way clutch, but rather a regular clutch such as clutch D in the Ford FWD six-speed AT, the off-going clutch torque will not be zero when the 1–2 shift enters the inertia phase. It is pointed out here that torque T_F is kept in Eq. (5.89) for generality. If clutch D is used as the reaction clutch in first gear, then as with the off-going clutch in the 1–2 shift, its torque is not equal to zero during the inertia phase and T_F in Eq. (5.89) is replaced by T_D as a control variable.

The specific state variable equation system for the whole vehicle is formed by combining Eqs (5.89–5.91) with the vehicle equation of motion, i.e. Eq. (5.82). If the transmission input torque T_{in} and the control variables T_C and T_D are given, then all state variables, including $\dot{\omega}_{in}$, $\dot{\omega}_{out}$, T_E, and T_{out}, can be solved from the specific state variable equation system. By variable substitution, state variables T_E, $\dot{\omega}_{in}$, and T_{out} can be solved in terms of T_{in}, T_C, T_F, and $\dot{\omega}_{out}$ directly from Eqs (5.89–5.91) in the form:

$$T_E = 1.3131T_{in} - 1.3131I_{in}\dot{\omega}_{in} - 0.5856I_{C2R3}\dot{\omega}_{out} \tag{5.92}$$

$$\dot{\omega}_{in} = \frac{2.1707T_{in}}{(2.1707I_{in} + 6.9441I_{S1} + 0.4607I_{C1R2})} - \frac{3.8824T_C + T_F}{(2.1707I_{in} + 6.9441I_{S1} + 0.4607I_{C1R2})} + \frac{(19.9427I_{S1} + 2.0655I_{C1R2})}{(2.1707I_{in} + 6.9441I_{S1} + 0.4607I_{C1R2})}\dot{\omega}_{out} \tag{5.93}$$

$$\begin{aligned} T_{out} = {} & 4.4837T_{in} - 2.8824T_C - \frac{(5.1555I_{S1} + 4.4837I_{in})}{(2.1707I_{in} + 6.9441I_{S1} + 0.4607I_{C1R2})}(2.1707T_{in}) + \\ & \frac{(5.1555I_{S1} + 4.4837I_{in})}{(2.1707I_{in} + 6.9441I_{S1} + 0.4607I_{C1R2})}(3.8824T_C + T_F) + \\ & \frac{(5.1555I_{S1} + 4.4837I_{in})(19.9427I_{S1} + 2.0655I_{C1R2})}{(2.1707I_{in} + 6.9441I_{S1} + 0.4607I_{C1R2})}\dot{\omega}_{out} \\ & - (2.0I_{C2R3} - 14.8006I_{S1} + I_{out})\dot{\omega}_{out} \end{aligned} \tag{5.94}$$

It is again emphasized here that torque T_F will be replaced by torque T_D if clutch D is applied in first gear instead of clutch F. Obviously, $\dot{\omega}_{out}$ can be solved by combining Eq. (5.94) and Eq. (5.82) in terms of T_{in}, T_C, and T_F, and all other state variables are then solved by back substitution. This will lead to the accurate solution of the specific state variable equations for all state variables. Shift response and smoothness primarily depend on the variation patterns of three variables: output torque T_{out}, the input angular acceleration $\dot{\omega}_{in}$, and the output angular acceleration $\dot{\omega}_{out}$ during the shift process. These variables can be qualitatively analyzed using Eqs (5.92–5.94) and the following observations can be made on the characteristics of the inertia phase of the 1–2 upshift.

- During the 1–2 shift inertia phase, the transmission ratio i_t starts to change toward the lower second gear ratio. The input angular velocity must start to drop quickly, i.e. $\dot{\omega}_{in}$ is negative. As can be seen in Eq. (5.93), the magnitude of $\dot{\omega}_{in}$ depends on the input torque T_{in} and the combination of the oncoming clutch torque T_C and the residual off-going torque T_D. For $\dot{\omega}_{in}$ to be negative, $3.8824T_C + T_D > 2.1707T_{in}$. For the Ford FWD six-speed AT which uses clutch F as the reaction clutch in first gear, $T_C > 0.5591T_{in}$. In order to complete the shift as quickly as possible, the oncoming clutch torque T_C must be ramped up to a high value quickly to decelerate the input angular velocity ω_{in}. To increase the magnitude of $\dot{\omega}_{in}$, it is necessary to reduce the input torque T_{in} in the inertia phase. In the practice of transmission control, the reduction of input torque T_{in} is realized by "spark retarding". The timing for spark retarding must be controlled at the end of the torque phase, otherwise the output torque will drop further in the torque phase.

- Before the 1–2 upshift, or any other power on upshifts, is initiated the vehicle is being accelerated in the current gear. To avoid a sudden drop of vehicle acceleration during shift, felt as "jerk" by the driver or passenger, the output torque should be controlled above a certain level and should not oscillate at large amplitude. This also means that the output angular acceleration $\dot{\omega}_{out}$ determined by Eq. (5.82) is positive and varies continuously during the shift process.
- During the shift process, the magnitude of the input angular acceleration is far larger than the output angular acceleration, i.e. $|\dot{\omega}_{in}| \gg |\dot{\omega}_{out}|$. In addition, the mass moment of inertia of the input assembly is much larger than other assemblies, i.e. $I_{in} \gg I_{S1} \sim I_{C1R2} \sim I_{C2R3} \sim I_{out}$. As can be observed in Eq. (5.94), the output torque T_{out} is dominated by the first four terms. For a qualitative analysis, the magnitude of the output torque can be approximated by dividing both the numerator and denominator of the third and fourth terms in Eq. (5.58) by I_{in} and then taking the limit when I_{in} approaches the infinity. This will lead to the approximation of the output torque as: $T_{out} \approx 5.1369 T_C + 2.0656 T_D$. Here T_D replaces T_F if clutch D is the reaction torque in first gear.
- As seen above, the magnitude of the output torque during the inertia phase is proportional to the oncoming clutch torque to a large degree. To decrease ω_{in} quickly, larger oncoming clutch torque T_C is needed. But on the other hand, an overly large oncoming clutch torque can cause an overshoot in the output torque during the inertia phase. This contradiction can be solved by the input torque reduction through spark retarding, as mentioned previously. For the clutch to one-way clutch 1–2 shift, the oncoming clutch torque is already equal to $0.5591 T_{in}$ at the end of the torque phase. As soon as the shift enters the inertia phase, the input torque is reduced by engine spark retarding. The oncoming clutch torque will be immediately larger than $0.5591 T_{in}$ and is controlled to increase continuously, resulting in an angular deceleration for the input assembly. As the input angular velocity decreases, the transmission ratio also decreases from the first gear ratio toward the second gear ratio. At some point toward the end of the shift, engine spark retarding is cancelled and the input torque is recovered to get close to the relation $T_C = 0.5591 T_{in}$. The perfect timing to end the input torque reduction would be such that when the transmission ratio reaches the target second gear ratio, the input torque is just recovered to the point at which $T_C = 0.5591 T_{in}$. This will provide a seamless transfer from the first gear to the second gear without any torque overshoot. In engineering practice, the timing of input torque reduction through engine spark retarding and the ramping up of oncoming torque are carefully calibrated. A small amount of output torque overshoot is allowable and is mostly damped by the torque converter equipped powertrain itself.
- If clutch D replaces clutch F in first gear as the reaction clutch in the Ford FWD six-speed AT, the off-going clutch torque T_D increases the overshoot of the output torque during the inertia phase. The residual off-going clutch torque should be brought to zero asoon as the inertia phase starts to lower the output torque overshoot. Meanwhile, the oncoming clutch torque T_C must be ramped up quickly to be greater than $0.5591 T_{in}$ in order to decrease the input angular velocity ω_{in}, otherwise "engine flare" will occur at the start of the inertia phase.

- Note that the observations above are based on the qualitative analysis on the 1–2 upshift in the Ford FWD six-speed AT. The observations can be generalized for all upshifts in other automatic transmissions. The behavior of shift quality related variables during the 1–2 shift are shown in Figure 5.22.

Criteria for shift quality: The time required to complete shifts and the variation of the vehicle acceleration during shifts are the two most important parameters for the shift quality of automatic transmissions. The shift time depends on how quickly and timely the off-going clutch is released and the oncoming clutch is engaged. The variation of vehicle acceleration depends largely on the behavior of transmission output torque during the shift process. The size of the torque hole shown in Figure 5.22 is directly proportional to the vehicle acceleration drop and must be minimized to enhance the shift quality. The torque overshoot during the inertia phase causes sudden acceleration jerks during shifts. For the 1–2 shift of the Ford FWD six-speed AT which uses one-way clutch F in first gear, the oncoming clutch torque T_C must be ramped up quickly to be around a magnitude of $0.5591T_{in}$, at which the one-way clutch starts to slip by itself and the shift enters the inertia phase. If the oncoming torque T_C is controlled to be close to $0.5591T_{in}$, that is, the reaction torque of clutch C in second gear at the end of the torque phase, then the corresponding output torque should be close to the target value in second gear and the depth of the torque hole will be minimized. As the shift enters the inertia phase, the oncoming clutch torque should be ramped up at a lower rate because excessive oncoming clutch torque in the inertia phase causes the output torque to overshoot. Meanwhile, the oncoming clutch torque T_C. must be high enough to act against the input torque T_{in} to decelerate the input angular velocity ω_{in}. To overcome this contradiction, it is necessary to reduce the input torque through engine spark retarding, as shown in Figure 5.22 as soon as the inertia phase starts. When the input angular velocity is reduced to be the target value corresponding to the second gear ratio, the spark retarding is cancelled to recover the input torque. Simultaneously, the oncoming clutch torque is further increased to complete the shift and secure the engagement of the oncoming clutch. Theoretically, it is possible to achieve a perfect 1–2 shift as illustrated in Figure 5.23, with a seamless output torque transfer during the shift.

If clutch D replaces the one-way clutch F in first gear of the Ford FWD six-speed AT, the characteristics of key variables during the clutch to clutch 1–2 shift are similar to those shown in Figure 5.23, but the torque profiles of the off-going clutch and the oncoming clutch are different, as shown in Figure 5.24. In this case, the torque capacity of the off-going clutch D must be ramped down quickly to the threshold of slippage before the actual torque in clutch D reaches zero. This means that the torque phase in the clutch to clutch 1–2 shift will end sooner than in the clutch to one-way clutch 1–2 shift. For the same input torque, the torque in the oncoming clutch is lower at the time when the off-going starts to slip. As the shift enters the inertia phase, there is still a torque in the off-going clutch and this torque must be brought to zero as quickly as possible. Because the torque in the oncoming clutch is lower at the beginning of the inertia phase, it must be ramped up more quickly during the inertia phase in order to act against the input torque to decelerate the input angular velocity. Through the control of spark retarding timing and oncoming clutch torque, it is possible to achieve near perfect shifts with smooth output torque transfer similar to that shown in Figure 5.23 for the clutch to one-way clutch shift.

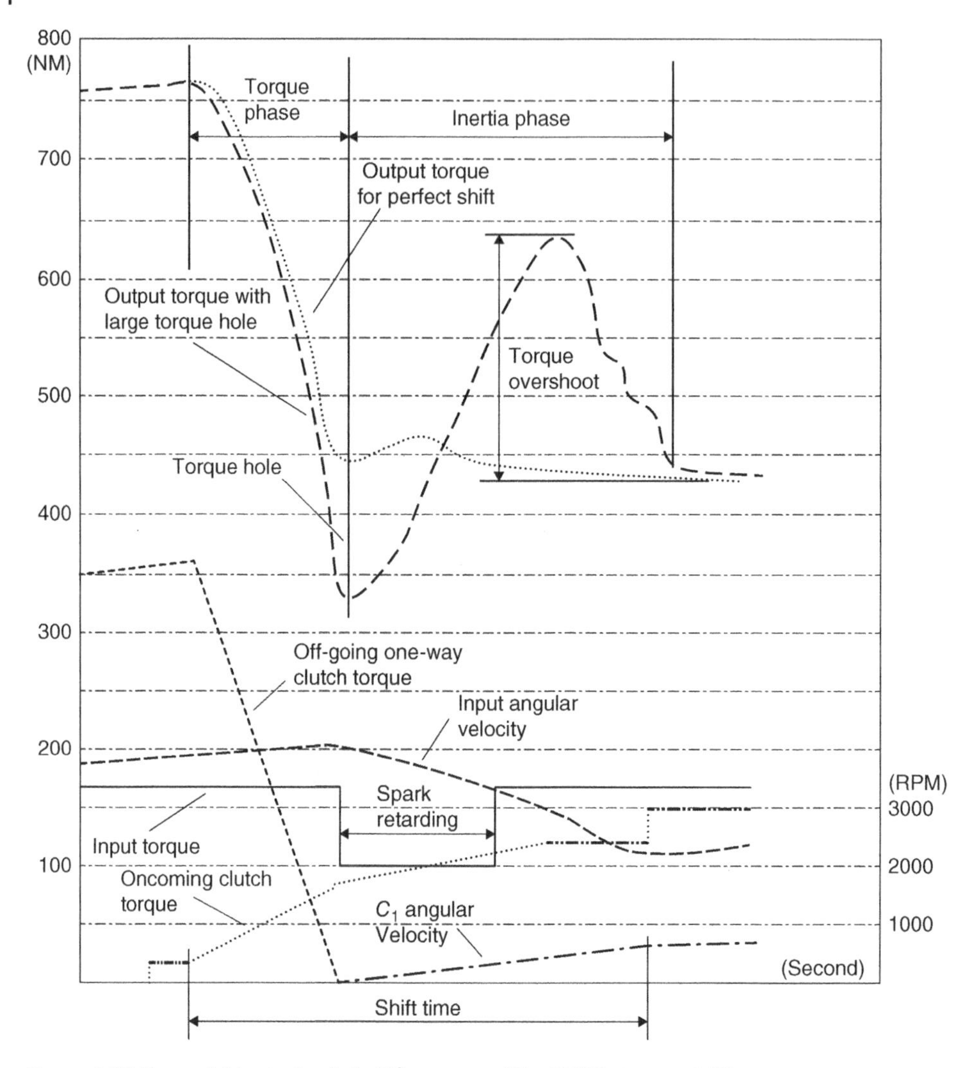

Figure 5.23 Key variables in the 1–2 shift process of Ford FWD six-speed AT.

Power-on downshifts: Power-on downshifts are initiated during vehicle operations such as passing, slope climbing, and other deep throttle maneuvers either triggered by the driver or commanded by the transmission controller under certain road conditions. These types of downshifts occur at relatively high vehicle speed, with the transmission operating in higher gears such as fourth, fifth, and sixth. As higher vehicle acceleration is the main intention, power-on downshifts must be implemented expeditiously and some harshness during such shifts is not a major concern. If a downshift is triggered by the driver, then the driver will anticipate it as an oncoming event and will feel the shift harshness with lessened unpleasantness. A power-on downshift also consists of the inertia phase and torque phase. However, power-on downshifts start with inertia phase and end with torque phase. The order of the two phases is opposite

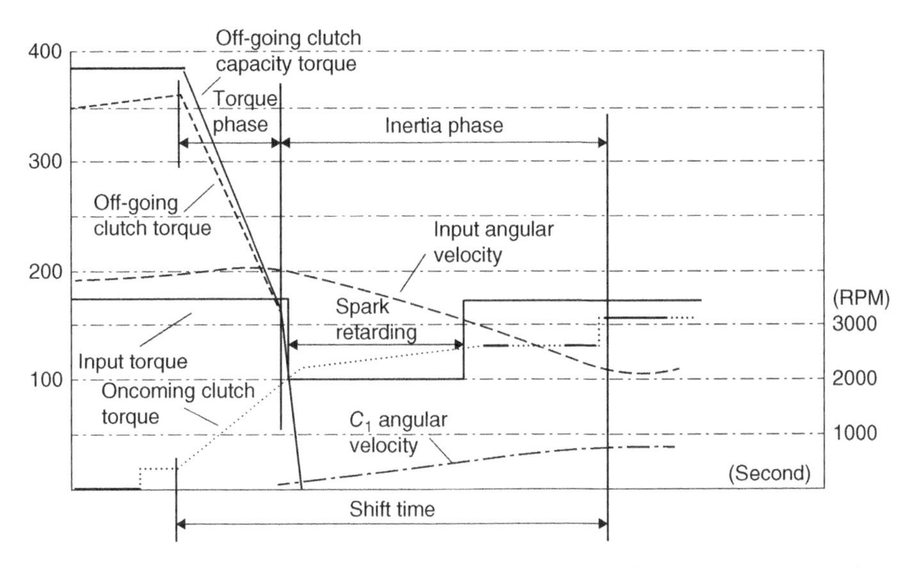

Figure 5.24 Clutch torque profiles in the clutch to clutch 1–2 shift of Ford FWD six-speed AT.

to that of power-on upshifts discussed previously. This means that the off-going clutch must be controlled to slip before the oncoming clutch starts to apply a clutch torque.

Shift transients during downshifts can be analyzed qualitatively similar to the previous case study which analyzes the 1–2 upshift of the Ford FWD six-speed AT. Let's consider the 4–3 downshift of the Ford FWD six-speed AT in a simple qualitative analysis. In this downshift, clutch A is off-going and clutch B is oncoming. When the transmission operates in fourth gear, the angular velocity of sun gear S_1, ω_{S1}, is higher than the input angular velocity ω_{in} as shown the transmission ratio analysis. Clutch B, when applied, couples the input assembly with the sun S_1 assembly, as shown in Figure 5.17. If clutch B is applied before clutch A starts to slip in the 4–3 downshift, it will cause backward power recirculation, or "clutch tie-up", because the transmission still operates in the fourth gear and thus sun gear S_1 turns fast than the input assembly. To avoid this backward power recirculation, the off-going clutch A must be brought to slip first before any clutch torque exists in the oncoming clutch B. As soon as clutch A slips, the transmission ratio will start to increase from the fourth gear ratio as the input angular velocity ω_{in} increases due to the reduced torque in clutch A and the 4–3 downshift starts its inertia phase immediately. After the inertia phase starts, the torque in the oncoming clutch B is then ramped up from zero rapidly and the torque in the off-going clutch A is further ramped down. The inertia phase is finished at the point when the target third gear ratio is reached. At this point, the pressure in clutch B is stepped up quickly to secure its engagement and the 4–3 downshift enters the torque phase. The residual torque in clutch A is now a drag torque on the transmission and must be rapidly brought to zero to end the torque phase. Note here that if the torque in the off-going clutch A is brought to zero before the target third gear ratio is reached, then the 4–3 downshift will be completed without the torque phase.

A qualitative analysis based on the specific state variable equation system during a 2–1 downshift can be performed similarly to the previous case study for the 1–2 upshift of the Ford FWD six-speed AT. The two sets of specific state variable equation systems are the same for both the upshift and downshift involving the same two gears. Equations (5.84–5.86) and (5.93–5.95) apply for both the 1–2 upshift and 2–1 downshift. Without losing generality, let's consider that clutch D is used instead of the one-way clutch F in first gear. Suppose a 2–1 power-on downshift is to be implemented, then the clutch torque capacity of the off-going clutch C must be reduced so as to start the slippage in clutch C before clutch D is pressured, otherwise clutch D would apply a drag torque on the transmission. The clutch torque capacity is reduced by the reduction of ATF pressure in the clutch piston chamber. As clutch C starts to slip, the 2–1 downshift enters the inertia phase immediately and the gear ratio starts to increase toward the first gear ratio. Meanwhile, the torque in the oncoming clutch D is ramped up gradually. As can be observed in Eq. (5.93), with T_{in} maintained or even increased in a power-on downshift, $\dot{\omega}_{in}$ can be increased rapidly if the off-going clutch torque T_C is decreased rapidly while the oncoming clutch torque is still low during the inertia phase. When the transmission ratio determined by $\frac{\omega_{in}}{\omega_{out}}$ reaches the first gear ratio, the oncoming clutch torque T_D is stepped up quickly to close the clutch, and the 2–1 downshift then enters the torque phase during which the residual torque in the off-going clutch is brought to zero rapidly. If the one-way clutch F is used in the first as in the Ford FWD six-speed AT, it will be actuated by itself to the holding status when the ratio $\frac{\omega_{in}}{\omega_{out}}$ reaches the value of the first gear ratio. Note here that the specific state variable equations for the 2–1 downshift are used here for the analysis of downshifts in general. In real world driving, power-on 2–1 downshift is rarely implemented in transmission control.

This qualitative analysis can be generalized for all power-on downshifts. The torque profiles of the off-going clutch and the oncoming clutch, as well as other variables in such shifts, are illustrated in Figure 5.25. As observed in this figure, the actual downshift process does not start immediately after the shift is initiated by the transmission control system. There is a small amount of time between the shift initiation and the point at which the off-going clutch starts to slip. During this time, the off-going clutch torque capacity is ramped down to the threshold for slippage, which is equal to the value of the off-going clutch torque determined by the system dynamics. In a power-on downshift, the input torque increases after the shift is initiated and the output torque increases accordingly before the actual shift starts. As the shift enters the inertia phase, the output torque will oscillate somewhat due to the slippage of the off-going clutch. At this time, the oncoming clutch torque starts to ramp up but is still small. The input angular velocity is therefore accelerated by the input torque which works against both the off-going clutch torque and the oncoming clutch torque. As the input angular velocity increases, the transmission ratio is gradually changed to the target value of the lower gear. As shown in Figure 5.25, the torque phase is much shorter in time than the inertia phase in a typical power-on downshift. If the torque profiles of the off-going clutch and the oncoming clutch are controlled accurately and timely, power-on downshifts can also be made with optimized responsiveness and smoothness, as indicated by the shift time and output torque behavior shown in Figure 5.25.

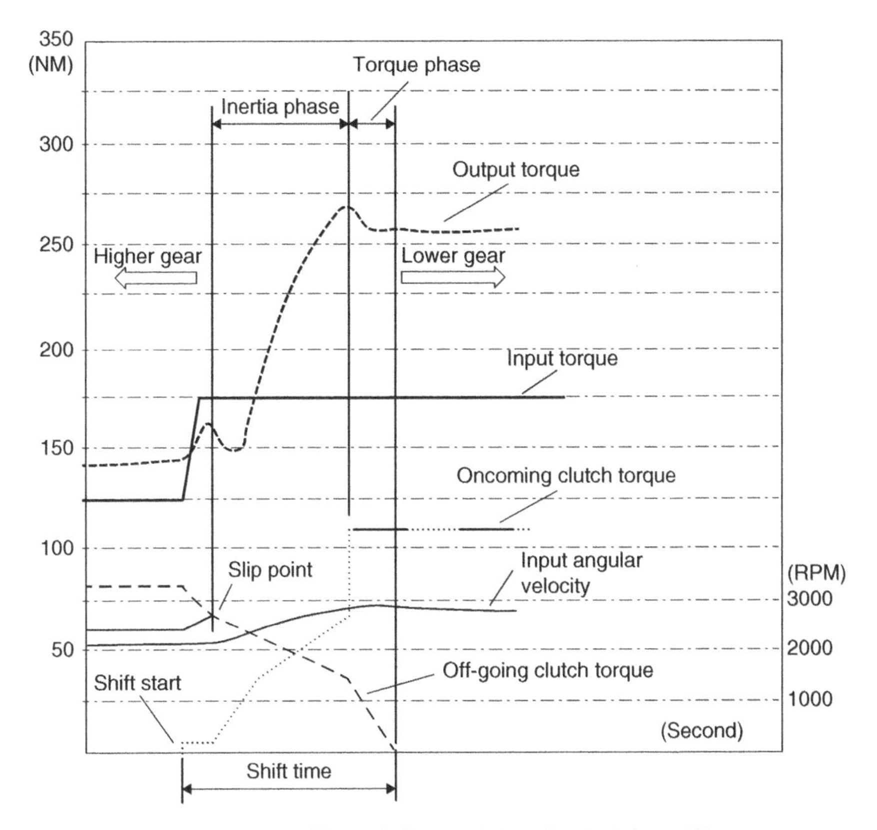

Figure 5.25 Clutch torque profiles and characteristics of typical downshifts.

In addition to power-on upshifts and downshifts, power-off upshifts and downshifts are also frequently implemented in daily driving. Typically, **power-off upshifts** are used when the vehicle has been accelerated to a relatively high speed and the driver decides not to accelerate the vehicle further. A shift like this is triggered when the driver releases the gas pedal. In such shifts, it is the wheels that drive the powertrain system since the input torque is almost zero. In order to make quick and smooth power-off upshifts, the off-going clutch should be controlled to slip first before the oncoming clutch is pressured. In other words, the inertia phase comes before the torque phase in power-off upshifts. Since there is almost no input torque, the control of such shifts is less challenging in comparison with power-on upshifts.

Power-off downshifts are used when the vehicle is being coasted to low speeds or a full stop. Such shifts occur when the vehicle speed is relatively low. Because the transmission input is the driven side in such shifts, the torque phase comes before the inertia phase. The oncoming clutch torque should be ramped up before the off-going clutch starts to slip in power-off downshifts for optimized shift feel. In summary, power-off upshifts are used at high vehicle speed and power-off downshifts are used at low vehicle speed. These two types of shifts are implemented without power input to the transmission and are less technically challenging in comparison with power-on shifts.

5.6 General Vehicle Powertrain Dynamics

Vehicle drivetrain dynamics, especially transmission shift dynamics, have been the subject of in-depth studies in the automotive industry [5–7]. In this section, the ZF RWD eight-speed AT shown in Figure 5.14 will be used as an example for the study of vehicle powertrain dynamics. The example can be readily extended to all other automatic transmissions in general. For this example transmission, the general state variable equation system consists of Eqs (5.71–5.78), which were derived in Section 5.4. The vehicle general state variable equation system is formed by combining Eqs (5.71–5.78) and the vehicle equation of motion represented by Eq. (5.82). The transmission input torque T_{in} is considered as a given variable in the state variable equation system. Note that T_{in} depends on the engine output and the torque converter characteristics. The integrated vehicle system will incorporate the engine, the torque converter, and the transmission, as will be discussed later in this section.

The vehicle general state variable equation system must be combined with the clutch constraints that exist in fixed gear operations or during shifts. Any shift involves two separate statuses: torque phase and inertia phase. Not counting the reverse gear, the ZF RWD eight-speed AT has eight forward speeds and 36 direct shifts, as mentioned previously. Therefore, there are a total of 80 operation statuses: eight fixed gear operations, and 36 torque phases and 36 inertia phases for the 36 direct shifts. Each of these 80 operation statuses requires a set of specific state variable equations to uniquely describe the system dynamics. Note that the specific state variable equations are the same for both the upshift and the downshift in any of the 36 direct shifts. That is, the state variable equations for the torque phase of the 1–2 upshift are the same as for the torque phase of the 2–1 downshift, and the state variable equations for the inertia phase of the 1–2 upshift are also the same as for the inertia phase of the 2–1 downshift.

In the vehicle general state variable equation system, formed by combining Eqs (5.71–5.78) and Eq. (5.82), there are a total of nine equations and 14 state variables, with the input torque T_{in} as a given variable. Of these 14 state variables, eight are angular accelerations: $\dot{\omega}_{in}$, $\dot{\omega}_{S1}$, $\dot{\omega}_{R1}$, $\dot{\omega}_{C1}$, $\dot{\omega}_{R2}$, $\dot{\omega}_{R3}$, $\dot{\omega}_{C3}$, and $\dot{\omega}_{out}$; five are clutch torques: T_A, T_B, T_C, T_D, and T_E; and one is the output torque T_{out}. In fixed gear operations, three clutches are applied, providing three constraints on the angular accelerations and reducing the number of angular accelerations as state variables to five. The torques in the two open clutches are both zero, reducing the torque state variables to four. Therefore, the total number of unknown state variables, including five angular accelerations, three clutch torques, and one output torque, is nine. With the input torque T_{in} given, these nine state variables are uniquely determined by the vehicle state variable equation system. In the torque phase of shifts, the clutch constraints are the same as in fixed gear operations and the number of unknown state variables is also equal to nine. The oncoming clutch torque is a given control variable provided to the transmission system. In the inertia phase of shifts, only two clutches are applied and thus the number of clutch constraints is only two. But in this case, both the oncoming and off-going clutch torques are given control variables provided to the transmission system. Therefore, the number of unknown state variables during shifts is still equal to the number of state variable equations and the dynamic status of the powertrain system is uniquely determined during all shifting operations.

5.6.1 General State Variable Equation in Matrix Form

The vehicle state variable equation system formed by Eqs (5.71–5.58) and Eq. (5.82) can also be expressed in matrix form. Firstly, the eight angular accelerations and the seven torque variable, including the five clutch torques, the input torque and the output torque, are represented by two separate column vectors:

$$[X]=[x_1,x_2,x_3,x_4,x_5,x_6,x_7,x_8]^T=[\dot{\omega}_{in},\dot{\omega}_{S1},\dot{\omega}_{R1},\dot{\omega}_{C1},\dot{\omega}_{R2},\dot{\omega}_{R3},\dot{\omega}_{C3},\dot{\omega}_{out}]^T \tag{5.95}$$

$$[Y]=[y_1,y_2,y_3,y_4,y_5,y_6,y_7]^T=[T_{in},T_A,T_B,T_C,T_D,T_E,T_{out}]^T \tag{5.96}$$

where the column vectors $[X]$ and $[Y]$ contain respectively the eight angular accelerations and the seven torques. In terms of variable $x_i\ (i=1,2,\ldots,8)$ and $y_i\ (i=1,2,\ldots,7)$, Eqs (5.71–5.78) and Eq. (5.82) can be readily rewritten in the following form:

$$x_2+2.0x_3-3.0x_4=0 \tag{5.97}$$

$$-3.0x_1+x_2+2.0x_5=0 \tag{5.98}$$

$$x_5+1.61x_6-2.61x_7=0 \tag{5.99}$$

$$x_6+3.696x_4-4.696x_8=0 \tag{5.100}$$

$$-0.3333I_2x_1-I_{S1S2}x_2+0.5I_{R1}x_3+0.3333y_1+y_2-0.5y_3-0.3333y_4=0 \tag{5.101}$$

$$1.5I_{R1}x_3+I_{C1R4}x_4+0.787I_{C4}x_8-1.5y_3-0.787y_5+0.787y_7=0 \tag{5.102}$$

$$-0.6667I_{in}x_1-I_{R2S3}x_5-0.3831I_{C3}x_7+0.6667y_1 \\ -0.6667y_4-0.3831y_5-y_6=0 \tag{5.103}$$

$$-I_{R3S4}x_6-0.6168I_{C3}x_7-0.2129I_{C4}x_8+y_4-0.4039y_5+y_6-0.2129y_7=0 \tag{5.104}$$

$$\frac{Wr}{gi_a}x_8-\frac{\eta i_a}{r}y_7=-R \tag{5.105}$$

These equations can be written in matrix form:

$$[A_{9,8}\ \ B_{9,7}]_{9,15}\begin{bmatrix}X_8\\Y_7\end{bmatrix}_{15,1}=[R]_{9,1} \tag{5.106}$$

In this matrix equation, the coefficient matrix, formed by partitioning matrices A and B, has a dimension of 9 × 15 since there are nine state variable equations and 15 state variables, as shown in Eqs (5.95) and (5.96). Matrix A is 9 × 8 and its elements are multipliers of the angular accelerations or variables $x_i\ (i=1,2,\ldots,8)$ in the state variable equations. Matrix B is 9 × 15 and its elements are the multipliers of torques or variables $y_i\ (i=1,2,\ldots,7)$. The column vector on the right-hand side has 15 elements, with the ninth element equal to $(-R)$ and all the other eight elements equal to zero. The general state variable equations are the characteristics of the vehicle powertrain system and apply for all vehicle operations. This means that the elements of the coefficient matrix in Eq. (5.106) are constants for a given transmission regardless of operation status. The elements of matrices A and B for the ZF RWD eight-speed AT are represented in the following:

$$[A]_{9,8} = \begin{bmatrix} 0 & 1.0 & 2.0 & -3.0 & 0 & 0 & 0 & 0 \\ -3.0 & 1.0 & 0 & 0 & 2.0 & 0 & 0 & 0 \\ 0 & 0 & 0 & 0 & 1.0 & 1.61 & -2.61 & 0 \\ 0 & 0 & 0 & 3.696 & 0 & 1.0 & 0 & -4.696 \\ -3.3333I_I & -I_{S1S2} & 0.5I_{R1} & 0 & 0 & 0 & 0 & 0 \\ 0 & 0 & 1.5I_{R1} & I_{C1R4} & 0 & 0 & 0 & 0.787I_{C4} \\ -0.6667I_I & 0 & 0 & 0 & -I_{R2S3} & 0 & -0.3931I_{C3} & 0 \\ 0 & 0 & 0 & 0 & 0 & -I_{R3S4} & -0.6168I_{C3} & -0.21294 \\ 0 & 0 & 0 & 0 & 0 & 0 & 0 & Wr/gi_a \end{bmatrix}$$

$$[B]_{9,7} = \begin{bmatrix} 0 & 0 & 0 & 0 & 0 & 0 & 0 \\ 0 & 0 & 0 & 0 & 0 & 0 & 0 \\ 0 & 0 & 0 & 0 & 0 & 0 & 0 \\ 0 & 0 & 0 & 0 & 0 & 0 & 0 \\ 0.3333 & 1.0 & -0.5 & 0.3333 & 0 & 0 & 0 \\ 0 & 0 & -1.5 & 0 & -0.787 & 0 & 0.787 \\ 0.6667 & 0 & 0 & -0.6667 & -0.3831 & -1.0 & 0 \\ 0 & 0 & 0 & 1.0 & -0.4039 & 1.0 & -0.2129 \\ 0 & 0 & 0 & 0 & 0 & 0 & -\eta i_a/r \end{bmatrix}$$

5.6.2 Specific State Variable Equation

The specific state variable equation system is formed by superimposing the clutch constraints upon the general state variable equation system expressed by Eq. (5.106) in each operation. For the convenience of matrix operation and linear algebra, the specific state variable equation system is formed by augmenting the general state variable equation system with the clutch constraint equations rather than by direct variable substitutions. This means that constraint equations are just treated as state variable equations and thus the number of rows in the coefficient matrix in Eq. (5.106) increases by the number of constraint equations. The clutch constraint equations are obtained from the clutch table directly, in accordance with the three operation statuses: fixed gear operation, torque phase, and inertia phase.

Fixed gear operation: Each applied clutch corresponds to a constraint equation on the angular acceleration. If a reaction clutch is applied, the constraint equation is simply formed by making the related angular acceleration equal to zero. If a coupling clutch is applied, then the constraint equation is formed by making the subtraction of the two angular accelerations of the two coupled components equal to zero. For the ZF RWD eight-speed AT, there are three such constraint equations in each gear and the number of equations in the specific state variable equation system is 12. The two torques in the two open clutches are equal to zero and are dropped from the specific state variable equation system.

Torque phase of a shift: During the torque phase of a shift, the clutch constraint equations are the same as for the fixed gear operation. However, the oncoming clutch torque is now an input control variable that appears on the right side of the specific state variable equation system, as illustrated later.

Inertia phase of a shift: During the inertia phase of a shift, the clutch constraint equation caused by the off-going clutch is released. For the ZF RWD eight-speed AT, there are still two clutch constraint equations during the inertia phase of a shift. The torque in the open clutch is equal to zero and is dropped from the specific state variable equation system. The torques in both the off-going clutch and the oncoming clutch are control variables. The number of equations in the specific state variable equation system is 11.

In what follows, the first gear operation, the torque phase and the inertia phase of the 1–2 shift in the ZF RWD eight-speed AT are used as examples in formulating the specific state variable equation systems. These examples can be readily extended to all other operations and for all other automatic transmissions.

First gear operation: In the first gear of the ZF RWD eight-speed AT, clutches A, B, and C are applied, resulting in three clutch constraints: $\dot{\omega}_{S1} = 0$, $\dot{\omega}_{R1} = 0$, and $\dot{\omega}_{in} = \dot{\omega}_{R3}$. In terms of the variables of the column vector $[X]$ expressed in Eq. (5.95), the three constraint equations are written in the following form:

$$x_2 = 0 \tag{5.107}$$

$$x_3 = 0 \tag{5.108}$$

$$x_1 - x_6 = 0 \tag{5.109}$$

These three constraint equations are combined with the general state variable equations from Eqs (5.97–5.105) to form the general state variable equation system, which then has 12 equations and is represented in matrix form as follows:

$$\left[A_{12,8} \;\; B_{12,7}\right]_{12,15} \begin{bmatrix} X_8 \\ Y_7 \end{bmatrix}_{15,1} = [R]_{12,1} \tag{5.110}$$

Here the coefficient matrix is 12 × 12 since it has been augmented by Eqs (5.107–5.109). The right-hand side is augmented from $[R]_{9,1}$ in Eq. (5.106) by three zero elements. Matrix $[A]_{12,8}$ and $[B]_{12,7}$ are augmented from $[A]_{9,8}$ and $[B]_{9,7}$ in the general state variable equation system expressed by Eq. (5.106), in the following partitions:

$$[A]_{12,8} = \begin{bmatrix} A_{9,8} \\ A^a_{3,8} \end{bmatrix}; [B]_{12,7} = \begin{bmatrix} B_{8,7} \\ B^a_{3,7} \end{bmatrix} \tag{5.111}$$

In these matrix augmentations, $[A]_{9,8}$ and $[B]_{8,7}$ are the same as in Eq. (5.106) and are characteristic of the transmission, i.e. these two matrices are always the same regardless of transmission operations. The two augmenting matrices, $[A]^a_{3,8}$ and $[B]^a_{3,7}$, depend on the transmission operation status and are represented for the first gear operation of the ZF RWD eight-speed AT as follows:

$$[A]^a_{3,8} = \begin{bmatrix} 0 & 1 & 0 & 0 & 0 & 0 & 0 \\ 0 & 0 & 1 & 0 & 0 & 0 & 0 \\ 1 & 0 & 0 & 0 & -1 & 0 & 0 \end{bmatrix}; \quad [B]^a_{3,7} = \begin{bmatrix} 0 & 0 & 0 & 0 & 0 & 0 & 0 \\ 0 & 0 & 0 & 0 & 0 & 0 & 0 \\ 0 & 0 & 0 & 0 & 0 & 0 & 0 \end{bmatrix}$$

Note that, in general, the augmenting matrix $[B]^a_{3,7}$ is always a null matrix, with the number of columns equal to the number of torques and number of rows equal to the number of constraint equations. The numbers of columns and rows of matrix $[A]^a_{3,8}$ are equal to the number of angular accelerations and the number of constraint equations. In first gear, clutch torques T_D and T_E, or variables y_5 and y_6, are equal to zero and drop out of the equation system. The input torque T_{in} or variable y_1 is considered as a given variable and is moved to the right-hand side of the equation. The remaining 12 unknown state variables can be uniquely solved from the following linear equation system:

$$[A_{12,8} \;\; B_{12,4}]_{12,12}[X^*]_{12} = [R]_{12,1} - \sum_{i=1}^{12} B_{i1}y_1 \tag{5.112}$$

In this equation system, the coefficient matrix is 12×12 since the two columns corresponding to y_5 and y_5 are dropped out from matrix $[B]_{12,7}$ in Eq. (5.111) and the column corresponding to y_1 is moved to the right-hand side. The unknown column contains 12 variables and $[X^*]_{12} = [x_1, x_2, x_3, x_4, x_5, x_6, x_7, x_8, y_2, y_3, y_4, y_7]^T$. The dynamic status of the whole vehicle powertrain system is uniquely defined by these variables.

Torque phase of 1–2 shift: As mentioned previously, the torque phase in a 1–2 shift differs from the first gear operation only in that the oncoming clutch torque T_E, or variable y_6, is a control variable rather than equal to zero. The corresponding specific state variable equation system is therefore represented as follows:

$$[A_{12,8} \;\; B_{12,4}]_{12,12}[X^*]_{12} = [R]_{12,1} - \sum_{i=1}^{12} B_{i1}y_1 - \sum_{i=1}^{12} B_{i6}y_6 \tag{5.113}$$

It is observed here that Eq. (5.113) only differs from Eq. (5.112) on the right-hand side in an additional term reflecting the effect of the oncoming clutch torque or variable y_6. The specific equation system (5.113) is used to solve the state variables until the end of the torque phase, which is determined at the time when the off-going clutch starts to slip. The threshold of the clutch slippage is at the point when the clutch torque capacity becomes smaller than the clutch torque determined as a state variable. For the ZF RWD eight-speed AT, the end of the 1–2 shift torque phase is judged by the following inequality:

$$T^c_C(\mu, p) \le T_C = y_4 \tag{5.114}$$

In this inequality T^c_C is the torque capacity of clutch C. For a given clutch, the clutch torque capacity mainly depends on the hydraulic pressure p in the clutch piston chamber and the clutch friction coefficient μ, which varies with clutch temperature. T_C is the clutch torque in clutch C determined as the state variable y_4. As every time step in the solution of Eq. (5.113), the clutch torque capacity T^c_C is evaluated in terms of the clutch hydraulic pressure and compared with the clutch torque T_c solved as the state variable y_4. As soon as the inequality above is satisfied, the 1–2 shift will enter the inertia phase.

Inertia phase of 1–2 shift: In the inertia phase of the 1–2 shift in the ZF RWD eight-speed AT, clutches A and B are still applied, but clutch C is slipping. There are now only two constraint equations, represented by Eqs (5.107) and (5.108). These two constraint equations are combined with the general state variable equations from Eqs (5.97–5.105) to form the general state variable equation system for the inertia phase, which then has 11 equations and is represented in matrix form as follows:

$$[A_{11,8} \;\; B_{11,7}]_{11,15} \begin{bmatrix} X_8 \\ Y_7 \end{bmatrix}_{15,1} = [R]_{11,1} \tag{5.115}$$

Here the coefficient matrix is 11 × 11 after being augmented by Eqs (5.107–5.108). The right-hand side is augmented from $[R]_{9,1}$ in Eq. (5.106) by two zero elements. Matrix $[A]_{11,8}$ and $[B]_{11,7}$ are augmented from $[A]_{9,8}$ and $[B]_{9,7}$ in the general state variable equation system expressed by Eq. (5.106), in the following partitions:

$$[A]_{11,8} = \begin{bmatrix} A_{9,8} \\ A^a_{2,8} \end{bmatrix}; [B]_{11,7} = \begin{bmatrix} B_{8,7} \\ B^a_{2,7} \end{bmatrix} \tag{5.116}$$

In this matrix augmentations, $[A]_{9,8}$ and $[B]_{8,7}$ are the same as in Eq. (5.106) and are characteristic of the transmission, as mentioned previously. The two augmenting matrices, $[A]^a_{2,8}$ and $[B]^a_{2,7}$, depend on the transmission operation status and are represented for the 1–2 shift inertia phase of the ZF RWD eight-speed AT as follows:

$$[A]^a_{2,8} = \begin{bmatrix} 0 & 1 & 0 & 0 & 0 & 0 & 0 \\ 0 & 0 & 1 & 1 & 1 & 1 & 1 \end{bmatrix}; \quad [B]^a_{2,7} = \begin{bmatrix} 0 & 0 & 0 & 0 & 0 & 0 & 0 \\ 0 & 0 & 0 & 0 & 0 & 0 & 0 \end{bmatrix}$$

In general, the augmenting matrix $[B]^a_{2,7}$ is always a null matrix, with the number of columns equal to the number of torques and number of rows equal to the number of constraint equations in the inertia phase. The numbers of columns and rows of matrix $[A]^a_{2,8}$ are equal to the number of angular accelerations and the number of constraint equations in the inertia phase. In the inertia phase of the 1–2 shift, clutch torque T_D or variable y_5 is equal to zero and drops out of the equation system. The input torque T_{in} or variable y_1, the clutch torques T_C and T_E, or variables y_4 and y_6, are considered as a given variables and are moved to the right-hand side of the equation. The remaining 11 unknown state variables can then be uniquely solved from the following linear equation system:

$$[A_{11,8} \;\; B_{11,3}]_{11,11}[X^*]_{11} = [R]_{11,1} - \sum_{i=1}^{11} B_{i1}y_1 - \sum_{i=1}^{11} B_{i4}y_4 - \sum_{i=1}^{11} B_{i6}y_6 \tag{5.117}$$

In this equation system, variables y_4 and y_6 represent the clutch torques in the off-going clutch C and the oncoming clutch E. Each clutch torque is controlled by the hydraulic pressure in the piston chamber, formulated in the following according Eq. (2.7) in Chapter 2.

$$T_C = y_4 = T_C^c(\mu, p_c) \tag{5.118}$$

$$T_E = y_6 = T_E^c(\mu, p_E) \tag{5.119}$$

At every time step, the right-hand side of Eq. (5.117) is given and the 11 unknown state variables, i.e. $[X^*]_{11} = [x_1, x_2, x_3, x_4, x_5, x_6, x_7, x_8, y_2, y_3, y_7]^T$, are uniquely solved for the determination of powertrain system dynamic status during the inertia phase of the 1–2 shift.

Note that the same procedure can be readily extended to all transmission operations for all automatic transmissions. It is just a matter of systematically augmenting the general state variable equation system by the augmenting matrices corresponding to the constraint equations in the specific operation.

5.6.3 Solution of State Variables by Variable Substitution

As shown in the qualitative analysis on the 1–2 shift of the Ford FWD six-speed AT, the state variables in the vehicle powertrain system can be solved directly by variable substitution from the specific state variable equations in terms of the transmission input torque and the torques of the off-going and oncoming clutches. For first gear operation, the torque phase and the inertia phase of the 1–2 shift of the Ford FWD six-speed AT, and the solution of state variables by variable substitution are represented by Eqs (5.79–5.81), Eqs (5.83–5.85), and Eqs (5.92–5.94) respectively. The same procedure can be performed on the specific state variable equations for the ZF RWD eight-speed AT, and all other transmissions in general. Without losing generality, the direct solution of state variables for the first gear operation, torque, and inertia phases of the 1–2 shift for the ZF RWD eight-speed AT are illustrated in Figure 5.26.

For all transmission operations, including fixed gear operations and direct shifts, the state variables can be solved from the corresponding specific state variable equations by variable substitution in terms of the transmission input torque and the torques in the off-going clutch and oncoming clutch, as illustrated in Figure 5.26. But this can be a very lengthy process for transmissions with multiple direct shifts. For example, the ZF RWD eight-speed AT has eight fixed gear operations and 36 direct shifts. It can be a challenging task to solve the state variable equations manually for so many operations without making any mistakes. In comparison, the matrix form representation of the state variable equation systems is more convenient for programming in the modeling of vehicle powertrain system dynamics.

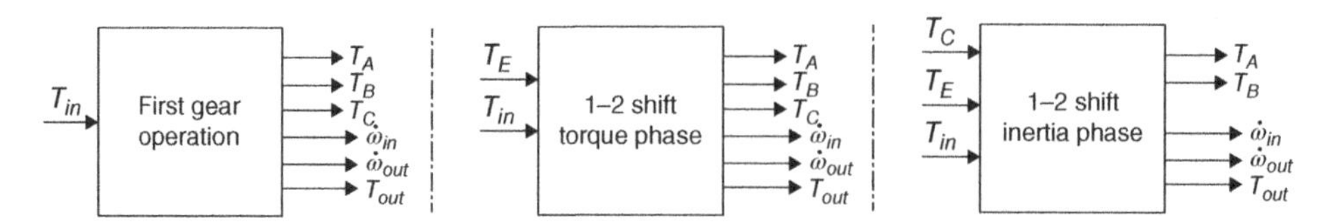

Figure 5.26 Direct solution of state variables.

5.6.4 Vehicle System Integration

The automatic transmission must be matched with the engine and the torque converter for the determination of the joint operation status of the overall vehicle powertrain system. The joint operation of the engine, the torque converter, and the transmission was illustrated in Figure 4.11 of Chapter 4, which details the characteristics of the torque converter coupling the engine and the transmission. The transmission system discussed so far in this chapter can be considered as a subsystem module to be integrated to the overall vehicle system. Model integration depends on the software platform used and the programming language. In the Matlab/Simulink environment, the vehicle system model can be structured based on the following block diagram in Figure 5.27.

In Figure 5.27, the engine torque map describes the engine output torque in terms of the engine RPM and throttle opening, as discussed in Chapter 1. The torque converter block describes the converter input and output characteristics as discussed in Chapter 4. The shift schedule block provides shift decisions to the transmission system based on vehicle speed and engine throttle opening. The transmission system module is the core of the integrated vehicle powertrain system and is built based on the approaches presented in this chapter. Once the specific state variable equation systems of the transmission are established, either using matrix presentation or by variable substitution, the integrated vehicle powertrain system can be conveniently established on a Matlab/Simulink platform based on the structure shown in Figure 5.27. Note that the engine angular acceleration is treated as a separate state variable by a sub-block in Figure 5.27, which implements the equation of motion for the engine flywheel and torque converter impeller assembly represented as follows:

$$T_e - T_i = I_i \dot{\omega}_e \tag{5.120}$$

where $\dot{\omega}_e$ is the engine angular acceleration, T_e is the engine output torque, and T_i is the impeller torque. I_i is the mass moment of inertia of the flywheel and impeller assembly. The transmission input torque T_{in}, or the turbine torque T_t, is related to the impeller

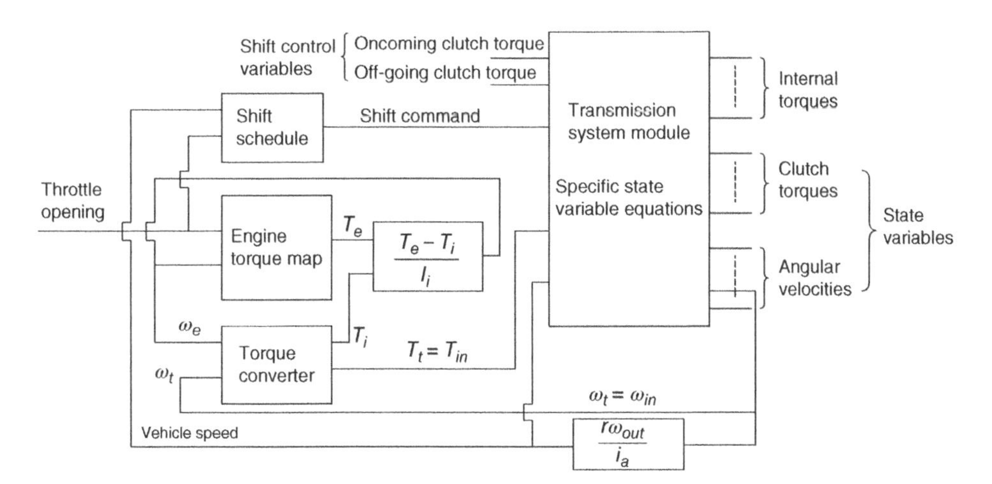

Figure 5.27 Block diagram of integrated vehicle powertrain system.

torque by the torque converter characteristics, which are discussed in Chapter 4 and reiterated in the following:

$$i_q = f_1(i_s) \tag{5.121}$$

$$K = f_2(i_s) \tag{5.122}$$

$$T_i = \frac{n_e^2}{K^2} \tag{5.123}$$

$$T_t = T_{in} = i_q T_i \tag{5.124}$$

The torque ratio i_q and the torque capacity factor K are functions of the speed ratio i_s, which is defined as the ratio between the angular velocities of the turbine and the impeller. The torque converter block in Figure 5.27 implements the four equations above with the turbine torque T_t as the transmission input torque T_{in}.

The vehicle powertrain system itself has certain torsional elasticity and damping characteristics that can affect system dynamics, particularly during launch and shift operations. To account for these system attributes, the transmission input and output shafts can be modeled as torsional spring dampers, as shown in Figure 5.28.

In the spring–damper model shown in Figure 5.28, the angular acceleration of the turbine $\dot{\omega}_t$ and the angular acceleration of the transmission input $\dot{\omega}_{in}$ are not the same because of the torsional spring damper between the turbine and the transmission input. Similarly, the angular acceleration of the transmission output $\dot{\omega}_{out}$ and the angular acceleration of the final drive input $\dot{\omega}_a$ are not the same because of the output spring damper. The introduction of the two torsional spring dampers will add two additional state variables: $\dot{\omega}_t$ and $\dot{\omega}_a$. These two state variables are governed respectively by the equation of motion for the turbine assembly with mass moment of inertia I_t and the vehicle equation of motion modified from Eq. (5.82) or (5.105), as:

$$I_t \dot{\omega}_t = T_t - T_{in} \tag{5.125}$$

$$\frac{Wr^2}{\eta\, i_a^2} \dot{\omega}_a = T_{out} - \frac{Rr}{\eta i_a} \tag{5.126}$$

The input torque and output torque depend on the stiffness and the damping coefficient of the respective spring damper, as formulated by:

$$T_{in} = k_i(\theta_t - \theta_{in}) + c_i(\omega_t - \omega_{in}) \tag{5.127}$$

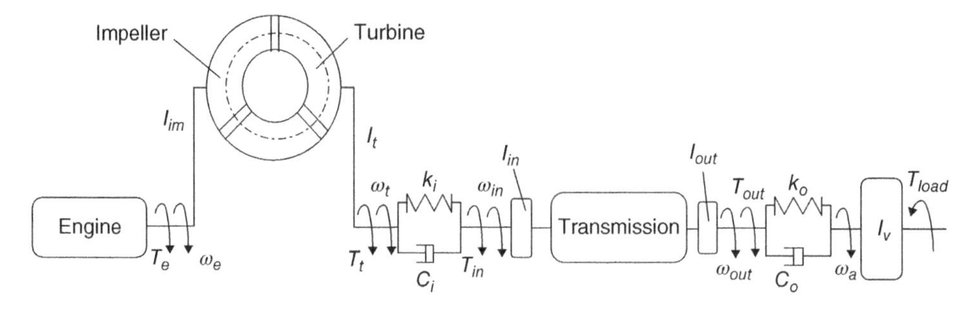

Figure 5.28 Spring–damper modeling for Input and output shafts.

$$T_{out} = k_o(\theta_o - \theta_a) + c_o(\omega_o - \omega_a) \tag{5.128}$$

With the input and output shafts modeled as spring dampers, one more state variable equation will be added to the vehicle state variable equation system and the last state variable equation will be replaced by Eq. (5.126). For the ZF RWD eight-speed AT, Eq. (5.125) can be added after Eq. (5.104) to become the new Eq. (5.105) and the old Eq. (5.105) will be replaced by Eq. (5.126) as the last equation in the new equation system. The number of equations in the general state variable system now becomes 10. The two state variables, $\dot{\omega}_t$ and $\dot{\omega}_a$, are included in the state variable equations, but the number of state variables increases by only one since T_{out} is no longer a state variable. Therefore, the specific state variable equation system in fixed gear and shift operations discussed previously will uniquely determine the state variables and the vehicle powertrain dynamics. Note that the elements of the new coefficient matrix, the state variable column, and the right-hand side must be modified accordingly in the state variable equation system to accommodate the introduction of the spring dampers. Both the input and output torques are not state variables and will be evaluated using Eqs (5.127) and (5.128) after twice integrating the four related state variables, $\dot{\omega}_t$, $\dot{\omega}_{in}$, $\dot{\omega}_{out}$, and $\dot{\omega}_a$.

5.7 Simulation of Vehicle Powertrain Dynamics

The formulations on the vehicle powertrain dynamics in the previous section are capable of simulating transmission performance during launch and shifting operations, or of vehicle performance in acceleration and fuel economy over a specified speed range. The former can then be termed as transmission simulation and the latter as range simulation. The powertrain model structure shown in Figure 5.27 needs be modified according to the intended objectives. The following list of data must be provided for transmission performance simulation and range simulation:

- Engine torque map and fuel map
- Torque converter characteristics
- Transmission data, including planetary gear train parameters, gear ratios and efficiency values
- Vehicle data, including weight, dimension, air drag coefficient, front projected area, tire radius
- Inertias of involved components, spring-damper parameters
- Torque profiles of oncoming and off-going clutches during shifts

Transmission simulation: The objectives of transmission simulation are obviously to assess the transmission performance during launch and shift operations based on shift quality criteria discussed previously. A large number of shifts under various vehicle operation conditions need to be implemented for automatic transmissions with multiple gears. The control and calibration of these shifts is a challenging and time-consuming process that often relies on trial and error. With transmission simulation, engineers can have an upfront prediction on the transmission shift performances with selected clutch torque profiles and thus avoid the lengthy trial and error process in optimizing the clutch torque control. This will shorten greatly the transmission control and calibration time and enhance transmission shift quality.

Initial condition: If the simulation starts from standstill, then the dynamic model will start to simulate the vehicle launching in first gear. The initial condition at launch for the engine and torque converter is defined by Eq. (4.29) in Chapter 4. The values of all other state variables at launch are determined based on powertrain kinematics and static equilibrium. Generally, all angular velocities, other than the engine angular velocity are equal to zero at time zero during vehicle launching operations. The initial conditions for all shifts are set by the values of the related state variables, which are determined by the simulation model at the point of time when the shifts are initiated. Similarly, the initial condition at the start of the inertia phase is defined by the state variable values at the end of the torque phase. If the simulation of a particular shift is of interest, then the vehicle dynamic status must be determined by pre-processing the simulation model up to the point at which the shift is initiated. The related state variable values at the point of shift initiation will define the initial condition for the particular shift.

Simulation output and applications: The simulation of transmission performance will provide output data that are directly related to shift quality:

- Transmission output torque during a shift. This shows how the output torque varies versus time during the shift process and provides the primary data for the judgment of shift quality as discussed in Section 5.5.
- Angular velocities of all rotational components, including the engine, turbine, gears, and clutches.
- Vehicle acceleration variations during the shift. This is also an indication for shift quality.
- Shift responsiveness as defined by shift time.
- Torque values for all clutches and gear members.

These output data provide a quantitative assessment of the shift quality resulting from the adopted control strategy and the related control variables which are primarily the torque profiles of the off-going and oncoming clutches. It is possible to use the simulation model interactively to optimize the clutch torque profiles for the shift quality. In addition, the effect of engine torque reduction on shift quality can also be simulated to obtain the optimized timing and reduction amount. These model optimized control variables are then used as the initial values in the process of transmission control and calibration.

Range simulation: A drive range is the speed–time relationship specified to emulate a traffic pattern in model simulations. There are various standard Environment Protection Agency (EPA) driving cycles designed to mimic real-world traffic conditions. For example, the speed-time relationship of the EPA urban dynamometer driving schedule (UDDS) range is shown in Figure 5.29. The vehicle in simulation is required to follow the speed–time relationship specified in the drive range.

For range simulation, a driver model must be provided in the integrated vehicle system model to emulate the throttle and brake controls. The driver model consists of two parts: throttle controller and brake controller. Each of the two controllers is based on the difference between the simulated vehicle speed and the speed specified in the drive range. PID controllers are used for both the throttle and the brake to determine the throttle angle TA and brake force F_B as shown in the following:

$$\Delta V = V_{spec} - V$$

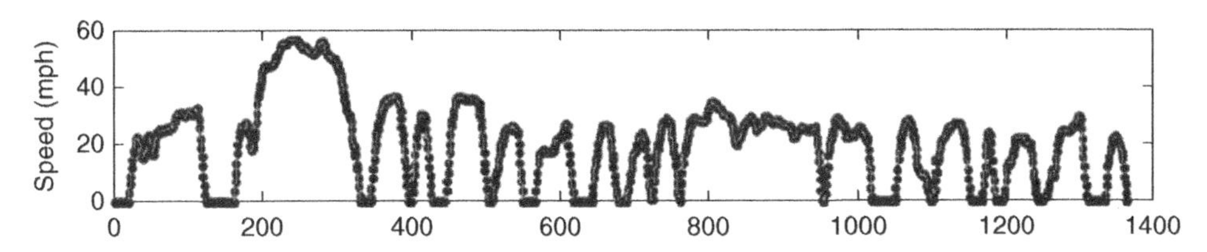

Figure 5.29 EPA UDDS drive range.

$$TA = k_{pt}\Delta V + k_{it}\int \Delta V dt + k_{dt}\frac{d(\Delta V)}{dt} \tag{5.129}$$

$$0 \le TA \le TA_{max}$$

$$\Delta V = max\{(V_{spec} - V), 0\}$$

$$F_B = k_{pb}\Delta V + k_{ib}\int \Delta V dt + k_{db}\frac{d(\Delta V)}{dt} \tag{5.130}$$

$$0 \le F_B \le F_{Bmax}$$

These two equations formulate respectively the throttle controller and the brake controller. In the two equations, V_{spec} and V are respectively the vehicle speed specified in the drive range and the vehicle speed obtained from model simulation, and k_p, k_i, and k_d are the proportion, integration, and differentiation factors of the PID controller, with a second subscript indicating the throttle and the brake. The PID throttle and brake controllers can be easily integrated into the vehicle powertrain system model shown in Figure 5.27.

Output of range simulation: Range simulation provides a broad spectrum of outputs on vehicle dynamics and fuel economy performance on specified traffic patterns:

- Vehicle speed versus time traces. This demonstrates that the vehicle in simulation follows the traffic patterns closely through the throttle controller and the brake controller.
- Engine torque and speed output. This depicts the engine operation status for the whole drive range.
- Shift frequency and operation time in each gear during the simulated range. These indicate how the transmission shift schedule matches the traffic pattern.
- Transients during shifts, including shift time, output torque, and vehicle acceleration variations. These are the data that judge the transmission shift quality.
- Fuel economy and emissions for the simulated drive range obtained by engine map interpolation.

These data from range simulation listed are useful for upfront vehicle validations on acceleration and fuel economy performances, even prior to vehicle prototyping. More importantly, based on these data, engineers can modify the existing transmission shift schedule so as to optimize it for the best fuel economy for the simulated drive range. It is also possible to make the shift schedule adaptive to traffic patterns for fuel economy optimization. The data on transmission shift transients in range simulation provide model-based validations for the transmission shift control strategy and control variables.

This helps engineers to shorten the time needed to achieve the optimum transmission control and calibration.

References

1 https://en.wikipedia.org/wiki/Automatic_transmission
2 Hydra-matic – Wikipedia https://en.wikipedia.org/wiki/Hydra-matic
3 Gott, P.G.: *Changing Gears: The Development of the Automotive*, Society of Automotive Engineers, 1991, ISBN 1-56091-099-2.
4 Gage, K.W. and Rhodes, P.J.: *The New General Motors Hydra-matic Transmission*, SAE Transactions, Vol. 65, 1957, p. 462.
5 Megil, T.W., Haghgooie, M., and Colvin, D.S.: *Shift characteristic of a four-speed automatic transmission*, SAE Paper # 1999-01-1060.
6 Fujii, E., Tobler, W.E., Clausing, E.M., Megli, T.W., and Haghgooie, M.: *Application of Dynamic Band Model for Enhanced Drivetrain Simulation, Journal of Automobile Engineering*, Vol. 216, No. 11, pp. 873–881.
7 Lee, S., Zhang, Y., Jung, D., and Lee, B.: 2014, *A Systematic Approach for Dynamic Analysis of Vehicles with Eight or More Speed Automotive*, ASME Journal of Dynamic Systems, Measurement, and Control, Vol. 136, No. 5, 051008(1-11), 2014.

Problems

1 A four-speed AT for a FWD vehicle is shown in the stick diagram. The clutch engagement schedule is shown in the table. The second gear ratio is designed to be 1.63. The chain ratio is 1.0 and the final drive PGT parameter β_3 is 1.90.

a) When the transmission makes a 3–4 upshift with the torque converter locked at a vehicle speed of 45 mph, the engine speed drops by 640 RPM. Find the gear ratios for all forward and reverse gear ratios.

b) Assuming that the turbine torque is T, find the reaction torque in the first, second, fourth and reverse gears respectively.

c) The transmission is in the process of making a 2–3 upshift. The sensors of the control system detect a vehicle speed of 35 mph and a turbine speed of 2200 RPM. Determine the RPM of sun (S1) and the RPM of carrier (C1).

d) The vehicle is being driven in first gear at a speed of 12 mph on a 5% slope and the vehicle acceleration is 5 ft/s^2. Determine the combined reaction torque in clutches B2 and R3.

Vehicle data:

Front axle weight: 1800 lb
Rear axle weight: 1500 lb
Center of gravity height: 14 inch
Wheel base: 108 in.
Air drag coefficient: 0.31
Frontal projected area: 21 sq.ft
Tire radius: 11.0 in.
Roll resistance coefficient: 0.02
Powertrain efficiency: 0.91

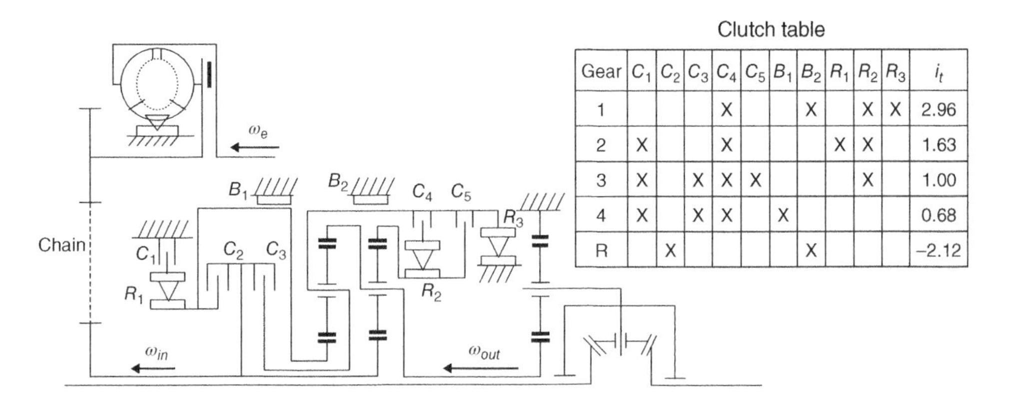

Clutch table

Gear	C_1	C_2	C_3	C_4	C_5	B_1	B_2	R_1	R_2	R_3	i_t
1				X			X		X	X	2.96
2	X			X				X	X		1.63
3	X		X	X	X				X		1.00
4	X		X	X		X					0.68
R		X					X				−2.12

2 The stick diagram of an eight-speed RWD automatic transmission and the clutch engagement are shown in Figure 5.12.

a) Derive the characteristic equations for the planetary gear trains and superimpose the structural constraints upon these equations.
b) Derive the formula for the transmission ratios for all gears in terms of the PGT parameters.
c) Draw the FBD for each of the assemblies in the transmission. The input, output, and reaction torque directions are shown in the stick diagram for the whole transmission and must be followed in your FBDs. Derive the equation of motion for each of the assemblies.
d) Derive the system of dynamic equations after eliminating the gear torques in the equations of motion.
e) Determine the clutch torques for all gears under static condition for unity input torque.
f) Derive the general state variable equation system in matrix form.
g) Derive the specific state variable equation for a 1–2 shift in matrix form.
h) Solve the output angular acceleration $\dot{\omega}_{out}$ for the 1–2 shift in (g) in terms of T_{in} and T_E.

Symbols for vehicle data:

Vehicle weight: W	Tire radius: R
Air drag coefficient: C_d	Frontal projected area: A
Final drive efficiency: η	Roll resistance coefficient: f

3 The stick diagram of the Ford 10-speed RWD automatic transmission for pickup trucks and the clutch engagement are shown in Figure 5.30.

a) Derive the formula for the transmission ratios for all gears in terms of the PGT parameters.
b) Draw the FBD for each of the assemblies in the transmission. The input, output, and reaction torque directions are shown in the stick diagram for the whole transmission and must be followed in your FBDs. Derive the equation of motion for each of the assemblies.

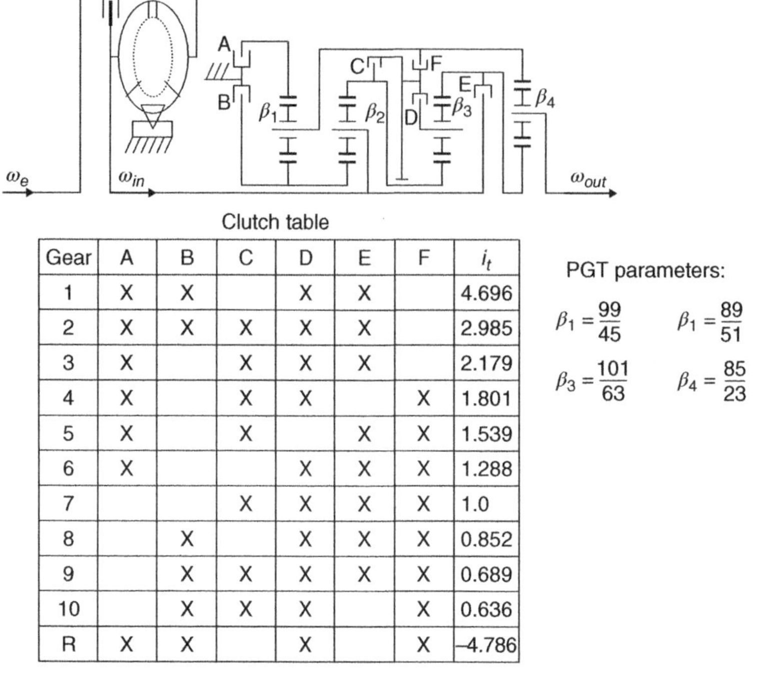

Gear	A	B	C	D	E	F	i_t
1	X	X		X	X		4.696
2	X	X	X	X	X		2.985
3	X		X	X	X		2.179
4	X		X	X		X	1.801
5	X		X		X	X	1.539
6	X			X	X	X	1.288
7			X	X	X	X	1.0
8		X		X	X	X	0.852
9		X	X	X	X	X	0.689
10		X	X	X		X	0.636
R	X	X		X		X	−4.786

Figure 5.30 Ford 10-speed RWD AT.

c) Derive the system of dynamic equations after eliminating the gear torques in the equations of motion.

d) Determine the clutch torques for all gears under static condition for unity input torque.

6

Automatic Transmissions

Control and Calibration

6.1 Introduction

As discussed in the previous chapter, automatic transmissions (AT) that use planetary gear trains (PGT) realize multiple gear ratios by alternately engaging and disengaging hydraulically actuated clutches according to the transmission clutch table. A gear shift in the transmission involves the engagement of an oncoming clutch and the release of an off-going clutch. Two technical issues naturally arise: (a) when should transmission shifts be initiated, (b) how should the clutches involved in a shift be controlled so that the shift is made quickly and smoothly? The primary functions of transmission control systems are therefore to address these two issues.

Clutches in automatic transmissions of early types were both hydraulically actuated and controlled [1]. A gear shift was initiated by a governor that was driven by the transmission output shaft [1,2]. The governor assembly is actually a centrifugal mechanism that has weights rotating with the governor and swinging about a pivot attached to the governor. As the governor rotates, the weights will swing outward due to the centrifugal effect. The position of the swinging weights depends on the governor rotational speed which is in turn linearly related to the transmission output speed. As the weight swings outward, it switches on a pilot hydraulic valve to trigger a shift corresponding to a specific vehicle speed. Once a shift is initiated, the responsiveness and smoothness depend on the hydraulic control circuit that controls the apply pressures of the involved clutches. Hydraulically controlled automatic transmissions had to be exclusively applied in the automotive industry right up to the 1970s.

Automatic transmissions with partial electronic control date back to the early 1970s [3]. In these early types, only shift point control, or shift schedule, were controlled electronically. In a transmission controlled fully electronically, a microcomputer is used to control both the shift schedule and the shift processes through hydraulic solenoids. Automatic transmissions fully controlled by electronic systems were developed by the automotive industry from the mid 1980s and were well received by the automotive market due to the enhanced vehicle fuel economy and drivability. After 1990, all newly developed automatic transmissions, typically with four speeds and lock-up torque converters, were controlled fully electronically. In the control systems of these transmissions, two or three shift solenoids are used as switches in the hydraulic circuits to realize transmission shift scheduling, and one pressure control solenoid is used to control the line pressure

Automotive Power Transmission Systems, First Edition. Yi Zhang and Chris Mi.

and clutch apply pressure [4–13]. This control system layout and its variations were extensively applied for automatic transmissions until the turn of the century, when current generation automatic transmissions were developed and started to enter production in the industry. In comparison with the previous generation, current generation ATs generally have more speeds – mostly with six and some with eight speeds [14,15] – and more advanced control technologies that provide independent control for clutch pressure and line pressure via pressure control solenoids (PCS) or variable force solenoids (VFS). In general, the control system of today's automatic transmissions possess the following functions:

- **Shift schedule control**: The transmission shift schedule is directly related to vehicle acceleration performance, fuel economy, and drivability. The transmission control unit (TCU), or the powertrain control module (PCM), which controls the whole power train system, makes the shift decisions with the input signals on the vehicle speed and the engine throttle opening, and other supplemental input signals on vehicle operation status, for the optimization of performance, fuel economy, and drivability. The shift schedule is established through intensive calibration and can be designed with adaptability and self-learning features to fit the driving style of individual drivers. It can also be designed with different modes, such as "normal", "economy", and "sport", to fit driver priorities and road conditions.
- **Shift process control**: Transmission shift response and smoothness depend on the accurate torque control for the oncoming and off-going clutches. Once the TCU sends out the command signal to make a shift, the apply pressure for the oncoming clutch must be ramped up, and release pressure in the off-going clutch must be ramped down, according to the designed control strategies validated by calibration. Technically, shift process control is the most challenging task in the development of a transmission control system. For transmissions with six or more speeds, there are already a large number of direct shifts – i.e. clutch to clutch shifts – based on transmission design. Each of these direct shifts corresponds to a specific vehicle operation condition and engine operation status and warrants separate clutch torque control and calibration in order to achieve optimized shift quality in the overall drive range.
- **Torque converter locking control**: The status of the torque converter concerns powertrain efficiency and transmission shift smoothness. An open converter functions as a fluid couple and is conducive to smooth shifts, but at the expense of efficiency which can be up to 3% fuel economy loss. To take full advantage of torque converter characteristics without sacrificing fuel economy, a torque converter locking schedule must be designed and calibrated as an important portion of the overall transmission control system. When the torque converter is locked or released, the lock clutch apply pressure or release pressure is ramped up or down using a pressure control solenoid to minimize powertrain transients.
- **Engine torque control during shift**: This is usually achieved by spark retarding in a selected number of engine cylinders. As discussed in the previous chapter, spark retarding during a transmission shift decreases the engine torque to lower the oncoming clutch torque required to complete the shift, thus enhancing shift response and smoothness. Obviously, spark retarding can only be made possible technically in vehicles that are equipped with electronically controlled automatic transmissions.

- **System diagnosis and failure mode management**: In case of component malfunctions, the control system will provide warning signals on the instrument panel and enable vehicle operation under safe conditions in the failure management mode. The vehicle can then be driven to a service station for diagnosis and repair. In addition, control and calibration variables may be modified for further improvements of transmission quality or may even need to be corrected after the vehicle model is already in the market. Therefore, the software of the transmission control system, including control codes and data, should have the capability of being updated if necessary.

The development of transmission control system is always focused on the fulfillment and improvement of the functions just described. This chapter naturally concentrates on the hardware and software technologies applied in the transmission control systems. Following this introduction section, Section 6.2 provides the functional descriptions for the hardware components, including pump, clutches and pistons, accumulators, sensors, various valves, and solenoids, in the transmission control systems. Section 6.3 presents the transmission control system configurations and the related design guidelines. Examples based on production transmissions of the last and current generations will be used to demonstrate the operation logic and functions of the control systems. The section will also provide an introduction to the development of control systems for automatic transmissions for the next generation. Section 6.4 will present concurrent transmission control technologies commonly applied in the automotive industry. The content of this section concentrates on the accurate control of clutch apply pressure during gearshifts. It also details the strategy for torque converter lock and release control. The transmission control system comes with a large number of variables and component characteristics that must be calibrated for the optimization of the control functions. The identification of control variables to be calibrated and calibration of transmission control system will be the topic of Section 6.5 at the end of this chapter.

6.2 Components and Hydraulic Circuits for Transmission Control

The control system of automatic transmissions is typically a synergetic combination of mechanical, hydraulic, and electronic subsystems or components. The mechanical aspects of automatic transmissions have been discussed in detail in Chapter 5. To better understand the functions of transmission control systems, it is necessary to have a general knowledge of the characteristics of individual components and hydraulic circuits. The following text provides the description on the functions of key hydraulic and electronic components, as applied in the transmission control system.

System ATF supply and line pressure control circuit: Driven by the engine, the hydraulic pump generates the ATF flow and pressure in the hydraulic circuit of the transmission control system. Gear pumps and vane type pumps have both been applied in automatic transmissions. The pump consumes power in proportion to its capacity and a fixed volume pump may pump out more ATF than necessary under a given transmission operation, resulting in fuel economy loss. Variable capacity pumps supply an ATF amount to the control system in proportion to the engine RPM to meet system requirements with improved pump efficiency. The design guidelines of hydraulic pumps can be found in open literature, such as SAE publications [2]. Typically, the hydraulic pump forms the system pressure supply circuit with a pressure regulator valve, a pressure limit valve, and a pressure control solenoid or variable force solenoid (VFS) for the supply and control of line pressure by regulating the AFT flow volume entering the main circuit, as shown in Figure 6.1. This circuit design and its variations are generally applied in the control systems of the last generation and current automatic transmissions [4,6,7], such as the GM four-speed Hydra-Matic 4 T80-E [16] and the GM six-speed AT Hydra-Matic 6 T40/45 [17]. It can be predicted also that the control system for future automatic transmissions with full direct clutch pressure control will also be based on designs similar to the line pressure control circuit shown in Figure 6.1. Since the current generation ATs do not use accumulators in the control system, the torque signal port to accumulator control valve is eliminated in the PSC in Figure 6.1.

Pressure regulator valve: The basic structure of this type of valve is illustrated in Figure 6.1. These valves can be designed with in-port and out-port variations. A pressure regulator valve regulates the ATF amount entering the main circuit, or pressure in equivalence, through the combination of valve spring and torque signal pressure, which is also termed throttle signal pressure. The sump port of the valve has a variable opening to let pumped ATF bleed through to the pump suction circuit. To boost the line pressure, the torque signal pressure controlled by a pressure control solenoid acts with the valve spring downward against the upward line pressure force on the valve body, diminishing the sump port opening and letting in more ATF. In the circuit shown in Figure 6.1, there is a reverse ATF port in the pressure regulator valve. When the transmission is shifted into the reverse gear, the ATF pressure in the reverse circuit is routed to the reverse port,

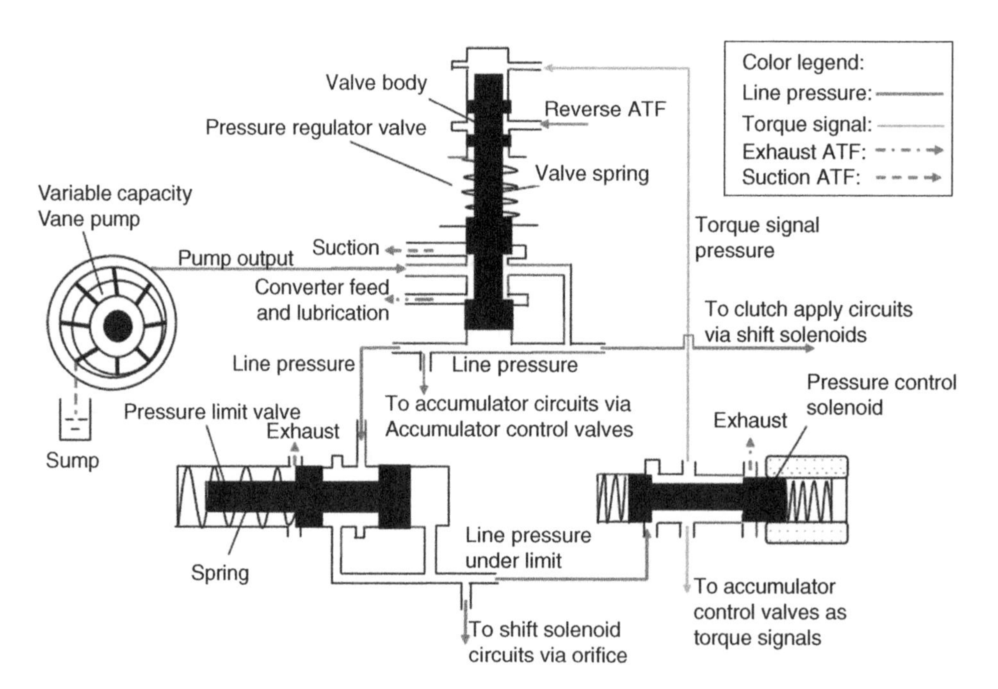

Figure 6.1 System ATF supply and line pressure control circuit.

pushing the valve body down with the valve spring to boost the line pressure that is needed to generate a large reaction torque in the reverse clutch. This also has the effect of filling the hydraulic system circuit quickly to make it ready for shift operations. Note that the pressure supply circuit shown in Figure 6.1 provides stable line pressure to the control system circuit with minimized pressure pulsation caused by the pump.

Pressure limit valve: As shown in Figure 6.1, this valve sets a limit on the line pressure in a very simple structure. The line pressure acts on the valve body against the valve spring. When the line pressure exceeds the limit allowed by the valve spring, the valve body will be pushed by the line pressure to open the exhaust port, thus limiting the ATF pressure passing through the valve. The ATF pressure after the limit valve is then routed to the pressure control solenoid, as shown in Figure 6.1, and to the shift solenoids that actuate transmission shifts. This pressure is therefore termed the actuator feed pressure.

Pressure control solenoid: This is the key valve that electronically controls the ATF pressure with high precision. The structure of a typical pressure control solenoid (PCS) is illustrated by a section view in Figure 6.2. There are two portions – an electronic portion and a hydraulic portion – in a pressure control solenoid. The electronic portion consists of coil assembly, armature, push rod, and spring housed inside the cylindrical frame. The hydraulic portion consists of valve core, spring, valve sleeve, and valve shell. There are three ports in the hydraulic portion on the valve shell: actuator feed fluid in-port from the pressure limit valve, as shown in Figure 6.1, torque signal pressure out-port, and exhaust port. In addition, the actuator feed fluid can also be exhausted from the variable bleed orifice between the valve sleeve and the armature. As can be observed in Figure 6.2, the function of the pressure control solenoid is to control the actuator feed fluid and convert it to the torque signal fluid pressure with high precision. The torque signal pressure is then routed to the pressure regulator valve and accumulator control valves as the pilot pressure for the control of line pressure and clutch apply pressure during shifts. Apparently, the pressure control solenoid is crucial for transmission shift feel since it is the primary, if not the only, component for the real time control of clutch apply pressure.

Pressure control solenoids (PCS), also called variable force solenoids (VFS) or variable bleed solenoids (VBS), work on the principle of pulse width modulation (PWM). In general, the operation of the pressure control solenoid can be explained by duty cycle and frequency. A duty cycle is the percent of time that electric current flows through the coil of the solenoid in a cycle whose time duration is defined by the frequency. In other words, a duty cycle is the percent of time in a cycle the TCU sends electric current through the coil. A pressure control solenoid works at a given specified frequency. For example, GM transmission Hydra-Matic 4 T80-E uses a frequency of 614 Hz for the pressure control solenoid [16]. A 20% duty cycle means that electric current flows in the solenoid coil for $0.2 \times (1/614)$ s (approximately 0.0003 s) in every cycle that lasts for 1/614 s (approximately 0.0016 s), as shown in Figure 6.2. The current in the solenoid coil generates a magnetic field that fills the center of the electronic portion in which the armature is located. The magnetic force, the force of the push rod spring, the hydraulic pressure force on the end of the valve core, and the spring force on the valve core interact with each other to vary the armature and valve core positions for the accurate control of the torque signal pressure.

The pressure control solenoid shown in Figure 6.2 is actually a variable bleed valve (VBS) and can be either "normally high" or "normally low". A normally high pressure control solenoid regulates torque signal pressure in reverse proportion to the current

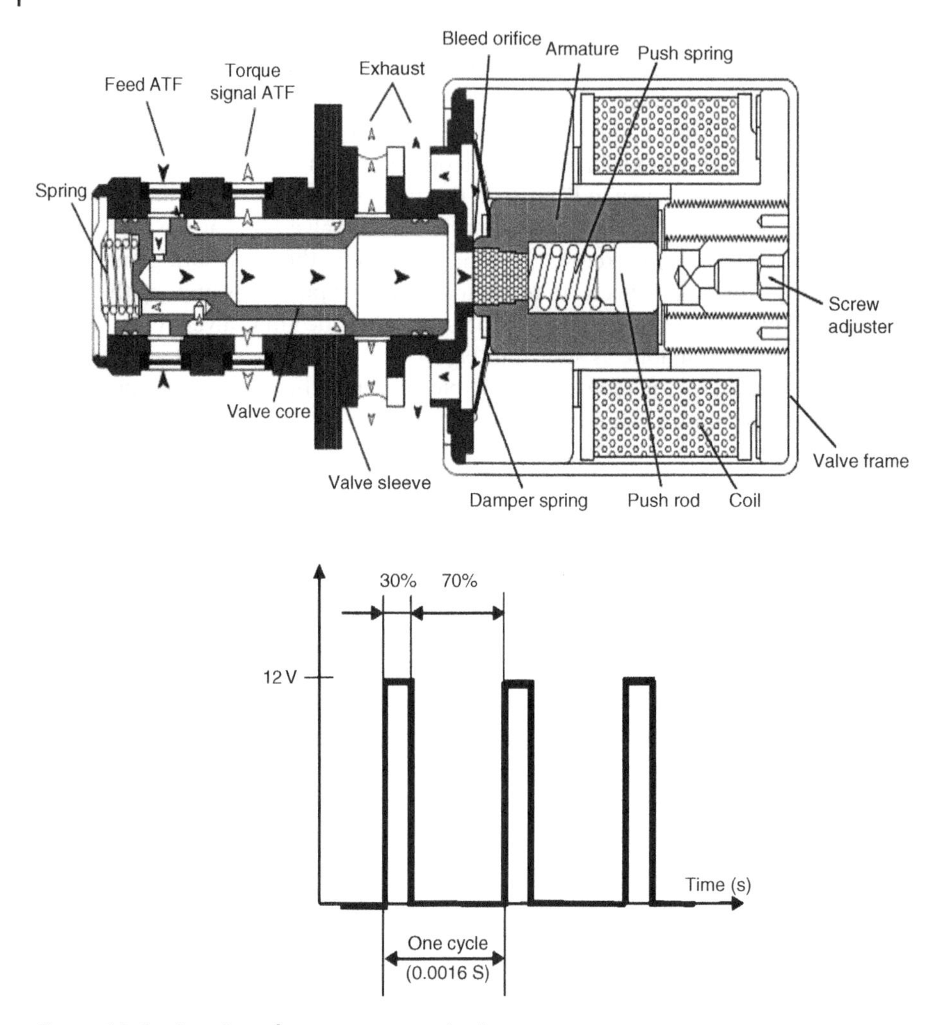

Figure 6.2 Section view of a pressure control valve.

in the coil. The line pressure control solenoid in Figure 6.1 is normally low. It controls the line pressure by bleeding more or less ATF from the valve as controlled by the TCU according to the engine load and vehicle operation conditions. The line pressure needed to secure transmission control functionality varies with the engine load which is reflected by the engine throttle opening. As engine throttle increases, the line pressure is also increased by bleeding less ATF from the pressure control valve through the decrease of duty cycles. By contrary, a normally high pressure control solenoid regulates the out-port pressure in direct proportion to the current in the coil; that is, to increase the out-port pressure, the TCU increases the duty cycles of the coil. Note that the PCS or VFS is energized by duty cycles through PWM by TCU, but the ATF pressure controlled is calibrated against the current through the solenoid coil. When applied in the transmission control system, the TCU monitors the current in the electric circuit from the solenoid to ground and uses this as the feedback to adjust the duty cycles for accurate control of torque signal pressure.

Shift solenoid: This is actually an electronically controlled switch valve in the transmission hydraulic circuit and only has two states: On or Off. A combination of On and Off states of shift solenoids defines a particular gear position. For example, two shift solenoids can be used to define uniquely the four gears in a four-speed AT through the four combinations of On and Off states. A sequential shift can then be triggered by the change of status of one of the two shift solenoids. The section view of a normally open shift solenoid is illustrated in Figure 6.3. The line pressure under limit shown in Figure 6.1 is routed to the shift solenoid via an orifice and acts on the check ball. When de-energized by the TCU, the combined force of the ATF pressure and the spring moves the plunger away from the metering ball, allowing the solenoid ATF from the pressure limit valve shown in Figure 6.1 to unseat the metering ball to be exhausted. When it is energized, current in the coil generates a magnetic field, which in turn generates a magnetic force on the plunger and moves it to seat the metering ball, blocking the solenoid fluid from the exhaust circuit. The solenoid AFT pressure is then used to control the positons of shift valves, which then route ATF pressure for clutch applications in different gears. Note that a pressure solenoid control valve can also function as a switch solenoid with On or Off states, in addition to its pressure control capability.

Shift valve: These valves are used in the hydraulic system circuit for the transmission control unit to route the ATF flow in various operations, including park, neutral, fixed gear operations, and shifts. Typically, one shift valve is used for the shift between two neighboring gears. For example, GM four-speed AT Hydra-Matic 4 T80-E uses three shift valves: 1–2 shift valve, 2–3 shift valve, and 3–4 shift valve in the hydraulic system [16]. A shift valve has multiple ports and two positions that are controlled by the solenoid ATF pressure from a shift solenoid. At each position, a shift valve will either block ATF from entering the clutch apply circuit or let ATF pressure through it to become clutch apply pressure in the clutch apply circuit. Note that the manual valve connected to the shift stick can also be considered as a shift valve manually controlled by the driver.

Clutch: Gear ratios of an automatic transmission are realized by applying different clutches according to the transmission clutch table, as discussed in Chapter 5. Three

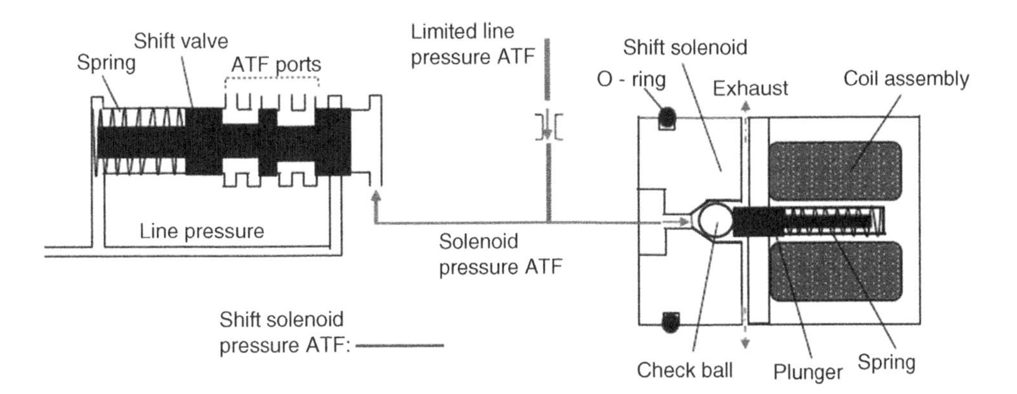

Figure 6.3 Shift solenoid and shift valve circuit.

types of clutches – multiple disk clutch, band clutch, and one-way clutch – are used in automatic transmissions. The one-way clutch is self-actuated and does not need to be controlled. Current generation ATs only use multiple disk clutches due to their compactness and characteristics in apply and release. Some six-speed ATs of the current generation still use a one-way clutch to facilitate the 1–2 shift as explained in Chapter 5. Multiple disk clutches can be used for both coupling and reaction. The basic structure of multiple disk clutch is shown in Figure 6.4. When applied, the two rotating components – the drum and the hub, – are coupled together by the friction generated between the friction disks splined to the hub and the steel plates splined to the drum. To apply the clutch, the clutch apply ATF pressure from a shift valve enters the piston chamber to fill the piston cavity before pressure is built up in the piston. The ATF then pushes the piston against the return spring and then the apply plate. The friction disks and the steel plates are then clamped against each other, and friction is generated on each of the contact faces between the friction disks and steel plates. As the apply pressure is ramped up via clutch pressure control solenoid, the clutch torque will be increased to the level for the full engagement of the clutch.

There are several important variables or characteristics associated with the clutch apply process. As shown in Figure 6.4, clutch apply ATF will fill up the clutch piston cavity for some time while the clutch pressure is built up, and the piston will not move

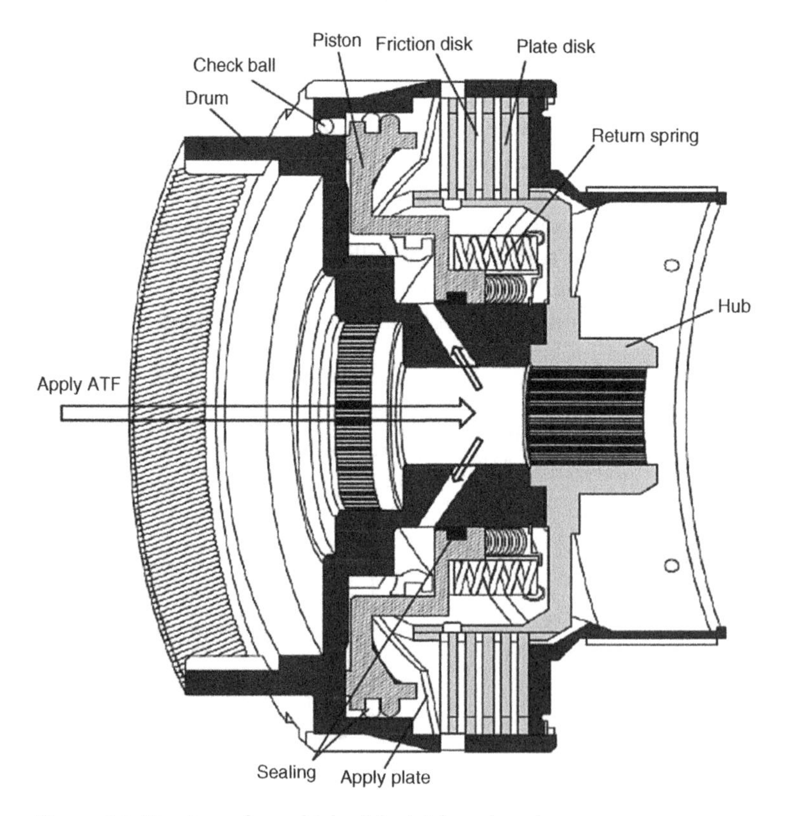

Figure 6.4 Structure of a multiple disk clutch and apply process.

until the pressure build-up is high enough to overcome the return spring force and push the apply plate afterwards. The clutch hardware designed this way in order to smooth out the harshness of the clutch apply. However, the initial clutch apply characteristics adversely affect shift timing and must be calibrated for the accurate control of clutch pressure ramp-up profile. To eliminate the effects of the clutch apply side cavity, the return spring side can be designed with a counter chamber or compensator chamber [8]. Low pressure ATF fills up the compensator chamber and the piston apply side cavity when the clutch is open. This improves the accuracy of clutch apply timing and clutch pressure as shifts are initiated. As shown in Section 2.4 of Chapter 2, if disk wear is assumed to be uniform, the torque capacity of a multiple disk is given by:

$$T_{CL} = n\frac{D+d}{2}\mu\left[p_c A - \left(F_s + kx_p\right)\right] \tag{6.1}$$

where n is the number of friction disks, D and d are the friction disk outer and inner diameters, μ is the friction coefficient, A is the clutch piston area, F_s is the return spring force when the clutch is released, and k and x_p are the spring stiffness and piston displacement respectively. Clutch apply pressure is denoted by p_c, which is controlled by pressure control solenoid during shifts through clutch pressure control circuits.

Clutch pressure control circuit with accumulator: In the control system of the previous generation of automatic transmissions [16], hydraulic accumulators were usually used to dampen clutch apply harshness in a circuit shown in Figure 6.5. ATF from the line pressure control circuit shown in Figure 6.1 is routed through an orifice to fill up the accumulator back side (spring side) and to the accumulator control valve counter side (left side as shown in Figure 6.5), when the transmission is in Park. The accumulator spring is stretched when the related clutch is not applied. Generally, the spring side of all accumulators is filled up by accumulator back pressure ATF in Park and stays this way in the whole range to be ready for clutch apply and release. During clutch apply, ATF from the line pressure control circuit shown in Figure 6.1 is routed to the accumulator valve and is directed by shift solenoids and shift valves to enter the clutch apply circuit. Meanwhile, torque signal pressure from the pressure control solenoid acts on the accumulator control valve against the accumulator back pressure, causing the accumulator piston to displace back and forth so as to control the clutch apply pressure. For the clutch

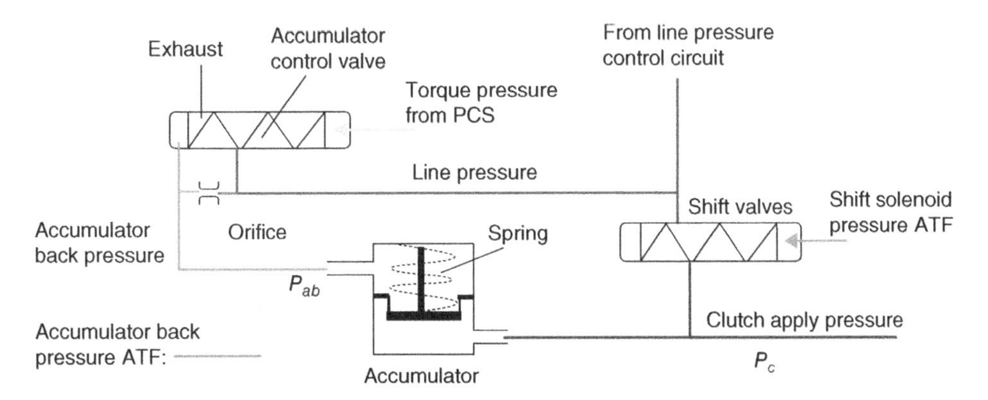

Figure 6.5 Clutch pressure control circuit with accumulator.

pressure control circuit shown in Figure 6.5, the clutch apply pressure p_c is related to the accumulator back pressure p_{ab} as:

$$p_c = \frac{A_{ab}p_{ab} - (F_{sa} - k_a x_a)}{A_{ac}} \tag{6.2}$$

where F_{sa} is the spring force that balances the accumulator back pressure before clutch apply is initiated, and x_a and k_a are the accumulator piston displacement and spring stiffness respectively. The accumulator piston back area A_{ab} is designed to be smaller than the area A_{ac} that receives the clutch apply pressure. During shift operations, the torque signal pressure from the pressure control solenoid acts against the accumulator back pressure to vary the opening in the accumulator control valve. This leads to the control of accumulator back pressure and thus the clutch apply pressure momentarily according to Eq. (6.2) during shifts to satisfactory accuracy. It is noted that p_c is lower than the line pressure during the clutch apply process. The circuit design shown in Figure 6.5 had been applied almost exclusively for clutch apply pressure control in the previous generation automatic transmissions. In general, a particular accumulator is attached to each clutch, and line pressure is routed by shift valves to the accumulator and apply circuit for the related clutch, as shown in Figure 6.5.

Clutch pressure control circuit with independent PCS: In typical current generation six-speed automatic transmissions, no accumulator is used in the hydraulic circuit in the control system. This reduces the number of hardware components, thus lowering the transmission overall size and weight, as well as simplifying the control system hydraulic circuit. Unlike previous generation ATs, which use, as illustrated in Figure 6.1 and Figure 6.5, only one PCS for the control of both line pressure and clutch apply pressure, current generation ATs generally use a PCS to control the apply and release of each clutch in the clutch pressure control circuit shown in Figure 6.6, and a separate PCS to control the system line pressure in a circuit similar to that shown in Figure 6.1.

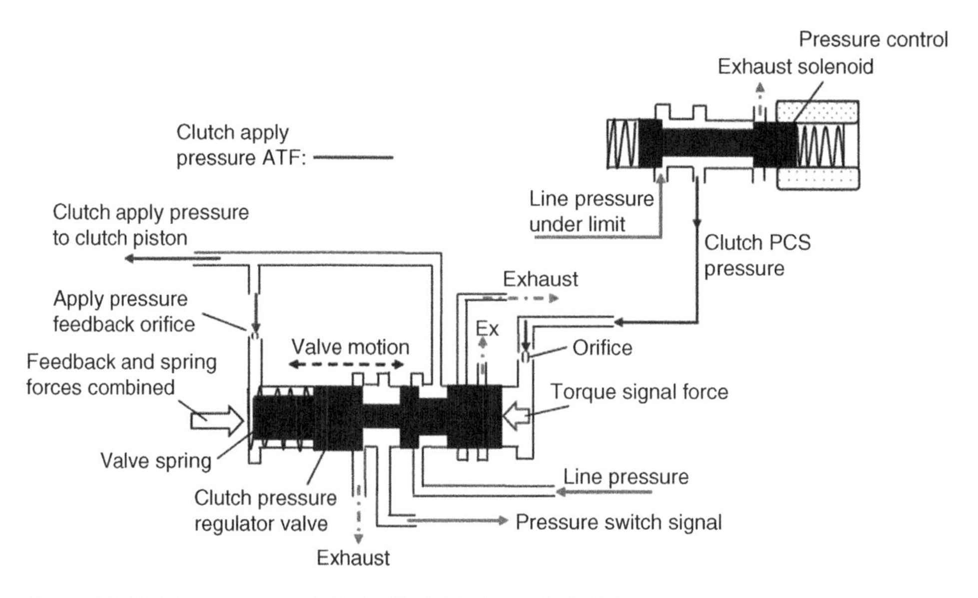

Figure 6.6 Clutch pressure control circuit with independent PCS.

The advantage of the clutch pressure control circuit shown in Figure 6.6 is obvious in comparison with the old version shown in Figure 6.5. Firstly, it eliminates the accumulator for each of the clutches in the transmission. For a transmission that has five clutches, eliminating five accumulators would mean significant reduction in overall transmission dimension and weight. More importantly, the new circuit allows independent clutch pressure control during shift operations. This is crucial for smooth shifts in current generation ATs that feature multiple speeds and clutch to clutch shifts. During shift operations, the ATF pressures in the oncoming clutch and in the off-going clutch are controlled by two separate solenoids. This leads to more accurate clutch pressure control and improved shift smoothness and response. In addition, the independent control of line pressure and clutch pressure control minimizes system disturbance and interference between clutches, which is also conducive to accurate clutch pressure control.

In general, each clutch pressure regulator valve corresponds to a particular clutch in automatic transmission with independent clutch pressure control. In addition to pressure regulation functionality, a pressure regulator valve also functions as a switch with On or Off status to route ATF flow. As shown in Figure 6.6, the clutch pressure regulator valve has multiple ports, including an in-port for line pressure, out-port for clutch apply pressure, exhaust port and pressure switch signal port. Torque signal pressure controlled by the respective clutch PCS is routed via an orifice to one side of the valve and acts against the valve spring force. When the PCS is energized to be at On status, the torque signal force overcomes the spring force and displaces the valve body toward the spring side, creating an opening between the valve shoulder and the line pressure in-port, as shown in Figure 6.6. Controlled by the PCS duty cycle, the torque signal force counteracts the spring force to move the valve body back and forth, varying the opening, or "restriction", for the line pressure to enter the clutch apply circuit.

When the PCS is at Off status, i.e. not energized, there is no torque signal force acting on the valve and the spring force moves the valve body toward the torque signal side and blocks the line pressure from entering the clutch apply circuit. Line pressure is then routed through the valve to become the pressure switch signal which indicates the status of the clutch pressure regulator valve. Switch signals from clutch regulator valves are also used for fault diagnosis and fail mode management function of the transmission control system. To release the clutch, the clutch PCS pressure will be reduced firstly so that the spring force will move the valve body to the right and block the entrance for the line pressure. As the valve body moves to the right, it also creates an opening to the exhaust circuit for the clutch apply circuit. The exhaust opening is controlled afterwards by the PCS to control the clutch release pressure.

As shown in Figure 6.6, the clutch apply pressure is usually fed back to the pressure regulator valve directly via an orifice. If necessary, the clutch pressure control circuit shown in Figure 6.6 can be augmented by a feedback loop on the clutch apply pressure via a clutch pressure boost valve as shown in Figure 6.7. In this setup, the clutch PCS pressure acts on an area differential in the boost valve against the spring to vary the exit opening of the clutch pressure feedback ATF, which is routed via an orifice to the spring side of the clutch pressure regulator valve in the circuit shown in Figure 6.6. As the clutch PCS pressure is controlled to reach a designated level, it will move the boost valve further to the right and let the feedback pressure exhaust.

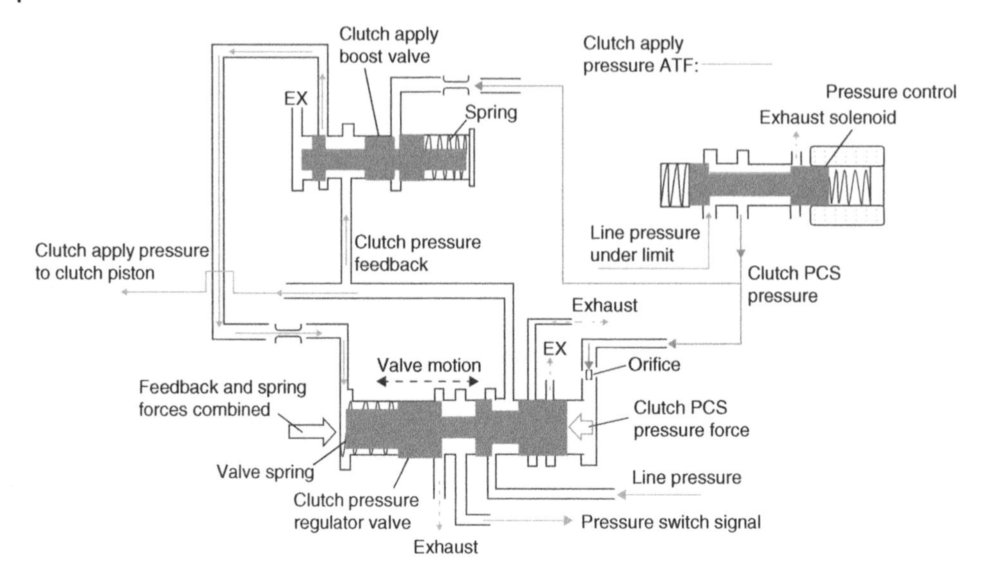

Figure 6.7 Clutch apply pressure circuit with boost valve.

Clutch compensator feed circuit: When a clutch is designed with a compensator chamber on the return spring side, ATF under a low pressure from the compensator feed valve fills the compensator and also the clutch apply side cavity so that the clutch is ready for apply, with minimal effects of the initial apply characteristics. As shown in Figure 6.8, the compensator feed ATF pressure depends on the spring force acting against the orificed compensator feed pressure feedback and is designed at a low value that is not sufficient to overcome the clutch return spring so that no clutch drag is created in the clutch. If the compensator feed pressure p_{cp} exceeds the level allowed by the spring, the valve body will be moved to the exhaust position. In addition to minimizing unwanted clutch initial apply attributes, clutch compensator pressure also increases the responsiveness of clutch actuation during shift operations. Note that the clutch compensator feed circuit is independent of other hydraulic circuits in the control system and its status is kept the same in the whole drive range.

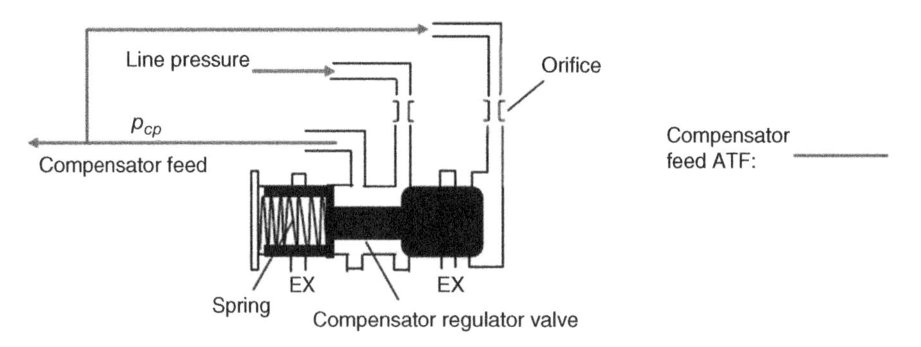

Figure 6.8 Clutch compensator feed circuit.

Torque converter circuit: The torque converter clutch (TCC) apply and release is controlled by the circuit shown in Figure 6.9 or its variations in various automatic transmissions [8,16]. The TCC control valve is basically a position valve controlled by the ATF pressure from the TCC pressure solenoid. When the TCC PCS is not energized, the spring force keeps the TCC control valve in the release position for the feed ATF from the line pressure regulator valve shown in Figure 6.1 to enter the torque converter and leave it to cool. The TCC apply port is blocked by the valve land at the release position. Meanwhile, the TCC regulator valve is actuated by its spring and is kept at the position that blocks the line pressure from flowing through. Therefore, the TCC regulator valve does not have any effect on the converter feed circuit when the TCC PCS is not energized. In some transmissions, such as the GM Hydra-Matic 6 T40/45, the TCC regulator valve also has a shift solenoid pressure port. Shift solenoid pressure is routed to the TCC regulator valve and acts on the shuttle, only in first gear, to close the TCC PCS pressure port, providing a redundant condition that the torque converter will not be locked in first gear. In the whole drive range, the shift solenoid pressure is not routed to the TCC regulator valve, which therefore does not have any effect on the converter feed circuit if the TCC PCS is not energized.

When the TCC PCS is energized, the TCC PCS pressure acts against the spring to move the TC control valve to the apply position, and meanwhile to move the TCC regulator valve to the pressure regulating position, as shown in Figure 6.9. In the pressure regulating position, line pressure ATF flows through the TCC regulator valve to enter the TCC apply pressure regulation circuit. During the torque converter locking up process, the TCC PCS provides the pilot pressure to the TCC regulator valve to act against the spring and the TC apply pressure feedback, as illustrated by Figure 6.9. The TCU varies the duty cycles of the TCC PCS to control the TCC PCS pressure, which in turn controls the valve opening for the line pressure ATF to enter the TCC apply regulation circuit. The TCC apply ATF then flows through the TCC control valve to the apply side or feed side of the torque converter pressure plate to lock up the clutch.

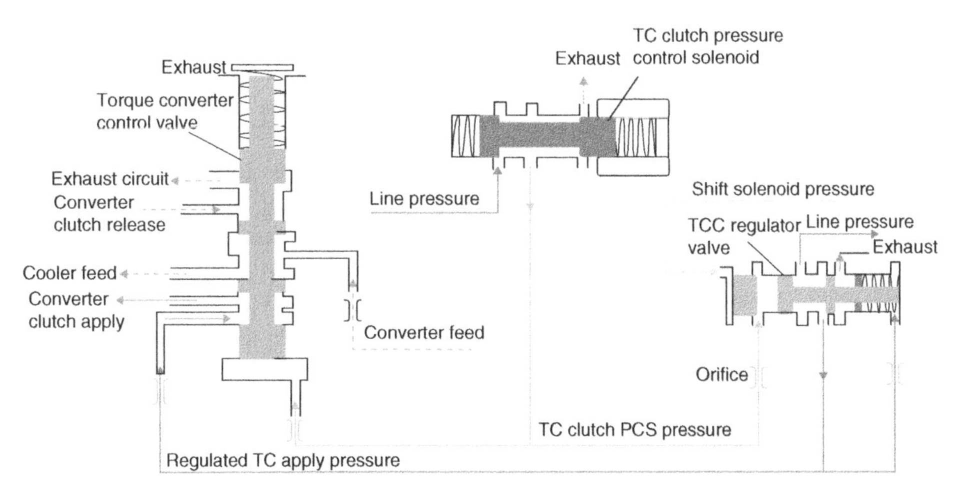

Figure 6.9 Torque converter clutch pressure control circuit.

To release the torque converter clutch, the TCU firstly decreases the TCC PCS pressure to a level enough for the spring force to move the TCC control valve and the TCC regulator valve both to the release position. The TCC apply ATF circuit is then linked with the cooler feed port of the TCC control valve, and the converter feed ATF port is linked to the TC release circuit. ATF is then released from between the pressure plate and the cover of the torque converter to the exhaust circuit, as discussed in Chapter 4.

Speed sensors: Hall-effect type speed sensors are used in transmission control systems to measure the angular velocities of the transmission input and output. A speed sensor is positioned to face a toothed wheel (called a reluctor wheel) that is mounted on, and turns with, the shaft. As the reluctor wheel turns, the speed sensor produces an electric signal at a frequency correlated to the number of teeth and the rotational speed of the shaft. Typically, there are three speed sensors in the control system of an automatic transmission: one on the transmission input, another on the transmission output, and the third on the engine output or the impeller of the torque converter. The TCU receives signals from these three speed sensors via the respective circuit and processes these signals for various control operations. Note that speed sensors are the most important sensors in the transmission control systems; they provide real time data for the TCU to make decisions for shift schedule control, shift point control, and torque converter lock and release control, as discussed in Section 6.4.

ATF temperature sensor: The ATF temperature sensor is basically a thermistor that changes resistance in reverse proportion to temperature changes. A reference voltage is supplied to the circuit of the temperature sensor. The TCU measures the voltage change in the circuit as a measure of the ATF temperature and it can use this information to make adjustments to the shift schedule control, shift point control, and the torque converter lock and release control.

Shift range sensor or switches: This sensor provides information on the driver's selection of drive range. For automatic transmissions with manual shift option, this sensor also sends a signal to the TCU on driver's intention to initiate a shift. For transmissions that use a shift switch instead of a shift stick, the shift switch provides the TCU with similar signals to those of the shift range sensor.

In addition to the sensors mentioned, the transmission control system also shares signals via control area network (CAN) from numerous sensors that are mounted separately for other vehicle systems, mainly the engine control system. The following is a list of these sensors and how signals from these sensors are used in the transmission control system.

Engine throttle opening sensor: The engine operation status is determined by the engine throttle opening and the engine RPM. The engine throttle position is one of the two main variables – vehicle speed and engine throttle position – which determine the shift schedule. Since the transmission control strategy is torque based, the signal from the engine throttle sensor is used by the TCU for almost every aspect of transmission control, including shift point control, shift process control, and torque converter lock and release control, as will be discussed in Section 6.3.

Other engine related sensors include **engine coolant temperature sensor**, **crankshaft position sensor**, and **manifold pressure sensor**. These sensors complement the engine speed and throttle sensors for accurate determination of engine operating status and provide engine related data to the TCU for the control of transmission shift schedule and shift processes.

The operation of the air conditioning system affects the engine load and net transmission input torque, especially for passenger vehicles equipped with small engines. An **air conditioner switch** provides the On or Off status of the air conditioning system to the TCU, which calculates the net transmission input torque and adjusts the line pressure and clutch pressure accordingly. Note too that the TCU interacts via CAN with other vehicle operations or systems, such as the ABS system, the vehicle stability program, and cruise control operation, for decision making in the transmission shift schedule.

Transmission control unit: An example TCU layout is shown in Figure 6.10. As the core of the transmission control system, the TCU receives and processes signals from various sensors described above, performs calculations with the software that implements control strategies for shift schedule and shift processes, and sends commands to shift solenoids and pressure control solenoids for the accurate control of line pressure, shift point, clutch pressures during shifts, and pressure for the torque converter lock and release clutch. The TCU of current generation ATs may use 32-bit or 64-bit microprocessors (CPU). For example, the Delphi TCM8 transmission controller uses a 32-bit, 80 MHz microprocessor with 1.5 MB flash memory and 56 kB RAM. Transmission controllers, such as the Delphi TCM8, must possess the high-speed processing capability required for real time transmission control and sufficient memory for the storage of the control software and database. A TCU with a 32-bit CPU is usually sufficient for handling all function requirements for ATs with up to six speeds. Typically, transmission controllers are equipped with built-in input–output devices or drives for signal reception and control command delivery. The Delphi TCM8 features a configurable pulse width modulation (PWM) and pressure control solenoid drive with current feedback.

Stored in the TCU as EEPROM, the transmission control software consists of two parts: programs and database. The programs include the CPU operating system, I/O interface, and driver functions, software code for transmission control functions, as well as diagnostics and failure mode functions. The database contains data in several groups: (a) attributes of powertrain subsystems, such as engine maps and torque converter characteristics; (b) transmission calibration data, such as line pressures and clutch pressures under various operation conditions, clutch torque profiles, and clutch friction coefficient look-up table; and (c) shift schedules for different transmission operation modes. Note that transmission TCUs are capable of software and database upgrading.

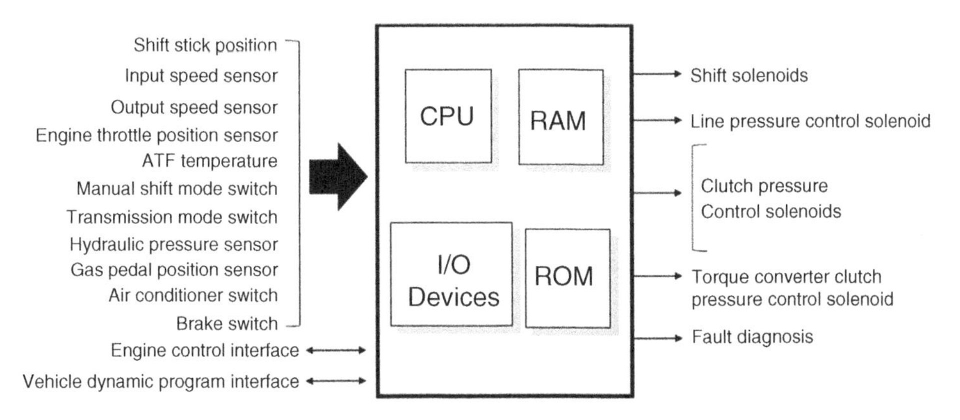

Figure 6.10 Configuration of transmission control unit (TCU).

6.3 System Circuit Configurations for Transmission Control

The system hydraulic circuitry for the control of automatic transmissions is the integration of the components and sub-circuits discussed in the previous section. In general, the sub-circuits for line pressure control, the torque converter lock, and the release control and clutch pressure control are similar in the control systems of various ATs. As discussed in Section 6.2, the status of certain sub-circuits, such as the accumulator back pressure circuit and the clutch compensator feed circuit (if applied), remains unchanged in all drive ranges. The main difference lies in the ATF routing circuits that are controlled by the shift solenoids via the shift valves. The following guidelines should be useful in the design and analysis of the hydraulic circuitry for automatic transmission control systems.

- The transmission gear position and shift logic are uniquely defined by the combinations of On and Off status of shift solenoids. In ATs with four or five speeds, two or three shift solenoids are sufficient to define all the transmission operation states, including fixed gear operations and shift operations. For automatic transmissions with all clutches independently controlled by pressure control solenoids (PCS), the number of PCSs is the same as the number of clutches, and the transmission operation state is naturally defined by the On or Off status of the related PCSs. A clutch to clutch shift would then correspond to the changes of status of the two PCSs that control the two concerned clutches respectively. If a one-way is used in a transmission, then a shift solenoid can be used, in addition to the PCSs that control all the other clutches, to define the transmission status and route the clutch apply ATF.
- The system hydraulic circuit should be designed to fulfill control functionality with minimized complexity. This is not only cost advantageous but also reduces unwanted attributes in the hydraulic circuit that worsen hydraulic pressure control response and accuracy.
- For automatic transmissions with clutches independently controlled by PCSs, the clutch pressure control circuits are laid out similarly to that shown in Figure 6.6, and if necessary a clutch apply pressure feedback circuit shown in Figure 6.7 can be applied for a particular clutch.
- Clutch compensator circuits are designed for clutches involved in shifts that are frequently made and most sensitive to the driver in response and harshness. As mentioned previously, compensator feed circuits are separate from the rest of the hydraulic system.

This section will briefly present the system designs for the hydraulic circuitry for the control of the previous generation of ATs, with the GM Hydra-Matic 4 T80-E as the example for illustrative purposes. The section will then focus on the control system configurations of the current generation ATs as exemplified by the GM Hydra-Matic 6 T40/45. The transmission control strategies to be covered in Section 6.4 will be discussed in reference to the current generation ATs. Finally, the section will provide an introduction to the control systems of automatic transmissions currently under development.

6.3.1 System Hydraulic Circuitry for the Previous Generation of ATs

As discussed in Section 5.2, the GM four-speed AT, Hydra-Matic 4 T80-E, uses five multiple disk clutches, two band clutches, and three one-way clutches to realize clutch to one-way clutch shift for all sequential shifts. For readers' convenience, the stick diagram with the clutch table and the status of the shift solenoids is shown in Figure 6.11. Clutch C_5 is only applied in third gear for engine braking during coasting. Two solenoids, A and B, are used in the transmission control system to define the gear position and to initiate shifts. The park position (P), reverse gear (R), neutral position (N), and first gear share the same solenoid status. The manual shift valve routes the ATF flow in these positions respectively. In the drive range, a gear position is defined respectively by a combination of "on" and "off" of the two solenoids. Because all sequential shifts are clutch to one-way clutches, each sequential shift is triggered by the change of status of one solenoid. A skip downshift would be triggered by the change of statuses of both solenoids, as can be observed in the clutch table.

Except for the coasting clutch C_5 and the band clutch B_2 that is applied only in P, R, N, and first gear, the apply circuit for the other five hydraulically actuated clutches features an accumulator respectively. Each of the five accumulators is used in the clutch apply pressure control circuit for the shift process involving the related clutch, in a layout shown in Figure 6.5. Two accumulator control valves are used for the control of clutch apply and release pressure for all sequential and skip shifts. For a sequential shift, one shift solenoid will change its status to control the position of a shift valve in

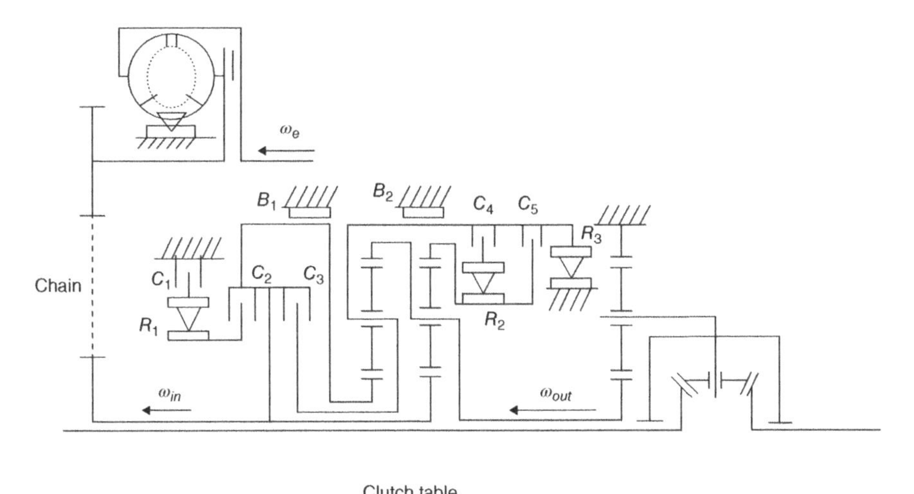

Clutch table

Gear	SOL A	SOL B	C_1	C_2	C_3	C_4	C_5	B_1	B_2	R_1	R_2	R_3	i_t
P	ON	OFF							X				
R	ON	OFF		X					X				−2.12
N	ON	OFF							X				
1	ON	OFF				X			X		X	X	2.96
2	OFF	OFF	X			X				X	X		1.63
3	OFF	ON	X		X	X	X				X		1.00
4	ON	ON	X		X	X		X					0.68

Figure 6.11 Stick diagram, clutch, and shift solenoids table for GM Hydra-Matic 4T80-E.

Figure 6.5 and route the line pressure to the accumulator apply side to close the clutch apply circuit. In a skip shift, the statuses of both solenoids will change to reposition the shift valves and route the line pressure to the two accumulators for the two clutches involved in the shift.

The system configuration of the hydraulic circuitry for the GM four-speed AT, Hydra-Matic 4 T80-E is illustrated in Figure 6.12. The sub-circuits for line pressure control, TC clutch apply and release control, and clutch pressure control have been discussed previously and are drawn as blocks in the system configuration. In addition to the manual valve, the two shift solenoids control the positions of the three shift valves, routing the line pressure to the respective clutch apply circuit for fixed gear and shifting operations.

As can be observed from Figure 6.12, there is only one PCS, namely the line PCS, which controls both the line pressure and the clutch apply pressure during shifts. When a sequential shift is to be initiated, the TCU signals one of the two solenoids, either A or B, and flips its status. Then the shift solenoid pressure acts against the spring of a related shift valve and repositions it. This will route the line pressure through the shift valve and connect it to the intended clutch pressure control circuit. The TCU then sends electric current to the line PCS by pulse width modulation (PWM) and controls the torque signal pressure. The torque signal pressure acts as a pilot pressure to regulate the clutch apply pressure in the corresponding clutch pressure control circuit. For a skip downshift, the TCU flips the statuses of both solenoids A and B, which reposition the shift valves. The ATF in the off-going clutch circuit is then connected to the exhaust circuit, while the oncoming clutch apply circuit is connected to the line pressure, both by the repositioned shift valves. The torque signal pressure from the line PCS then rapidly ramps up the oncoming clutch apply pressure to complete the downshift.

The shortcomings of the hydraulic circuitry shown in Figure 6.12 are obvious. Firstly, the clutch apply and release pressures are not controlled by PCS independently, making it difficult to accurately control apply pressure and shift timing. Secondly, since the line PCS controls both the line pressure and the clutch apply pressure during a shift, the pressure of other clutches that need to be applied during the shift may be affected. This may even result in slippage in these clutches if the line pressure fluctuates too much during the shift in the control of the oncoming clutch apply pressure. To overcome these shortcoming, engineers had adopted two approaches in the design and control of the previous generation ATs. On the design side, one-way clutches are widely used in the transmission to realize clutch to one-way clutch sequential shifts. This is typified by the GM Hydra-Matic 4 T80-E shown in Figure 6.11, which uses three one-way clutches and has all sequential shifts clutch to one-way clutch. On the control side, hydraulic accumulators are extensively used in the control systems for the clutch apply circuits. The adoption of these two approaches greatly enhances the shift smoothness of the previous generation ATs with four or five speeds, but of course at the expense of hardware cost and overall transmission weight and dimensions.

6.3.2 System Hydraulic Circuitry for ATs with Independent Clutch Pressure Control

ATs of the current generation in mass production typically have six speed and feature clutch to clutch sequential shifts, except for the 1–2 shift in some designs. There are two apparent differences in the clutch control circuits between the previous and current

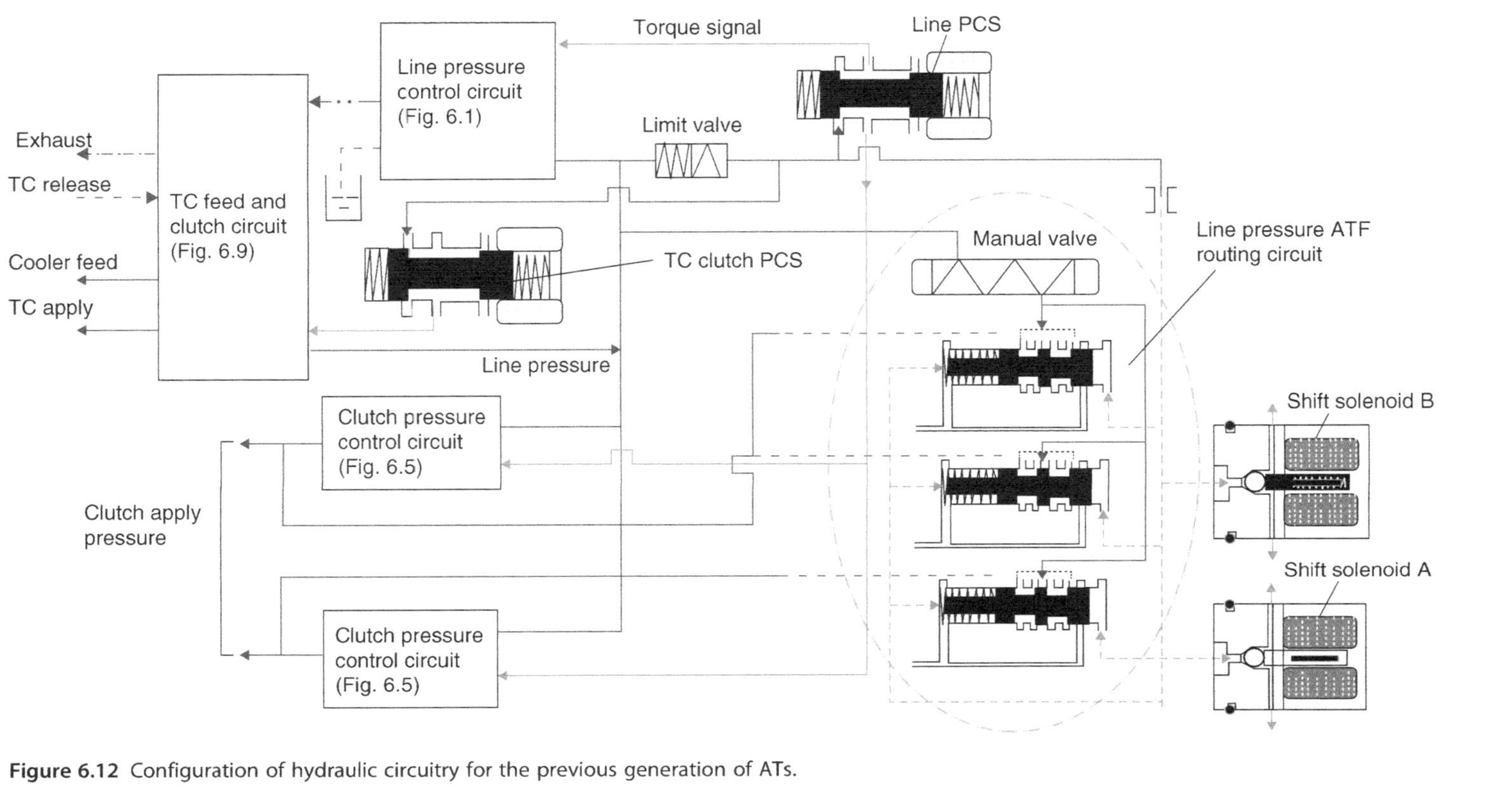

Figure 6.12 Configuration of hydraulic circuitry for the previous generation of ATs.

generation ATs, as can be observed in Figures 6.5 and 6.6. In the ATs of the current generation, accumulators are eliminated in the clutch pressure control circuit, and a PCS designated to a particular clutch controls the pressure regulator valve position to control the clutch apply pressure independent of other sub-circuits. This setup overcomes the shortcomings of the hydraulic circuitry of the previous generation ATs, with substantial reduction in overall transmission dimension and weight. The system hydraulic circuitry of ATs with independent clutch pressure control is illustrated in general form in Figure 6.13.

In the system hydraulic circuitry shown in Figure 6.13, the line pressure control circuit and the torque converter clutch control circuit are illustrated from Figures 6.1 and 6.9, as discussed in Section 6.2. The compensator feed circuit shown in Figure 6.8 is designed for clutches with compensator pistons, and the status of this circuit remains unchanged in the whole transmission operation range. The clutch pressure control circuit can be designed in the setup shown in Figure 6.6 or Figure 6.7, depending on the pressure control requirements for the specific clutch. The manual shift valve has five positions, Park, Reverse, Neutral, Drive, and Manual, which correspond to different ports, and serves as a switch to route the ATF under line pressure, respectively. In some designs, a position sensor provides the signal indicating the shift level position for the TCU to control the statuses of the clutch solenoids accordingly, eliminating the need for the manual shift valve to route line pressure ATF.

The hydraulic circuit design illustrated in Figure 6.13 and its variations can be applied in all ATs with clutch to clutch shifts and independent clutch pressure control, such as the Ford six-speed RWD AT discussed in Section 5.3 and shown in Figure 5.11. In the Ford six-speed RWD AT, there are five multiple disk clutches and no one-way clutch,

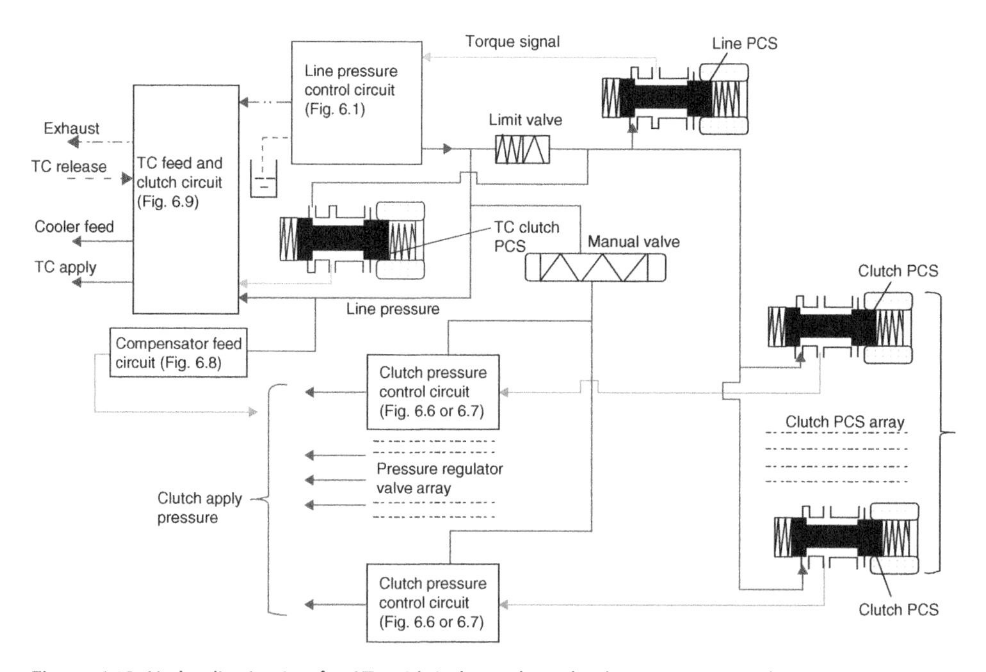

Figure 6.13 Hydraulic circuitry for ATs with independent clutch pressure control.

with all shifts clutch to clutch. Therefore, there will be five clutch pressure control solenoids (PCS) and five clutch pressure control circuits. Each clutch PCS controls independently one of the five pressure control circuits designated for a particular clutch. In addition to controlling clutch pressure, each clutch PCS also possesses the On or Off status. Therefore, all transmission operation statues, fixed gear operations or shift operations, are uniquely defined by the combinations of the On or Off status of the five clutch pressure control solenoids.

As discussed in Chapter 5, many current generation automatic transmissions with six or eight speeds – such as the Ford six-speed FWD AT and the Lexus eight-speed RWD AT shown in Figures 5.10 and 5.12 respectively – still use a one-way clutch in first gear as the reaction clutch and the 1–2 shift is therefore clutch to one-way clutch. In these transmissions, it is not necessary to use a PCS per each of the clutches. A clutch pressure control circuit can be shared by the clutch applied in first gear for engine braking and another clutch that is applied in higher gears, similar to the design shown in Figure 6.12, by using a shift solenoid for ATF routing and a PCS for clutch apply pressure control. The pressure control circuits for all other clutches are laid out as shown in Figure 6.13.

In the GM six-speed FWD AT, Hydra-Matic 6 T40/45, one-way clutch F serves as the reaction clutch in the first gear of the drive range and clutch D serves as the reaction clutch in the first gear with engine braking capability, as shown in Figure 6.14. When the vehicle is launched in first gear, clutch D is applied as the reaction clutch, and is then released after launch before the 1–2 shift is initiated. This design makes the 1–2 shift in the drive range clutch to one-way clutch, with engine braking capability at low vehicle speed in first gear. As shown in the clutch and solenoid status table, the control system uses four PCSs and one shift solenoid (SS). The shift solenoid status is On in P, R, N, and D1 EB (engine braking), and is Off in the whole drive range, flipping only once for the status change from D1 EB to D1. The P-R, R-N, and N-D changes are triggered by flipping the status of one PCS respectively, and 1–2 shift is triggered by the flipping of the status of the clutch C PCS. All other sequential shifts are triggered by flipping the status of two PCSs. It is also interesting to note that skip downshifts to first gear from 3rd and 4th gears are triggered by flipping one PCS.

The hydraulic circuit for the control system of the GM Hydra-Matic 6 T40/45, shown in Figure 6.15, is based on the configuration shown in Figure 6.13 with minor modifications. Clutch A, applied in the fourth, fifth, and sixth gears, and Clutch D, applied in the reverse and first gear (with engine braking), share the same PCS (PCS A), with a shift solenoid to control the shift valve position for line pressure ATF routing. The shift valve has two out-ports for clutch apply pressure, one for clutch A and another for clutch D. It also provides passages for the line pressure ATF in the drive range for other purposes, such as the line pressure input for the torque converter clutch pressure control circuit shown in Figure 6.9. The shift solenoid is energized as On in P, R, N, and D1 EB and the shift valve is positioned by the SS pressure acting against the valve spring. At this position, clutch D remains applied by the pressure regulated by the clutch A and D pressure control circuit. Meanwhile, the manual valve provides additional routing for the line pressure ATF. When the manual valve is at P, PCS A is energized to be On and all other PCSs are Off. Line pressure ATF is routed to the clutch A and D pressure control circuit to apply clutch D and to fill the compensator feed circuit. When the manual valve is switched to R from P, it routes the line pressure through the shift valve passage to enter

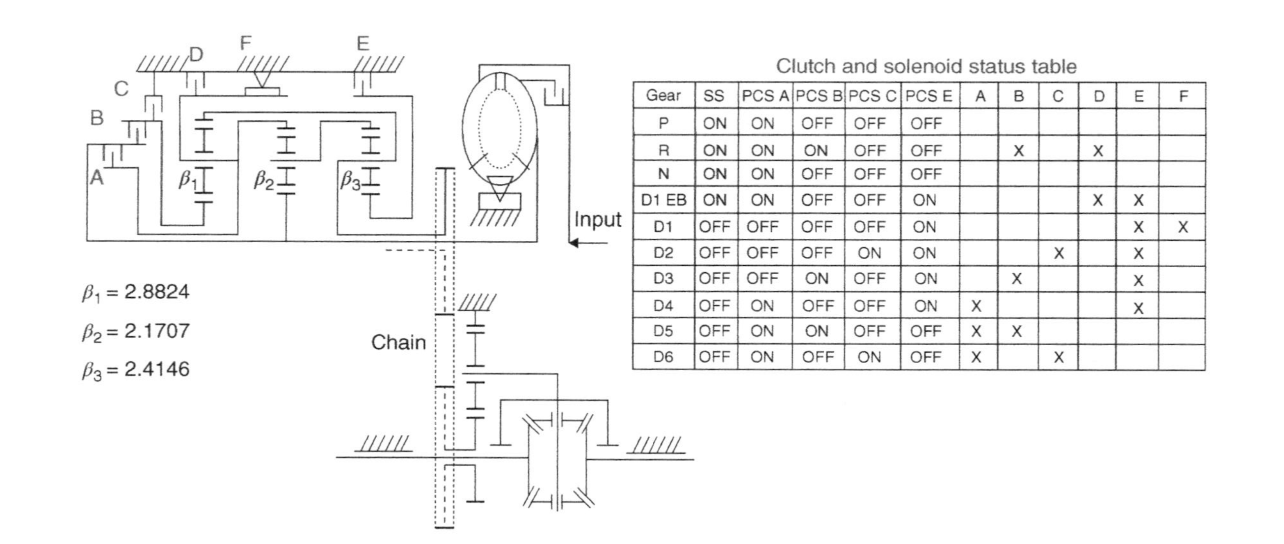

Gear	SS	PCS A	PCS B	PCS C	PCS E	A	B	C	D	E	F
P	ON	ON	OFF	OFF	OFF						
R	ON	ON	ON	OFF	OFF		X		X		
N	ON	ON	OFF	OFF	OFF						
D1 EB	ON	ON	OFF	OFF	ON				X	X	
D1	OFF	OFF	OFF	OFF	ON					X	X
D2	OFF	OFF	OFF	ON	ON			X		X	
D3	OFF	OFF	ON	OFF	ON		X			X	
D4	OFF	ON	OFF	OFF	ON	X				X	
D5	OFF	ON	ON	OFF	OFF	X	X				
D6	OFF	ON	OFF	ON	OFF	X		X			

Figure 6.14 Clutch and solenoid status table for GM Hydra-Matic FWD six-speed AT.

the clutch B pressure control circuit, and PCS B is energized to be On and controls clutch B apply. When the manual valve is switched from R to N after a short time of operation in R, the status of the hydraulic circuit returns to that of P, with PCS B flipping back to Off. In a similar fashion, when the manual valve is switched to D, it routes the line pressure ATF to the in-ports of the pressure control circuits of clutches B, C, and E, making them ready for apply as required by the shift schedule. At this time, only PCS E flips from Off to On to control clutch E apply via its pressure control circuit. The transmission will then operate in D1 EB (first gear with engine braking) for a short time until both the SS and PCS A flip from On to Off. Then, the SS pressure is exhausted and the shift valve is repositioned by the valve spring. The shift valve remains in this position in all gears in the drive range and serves only as a line pressure ATF router. Note that the transfer from D1 EB to D1 is completed by only releasing clutch D; the one-way clutch F will automatically take over the role as the reaction clutch in first gear from clutch D. Once the transmission operates in the drive range from D1 to D6, the gear positions and shifts are then defined by the statuses or status flips of the four PCSs, as illustrated in the clutch and solenoid status table in Figure 6.14.

6.3.3 System Hydraulic Circuitry for ATs with Direct Clutch Pressure Control

The hydraulic circuits of automatic transmissions with direct clutch apply pressure control are further simplified from the hydraulic circuits of the current generation ATs as shown in Figure 6.15. As the industry trend, next generation ATs with eight or more speeds will feature direct clutch pressure control technology. In this new transmission control technology, clutch apply and release pressures are directly controlled by large

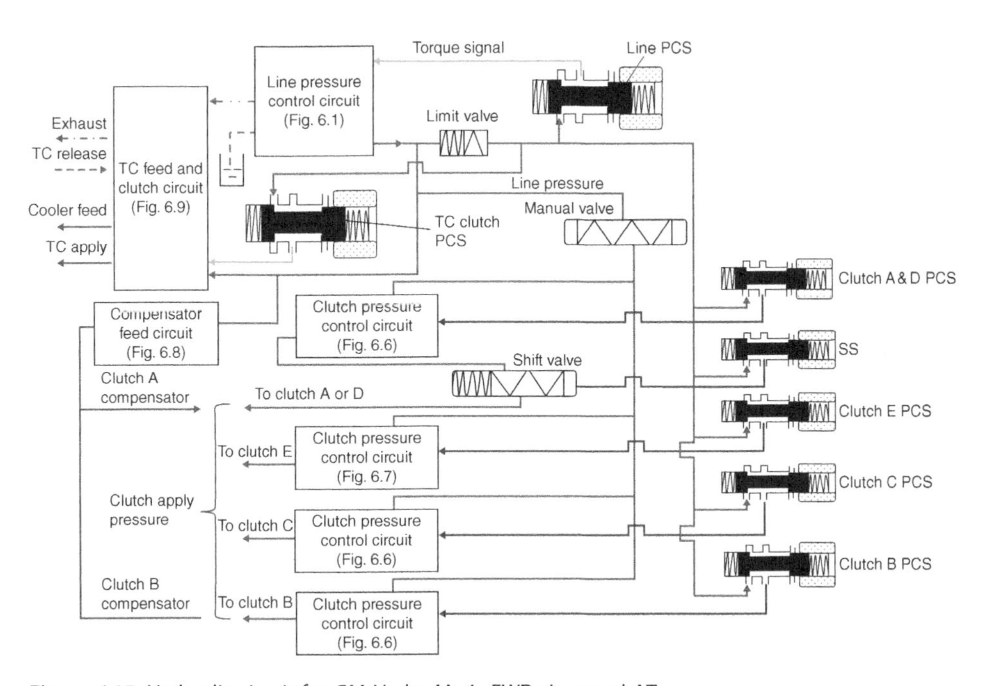

Figure 6.15 Hydraulic circuit for GM Hydra-Matic FWD six-speed AT.

volume pressure control solenoids (PCS), also called variable force solenoids (VFS) or variable bleed solenoid (VBS), without the pressure control circuits shown in Figures 6.6 and 6.7. Meanwhile, the manual valve, the shift valve, and the shift solenoid are also eliminated from the hydraulic system. The shift lever or shift nub only provides signals to the TCU on the positions indicating P, R, N, and the drive range D. Each clutch in the transmission corresponds to a specific PCS or VFS. Therefore, the apply pressure of the oncoming clutch and the release pressure of the off-going clutch are independently controlled by the respective VFS in all shifts. This results in enhanced accuracy in clutch pressure control and shift timing control, leading to optimized shift smoothness and shift response. In addition, the elimination of complex hydraulic circuits is conducive to overall transmission weight and cost reduction. Note that the hydraulic circuits for line pressure control and torque converter clutch control in ATs with direct clutch pressure control are the same as, or similar to, the circuits shown in Figures 6.1 and 6.9. If allowed by package space, clutches with dominating effects on shift smoothness and response are designed with compensators. The compensator feed circuit shown in Figure 6.8 is designed to fill the clutch piston compensator to minimize clutch initial apply attributes and to enhance shift response. The compensator feed circuit remains the same status in the whole drive range. The hydraulic system configuration implementing direct clutch pressure control is illustrated in Figure 6.16. This configuration or its variations can be used for the hydraulic circuit design for the control systems of next generation automatic transmissions. As an example, it can be well fitted to the control system of eight-speed ATs [14,15], such as the ZF RWD eight-speed AT that is shown in Figure 5.14 and

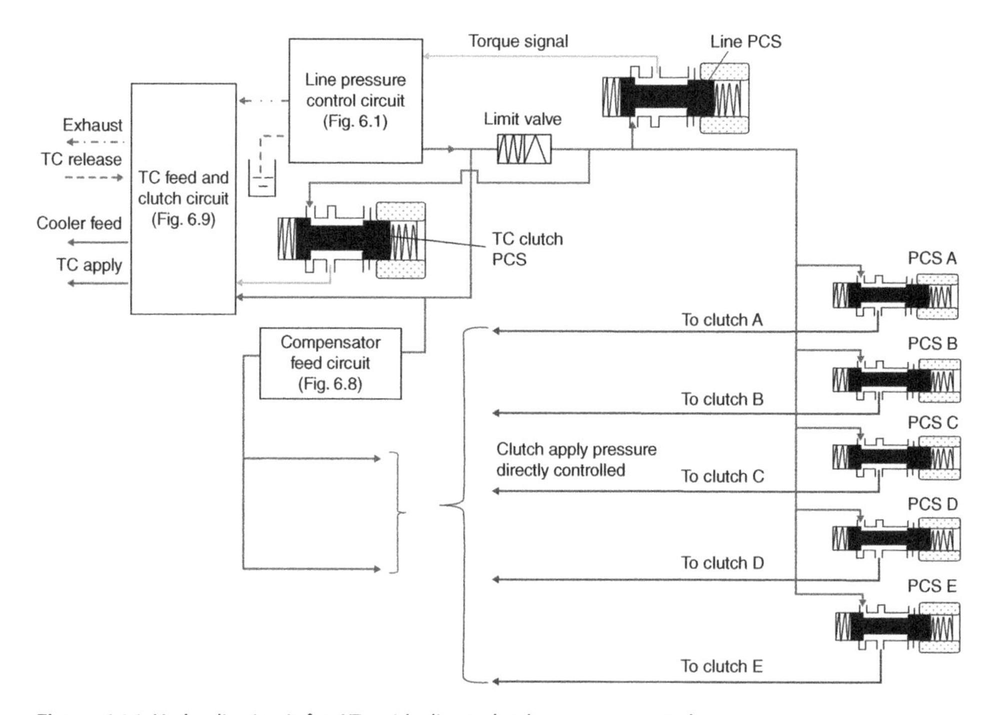

Figure 6.16 Hydraulic circuit for ATs with direct clutch pressure control.

analysed in Sections 5.3 and 5.4. For the ZF RWD eight-speed AT, the five clutches, A, B, C, D, and E are respectively and directly controlled by PCS A, PCS B, PCS C, PCS D, and PCS E. In any direct shifts, sequential or skip shifts, one PCS controls the apply pressure of the oncoming clutch, and another PCS controls the release pressure of the off-going clutch.

6.4 Transmission Control Strategy

As described in Section 6.1, the control system of an automatic transmission must possess five basic functions: (1) shift schedule control, (2) torque converter locking control, (3) engine torque control during shifts, (4) shift process control, and (5) system diagnosis and failure mode management. These functions are actuated by the transmission controller, based on the signals from relevant sensors and are executed according to predesigned and calibrated strategies or algorithms. The previous sections have provided a general description of the hardware components and hydraulic circuits for the transmission control systems. This section highlights the strategy and techniques for the implementation of the control system functions.

6.4.1 Transmission shift schedule

The transmission controller makes shift decisions according to the shift schedule based on two primary inputs – the transmission output speed and the engine throttle position – and other supplementary inputs such as ATF temperature and brake pedal depression. It is the shift schedule that defines the operating status of the vehicle power train system under any road condition. Therefore, the shift schedule is crucial to the vehicle characteristics such as fuel economy, dynamics performance, drivability, and pollutant emission levels. As discussed in Chapter 5, the fuel economy and acceleration performance of an automatic transmission vehicle with a specific shift schedule can be simulated by computer over various driving ranges, with the engine output data, the transmission data, and other vehicle data provided. These simulations can provide a model-based validation for the initial shift schedule in terms of fuel economy and performance. This initial shift schedule can then be used in test vehicles that undergo the intensive powertrain calibration process. The finalized shift schedule will be optimized in the calibration process as a well-balanced trade-off between fuel economy and dynamic performance. In some production transmissions [6,8,10], the shift schedules are designed with several modes, such as Economy, Normal, and Sport, which are chosen by the driver to meet different priorities and preferences.

An example shift schedule for a five-speed AT is shown in Figure 6.17, where the horizontal axis represents the vehicle speed and the vertical axis the engine throttle position. The solid lines are the thresholds for upshifts and dotted lines are for downshifts. As can be observed from Figure 6.17, some of the important attributes in an AT shift schedule are as follows:

- The current vehicle operation status is defined in Figure 6.17 by a point whose horizontal coordinate is the vehicle speed and vertical coordinate is the engine throttle position. Upshifts between two adjacent gears are made at higher vehicle speed than

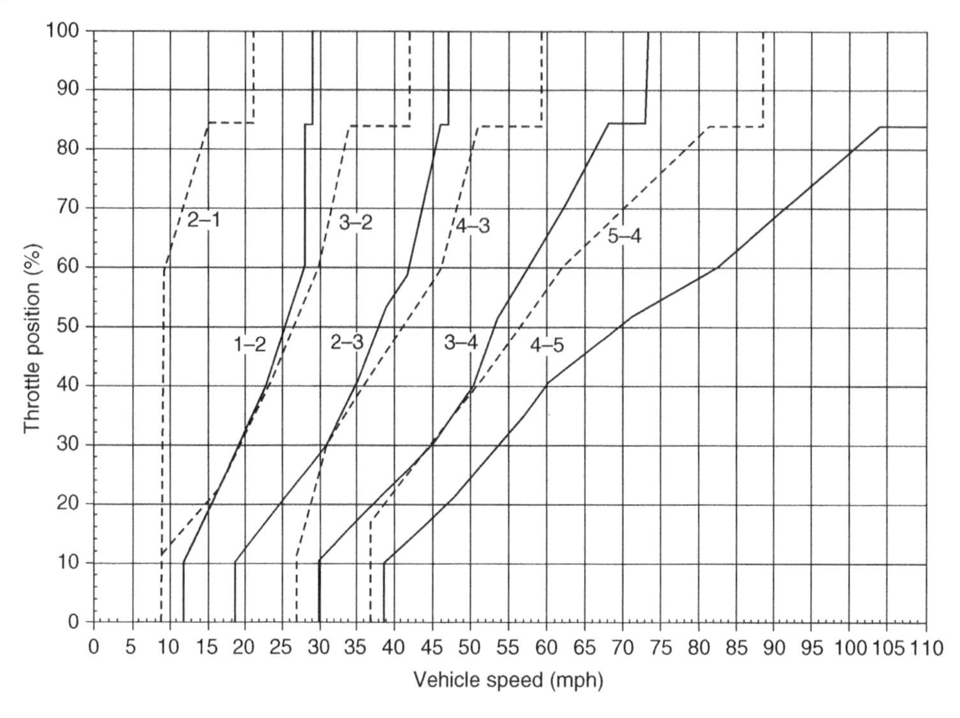

Figure 6.17 Example shift schedule for a five-speed automatic transmission.

downshifts. This means that there is a **buffer zone** between upshift and downshift. This buffer zone is designed for any two adjacent gears to keep the transmission from shifting back and forth along the shift threshold line. Without the buffer zone, a slight decrease of the vehicle speed right after an upshift is made may cause the status point to cross back over the downshift threshold line and trigger the downshift, resulting in frequent up/downshifts and poor drivability.

- The same upshift is made at higher vehicle speed at wider engine throttle opening, as indicated by the inclined upshift threshold lines in Figure 6.17. This design reflects the desire of the driver for acceleration performance. Deeper gas pedal depression will keep the vehicle running in the current gear for a longer time to achieve better acceleration. On the other hand, lighter throttle openings correspond to earlier upshifts and reach the high gear operation more quickly, resulting in better fuel economy as compared with aggressive gas pedal depressions.
- Each shift threshold line consists of several straight line segments and can be defined by the coordinates, i.e. the vehicle speed and the throttle position, at the intersection points of these segments. Therefore, the whole shift schedule is defined by these coordinates of the intersection points along the respective shift threshold lines. Adjustments or modifications on the shift schedule can be done by simply changing the coordinates of these intersection points.
- A shift is defined by a point on a shift threshold line. A number of points on each shift line are chosen to represent the corresponding shift conditions. These points include the points of intersection of the straight line segments mentioned above and may also

include intermediate points between the intersections, usually chosen at nearly even steps of engine throttle openings. For example, ten points might be chosen on the 1–2 shift line in Figure 6.17, six of them are at the intersections of the straight line segments and four are between the intersections for a nearly even throttle position interpolation. Theoretically, each one of such points constitutes a different shift and needs to be calibrated separately against the corresponding vehicle operation condition. Therefore, the number of shifts can become very large for an AT with multiple gear ratios.

- A transmission shift schedule works interactively with the torque converter clutch operation schedule. The apply and release of torque converter clutch closely follows the transmission shift schedule. As discussed later separately, the converter clutch operation schedule has two important effects on the transmission performance: shift smoothness and fuel economy.
- The shift schedule defined by threshold lines in Figure 6.17 is used for normal operation conditions. The shift thresholds are adjusted on a real time basis to better fit vehicle operation conditions, such as operation in low temperatures or when towing of a trailer. During a shift process, the controller will decide whether to proceed or discontinue while the driver depresses the brake.

Some vehicles are equipped with ATs that feature driver selectable shift modes, usually termed Economy, Normal and Power. Each of these modes is designed to fit an intended priority of the driver. For example, Figure 6.18 shows the shift schedule under the Normal mode and the Power mode respectively for a Toyota passenger car four-speed AT [8]. The Normal mode implements a balanced transmission shift schedule with the torque converter locked near the coupling point with the torque ratio approximately equal to 1.1. The Power mode is achieved by a power oriented shift schedule with an open

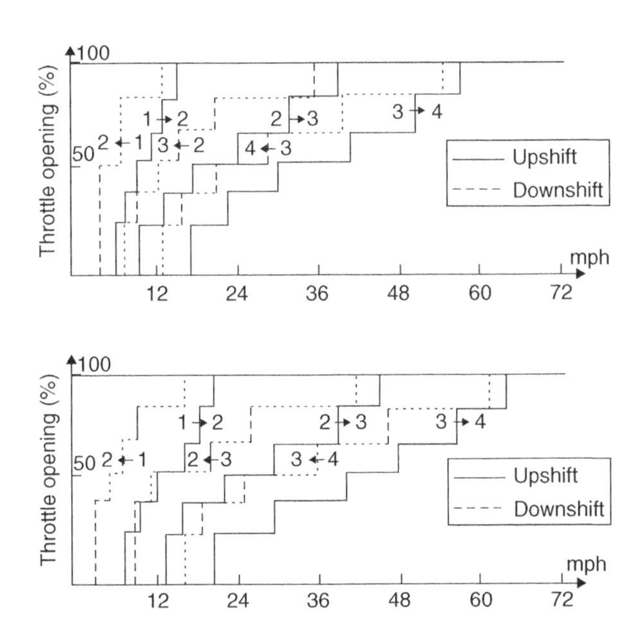

Figure 6.18 Shift schedules for Normal and Power modes.

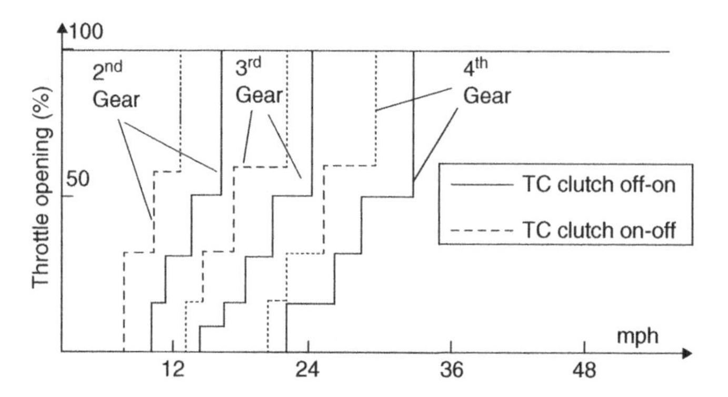

Figure 6.19 Torque converter clutch lock-up schedule optimized for fuel economy.

torque converter for the full advantage of torque multiplication. It can be observed from Figure 6.18 that the shift lines of the Power mode are shifted to the right as compared with those of the Normal mode, which keep the vehicle accelerating in lower gears for a longer time to reach the target vehicle speed. The Economy mode combines a fuel efficient shift schedule with a schedule for torque converter lock-up that is optimized for fuel economy and is shown in Figure 6.19. This schedule locks up the torque converter when the torque ratio becomes less than 1.3. According to the dyno test and actual vehicle test data obtained by Toyota engineers [8,9], the Economy mode achieves about 3% fuel economy improvement over the Normal mode, and the Power mode achieves a 3.6% improvement in 0–100 km/h acceleration performance over the Normal mode under WOT test conditions. These data support the necessity of having different shift modes to be selected by the driver for preferred priorities since the improvements in fuel economy and performance are not insignificant.

6.4.2 Torque Converter Lock Control

The objective of torque control lock control is mainly to achieve an optimized trade-off between efficiency and shift smoothness. As a torque multiplier and a fluid couple, an open torque converter functions as the vehicle launcher and as a damper for powertrain harshness during transmission shifts. However, there is a power loss in the operation of an open torque converter. This loss is more significant in the low speed range which corresponds to low torque converter speed ratios and may translate into about 3% fuel mileage loss. According to the model simulations and experiments conducted by Toyota researchers [8,9], vehicle fuel economy is improved in the whole speed range by locking up the torque converter, especially at low vehicle speed. In general, an earlier converter lock-up scheme, as shown in Figure 6.19, results in better fuel economy. In most ATs, torque converters start to be locked in second gear operation above some threshold vehicle speed and are locked afterwards in fixed gear operations. However, a locked torque converter loses its function as a fluid couple to dampen the transients and harshness during shift operations. In order to optimize shift quality, the torque converter is therefore released during shift operations for it to function as the harshness damper. Once a shift is completed, the torque converter will be locked again to enhance the powertrain efficiency. Note that torque converter locking or unlocking is a dynamic process by itself that may cause unwanted powertrain transient behavior and needs to be controlled for optimized driver and passenger feel. In summary, torque converter lock control concerns two technical issues: Lock-release schedule and lock-release operation control.

6.4.3 Lock-Release Schedule

The torque converter stays open in first gear operation and thus the lock-release schedule applies to transmission operations above second gear. Similar to the transmission shift schedule, the torque converter lock-release schedule makes the lock-release decisions based the vehicle speed and the engine throttle opening, as shown in Figure 6.19. A buffer zone is designed between the locking and unlocking threshold lines for each gear to keep the converter from locking and unlocking frequently. As mentioned previously, earlier locking-up results in better fuel economy and later locking-up corresponds to better acceleration performance. If fuel economy is the target, then the converter is locked at lower speed in each gear above second gear, as designed in the lock-release schedule. However, locking up the torque converter too early, i.e. at aggressively low vehicle speed, results in the loss of torque multiplication and may even excite powertrain vibration. This happens if the torque converter is locked too early in all gears, according to the model simulation and experiment as conducted by Toyota researchers [8,9]. Therefore, the threshold lines in the lock-release schedule must be designed for each gear such that the converter torque multiplication capability is used as much as possible, and the lock-up point is above the lowest speed that will avoid exciting powertrain vibration. This lowest speed can be determined initially by model simulation and validated by experimental means or test vehicle calibration.

The lock-release schedule shown in Figure 6.19 only determines the torque converter status during fixed gear operations. As mentioned previously, the torque converter is unlocked synchronously during shift operations, especially in low gears, since its functionality as a fluid couple is needed to dampen shift harshness. Since torque converter locking or unlocking is a process by itself, the timing of converter locking or unlocking control is crucial to transmission shift smoothness. The On or Off status of the torque converter during shifts follows the transmission shift schedule shown in Figure 6.17 or Figure 6.18. In addition, the timing of converter unlocking control depends on the operation conditions and type of the shift when it is initiated by the transmission control unit (TCU). In general, the control of converter unlocking during shifts is timed based on the following conditions:

- Threshold line that triggers the shift: this will identify both the type of shift – upshift, or downshift – and the gear positions before and after the intended shift.
- Power on or power off: this signal is from the accelerator pedal sensor.
- Current converter clutch status: this is determined by the converter clutch control solenoid status.
- Downshift triggered by deep depression of accelerator pedal: this is also power-on downshift.

The synchronization of torque converter clutch locking or release control during shifts is illustrated in Figure 6.20. A shift is signaled by the flip of a solenoid status, from On to Off, and the torque converter clutch status during shift is commanded also by the On or

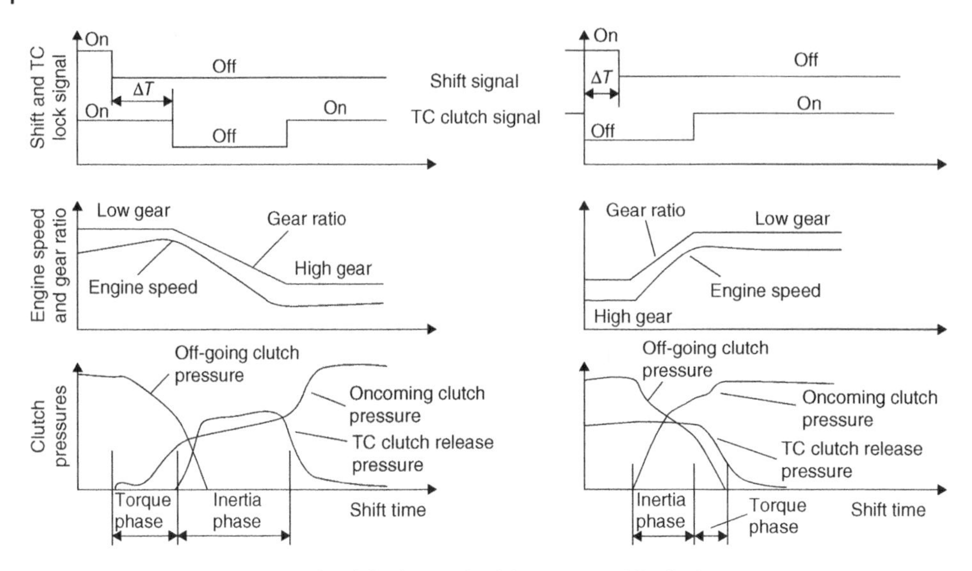

Figure 6.20 Torque converter clutch lock-up schedule optimized for fuel economy.

Off status of the related solenoid. The case for upshifts is shown on the left side in Figure 6.20. The shift is triggered by the flip from On to Off as indicated by the shift signal. The torque converter clutch signal On or Off corresponds to the locked or unlocked position. When a power-on upshift is to be made when the torque converter is locked in the current gear, releasing the converter clutch during the torque phase will give rise to engine flare because the mechanical coupling between the engine and the transmission input becomes momentarily a fluid coupling. The engine speed must be brought down in an upshift and engine flare should certainly be avoided since it elongates shift time and causes other undesirable shift attributes. On the other hand, if the torque converter clutch stays locked way into the inertia phase, shift harshness or even shift shocks may result due to the lack of converter damping effect. Therefore, it is critical to control the converter clutch release timing for the full advantage of fluid couple damping effect during the inertia phase. As shown in Figure 6.20, there is a time delay ΔT between the initiation of the upshift and the flip of the converter clutch signal. This time delay is designed to time the release of the torque converter clutch right at the transition from the torque phase to the inertia phase. Note that the value of time delay ΔT can be calibrated for all upshifts that involve torque converter clutch unlocking. After the upshift is completed, the transmission control unit (TCU) will command the torque converter clutch status from Off to On so as to lock up the torque converter in the high gear as shown in Figure 6.20, which is deemed appropriate by the schedule in Figure 6.19.

The order of torque phase and inertia phase in power-on downshifts are just reversed as compared with power-on upshifts. In a power-on downshift, the off-going clutch must be controlled to slip as soon as the downshift is initiated, so the engine speed will be brought up to the target value of the low gear to be downshifted. The torque converter clutch is usually locked at the time when a power-on downshift is commanded by the TCU, as indicated on the right side of Figure 6.20. As controlled by the TCU, the TC clutch signal flips from On to Off before the shift signal flips from On to Off by a time ΔT.

The value of ΔT can be preset using the transmission shift schedule as shown Figure 6.17 and can be validated through test vehicle calibration. For power-on downshift triggered by deep accelerator pedal depression, the TC clutch signal can be controlled to flip synchronously as the downshift is triggered. Since the converter clutch is released as soon as the power-on downshift is initiated, the engine speed will be synchronized to the low gear quickly to complete the shift. At the end of the downshift process, the status of the torque converter clutch is recovered to the status prior to the shift as deemed appropriate by the schedule shown in Figure 6.19, as illustrated on the right side of Figure 6.20. Note that the control logic for power-off upshifts is similar to that used for power-on downshifts for the avoidance of powertrain harshness.

6.4.4 Lock-Release Operation

The torque converter clutch pressure control circuit was shown in Figure 6.9. As mentioned previously, the TC clutch apply or release pressure is controlled by the TC clutch pressure control solenoid using pulse width modulation, i.e. via the duty cycle percentage. The control strategies may differ in specifics for different transmissions but they are all aimed at operation smoothness of the torque converter clutch during locking and release. As an example [17], the control of converter clutch apply and release processes is illustrated in Figure 6.21 for the GM Hydra-Matic six-speed AT shown in Figure 6.14. The apply pressure ramp-up profile is from point A to point H, and the release pressure ramp-down profile is from point I to point L respectively. It is noted that the electric current in the TC clutch pressure control solenoid follows the same profile due to its proportionality with the pressure.

When the vehicle is being driven in second or higher gear with an open torque converter and the transmission control unit (TCU) decides to lock up the torque converter according to the lock-up schedule similar to that shown in Figure 6.19, the status of the torque converter clutch control solenoid immediately flips from Off to On, as designed in the system control circuit shown in Figure 6.15. As detailed in the Technician's Guide for the GM Hydra-Matic six-speed AT [17], the torque converter clutch **locking-up process** is implemented in three steps:

1) The TCU ramps the TC clutch PCS pressure from point A to point B and holds it until point C, through the circuit shown in Figure 6.9. The pressure at point B is sufficient to position the TCC regulator valve for the line pressure ATF to enter the

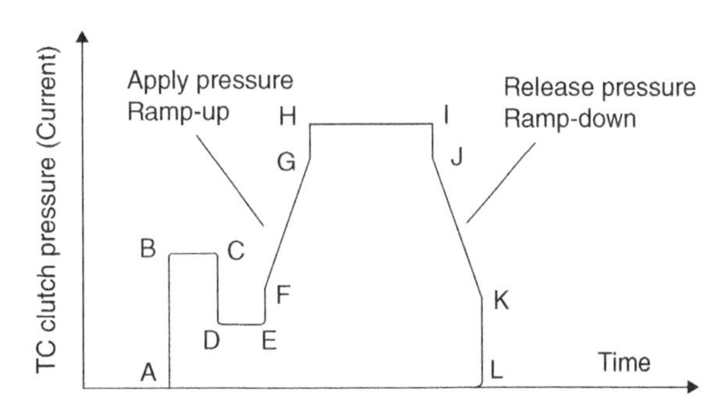

Figure 6.21 Pressure ramping for torque converter clutch control.

regulated TC apply circuit and to position the TCC control valve to allow the converter release ATF to exhaust. To avoid possible harshness caused by overly high initial apply pressure, the TCU lowers the TC clutch PCS pressure from point C to point D. The TCC PCS pressure at D is controlled to be still high enough to maintain the positions of the TCC regulator valve and the TCC control valve, but at this point, the converter clutch apply pressure from the TCC regulator valve is not yet high enough to apply the converter clutch.

2) The TCU increases the TCC PCS pressure from point E to point F in a small jump and further ramps it to point G following a designed slope. In this process, the regulated TC apply pressure from the TCC regulator valve enters the TCC apply circuit and becomes high enough to start applying the converter clutch. The converter slip rate – i.e. the speed difference between the impeller and the turbine – will decrease gradually as the TCC PCS pressure increases. At point G, the converter slip rate approaches zero.
3) To eliminate the residual slip in the torque converter and to secure the lock-up, the TCU further increases the TCC PCS pressure in a jump to point H. At this point, the torque converter becomes a mechanical couple that links the engine and the transmission.

When the vehicle is being driven with the torque converter locked-up and the transmission control unit (TCU) decides to release the torque converter in a shift event or according to the lock-up schedule, the status of the torque converter clutch control solenoid flips from On to Off. In a shift event, the time interval ΔT between the shift signal and the clutch signal depends on the shift type, as discussed previously and illustrated in Figure 6.20. The control of the release process is implemented following the release pressure ramp-down profile from point I to point L, as shown in Figure 6.21. In a shift event, the torque converter clutch needs to be released quickly in order to take full advantage of the damping effect of the converter fluid couple functionality. This is indicated by the steepness of the release pressure ramp-down profile from point I to point L. For the GM Hydra-Matic six-speed AT [17], the torque converter **release process** is controlled in two steps:

1) At the beginning of the release process, the TCU decreases the TCC PCS pressure by cutting down the current sent to the TC clutch control solenoid in the circuit shown in Figure 6.9 from point I to point J. The reduced TCC PCS pressure at point J allows the repositioning of the TCC control valve by the resultant pressure of ATFs from other circuits and diminishes the opening in the TCC control valve for the regulated TC apply ATF to enter the converter clutch apply circuit. As a result, the ATF pressure on the torque converter clutch apply side – i.e. the apply side of the converter pressure plate – is reduced and at some point the contact surface between the pressure plate and the converter cover will reach the threshold of slippage. This slippage threshold point can be calibrated per vehicle operation condition and the slip can be detected by the speed sensors as soon as it starts.
2) As the release proceeds, the TCU further decreases the TC clutch PCS pressure from point J to point K following the calibrated slope by decreasing the current sent to the TCC pressure control solenoid to zero. At point L, the residual electromagnetic effect of the TC clutch PCS is zero and the release ATF enters the space between the pressure plate and the converter cover from the release opening of the TC control valve as

shown in Figure 6.9. The torque converter is now fully open and provides the torque multiplication and damping functionalities as designed.

As mentioned previously, the torque converter locking and release processes are controlled in different ways for different transmissions. The control strategy presented above can be considered as typical and can be modified for general applications. In addition, the profiles shown in Figure 6.21 can also be trimmed for different transmissions and are validated in the calibration process of test vehicles.

6.4.5 Engine Torque Control During Shifts

As discussed in Section 5.5, the oncoming clutch torque drags on the transmission and decreases the transmission output torque during the torque phase of upshifts. During the inertia phase, the transmission output torque largely depends on the oncoming clutch torque and may have a large overshoot if the oncoming clutch torque is too large. The output torque drop forms the so-called torque hole that bottoms at the transition from torque phase to inertia phase, while the torque overshoot peaks toward the end of the inertia phase, as shown in Figure 5.22. Both the torque hole and the torque overshoot must be minimized for shift smoothness. Typical output torque variation patterns are shown in Figure 6.22 for clutch to one-way clutch upshifts without engine torque retarding and with engine torque retarding. As shown in Figure 6.22, engine torque reduction by spark retarding has a significant effect on the minimization of the output torque overshoot. For the best effect, engine torque reduction should be started as soon as the shift transfers to the inertia phase. If the engine torque reduction starts when the torque phase is yet to be finished, it will deepen the torque hole, resulting in unpleasant shift feeling. Therefore, the TCU must interact with the engine controller via the control area network to time the spark retarding accurately on a real time basis for power-on upshifts.

The effect of the engine torque reduction on shift quality has been analysed in detail in Section 5.5. In an upshift, the engine speed must be brought down quickly to be synchronized with the value corresponding to the high gear. If the upshift is clutch to one-way clutch, such as the 1–2 shift in the Ford FWD six-speed AT or the GM Hydra-Matic six-speed AT, the torque in the off-going clutch is zero after the inertia phase starts. Therefore, the magnitude of the engine angular deceleration in the inertia phase is only proportional to the oncoming clutch torque for a given engine output torque or transmission input torque, as shown in Eq. 5.94. The oncoming clutch torque T_C must be ramped up to a certain value in order to decelerate the transmission input speed or engine speed. A steep ramp-up profile for the oncoming clutch torque T_C results in rapid shift response and shortens shift time. But, on the other hand, larger oncoming torque magnitudes give rise to higher output torque overshoot and shift harshness, as analysed in Section 5.5. This contradiction is well addressed by reducing the engine torque, i.e. the transmission input torque, in the inertia phase as shown in Figure 6.22. As observed in Eq. 5.94, by reducing the engine torque via spark retarding, the transmission input torque is reduced proportionally, and therefore a smaller oncoming clutch torque is able to achieve the engine deceleration rate required to complete the shift in good time. When the engine speed is brought down near the target speed of the high gear toward the end of the inertia phase, spark retarding is cancelled by the engine controller and engine torque recovers to the normal level. The oncoming clutch torque is further ramped up in one

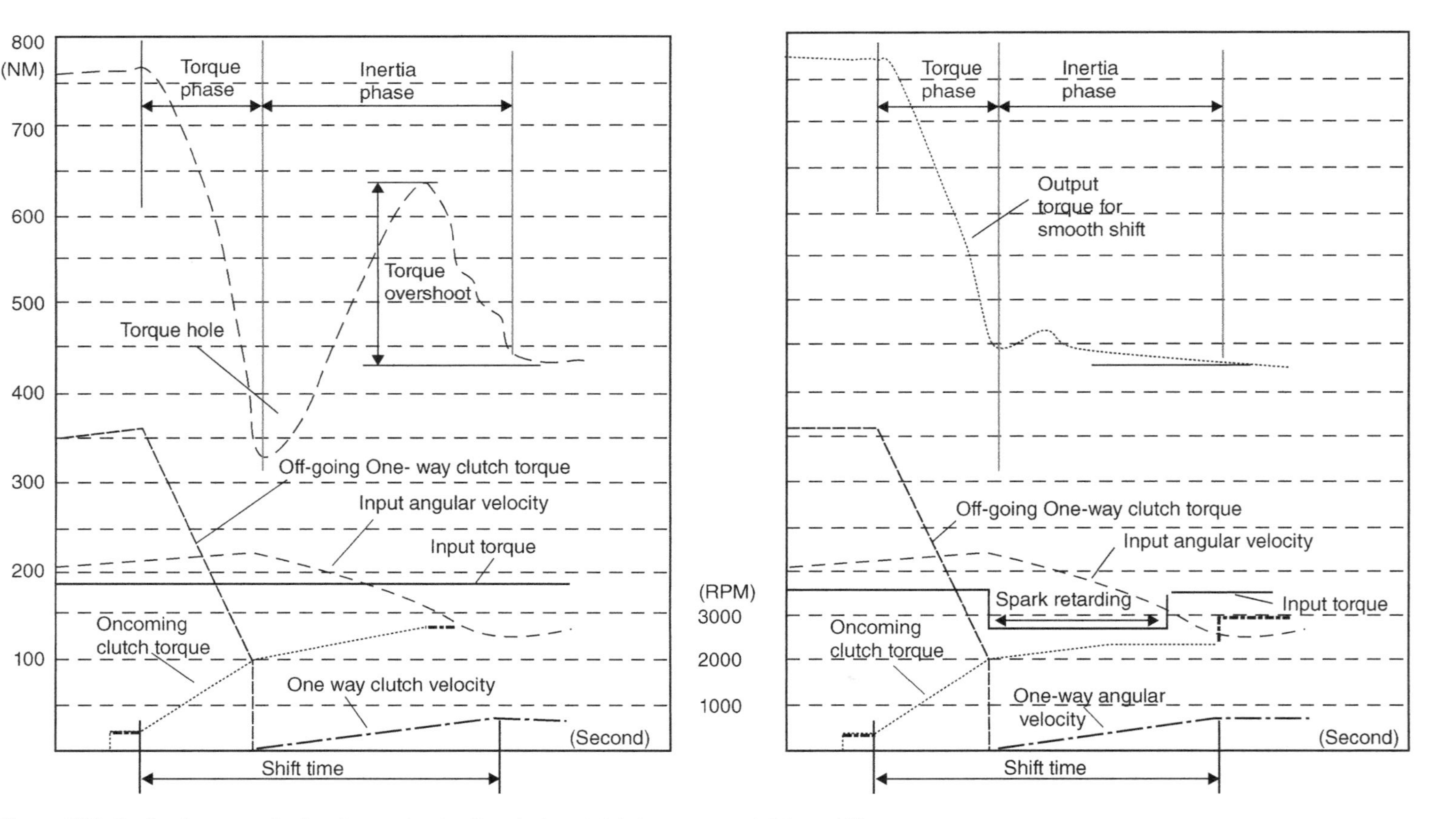

Figure 6.22 Engine torque reduction by spark retarding during clutch to one-way clutch upshifts.

step to secure the engagement of the oncoming clutch. If spark retarding is controlled in good time and the oncoming clutch torque ramp-up profile during inertia phase is controlled properly, it is possible to achieve near perfect upshifts, as shown in Figure 6.22 with minimized output torque overshoot.

As shown in Figure 6.23, the effect of engine torque reduction in clutch to clutch upshifts is similar to that for clutch to one-way clutch upshifts discussed earlier. In this case, the off-going clutch is a regular clutch whose torque capacity depends on the hydraulic pressure in the clutch piston chamber. In the torque phase, the hydraulic pressure in the off-going clutch piston chamber is controlled by the related hydraulic circuit to decrease rapidly for the off-going clutch to reach the slip threshold, at which point the off-going clutch torque capacity is equal to the off-going clutch torque determined by the system dynamic status, as shown in Figure 6.23. As the inertia phase starts, both the oncoming clutch torque and the residual torque in the off-going clutch act against the engine torque to decelerate the engine speed or transmission input speed, as observed in Eq. 5.94. The engine torque reduction starts after the torque phase immediately by timing the spark retarding accurately. Owing to the reduced engine torque, the oncoming clutch torque is controlled to follow a lowered profile for the minimization of the output torque overshoot, while still being high enough to act against the engine torque for the deceleration of the engine speed, as observed in Eq. (5.94). During the inertia phase, the TCU processes the data from the speed sensors on real time and notifies the engine controller to cancel spark retarding for engine torque recovery once it judges that the upshift is near completion. The oncoming clutch torque is then ramped up further in one step to securely engage the oncoming clutch. With the engine torque reduction controlled in good time and the oncoming clutch torque profile ramped up accurately, it is

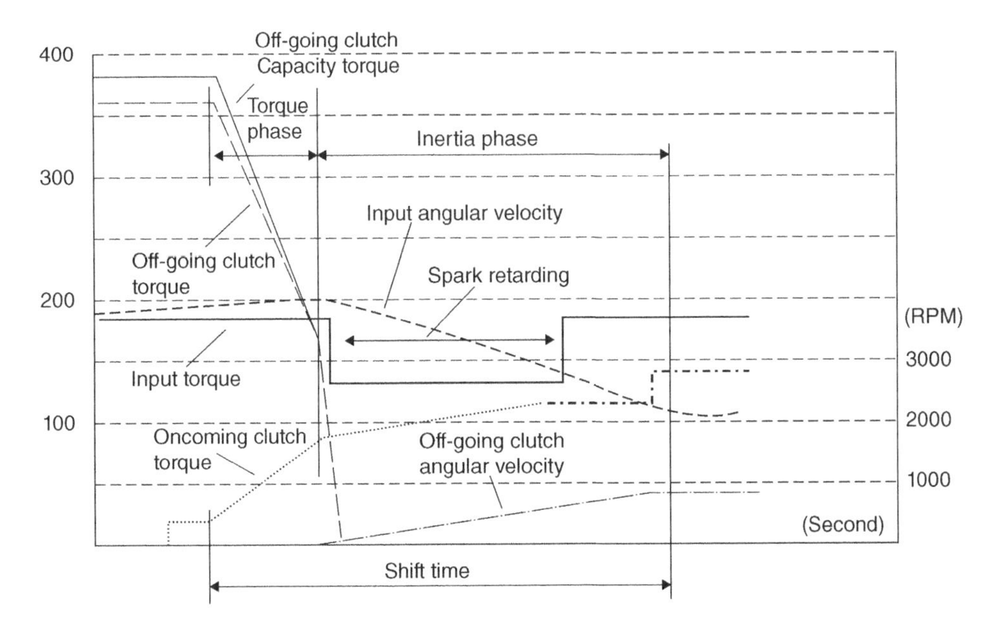

Figure 6.23 Engine torque reduction by spark retarding during clutch to clutch upshifts.

possible to achieve power-on clutch to clutch upshifts with minimized output torque overshoot similar to that for clutch to one-way clutch upshifts, as shown in Figure 6.22.

6.4.6 Shift Process Control

This is the key function of the transmission control system. Transmission shift response and smoothness primarily depend on the off-going clutch torque and the oncoming clutch torque, which are controlled by the clutch pressure control circuits illustrated in Figures 6.6 and 6.7. During shift operations, the hydraulic pressures in the piston chambers of the off-going clutch and the oncoming clutch are respectively controlled by these circuits using variable force solenoids (VFS). In addition, engine torque reduction and torque converter unlocking are introduced during shifts to enhance shift response and smoothness, as already discussed in detail. This subsection focuses on the techniques and strategies for the ramping-up and ramping-down of the ongoing and off-going clutch pressures. ATs currently in production or under development typically feature the following technical highlights for shift process control:

- The apply or release pressure of a clutch is independently controlled by a VFS in a circuit specific to the related clutch, as shown in Figures 6.6, 6.7, 6.15, and 6.16. This results in better accuracy and response in clutch apply pressure control.
- Torque based control is used for both the torque phase and the inertia phase during shifts. When a shift is initiated, the engine torque is estimated using the engine map and modified on a real time basis using an array of sensor-provided data, such as temperature, engine manifold pressure, and air conditioner status,The net transmission input torque is obtained by factoring in further the torque converter characteristics.
- Clutch torque profiles for various shifts optimized in transmission calibration are stored in the transmission control unit (TCU) as data. These torque profiles can be converted to clutch pressure profiles for real time shift control using a database on clutch friction coefficients.
- The attributes of clutch behavior during the initial stage of piston stroking are handled effectively by hardware design improvements and by pre-calibrated control software techniques. Counter piston design and compensator chamber design can be applied to eliminate the time delay caused by clutch piston chamber fill-up and minimize the effects of follow-up piston stroking transients. On the software side, a database can be established on the initial piston characteristics through calibration and is used to modify or compensate clutch pressures results on a real time basis [13].
- Open loop control is usually used for the torque phase, and closed loop is used for the inertia phase for power-on upshift. The feedback signals are mainly the speeds of the engine, transmission input, and output. Fine adjustments of clutch pressures are implemented based on the feedback, keeping shift performance consistent between vehicles and during extended service life.

Typical torque profiles during power-on clutch to clutch upshifts are shown in Figure 6.23. The pressure profiles of the two involved clutches follow the same pattern as the torque profiles but in a scaled proportion, as shown in Figure 6.24. In Section 5.4, the clutch torque magnitudes were determined for each gear in terms of the transmission input torque. The maximum torque magnitude required for a clutch when the

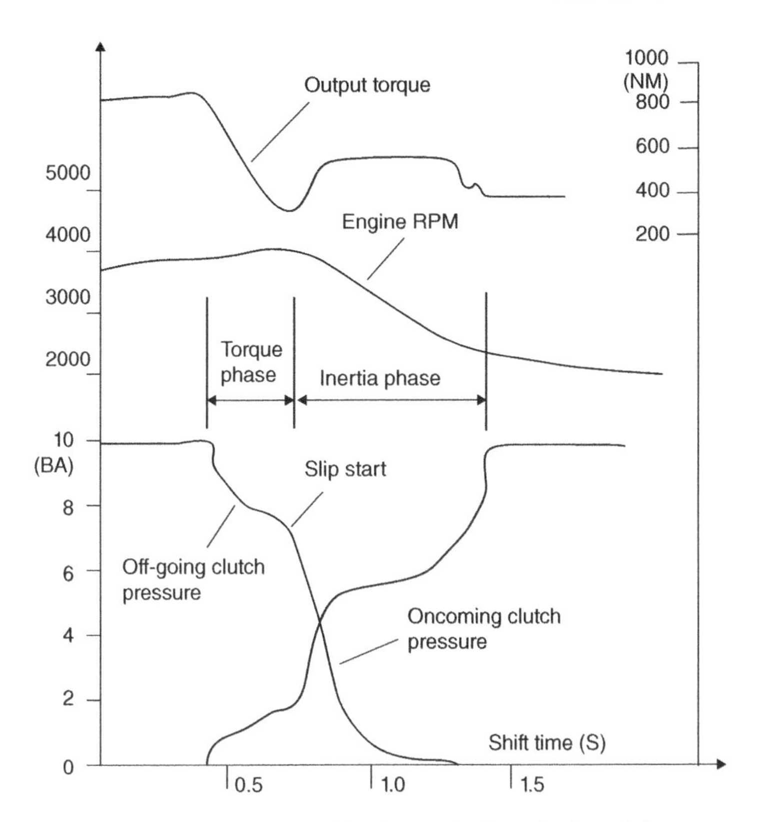

Figure 6.24 Clutch pressure profiles during clutch to clutch upshifts.

transmission operates in a fixed gear depends on the maximum transmission input torque, which in turn can be determined based on the engine torque map, torque converter characteristics, and transmission shift schedule, such as that shown in Figure 6.17. The maximum torque required of a clutch is used in clutch design in terms of sizing and the selection of the number of friction disks. For a clutch with given design parameters, the clutch torque only depends on the clutch apply pressure. These clutch design parameters are optimized such that the pressure required for full clutch engagement in each gear does not vary significantly between the various clutches in the transmission. Therefore, the system line pressure that is controlled according to vehicle operation conditions will fit all clutches for the control of clutch apply and release pressure profiles. When applied, the hydraulic pressure in the clutch piston chamber is equal to, or close to, the system line pressure. For a factor of safety, the system line pressure is thus controlled at a value somewhat higher than the pressure required to fully engage the clutches during fixed ratio operations. The pressure profiles of the off-going clutch and the oncoming clutch in a power-on upshift are shown in Figure 6.24, where the pressure unit is in BA. Usually, the clutch apply pressure in automatic transmission control systems does not exceed 10 BA or 1.0 MPa. The pressure profiles in Figure 6.24 generally refer to power-on upshifts under heavy engine loads with large to full engine throttle opening.

6.4.7 Initial Clutch Pressure Profiles

As already mentioned previously, the pressure profiles for clutch torque control during shifts must be validated and finalized in the calibration process of test vehicles. However, there must be an initial set of clutch pressure profiles for each of the shift events to be implemented by the transmission so that the calibration vehicle can be test driven. For transmissions with multiple gears, the number of shift events can be overwhelming. For example, the ZF eight-speed RWD AT analysed in Chapter 5 has 36 direct clutch to clutch shifts, and each of these direct shifts can be made under different operation conditions as dictated by the shift schedule, so the total number of shift events will then be in the hundreds. The initial clutch pressure profiles, if selected appropriately, not only significantly shorten calibration time, but also lead to better calibration results in terms of shift response and smoothness. Model simulation, as discussed in Chapter 5, is the main tool for the selection of these initial clutch pressure profiles. Firstly, the torque values of the applied clutches under static condition in each gear can be calculated using the clutch torque table of the transmission for each shift event according to the shift schedule. Secondly, these clutch torque values are converted to the clutch pressure, with a safety factor on the clutch torque capacity, using the clutch torque capacity formula as shown by Eq. 6.1. Thirdly, pressure profiles that follow the patterns illustrated in Figure 6.24 are used as input control variables for the vehicle system model shown in Figure 5.26 to simulate the transmission shift process for the assessment of shift quality. The pressure profiles that are optimized by model simulation can then be used in the test vehicles in the calibration process for validation and finalization. Note that the pressure profiles in Figure 6.24 need to be converted to current signals sent to the related variable force solenoids during transmission shifts, as shown in Figure 6.25.

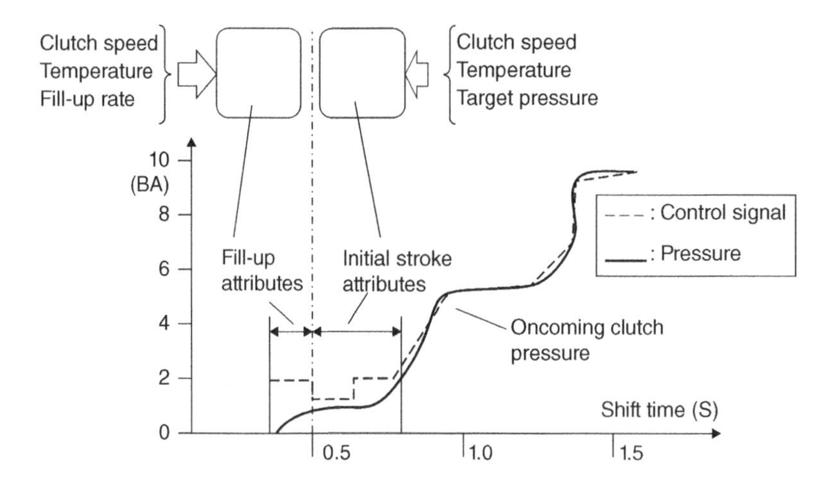

Figure 6.25 Clutch piston initial stroke attributes.

6.4.8 Initial Piston Stroke Attributes

For clutches that are not designed with counter pistons, there will always be a time delay for the clutch pressure to respond to the control signal at the beginning of the shift, as shown in Figure 6.25. This is because the clutch pressure can only build up after the piston chamber is filled up with ATF. After the piston chamber is filled up, the piston will move against the return spring and the wave disk (if present), to eliminate the backlash in the friction disks. The clutch pressure will then follow the control signal, as shown by the solid line and dotted line respectively in Figure 6.25. It is critical to quantify the initial piston stroke attributes to minimize the time delay caused by them and to control the clutch pressure accurately as soon as possible after the shift starts. Although fill-up and initial stroke attributes are difficult to model analytically, both of them can be calibrated in a laboratory setup or in a test vehicle. In the calibration setup, a pressure sensor is installed in a location close to the piston chamber and measures the pressure build-up process in response to the current signal sent to the related VFS. In the control system of some ATs, databases are established for piston fill-up attributes and initial stroke attributes based on the calibration data [12,13]. During transmission shifts, VFS control signals are interpolated in real time from the database by the transmission control unit (TCU) using an array of sensor-provided inputs, such as clutch speed, ATF temperature, and system line pressure. By using this technique, it is possible to correlate the clutch response in the initial stage of the shift to the control signal, even though the correlation is not in linear proportion. More importantly, it is therefore technically possible to control the duration of the initial piston attributes so that the clutch pressure can be controlled accurately and in good time.

6.4.9 Feedback Shift Control

Feedback control is only used for the inertia phase during shifts since the transmission ratio is not changed in the torque phase. When a shift is deemed necessary by the TCU, it will also figure out the desired ratio change rate and the clutch pressure profile based on the vehicle operation condition. Pressure control signals are sent to the VFS for the off-going and oncoming clutches respectively. This will cause the change of dynamic status of the powertrain system immediately. This change is monitored by the TCU via various sensors, such as the engine speed sensor, and the transmission input and output speed sensors. As the oncoming clutch pressure ramps up and the off-going clutch ramps down, the off-going clutch will start to slip at some point, which is detected by the sensors as soon as it happens. Closed loop control is then used for the control of the oncoming clutch pressures, with the off-going clutch pressure brought to zero as quickly as possible. In the inertia phase, the transmission input and output speeds and the engine speed are measured by speed sensors on a real time basis at designed sampling time intervals. The TCU calculates the ratio change rate based on the transmission input and output speeds and compares it to the desired ratio change rate, as shown in Figure 6.26. Clutch pressure is then refined upon the base pressure profile and controlled by the VFS to achieve the desired ratio change rate. In addition to the desired ratio change rate, other target variables such as the engine speed and the vehicle acceleration can also be used as the control reference for the feedback control in similar configurations to those

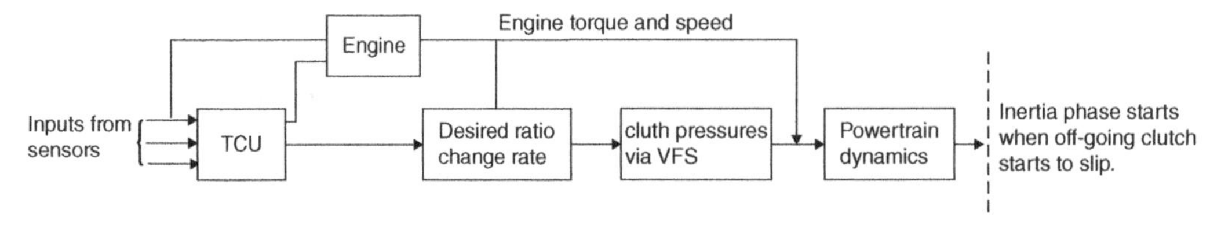

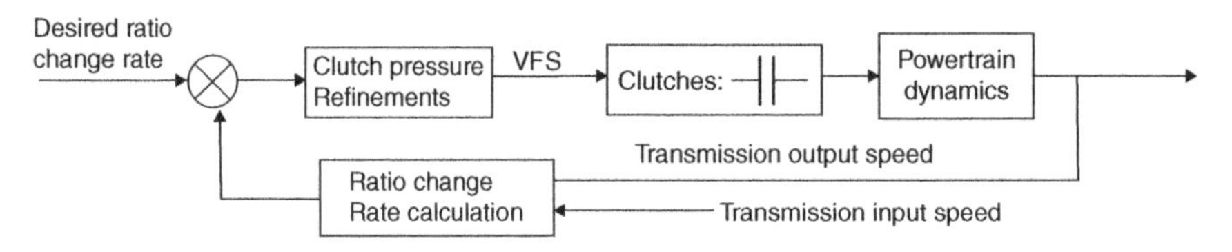

Figure 6.26 Control loops for torque phase and inertia phase.

shown in Figure 6.26 [8,13,14]. Generally, feedback control for transmission shifts possesses characteristics and benefits described in the following:

- Shift quality is further improved through fine adjustments of clutch pressure profiles in real time control.
- Variations in properties and attributes always exist for production vehicles of the same model on both component and system levels, leading to transmission shift quality variations from vehicle to vehicle. Feedback shift control minimizes the effects of these variations on shift quality and achieves better consistency in transmission shift quality and driver feel among production vehicles of the same model.
- The transmission properties and attributes on both component and system levels change as the vehicle ages, leading to a worsening in shift performance. Feedback shift control minimizes this worsening and achieves consistency in shift performance over the product service life.
- Feedback shift control is implemented by control software at no hardware cost increase.

The effects of feedback control for transmission shifts are illustrated in Figure 6.27, where the control reference is the engine speed or transmission input speed. During a shift, a target engine speed profile in the shift inertia phase is pre-designed and the oncoming clutch pressure is controlled so that the engine speed will follow the target profile. The feedback control starts at a point after the transfer from the torque phase to the inertia phase since the feedback signals can only be obtained after the system responds to the initial inertia phase control. In comparison with shift control without feedback, the overshoot in the transmission output torque during the inertia phase can be reduced for shift quality enhancements.

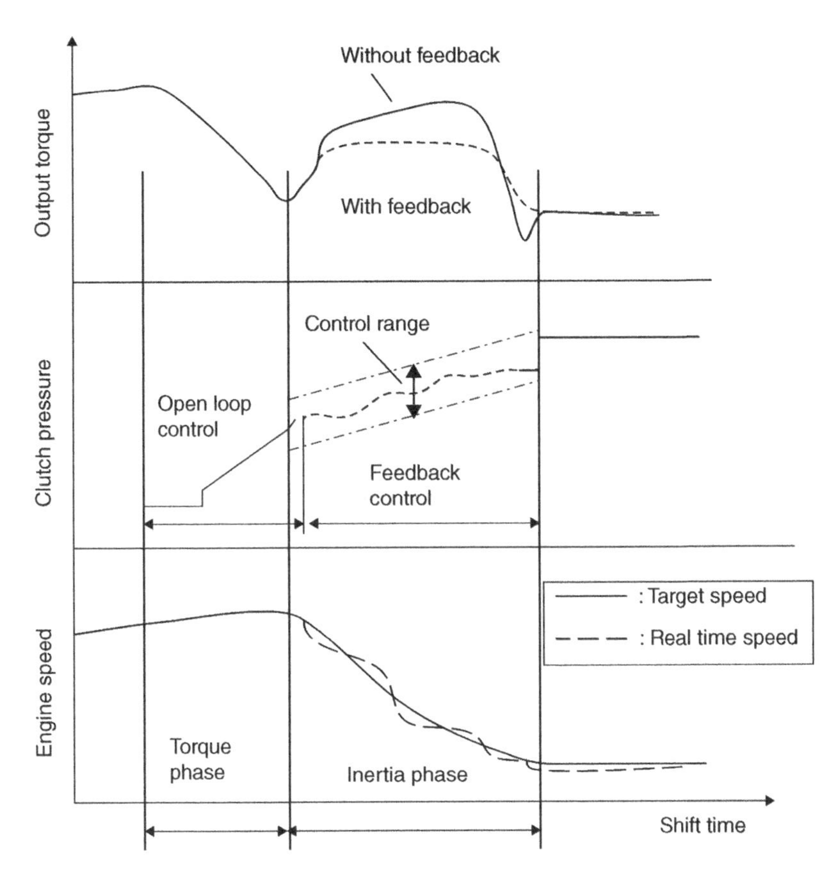

Figure 6.27 Shift control in inertia phase with feedback on engine speed.

6.4.10 Torque Based Shift Control

The advantage of torque based shift control is that it provides accurate clutch torque control in real time during shifts in accordance with the transmission input torque when shifts are initiated. This leads to shift quality optimization based on the powertrain operation status. There are three technical issues that need to be addressed for the implementation of torque based shift control:

- accurate estimation of transmission input torque
- determination of the torque profiles of the off-going and oncoming clutches during shifts
- accurate conversion of the clutch torque profiles to the clutch pressure profiles

The estimation of transmission input torque is straightforward and is mainly based on the engine map, engine operation condition, status of auxiliary systems, and torque converter characteristics. Firstly, the nominal engine torque value is interpolated from the engine torque map in terms of the engine throttle opening, RPM, and intake air pressure. This nominal torque is then modified based on signal inputs from related sensors, such as temperature, atmospheric condition, air/fuel mix ratio, and any other inputs that may affect engine torque output. The net torque applied by the engine to the torque converter impeller is obtained by subtracting the loads of the accessary systems, such as air conditioner and power steering pump, from the modified nominal torque. The transmission input torque is then determined by considering the mass moment of inertia of the flywheel–impeller assembly and the torque converter characteristics, using Eqs (5.122–5.125).

Shift operations are completed by controlling the off-going and oncoming clutch torque profiles with the objective of achieving a smooth transfer of the transmission output torque from the current gear to the target gear. As detailed in Chapter 5 and previous sections of this chapter, the variation of the transmission output torque during transmission shifts is critical for shift quality. This is because the transmission output torque is linearly related to the vehicle longitudinal acceleration or the G value to which the driver or passenger is most sensitive. Various techniques can be applied to determine the base torque profiles, which are then modified by model simulation detailed in Chapter 5 and finalized through the calibration process. For example, Honda engineers proposed a so-called "G design" method for the determination of clutch torque profiles [12,13]. A pre-selected G value that is validated to be conducive to shift quality is designed for each shift. Based on the transmission input torque that is determined by the steps described previously, the clutch torque profiles are then converted from the target G values, as illustrated in Figure 6.28. Note that the selection of G value in the G design

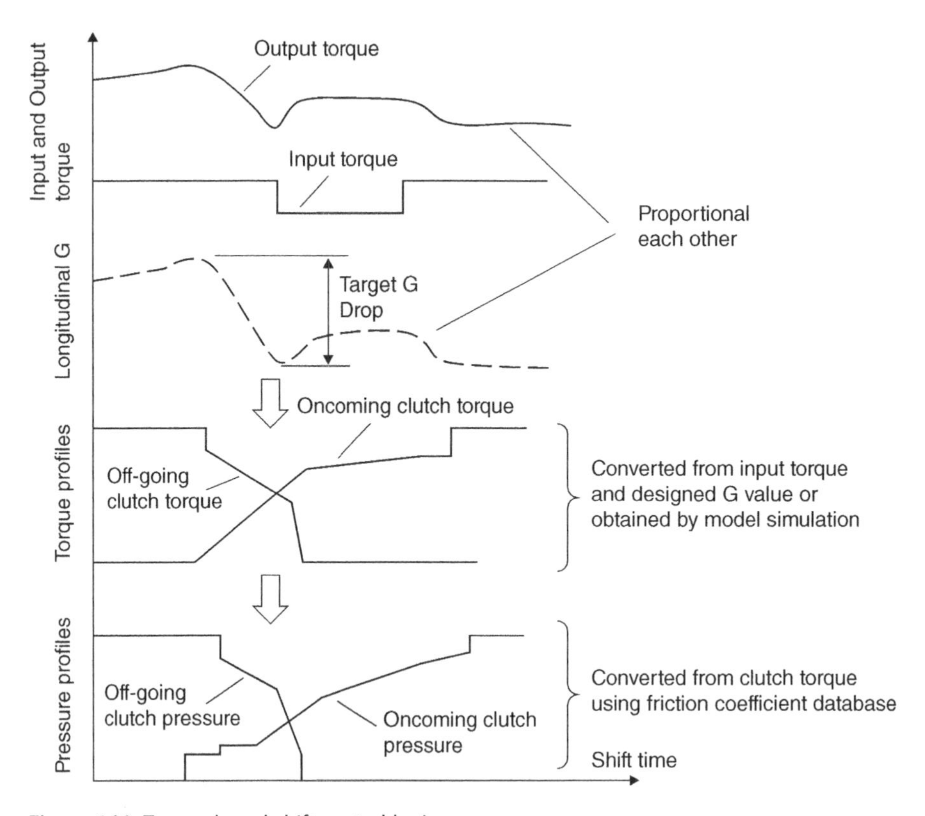

Figure 6.28 Torque based shift control logic.

method is equivalent to the minimization of torque hole and overshoot of the transmission output torque because of the proportionality between the longitudinal acceleration and the output torque. Using model simulation detailed in Chapter 5, it is possible to determine the clutch torque profiles that minimize the torque hole and overshoot for all shifts with the transmission input torque as a given variable.

The clutch torque profiles obtained in this way, as illustrated in Fig. 6.28, must be converted to the corresponding clutch pressure profiles in the shift control processes. This is the most difficult of the three technical issues mentioned previously. There are two main difficulties here: the correlation of the hydraulic pressure in clutch piston chamber and clutch torque; and the correlation of clutch pressure and the current signal sent by the TCU to the VFS in the clutch apply circuit. The first difficulty is mostly overcome if the friction coefficient of the clutch disks can be estimated with accuracy on a real time basis. As shown in Eq. 6.1, for a given clutch, its torque only depends on the pressure in the piston chamber and the friction coefficient. To convert the clutch torque into the control variable, i.e. the clutch pressure, there must be an effective method of estimating the friction coefficient under real time condition. Each clutch is designed with a given nominal friction coefficient for the friction disks but the friction coefficient variates around the nominal value with respect to clutch temperature and slippage. Therefore, it is necessary to establish a database of the variation of friction coefficient under various clutch temperatures and slippage deemed possible by vehicle operation conditions. Such databases have been successfully applied in transmission control systems, such as the control system of the Honda five-speed clutch to clutch AT [13]. The second difficulty can be solved by improving the accuracy of the variable force solenoid and design optimization of the clutch apply circuit. In automatic transmissions currently in production or under development, each clutch is assigned a VFS in the hydraulic circuit so that the clutch pressure is controlled directly with minimum interference from other clutches or actuators. Accurate correlation between pressure and current signals (in terms of PWM) can therefore be achieved to control the clutch torque during shifts.

In summary, torque based transmission shift control can be implemented in steps that are illustrated in Figure 6.29. When a shift is commanded by the TCU, the nominal engine torque is firstly interpolated from the engine map in terms of the engine speed and throttle opening and is then modified using sensor-provided data such as intake air pressure, atmospheric condition, and fuel/air ratio. The torque on the converter impeller is then obtained by subtracting the loads of auxiliary systems from the modified engine torque. The net transmission torque is the result after considering the torque converter dynamics and characteristics. Knowing the transmission torque, various techniques can then be used to determine the clutch torque for shift control. The G design method is one of these techniques. The clutch torque can also be determined through model simulation with designed transmission output torque patterns. The clutch apply pressure that produces the clutch torque is then interpolated from the friction coefficient database using sensor provided data on ATF temperature, clutch clamping force, and slip rate. Afterwards, the clutch apply pressure is converted to VFS signal by the TCU which controls the clutch torque via the related circuit. In the initial piston stroke stage, the control signal is interpolated from the piston attributes database to minimize the delay of clutch actuation. As the transmission system responds to the VFS signal during a shift, as shown in Figure 6.29, open loop control and feedback control are then used for the control of the torque phase and inertia phase during the shift, as illustrated in Figure 6.26.

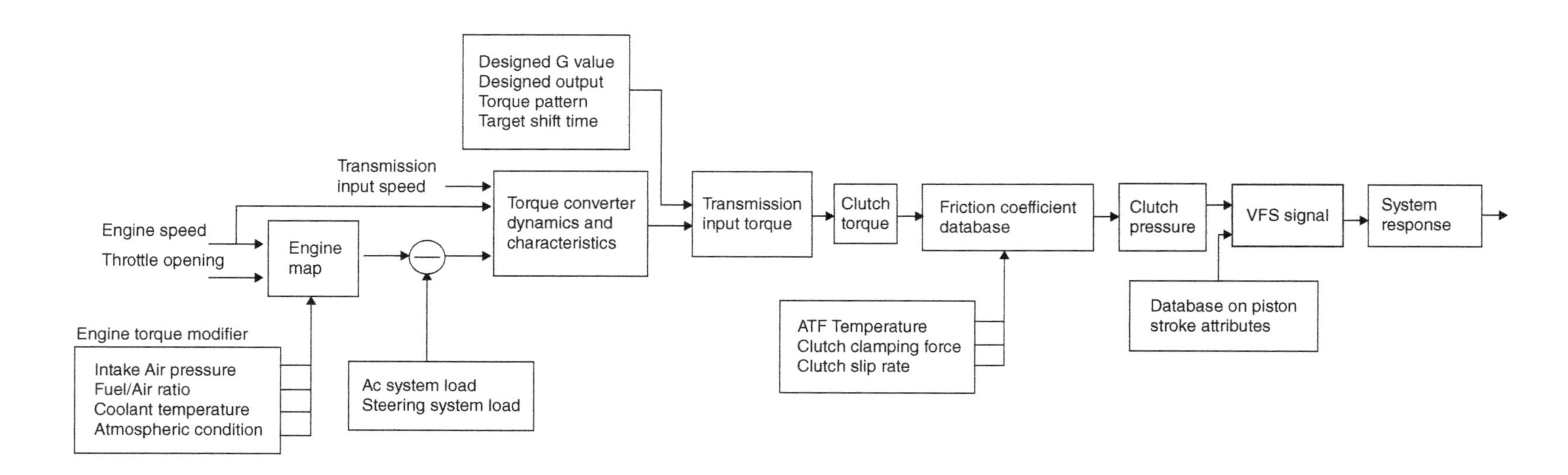

Figure 6.29 Torque based shift control logic.

6.4.11 System Diagnosis and Failure Mode Management

This function of the transmission control system is product specific and is realized by a combination of hardware and software. For example, pressure switches can be used in clutch pressure control circuits, as shown in Figures 6.15 and 6.16, to detect the pressure build-up. The status of these switches will tell whether or not the circuit is working properly. In case of malfunctions, the pressure switch will detect abnormal pressure values and provide relevant signals to the TCU, which then processes these signals to identify the root cause of the malfunctions. Generally, ATs today have the following features in system diagnosis and failure mode management:

- Control software and database can be updated after the sale of the vehicle. The OEM or transmission system suppliers develop and improve transmission control technologies continuously, the software or the database in existing vehicles can be replenished with updated versions at no cost to the vehicle owner. This allows existing vehicles to benefit from newly developed technologies and, more importantly, provides an easy fix for software issues in the transmission control system.
- Wherever possible, the TCU will send out a warning signal and display it on the instrument panel when malfunctions occur or an abnormal behavior, at component or system level, is detected. This urges the driver to have the vehicle inspected and serviced accordingly in the dealership or repair shop.
- With safety guaranteed, the control system allows the driver to drive the vehicle to a repair shop in case of a malfunction under the failure mode management scheme. This is usually realized by the default setup of the shift solenoids and hydraulic circuits in the transmission control system. For example, the transmission can be set in first gear in case of malfunction to allow the driver to drive the vehicle at low speed to the nearest repair shop.
- Transmission control systems can be checked on a regular basis after specified mileage intervals using specific inspection tools. This allows prognosis of potential malfunctions caused by component aging or function degradation by replacing problematic components ahead of their gross failures.

6.5 Calibration of Transmission Control System

There are many variables in the transmission control system, some of which are related to software that implements control strategies or algorithms, and others are related to hardware such as valves, clutches, and sub-circuits. These variables affect the transmission system response interactively. Moreover, the attributes or characteristics of control hardware components vary with respect to the system operation condition. Since each transmission shift corresponds to a specific system operation status, transmission control calibration needs to be conducted per each shift event for the optimization of shift quality in all vehicle operation ranges. Due to the large number of shift events in ATs with multiple gear ratios, transmission control calibration is a time consuming and expensive process in terms of labor cost. The calibration process is by nature highly experimental and is based on trial and error techniques. Engineer experiences and effective management are critical for shortening the powertrain calibration process of new vehicle models. Since the transmission works together with the engine and torque converter in the vehicle powertrain, transmission control calibration results directly depend on accurate and reliable data on engine output and converter characteristics which are obtained by engine and torque converter calibrations.

Transmission control calibration can be conducted on both component level and system level, in a laboratory environment or on test vehicles. At the component level, the main objectives are to obtain accurate data on the performance and attributes of key components in the transmission control system, such as clutch pistons, friction elements, and hydraulic circuits. Component level calibration can be mostly conducted in a laboratory setup that emulates conditions that the component is subject to when applied in the transmission system. Shift process control, including torque converter clutch control and engine spark retarding, is mainly calibrated at system level on test vehicles. Note that shift control can also be conducted at the initial calibration stage under laboratory environment, as shown in Figure 6.30. In some respects, the laboratory setup has advantages in emulating extreme transmission operation conditions and off-season weather patterns. In addition to the various sensors that are to be used in the control system of production vehicles, sensors used only for calibration purposes are installed at key positions in the setup shown in Figure 6.30 for a laboratory test or in the transmission of the test vehicles. For example, torque sensors, which are not used in production vehicles, may be used for transmission calibration to measure the torque values at key locations, such as the input and output shafts. An array of data is collected and processed by calibration hardware and software tools from sensors on various variables:

- Pressure values and profiles at key locations in the hydraulic circuit of the transmission control system. This indicates the response of the clutch apply or release pressure to control signals and is the main task in transmission calibration.
- Speeds of the engine, torque converter turbine, and transmission output. Additional speed sensors may be used in calibration to measure clutch slippage in shift operations. An accelerometer can be used in calibration vehicles to measure vehicle longitudinal acceleration in real time.
- ATF and environment temperatures.
- Engine intake pressure, throttle position, air/fuel ratio, etc.
- Torque on transmission output shaft or haft shaft.

6.5.1 Component Level Calibration

This can be conducted in a laboratory setup for the component concerned, and the main purpose is to obtain accurate quantitative data on the performance specifications and attributes when the component is applied in the transmission control system. As discussed previously, these data are critical for the correlation between

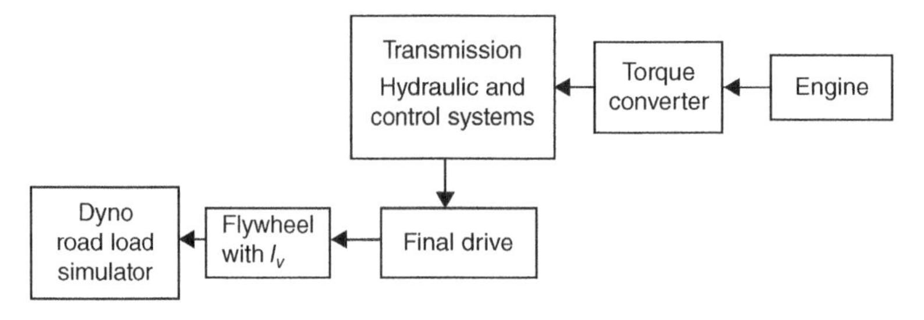

Figure 6.30 Transmission control system testing set-up.

control signals and clutch pressure profiles in the torque based shift control for transmission shifts. The calibrations on clutch piston attributes, clutch disk friction coefficient, and response of hydraulic circuits for line pressure control and clutch apply pressure control are most important on the component level, as described in the following.

Clutch piston attributes: As illustrated in Figure 6.25, for clutches that are not designed with counter pistons, there is a time delay for the piston chamber pressure to respond to the control signal. This time delay can be minimized by minimizing the piston chamber cavity in hardware design. For a given clutch, the time delay can be effectively handled if quantitative and reliable data on the attributes of clutch initial actuation are available. These data can be obtained by calibrating the clutch piston pressure response to the control signal (i.e. VFS current) in a laboratory setup as shown in Figure 6.30 or in test vehicles. In such calibrations, a pressure sensor is installed at a location close to the piston chamber to measure the pressure build-up when control signals are sent to the related VFS, while other sensors record the variables concerned with transmission operation status, such as temperature, clutch slip, and line pressure. A database is established using the calibration data on the clutch piston attributes during filling-up and initial stroke. This allows accurate timing for the actuation and pressure ramping-up of the oncoming clutch during shifts, as discussed previously.

Clutch disk friction coefficient: The friction coefficient of clutch disks depends on clutch temperature, clamping force, and clutch slippage. Accurate estimation of the friction coefficient is necessary for accurate clutch pressure control for the target clutch torque values during shifts. The friction coefficient variations of multiple disk clutches can be quantified under laboratory conditions that emulate transmission operations. In the experimental setup, torque sensors are installed on the shafts on both sides of the multiple disk clutch. The torque values are converted to the friction coefficients based on the clutch design parameters using Eq. 6.1. This leads to the establishment of the friction coefficient database shown in Figure 6.29.

Line pressure and clutch pressure circuits: The circuits for line pressure control and clutch pressure control are shown in Figures 6.1 and 6.5–6.7 respectively in various design variations. Each of these circuits is a subsystem consisting of valves, orifices, and hydraulic passages. The target pressure, whether it is line pressure or clutch piston chamber pressure, is controlled by the VFS in the related circuit. How the target pressure responds to the VFS control signal depends on the circuit attributes. It is therefore important to know the accurate correlation between the target pressure and the VFS control signal. This correlation can be calibrated either on test vehicles or in a laboratory setup. Pressure sensors are installed in the related circuits to measure the pressure values under real time conditions or conditions emulating real time transmission operation. The calibrated correlation can then be used to control the target line pressure or clutch pressure via the respective variable force solenoids.

6.5.2 System Level Calibration

The ultimate target of transmission calibration is to optimize the shift schedule and the quality of all shifts in the schedule that covers the whole vehicle operation range. Therefore, system level calibration concerns two mutually related issues: the calibration of the shift schedule and the calibration of all shifts in the schedule. As discussed in Chapter 5, model simulation can be used to predict the performance and fuel economy of an automatic vehicle with a given shift schedule. The simulation results and experiences on existing vehicles are used to establish an initial shift schedule for the optimized trade-off on fuel economy and performance. This initial shift schedule is then used for test vehicles in the calibration process and will be further adjusted to achieve the optimized trade-off on fuel economy, performance, and drivability. As for the control of the shifts in the shift schedule, model simulation can also be used to obtain initial clutch torque profiles that serve as the baseline shift control for test vehicles in the calibration process.

System level calibration can also be conducted in both the laboratory setup shown in Figure 6.30 and on test vehicles. Of course, laboratory calibration can only provide preliminary and complementary results on transmission shift control quality. The shift schedule and shift control for production vehicles have to be finalized through a painstaking calibration process on the test vehicles. There are several technical issues to be addressed in the vehicle calibration process, as described in the following.

Selection of shift events: As discussed previously, there are dozens of direct shifts, which only involve an off-going clutch and an oncoming clutch, in a multi-ratio automatic transmission. Each of these shifts can be made at different engine throttle opening and vehicle speeds along the particular shift threshold line in Figure 6.17. A number of points are selected on the shift threshold line and shift control calibration is conducted per each shift defined by one of these points. For example, if 10 points are chosen on the 1–2 upshift threshold line in Figure 6.17, there will be 10 1–2 upshift events that need to be separately calibrated. Calibration data on control variables, such as clutch pressure, line pressure, engine spark retard timing, and VFS signal current, are recorded for each of these shift events in the database of the transmission control system. If a shift is to be made between the selected points, the shift control variables are then determined by interpolation from the database. Counting all upshifts and downshifts as well as kick-down power-on downshifts, the total number of shift events to be calibrated for the shift schedule in Figure 6.17 will be in the hundreds. Note that the selection of shift points on the shift threshold line must reflect the most frequently used vehicle operation patterns. The shift threshold itself, as defined by the shift events, is a calibration target and is refined in the calibration process for the optimization of vehicle shift quality and drivability.

Converter clutch control calibration: The torque converter works together with the engine and transmission in the vehicle powertrain and its operation status is determined by system dynamics. The torque converter clutch control therefore needs to be calibrated on test vehicles. The converter clutch is applied or released according to the schedule shown in Figure 6.19. The calibration of converter clutch control concerns two issues: clutch release timing ΔT during shifts, as shown in Figure 6.20; and clutch apply and release process control, as shown in Figure 6.21. As discussed previously, the torque converter clutch release timing ΔT shown in Figure 6.20 directly affects the shift smoothness of upshifts and downshifts. This timing, as a delay after upshift initiation or as an amount of time ahead of downshift initiation, needs to be calibrated in each shift for the optimized damping effects of the torque converter on the shift process. This timing amount is just the difference between the points of time at which shift signal and converter release signal are sent by the TCU to the respective VFS. The process of converter clutch apply and release is controlled by the circuit shown in Figure 6.9. Pulse width modulation is used to control the converter clutch pressure during apply or release in a profile illustrated in Figure 6.21. The duty cycles in each step during apply and release are to be calibrated following the profile shown in Figure 6.21 for optimized smoothness when the converter clutch is released at the beginning of a transmission shift or applied after a shift in fixed gear operation.

Engine spark retarding timing calibration: As discussed in Chapter 5 and early in this chapter, transmission input torque reduction reduces the oncoming clutch torque required to complete the shift and lowers the transmission output torque overshoot. The transmission input torque reduction should be timed to start right after the inertia phase starts in upshifts. To control this timing, the engine controller should retard the spark to stop firing the selected cylinders at the exact crankshaft rotational position upon receiving the command from the TCU via control area network (CAN). The amount and timing of engine output torque reduction depend on the response to the engine controller and the engine characteristics. The key to the perfect engine spark retard timing is for the engine controller to send out spark retard commands in good time, so the engine responds with the reduced output torque as soon as the inertia phase starts. Through test vehicle calibration, engine spark retard timing can be determined for all shifts to minimize the torque hole and torque overshoot for the optimized shift quality.

In summary, calibration of transmission control is a lengthy trial and error practice where experience and existing data on similar products are critical in enhancing the effectiveness and shortening the process. This section only provides a guideline on control variables to be calibrated and the basic techniques used in transmission calibration. Publications are hard to find in the public domain in transmission control and calibration areas. Transmission control and calibration engineers are often internally trained at work by OEMs or suppliers through technical sessions or by technical manuals that do not circulate outside.

References

1 Gott, P.G.: *Changing Gears: The Development of the Automotive*, Society of Automotive Engineers, 1991, ISBN 1-56091-099-2.

2 ASE Transmission/Axle/Driveline Forum Committee, *Design Practices: Passenger Car Automatic Transmissions*, Third Edition, AE-18, SAE Publication, 1994, ISBN 1-56091-506-4.

3 Schwab, M.: *Electronically-Controlled Transmission Systems – Current Position and Future Developments*, SAE Paper No. 901156.

4 Taga, Y., Nakamura, K., Ito, H., and Taniguchi, T.: *Toyota Computer Controlled Four-Speed Automatic Transmission*, SAE Paper # 820740.

5 Wilfinger, E. and Thompson, J.: *Borg-Warner Australia Model 85 Automatic Transmission*, SAE Paper No. 880480.

6 Shinohara, M, Shibayama, T., Ohtsuka, K., Nawata, K, Ishii, S., and Yoshizumi, H.: *Nissan Electronically Controlled Four-speed Automatic Transmission*, SAE Paper # 890530.

7 Kondo, T., Iwatsuki, K., Taga, Y., Tanguchi, T., and Taniguchi, T.: *Toyota "ECT-I": a New Automatic Transmission with Intelligent Electronic Control System*, SAE Paper # 900550.

8 Taniguchi, H. and Ando, Y.: *Analysis of a New Automatic Transmission Control System for Lexus LS400*, SAE Paper # 910639.

9 Narita, Y.: *Improving Automatic Transmission Shift Quality by Feedback Control with a Turbine Speed Sensor*, SAE Paper # 911938.

10 Hojo, Y., Iwatsuki, K., Oba, H., and Ishikawa, K.: *Toyota Five-speed Automatic Transmission with Application of Modern Control Theory*, SAE Paper # 920610.

11 Taga, Y., Nakamura, K., Ito, H., ad Taniguchi, T.: *Toyota Computer Controlled Four-Speed Automatic Transmission*, SAE Paper # 820740.
12 Ohashi, T., Asatsuke, S., Moriya, H. and Nobie, T.: *Honda's 4-speed All clutch to Clutch Automatic Transmission*, SAE Paper # 980819.
13 Wakamatsu, H., Ohashi, T., Asatsuke, S., and Saitou, Y.: *Honda's 5-speed All clutch to Clutch Automatic Transmission*, SAE Paper # 2002-01-0932.
14 Konda, M., Hasegawa, Y., Takanami, Y., Arai, K., Tanaka, M., Kinoshita, M., Ootsuki., T., Yamaguchi, T., and Fukatsu, A.: *Toyota AA80E 8-speed Automatic Transmission with Novel Powertrain Control System*, SAE Paper # 2007-01-1311.
15 Lee, S., Zhang, Y., Jung, D. and Lee, B.: 2014, *A Systematic Approach for Dynamic Analysis of Vehicles with 8 or More Speed Automotive*, ASME Journal of Dynamic Systems, Measurement, and Control, Vol. 136, No. 5, 051008(1–11), 2014.
16 General Motors Corporation, *Hydra-Matic 4T80-E, Technician's Guide*, Second Edition, *Copyright 1994 Powertrain Division.*
17 General Motors Corporation, *Hydra-Matic 6T40/45, Technician's Guide, Copyright 2008 Powertrain Group.*

Problem

1 The stick diagram of the Ford 10-speed RWD automatic transmission for pickup trucks is as shown. Construct the framework hydraulic circuit similar to Figure 6.16 for the transmission control system.

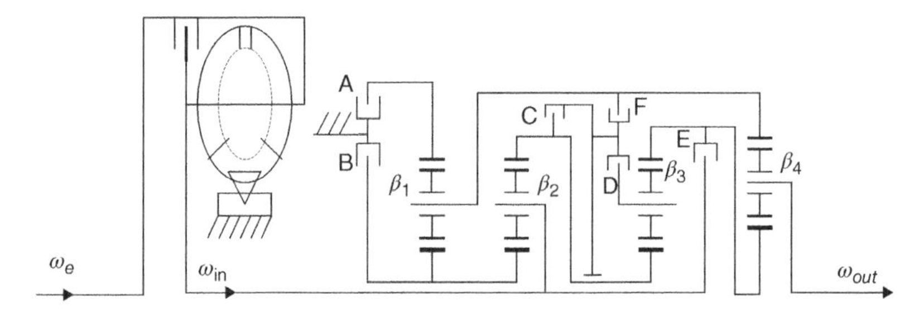

Clutch table

Gear	A	B	C	D	E	F	i_t
1	X	X		X	X		4.696
2	X	X	X	X			2.985
3	X		X	X	X		2.179
4	X		X	X		X	1.801
5	X		X		X	X	1.539
6	X			X	X	X	1.288
7			X	X	X	X	1.0
8		X		X	X	X	0.852
9		X	X		X	X	0.689
10		X	X	X		X	0.636
R	X	X		X		X	–4.786

PGT parameters:

$$\beta_1 = \frac{99}{45} \qquad \beta_2 = \frac{89}{51}$$

$$\beta_3 = \frac{101}{63} \qquad \beta_4 = \frac{85}{23}$$

7

Continuously Variable Transmissions

7.1 Introduction

As discussed in Chapter 1, continuously variable transmissions (CVT) are the ideal transmission that theoretically optimize engine matching. The idea of CVT dates back to the 15th century, the original design being credited to Leonardo da Vinci [1], who invented a mechanism with CVT functionality in 1490. CVTs have found applications in machinery that operates at relatively low power, but their applications in the automotive industry have gone through many setbacks and it is only in recent decades that the CVT vehicle market share has become significant. The British company Clyno was the first to develop a passenger car equipped with a CVT, in 1923 [1]. Not much happened for automotive CVTs since that, until, in 1961, the Dutch company DAF developed a production CVT for a small passenger vehicle with the brand name Daffodil [2]. Although this rubber CVT was not successful in the automotive industry due to its low efficiency, low torque capacity, and low reliability, it regenerated industrial interest in CVT development and inspired engineers in the design, manufacturing, and ratio control of this type of transmission. In 1987, Subaru developed a CVT vehicle named Justy which was well received by the market [3]. The Justy was a subcompact car with a 1.0 or 1.2 litre gas engine. In the same year, Ford and Fiat also launched compact cars of similar engine size equipped with CVTs. Due to limited torque capacity and reliability issues, CVT applications before the turn of the century were not widespread in the industry. The purchase of the patent of the CVT belt from DAF by Busch in 1995 was probably a major event in the transmission industry. As the leading automotive supplier, Busch played a leading role both in CVT technology and its market development.

The advancements in materials, manufacturing, and control technologies made possible CVT applications in vehicles with high engine outputs. As the leading CVT developer, Jatco supplies CVTs for passenger vehicles with engine sizes of 1.0–3.5 litres, covering most vehicle models. Nissan, as the industry leader in CVT automotive applications, uses CVTs across its product line from subcompact cars to full size SUVs. Other OEMs, such as Honda, Toyota, and Subaru, are also marketing popular cars and SUVs that are equipped with CVTs. It can be safely stated that CVTs today offer a service life as long as conventional ATs. Drivers can drive a CVT vehicle as long as they want to keep it, just like a vehicle equipped with a manual or conventional AT.

Automotive Power Transmission Systems, First Edition. Yi Zhang and Chris Mi.

The status of CVT technologies today is the result of intensive research and development by researchers and engineers both in the academic community and in the automotive industry. In a paper published in *Mechanism and Machine Theory*, Srivastava and Haque provided a comprehensive review of the development of CVT related technologies, with a long list of technical papers in the reference [4]. In this paper, CVT research and development was summarized in detail according to key CVT technical areas, including dynamic modeling of both belt and chain CVTs, as well as CVT ratio control. In the area of dynamic modeling, major contributions are credited to G. Gerbert for his fundamental work in the kinematics and mechanics of belt CVTs [5,6]. Miloiu, Worley, and Dolan proposed closed-form solutions that closely approximate the CVT belt slippage and forces involved under various CVT operations [7–9]. Belt CVT Transient behaviors and dynamics during ratio changing operations were investigated by Srivastava and Haque [10–12] and Carbone et al. in a series of papers [13–15]. Micklem et al. modeled the torque transmission mechanism of belt CVT based on elastohydrodynamic theory instead of Coulomb friction and studied the power transmission losses [16]. Pfeiffer and his co-workers conducted in-depth investigations into chain CVT dynamics and performance using multibody and FEM modeling [17,18]. In a series of papers [19–22], Fujii and Kurokawa presented valuable research results by analytical and experimental approaches on important technical issues for belt CVT design and control, including relationships between torque transmission and pulley thrust, compression of metal blocks and ring tension, and forces acting on metal blocks during fixed ratio operations and ratio changing operations. Fujii and Kurokawa's work sets up the practical guidelines for belt CVT design and control, and greatly contributes to the application of belt CVTs in vehicles with various engine sizes. In the area of CVT control, research efforts were mainly concentrated on continuous CVT ratio control with the objective of optimizing vehicle fuel economy, even though CVTs in today's vehicles are often controlled with multiple stepped ratios that emulate conventional ATs. For fuel economy optimization, the engine is controlled to always operate along the optimal operating line in the fuel map. This can be realized through engine–CVT integrated control where both the engine speed and the CVT ratio are controlled simultaneously [23,24].

Note that the research works referenced above are only the highlights of CVT research and development in the published domain. For a more thorough literature review, readers are recommended to read Srivastava and Haque's paper [4]. In addition, interested readers are also recommended to see the publications on the research and development on toroidal CVTs [25–28]. This chapter will mainly concentrate on the design and control of belt type CVTs. Readers may refer to publications that are mainly relevant to the contents in this chapter, such as the paper series by Fujii et al. [19–22].

Following this introduction section, the chapter will continue with Section 7.2 on the structural layouts of CVT systems and key components, including the basic CVT kinematics and operation principles. Section 7.3 will then concentrate on force analysis during CVT operations, and the mechanisms for torque transmission and ratio changes. Section 7.4 will look at control system design and the analysis of the control of ratio changing processes. Section 7.5 will present CVT system control strategies, including continuous ratio control, stepped ratio control, and system line pressure control.

7.2 CVT Layouts and Key Components

In a belt type or chain type continuously variable transmission, the belt–pulley or chain–pulley assembly realizes the ratio change functions. Overall transmission ratio is the multiplication of the variable CVT ratio with the stepped ratios of the gear sets in the transmission layouts. Typical belt or chain CVT systems are illustrated in Figure 7.1.

As shown in Figure 7.1, forward and reverse gears of the transmission are achieved by a simple planetary gear set through a forward clutch and a reverse clutch that are hydraulically actuated. A torque converter is used in most CVTs as the vehicle launcher, even though it is possible to use the forward clutch as the launcher if no torque converter is provided. There is no fundamental difference in the kinematics shown in Figure 7.1 between RWD CVTs and FWD CVTs Figure.

The planetary gear set can be placed either before the input (primary) pulley or after the output (secondary) pulley, as shown in the figure. For the case when the planetary gear set is placed before the input pulley as shown on the left in Figure 7.1, the input pulley rotates in opposite directions for forward and reverse gears, and the final drive usually consists of two gear sets, since the planetary gear set does not contribute to the overall drive line ratio. If placed after the output pulley as shown on the right in Figure 7.1, the planetary gear set provides an additional gear ratio in forward gears. This additional gear ratio is equal to the planetary gear train parameter β, and the overall drive train ratio is then equal to $(i_{cvt}\beta i_a)$, with i_{cvt} and βi_a as the CVT ratio and the final drive ratio respectively. In this layout, only one set of gears is needed in the final drive, and the input pulley always rotates in the same direction.

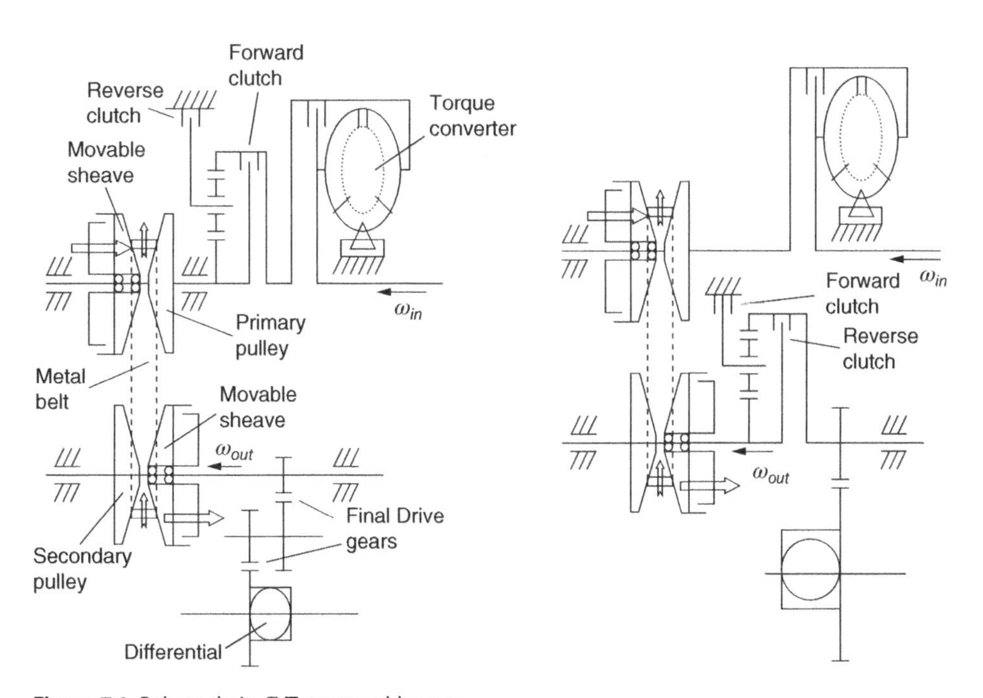

Figure 7.1 Belt or chain CVT structural layouts.

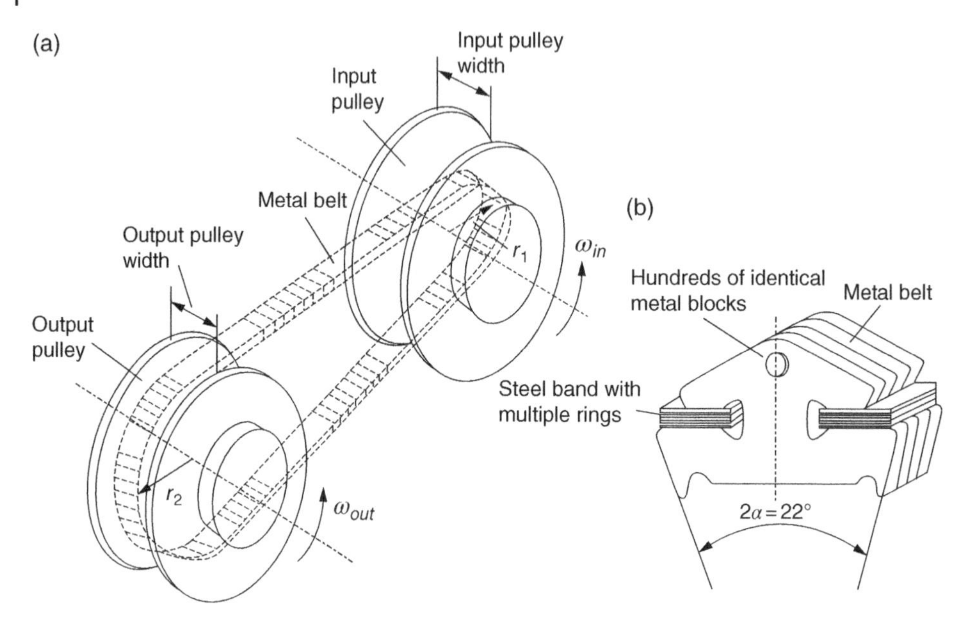

Figure 7.2 Structure of metal belt.

7.2.1 Belt Structure

The metal belt is the key CVT component and consists of many pieces of identical thin metal blocks and two steel bands, as shown in Figure 7.2. Each of the bands contains multiple layers (usually 9–12 layers) of thin steel rings (about 0.2 mm in thickness) laminated one on top another as shown in Figure 7.2b. The metal block has two slots, one on each side, into which the band is positioned. Hundreds of metal blocks – up to 400, and each with a thickness of about 2 mm – are strung along the two bands as shown in the figure. When assembled, the two bands (or rings as also commonly termed in the industry) will be in tension and the metal blocks will be under compression. When placed in the pulleys, as shown in Figure 7.2a, the side surfaces of each metal block contact the conical surface of the pulleys and it is the friction at this contact that transmits torque from the input pulley to the output pulley. The wedge angle of the pulleys is usually designed to be 22°, as indicated in Figure 7.2b. This gives a half wedge angle or groove angle of 11°, which optimizes the torque transmission and the lifting of the belt in the pulley wedges during ratio changing operations. The mean contact radius between the side surface of a metal block and a pulley groove surface is called the pitch radius, which is denoted as r_1 on input side and r_2 on the output side. The CVT ratio is then defined as $i_{cvt} = \frac{r_2}{r_1}$.

7.2.2 Input and Output Pulleys

Each of the two pulleys consists of two sheaves, one movable along the shaft and the other fixed on the shaft. As shown in Figure 7.3, the movable sheaves are supported on the input and output shafts by ball splines. These ball splines minimize the friction between the shaft and the movable sheave on it, during ratio changing processes, while keeping the movable

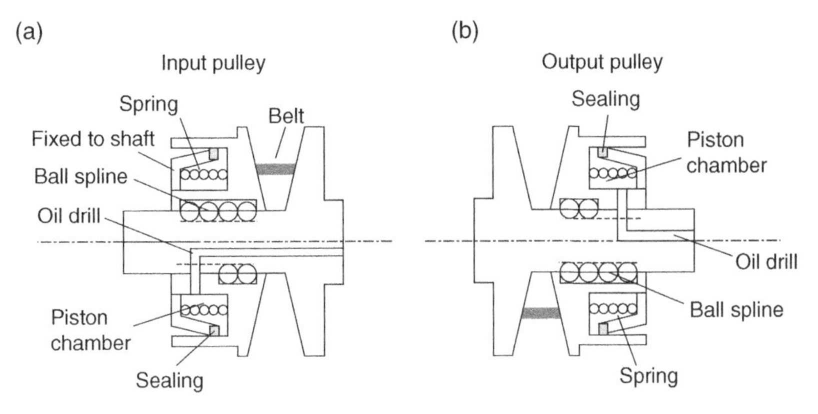

Figure 7.3 Input and output pulleys.

sheave rotating with the shaft. The axial motion of the two movable sheaves is actuated hydraulically, with the moveable sheaves also serving as the hydraulic pistons, as shown in Figure 7.3. The return spring in the piston chamber dampens the initial stroking force of the hydraulic piston. During ratio changing processes, the movable sheaves are actuated to move along the respective shafts, forcing the metal belt to rise or drop in the grooves of the input and output pulleys. In an upshift, the metal belt rises in the input pulley groove and drops in the output pulley groove, making r_1 larger and r_2 smaller so as to decrease the CVT ratio i_{cvt}. As can be observed in Figures 7.2 and 7.3, the width of the input pulley decreases whiles the width of the output pulley increases during upshifts. On the contrary, the metal belt rises in the output pulley groove and drops in the input pulley groove during downshifts, making r_2 larger and r_1 smaller so as to increase the CVT ratio i_{cvt}. This also increases the input pulley width while decreases the output pulley width.

As the movable sheaves move axially during shift operations, the central line or the line of symmetry of the input pulley and the output pulley will be displaced unequally along the respective shaft. As a result, the central line of the metal belt will be tilted as shown in Figure 7.4, causing unfavorable loading conditions on the belt. Belt tilt caused by ratio change can be minimized by arranging the movable sheaves of the input and the output pulleys on opposite sides, as shown in Figure 7.3. In this arrangement, both the input and output sides of the belt move in the same direction during ratio change operations, thus keeping the belt tilt angle at a negligible level. As shown in Figure 7.4, the displacement of the movable sheave and the belt displacement in the pulley groove are related as $\Delta_s = \Delta_r \tan\alpha$. The belt tilt angle is then determined as: $\sin^{-1}\frac{|\Delta_{s1} - \Delta_{s2}|}{\sqrt{E^2 - (r_2 - r_1)^2}}$.

7.2.3 Basic Ratio Equation

The length of the metal belt at assembly is equal to the metal block thickness multiplied by the number of the metal block and is a near constant since the compressive deflection of the metal blocks is negligible. A pitch line is defined on the metal belt to relate the angular velocities of the input and output pulleys. This pitch line has the same length as the metal belt and consists of the contact arcs on the pitch circles of the two pulleys with pitch radii r_1 and r_2, and the two identical tangents to the two pitch circles, as shown

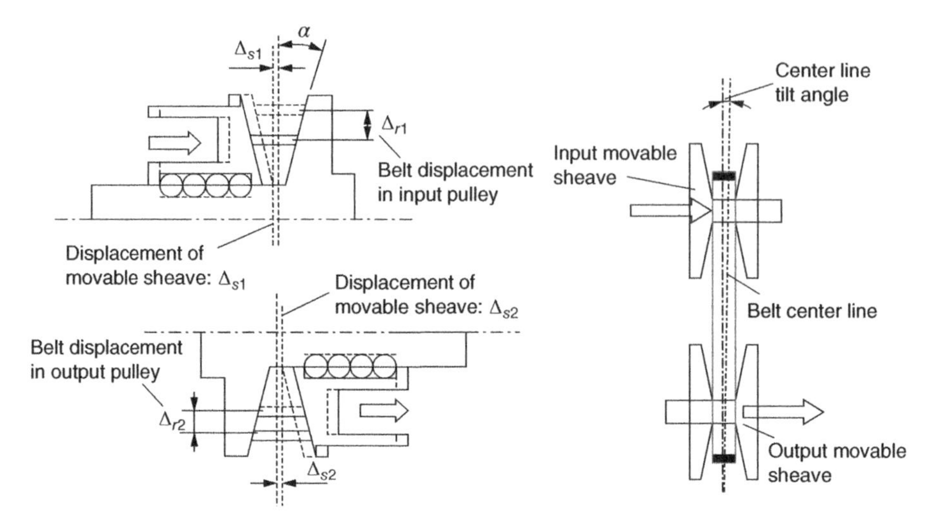

Figure 7.4 Belt central line tilt.

in Figure 7.5. Theoretically, there is no slippage along the pitch line between the side surfaces of the metal blocks and the pulley surfaces. In kinematics, it can be stated that the two pitch circles are defined by the CVT ratio i_{cvt}.

As shown in Figure 7.5, O_1 and O_2 are the centers of the input and output pulleys respectively and E denotes the center distance. Angle β is measured between the common tangent to the two pitch circles and line BC that is parallel to the center line O_1O_2. It is apparent from Figure 7.5 that the contact angles ϕ_1 and ϕ_2, which are the angles that the contact arcs span on the pitch circles of the input and output pulleys respectively, are equal to $(\pi - 2\beta)$ on the input pulley and $(\pi + 2\beta)$ on the output pulley. Since the length of the pitch line is a constant, the following equations can be readily derived by trigonometry using Figure 7.5:

$$\left(\frac{\pi}{2} - \beta\right) r_1 + \left(\frac{\pi}{2} + \beta\right) r_2 + \sqrt{E^2 - (r_2 - r_1)^2} = \frac{L}{2} \tag{7.1}$$

$$i_{cvt} = \frac{r_2}{r_1} \tag{7.2}$$

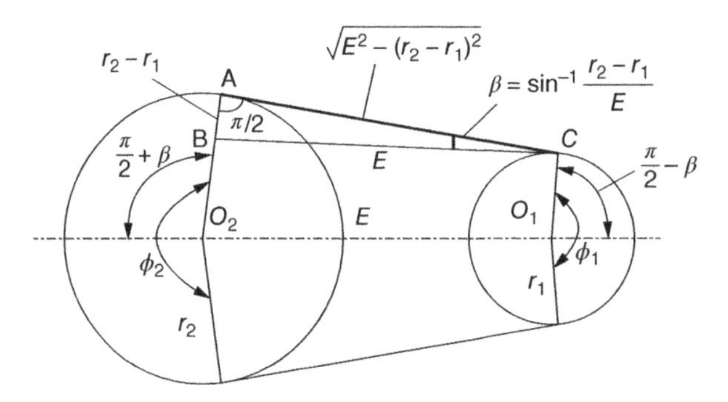

Figure 7.5 Belt pitch line and contact arcs.

where L is the length of the pitch line, which is almost a constant. For any specified value of the CVT ratio i_{cvt}, the two corresponding pitch radii r_1 and r_2 can be uniquely determined by solving Eqs (7.1) and (7.2) together. The two contact angles, ϕ_1 and ϕ_2 are then determined readily as shown in Figure 7.5. Note that the trigonometry shown in Figure 7.5 is valid when the CVT ratio is larger than 1.0. When the CVT ratio is smaller than 1.0, the subscripts for the pitch radii in Eq. (7.1) should be switched.

7.3 Force Analysis for Belt CVT

As mentioned in the introduction section, force analysis and the relation between various forces during belt CVT operations has been the subject of intensive research. This section presents the analysis, as quantitatively as possible, for the forces acting on individual metal blocks, pulleys, and the whole belt. The distribution of block compressive force and ring tension force will be analysed according to the CVT ratio and load conditions. The focus of the section is on the torque transmitting mechanism and the relations between the various forces involved in CVT operations, especially the relation between the thrust forces on the input pulley and the output pulley. Since the ratio between the pulley thrust forces is critical for belt CVT design and control, it has been the main subject of study in the paper series [19,20] by Fujii et al. as highlighted previously.

7.3.1 Forces Acting on a Metal Block

The free body diagram for an individual metal block inside the input or output pulley groove is shown in Figure 7.6. On the friction surface of the metal block, there exist three forces: the normal contact force N, the radial friction force F_r, and the tangential friction force F_t. The band normal force N_B and the band friction force F_b exist on the contact

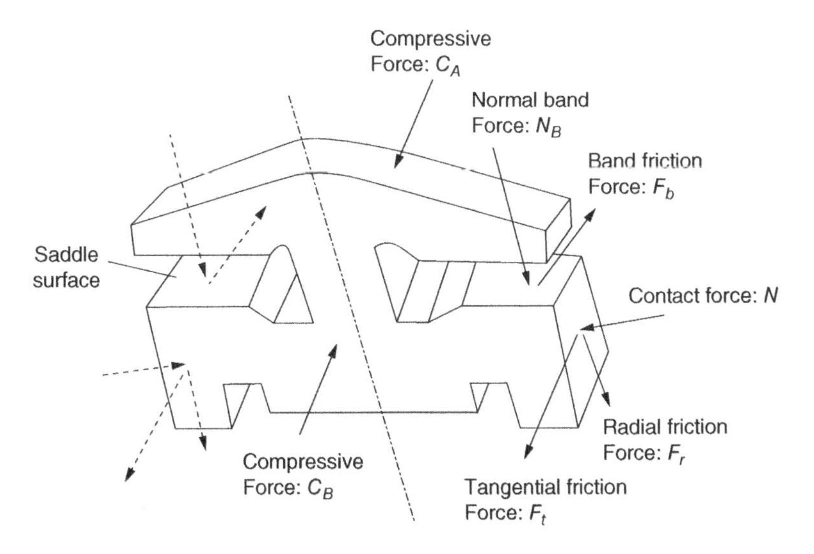

Figure 7.6 Forces acting on a metal block.

surface (saddle surface) between the steel band and the metal block. These forces are symmetrical on both sides of the metal block. The metal block is also subject to compressive forces (pushing forces) C_B and C_A, that are applied by the neighboring blocks before and after respectively. The friction forces are related to the normal contact forces as follows, based on Coulomb friction theory:

$$F_t = \mu_t N \tag{7.3}$$

$$F_r = \mu_r N \tag{7.4}$$

$$F_b = \mu_b N_B \tag{7.5}$$

where μ_t and μ_r are the tangential and radial friction coefficients between the block side surface and the pulley groove surface respectively, while μ_b is the friction coefficient between the band and the block slot surface. During CVT operation, torque is transmitted from the input pulley to the output pulley by the tangential friction force F_t. For normal CVT power transmission, there must be no slippage between the belt and the pulleys. This requires that the tangential friction coefficient μ_t in Eq. (7.3) must not exceed the maximum friction coefficient μ_{max}, i.e. $\mu_t \le \mu_{max}$. The radial friction force F_r is opposite in direction to the radial displacement of the metal belt in the pulley grooves during ratio changing operations. For example, the radial friction force is toward the input pulley axis during upshifts and outward during downshifts. Based on the equilibrium of the metal block, there exist the following relations between the forces shown in Figure 7.5:

$$N_B = N \sin\alpha \mp \mu_r N \cos\alpha \tag{7.6}$$

$$C_B - C_A = \mu_t N \mp \mu_b N_B \tag{7.7}$$

In Eq. (7.6), the minus sign is for upshifts and the plus sign is for downshifts for the radial friction force in the input pulley. The signs for the radial friction force in the output pulley are opposite to those for the input pulley. Note that the quantitative analysis on the band friction force is very difficult since Eq. (7.7) is highly indeterminate as the compression of metal blocks and the band slippage are dependent upon CVT operations.

7.3.2 Forces Acting on Pulley Sheaves

The normal contact force and the radial friction force acting on the metal block are shown in Figure 7.7a on the axial section of the input pulley when the metal belt is moving outward. These forces are symmetrically applied on each side of the metal block. The normal contact force and the radial friction force acting on the two sheaves of the input pulley are shown in Figure 7.7b. The thrust force P is applied respectively on the two sheaves along the axial direction. For the movable sheaves of the two pulleys, the thrust force P is applied by the respective hydraulic piston as shown in Figure 7.3a. As mentioned previously, the movable sheave is supported on the shaft by ball splines and the resultant reaction from the balls to the movable sheave is denoted by R in Figure 7.7b. Altogether, there are five forces applied to the movable sheave: four act on the axial section as shown in Figure 7.7b, the fifth force is the tangential friction force F_t that acts perpendicular to the axial section and is not shown in Figure 7.7b. Note that the resultant axial load on the input or output shaft is zero since the thrusts on the two

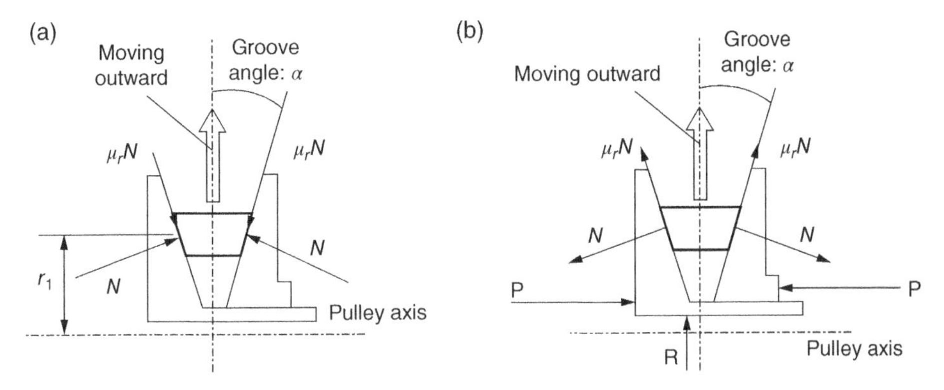

Figure 7.7 Forces acting on a metal block and on the input pulley on the axial section.

sheaves cancel each other. In addition, the groove angle α must be designed to be larger than the friction angle $\gamma = \tan^{-1}\mu_r$ since the metal block must be pushed outward by the thrust force during ratio changes.

Forces acting on pulley axial section are shown in Figure 7.7. The thrust force P is applied on each half of the pulley along the pulley axis and R is the reaction force applied on the movable sheave in the direction that is collinear with the CVT center line O_1O_2 shown in Figure 7.5. The contact force N and the radial friction force $\mu_r N$ of all metal blocks, shown in Figure 7.7, on the axial section for one block, have the same projections respectively upon the pulley axis. The sum of the projections of all forces on a movable sheave in the direction of the pulley axis must be equal to zero for equilibrium. This results in the following equations that relate the thrust force P, the contact force N and the radial friction force $\mu_r N$ for the movable sheaves of the input and output pulley respectively:

$$P_{in} = \cos\alpha\sum_{i=1}^{n} N_i \pm \sin\alpha\sum_{i=1}^{n}\mu_{r1}N_i = (\cos\alpha \pm \mu_{r1}\sin\alpha)\sum_{i=1}^{n} N_i \tag{7.8}$$

$$P_{out} = \cos\alpha\sum_{j=1}^{m} N_j \mp \sin\alpha\sum_{j=1}^{m}\mu_{r2}N_j = (\cos\alpha \mp \mu_{r2}\sin\alpha)\sum_{j=1}^{m} N_j \tag{7.9}$$

where P_{in} and P_{out} are the thrust forces applied by the hydraulic pistons, or the thrust forces generated by other means on the movable sheaves of the input and output pulleys respectively, and n and m are the numbers of metal blocks along the contact arcs on the two pulleys respectively. The plus sign is for the case when the block in the input pulley is moving outward and the minus sign is for the case when the block is moving inward. It is assumed in the equations that the radial friction coefficient μ_r is the same for all metal blocks in the same pulley. Assuming the tangential friction coefficient μ_t to be the same for all metal blocks in the same pulley, the resultant torque made by all the tangential friction forces on the input or output pulley is then given by the following two equations:

$$T_{in} = 2\mu_{t1}r_1\sum_{i=1}^{n} N_i \tag{7.10}$$

$$T_{out} = 2\mu_{t2}r_2\sum_{j=1}^{m} N_j \tag{7.11}$$

where r_1 and r_2 are the pitch radii of the input and output pulleys respectively. Clearly, the input torque and the output torque are related by the CVT ratio, i.e. $T_{out} = i_{cvt} T_{in} = \frac{r_2}{r_1} T_{in}$. The relation between thrust force and torque can be readily derived by combining Eq. (7.8) with Eq. (7.10) and Eq. (7.9) with Eq. (7.11), as follows:

$$P_{in} = \frac{(\cos\alpha \pm \mu_{r1} \sin\alpha)}{2\mu_{t1} r_1} T_{in} \tag{7.12}$$

$$P_{out} = \frac{(\cos\alpha \mp \mu_{r2} \sin\alpha)}{2\mu_{t2} r_2} T_{out} \tag{7.13}$$

Note that the derivation of the two equations above does not assume even distribution of the contact force N between the pulley groove surface and the metal blocks. Since the groove angle α is small and the term ($\mu_r \sin \alpha$) is negligibly small, these two equations can be simplified for the relation between thrust force and torque:

$$P_{in} = \frac{\cos\alpha}{2\mu_{t1} r_1} T_{in} \tag{7.14}$$

$$P_{out} = \frac{\cos\alpha}{2\mu_{t2} r_2} T_{out} = \frac{\cos\alpha}{2\mu_{t2} r_1} T_{in} \tag{7.15}$$

Note that the four friction coefficients, (μ_{r1}, μ_{t1}) and (μ_{r2}, μ_{t2}), are highly indeterminate during CVT operation. Practically, this means that the relation between the input thrust force P_{in} and the output thrust force P_{out} cannot be readily derived using Eqs (7.12) and (7.13). This relation is critical for belt CVT design and control and will be discussed later in detail. To avoid slippage between belt and pulley, both the real time tangential friction coefficients μ_{t1} and μ_{t2} must be smaller than the maximum friction coefficient μ_{tmax}. The maximum input torque that can be transmitted by belt CVT is then defined using Eq. (7.15) as:

$$T_{max} = \frac{2\mu_{max} r_1}{\cos\alpha} P_{out} \tag{7.16}$$

This simple equation is important for belt CVT design and control. The thrust force on the output pulley, P_{out} is usually controlled as the active force, with the thrust force on the input pulley P_{in} as the reaction force, as will be detailed later. For a given CVT, the value of the maximum friction coefficient μ_{max} can be estimated with decent accuracy experimentally. Even without knowing the exact μ_{max} value, the relation between the maximum input torque T_{max} and the thrust on the movable sheave of the output pulley can be determined by experiments on a CVT test rig. In such experiments, a certain thrust force P_{out} is set and the CVT under test is controlled to run at a certain CVT ratio. The input torque is then gradually ramped up until the belt starts to slip. The input torque recorded at the threshold of slippage is then the maximum input torque T_{max} for the specific thrust force P_{out}. For CVT design and control, the ratio between the real time input torque that is being transmitted and the maximum input torque determined by Eq. (7.16), is defined as the torque ratio, i.e. $i_t = \frac{T_{in}}{T_{max}}$. The torque ratio is a measure of the load condition and of how much the CVT operation status is below the block–pulley slippage threshold and thus can be considered as a safety factor. The torque ratio i_t is one of the important factors that determine the mechanism of torque transmission and the relation between the forces involved in CVT operations.

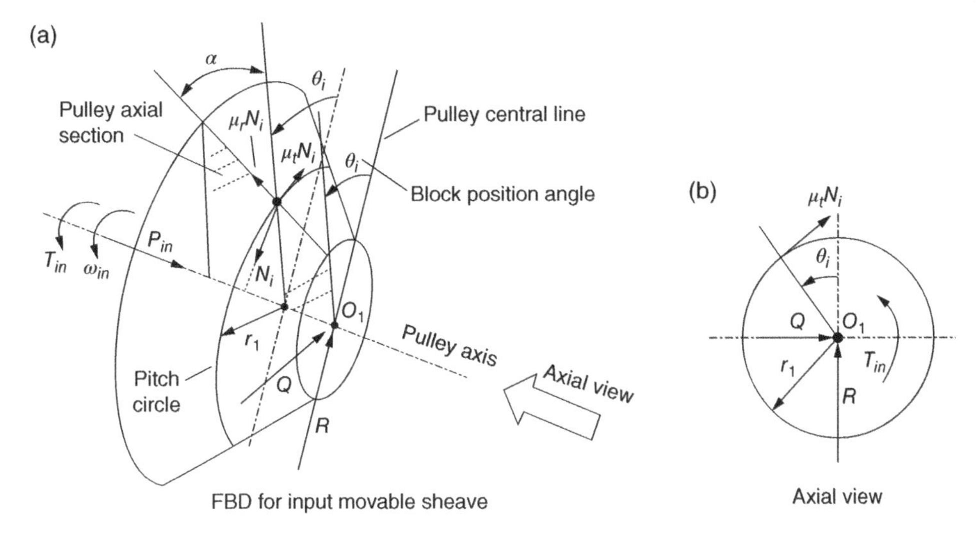

Figure 7.8 Free body diagram of the movable sheave of the input pulley.

As mentioned previously, Figure 7.7b only shows the forces applied by the metal block to the input pulley on the pulley axial section. In general, the free body diagram of the movable sheave of the input pulley is shown in Figure 7.8. In this figure, the pulley central line O_1O_2 bisects the input pulley. The position of a particular metal block is defined by the block position angle θ_i, and R and Q are, respectively, the reaction force applied by the ball splines shown in Figure 7.3 to the movable sheave in the direction along the CVT center line and along the direction perpendicular to the plane formed by the two pulley axes.

The block position angle θ_i in Figure 7.8 is related to the contact angle ϕ_1 defined in Figure 7.5, the pitch radius r_1 and the thickness of the metal block, denoted as t. It is apparent that the number of metal blocks on the input pulley and the position angle of each block along the contact arc can be determined by:

$$n = \frac{r_1\phi_1}{t} \tag{7.17}$$

$$\theta_i = \frac{t}{r_1}i \qquad i = \pm 1, 2, 3, \ldots, \frac{n}{2} \tag{7.18}$$

The number of blocks denoted as n in Eq. (7.17) can be rounded to the nearest even number without losing notable accuracy. The equilibrium condition along the pulley axis has been used to derive Eq. (7.8) for the determination of the thrust force P_{in}. The reaction forces R and Q can be determined respectively in terms of the contact force N_i and the friction forces applied by the belt to the movable sheave by considering the equilibrium conditions along the other two directions. Reaction force Q is only related to the load on the ball splines that support the movable sheave. The reaction force R, that acts along the CVT center line is more relevant to CVT torque transmission and ratio control and is determined by:

$$R = \sin\alpha\sum_{i=1}^{n}\sin\theta_i N_i \mp \cos\alpha\sum_{i=1}^{n}\cos\theta_i\mu_{r1}N_i - \sum_{i=1}^{n}\sin\theta_i\mu_{t1}N_i \tag{7.19}$$

The minus sign in this equation is for the case when the block is moving outward in the input pulley. The free body diagram for the movable sheave of the output pulley can be drawn similar to Figure 7.8 and the reaction force to the output pulley movable sheave can be represented by an equation similar to Eq. (7.19).

7.3.3 Block Compression and Ring Tension

As mentioned previously, the steel band or ring is in tension while the metal blocks are under compression when the CVT is assembled. However, the status of ring tension and block compression will change according to the CVT's operation conditions. The CVT ratio and the torque transmitted have the most effect on the status of ring tension and block compression, as analysed in the paper series published by Fujii et al. [19,20]. As shown in Figure 7.9, C_1 and C_2 are respectively the block compressive forces on side 1 and side 2, K_1 and K_2 are the respective ring tension forces. Away from the contact arcs, the block compressive force and the ring tension force on either side of the belt are constant. According to the analysis by Fujii and Kurokawa, there exists sliding between the ring and the block on the saddle surface (i.e. the contact surface between the ring and the block, as shown in Figure 7.6) at the pulley with smaller pitch radius. This is because the so-called saddle speed, that is the speed of the pulley at the radius of the saddle surface, is higher in the pulley with smaller pitch radius than that in the pulley with the larger pitch

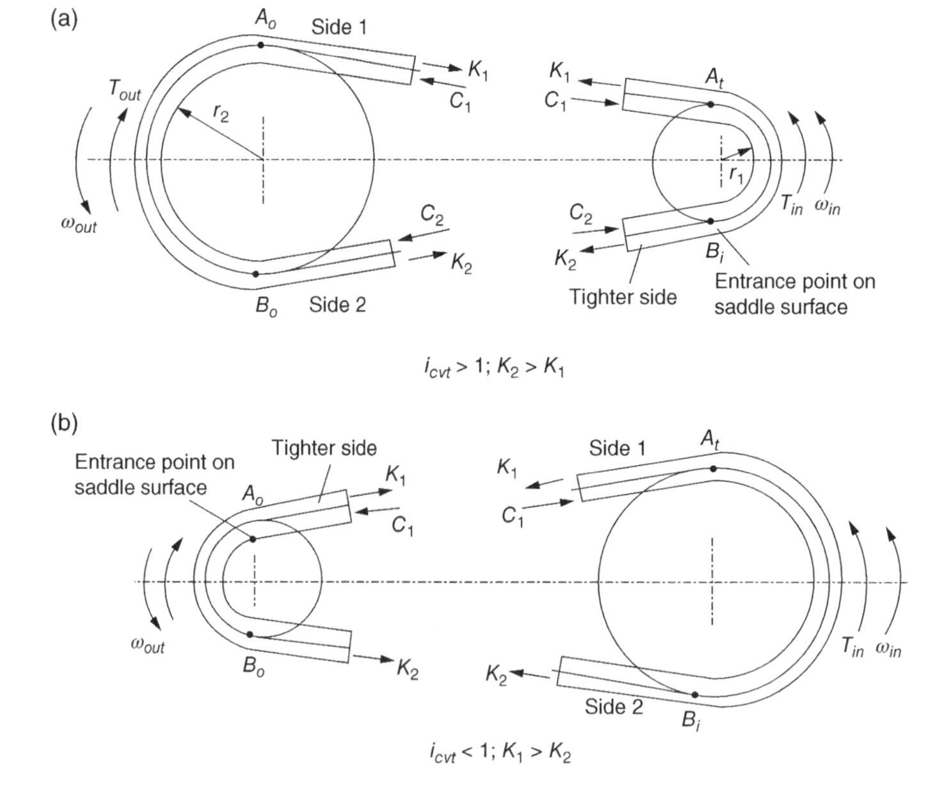

Figure 7.9 Ring tension and block compression forces.

radius. This speed difference is larger than ring elasticity can compensate for. As shown in Figure 7.9, ring sliding on the saddle surface occurs on the input pulley when $i_{cvt} > 1$ and on the output pulley when $i_{cvt} < 1$. As a result of this, the pulley with smaller pitch radius is always ahead of the belt on it, tightening up the belt side that moves toward it and slackening the other side that moves away from it, as shown in Figure 7.9. Ring tension exists under all CVT operation conditions on both sides, but block compressive force only exists on one side, as determined by the CVT ratio and the torque ratio defined previously. As shown in Figure 7.9, block compressive force only exists on Side 1 of the ring when $i_{cvt} < 1$, and may exist on either side of the ring as warranted by the torque ratio i_t when $i_{cvt} \geq 1$.

7.3.4 Torque Transmitting Mechanism

By considering the equilibrium of the input or output pulley in Figure 7.9, it is apparent that both the block compressive force and the ring tension force participate in the transmission of torque between the input pulley and the output pulley. This might be in contrast to the common conception that torque is only transmitted by the pushing force on the blocks from the input to the output. The qualitative effect of ring tension on torque transmission can be readily observed from Figure 7.9, which shows that the ring tension aids the block compressive force in transmitting torque from the input pulley to the output pulley when $i_{cvt} > 1$ since $K_2 > K_1$, but when $i_{cvt} < 1$, the ring tension acts against the torque transmission from the input pulley to the output pulley. When $i_{cvt} = 1$, the ring tension status is similar to the case for $i_{cvt} > 1$ as shown in Figure 7.9a. This is because the friction force between the pulley saddle surface and the ring, that is force F_b shown in Figure 7.6, tightens up Side 2 of the ring which moves toward the input pulley in the absence of the sliding of the ring on the saddle surface. A peculiar situation arises when $i_{cvt} > 1$ and there is no load on the output pulley (i.e. $T_{out} = 0$). Since ring tension always exists, a torque with its magnitude proportional to $(K_2 - K_1)$ is transmitted from the input pulley to the output pulley as shown in Figure 7.9a. To maintain the equilibrium of the output pulley, there must exist block compressive force C_2 on Side 2 of the ring to counterbalance the effect of the ring tension. Therefore, when $i_{cvt} > 1$ and there is no load on the output pulley (i.e. torque ratio $i_t = 0$), there exists block a compressive force on Side 2 of the ring. Since the compression status of the metal blocks can only be changed continuously as the transmitted torque or torque ratio i_t increases, block compressive force on Side 2 of the ring will maintain its existence until the torque ratio increases from zero to a certain threshold value i_{tc}. In summary, the input torque that is transmitted by the block compressive force and the ring tension force is formulated by the following equations:

$$T_{in} = C_e r_1 = \begin{cases} (K_2 - K_1) r_{s1} - C_2 r_{c1} & \text{when } i_{cvt} \geq 1 \text{ and } i_t < i_{tc} \\ (K_2 - K_1) r_{s1} + C_1 r_{c1} & \text{when } i_{cvt} \geq 1 \text{ and } i_t \geq i_{tc} \\ C_1 r_{c1} - (K_1 - K_2) r_{s1} & \text{when } i_{cvt} < 1 \end{cases} \quad (7.20)$$

where r_1 is the pitch radius of the input pulley; r_{s1} is the mean radius of the arc formed by the ring in the input pulley; r_{c1} is the radius of the point of concentration of the compressive force on the block face and can be approximated as the input pulley pitch radius r_1. C_e is the equivalent torque transmitting force as if it acts at the pitch radius. In Eq. (7.20), the ring tension difference inside the pair of parenthesis is always positive; the block compressive force C_1 or C_2 does not exist simultaneously on both sides of the ring. The following conclusions can be made on the belt CVT torque transmission mechanism based on Eq. (7.20):

- When $i_{cvt} \geq 1$ and torque ratio is lower than a certain value i_{tc} (i.e. when load is low), torque is transmitted by the ring tension force positively from the input to the output, but block compressive force acts against positive torque transmission, i.e. it acts against torque transmission from the input to the output. This is the only situation where the block compressive force exists on Side 2 of the ring and acts against torque transmission, as shown in Figure 7.9a.
- When $i_{cvt} \geq 1$ and torque ratio is higher than the threshold value i_{tc}, both the ring tension force and the block compressive force transmit torque positively from the input to the output. This is the most efficient belt CVT torque transmission mechanism and is also the most conducive to CVT durability. In order to keep the torque ratio i_t around the optimized value, it is necessary to control the thrust force on the output pulley P_{out} in Eq. (7.15) just above the value needed to transmit the required input torque. The torque ratio $i_t = 0.77$ or above is suggested in the paper series by Fujii et al. [19,20] for CVT operations corresponding to the most driven traffic conditions. In addition, lowering the thrust force P_{out} to be just above the value necessary for normal torque transmission reduces the loads on all components of the CVT system, including the CVT unit itself, pumps and hydraulic components, gears, bearings, and shaft, and is beneficial for CVT system efficiency, reliability, and durability.
- When $i_{cvt} < 1$, the ring tension force acts against positive torque transmission from the input to the output. The block compressive force always transmits torque positively from input to output when $i_{cvt} < 1$ regardless of the torque ratio.
- The equivalent torque transmitting force C_e is defined in Eq. (7.20) to reflect the resultant effect of the block compressive force and the ring tension on torque transmission. This is a virtual force that acts at the pitch radius of the input pulley.
- When $i_{cvt} \geq 1$, it is correct to say that a certain percentage of the input torque is transmitted from input to output by the ring tension force since it contributes positively to the equivalent torque transmitting force C_e. *However, when* $i_{cvt} < 1$, *it is not correct to say that a torque with a magnitude of* $[i_{cvt}(K_1 - K_2)r_{s1}]$ *is transmitted from the output pulley backward to input pulley.* Or otherwise, there would exist a significant amount of power recirculation between the input and the output pulleys, causing significant efficiency loss, overheating, and other detrimental issues.

The block compressive force and the ring tension force were investigated experimentally in depth by Fujii et al. [19–22]. As shown in Figure 7.9 and Figure 7.10, B_i and A_i are respectively the entrance and exit points on the input pulley contact arc; A_o and B_o are respectively the entrance and exit points on the output pulley contact arc. A_iA_o is belt Side 1 and B_iB_o is belt Side 2. The distribution of block compression and ring tension is shown in Figure 7.10 when $i_{cvt} \geq 1$ and the torque ratio i_t is higher than 0.5. As shown in the figure, ring tension on Side 2 is higher than on Side 1 ($K_2 > K_1$), and block compressive force C_1 exists on belt Side 1. A convincing interpretation of the experimental data was provided in [19–22] regarding the distribution of block compression and ring

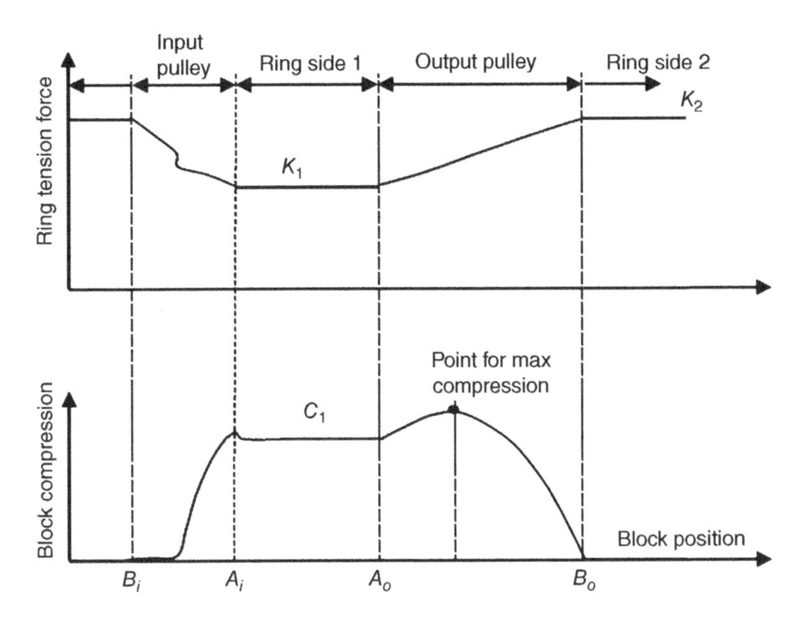

Figure 7.10 Ring tension and block compression distribution when $i_{cvt} \geq 1$ and $i_t \geq i_{tc}$.

tension shown in Figure 7.10. When the CVT under testing runs in stable condition, ring Side 2 B_iB_o is tightened up while Side 1 A_iA_o is slackened. There exists a small amount of backlash between blocks on Side 2 under this condition. Therefore, there is still no compression on the blocks after they have entered the input pulley on Side 2 at point B_i, because compression cannot build up until the backlash is totally eliminated by the tangential friction force between the blocks and the input pulley. The tangential friction force has the effect of pushing the blocks toward the input pulley exit A_i. After the block backlash is eliminated, the block compression then rises quickly in the input pulley. Theoretically, the compressive force of the block that leaves the input pulley exit should be equal to the compressive force C_1 on Side 1. The spiky transition at point A_i is caused by the tilt of the block when it leaves the input pulley exit. For the ring contained in the input pulley, the tension gradually decreases from the value K_2 on Side 2 toward the input pulley exit. It jerks somewhat at the position in the contact arc where the block compression starts to build up and is equal to the ring tension force K_1 on Side 1. After the blocks enter the output pulley at the entrance A_O, the block compressive force ramps up to a maximum value and then quickly ramps down to be zero at the exit B_o. This phenomenon was not explained by Fujii et al. and can be analysed as follows. After entering the output pulley, a block is acted on by four forces: two compressive forces that act on the front and rear of the block, denoted as C_B and C_A respectively in Figure 7.6, the tangential friction force F_t and ring or band friction force F_b. The ring friction force increases the block compression while the tangential friction force decreases it as indicated by Eq. (7.7). The effect of the ring friction force is more than that of the tangential friction force along the contact arc until the block compression reaches its maximum. Then, the effect of tangential friction dominates and reduces the block compression to zero quickly toward the exit B_o.

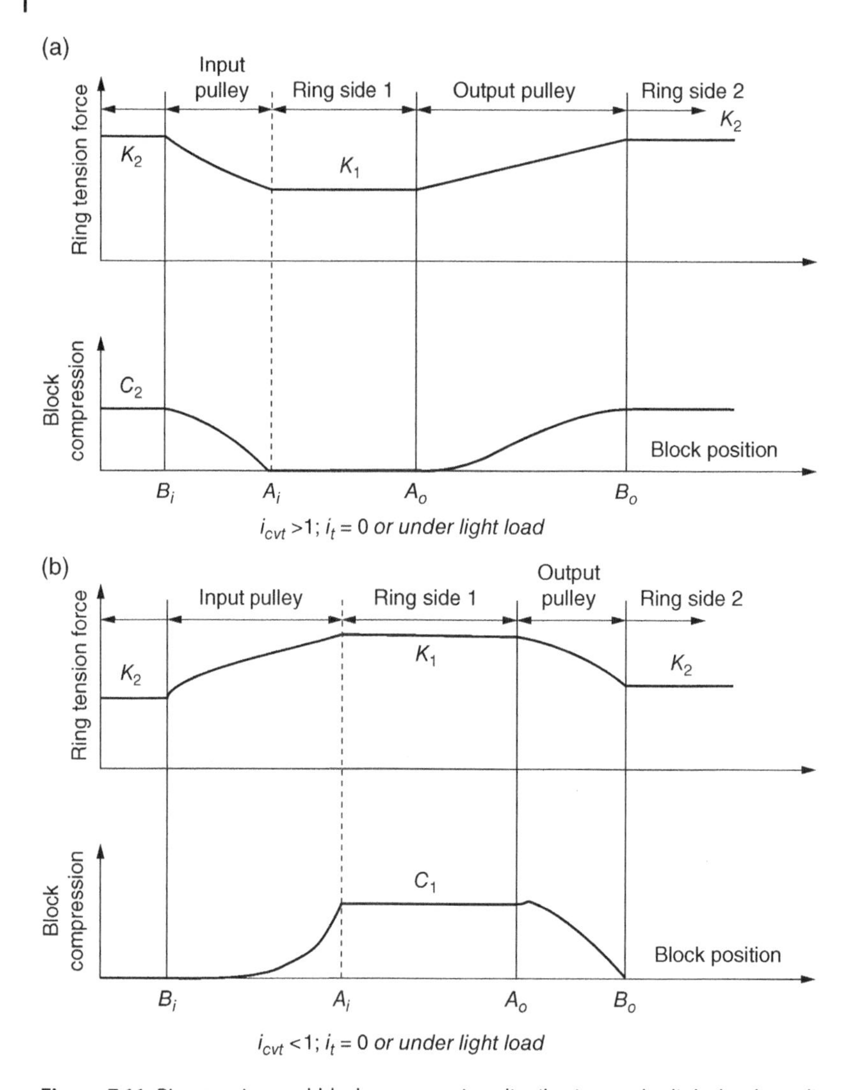

Figure 7.11 Ring tension and block compression distribution under light load conditions.

The distribution of block compression and ring tension is shown in Figure 7.11 when the CVT runs under light load conditions, with Figure 7.11a showing the case when $i_{cvt} \geq 1$ and Figure 7.11b showing the case when $i_{cvt} < 1$. As shown in Figure 11a, block compressive force C_2 acts on belt Side 2 and acts against torque transmission from the input to the output. There is no block compressive force on belt Side 1. Under this condition, it is the ring tension that overcomes the block compressive force C_2 and acts as the torque transmitting force. This is the only condition under which the block compressive force ramps down to be zero in the input pulley from entrance to exit and ramps up in the output pulley from zero at the entrance to C_2 at the exit. When $i_{cvt} < 1$, the ring tension always acts against torque transmission since $K_1 > K_2$, as shown in Figures 7.9b and 11b. Based on the experiment data in the paper series [19,20], the ring tension force difference $|K_1 - K_2|$ remains almost unchanged for all torque ratios at a given CVT ratio except when $i_{cvt} = 1$. This requires that the block compressive force C_1 keeps increasing as the torque ratio i_t increases. This is in contradiction to the analysis in the papers [19,20], which states that the block compressive force C_1 remains a constant when the torque ratio is below some threshold value i_{tc}.

7.3.5 Forces Acting on the Whole Belt

The free body diagram of the belt assembly as a whole body is shown in Figure 7.12. In this free body diagram, the contact force N, the radial friction force $\mu_r N$ and the tangential friction force $\mu_t N$ are applied by the input and output pulleys respectively to the belt. When the CVT runs at a fixed ratio with a constant input angular velocity, the belt is in a dynamic equilibrium. However, blocks in the pulleys rotate with the pulley and there is a centrifugal acceleration for each block along the contact arc. Therefore, an inertia force, that is equal to $m_i r_1 \omega_{in}^2$ for a block in the input pulley and $m_j r_2 \omega_{out}^2$ for a block in the output pulley, is added in the free body diagram based on the D'Alembert principle. Here, m_i is the mass of each block.

The reaction R that is applied by the ball splines to the movable sheave has been determined by Eq. (7.19). The magnitude of force R is equal to the sum of projections upon the

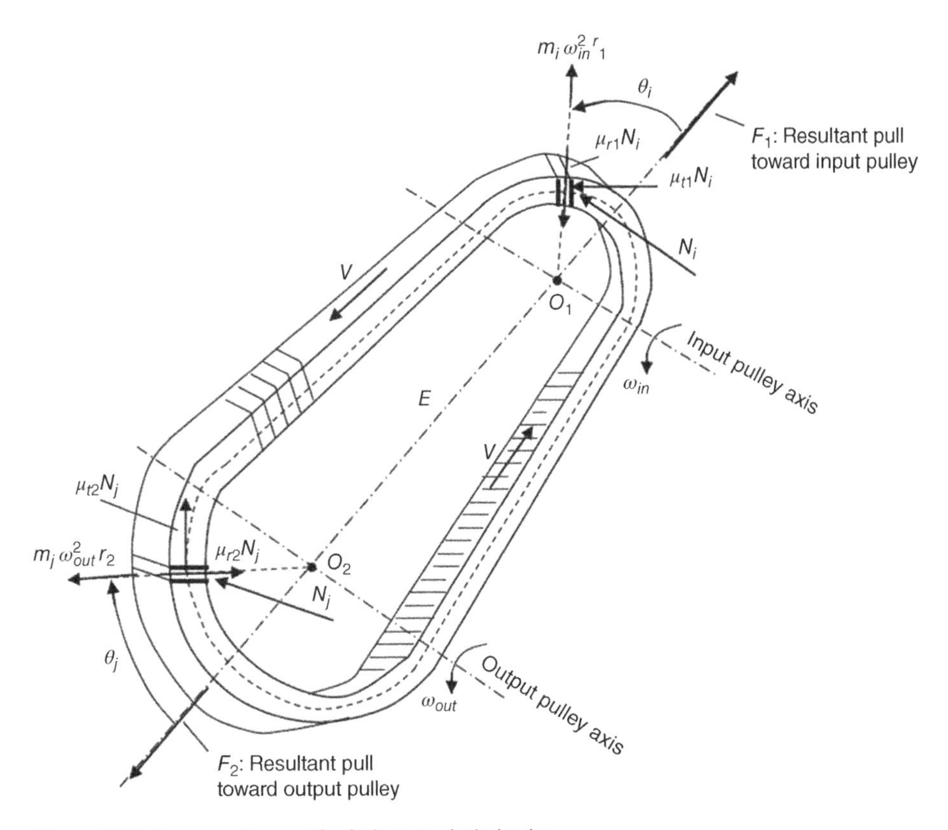

Figure 7.12 Forces acting on the belt as a whole body.

CVT center line O_1O_2 of the contact force N, the radial friction force $\mu_r N$, and the tangential friction force $\mu_t N$ applied on the belt by one side of the input or output pulley. Therefore, the sum of projections of all forces applied on the belt, including the inertia force, upon the center line O_1O_2 is represented for the input side and output side respectively:

$$F_1 = 2\sin\alpha\sum_{i=1}^{n}\sin\theta_i N_i \mp 2\cos\alpha\sum_{i=1}^{n}\cos\theta_i\mu_{r1}N_i - 2\sum_{i=1}^{n}\sin\theta_i\mu_{t1}N_i + \sum_{i=1}^{n}m_i r_1\omega_{in}^2\cos\theta_i \tag{7.21}$$

$$F_2 = 2\sin\alpha\sum_{j=1}^{m}\sin\theta_j N_j \mp 2\cos\alpha\sum_{j=1}^{m}\cos\theta_j\mu_{r2}N_j - 2\sum_{j=1}^{m}\sin\theta_j\mu_{t2}N_j + \sum_{j=1}^{m}m_j r_2\omega_{out}^2\cos\theta_j \tag{7.22}$$

where F_1 and F_2 are respectively the resultant force applied to the belt in the direction of the CVT center line on the input side and the output side. These two forces have the effect of pulling the belt toward the respective pulleys and are termed the resultant pull as denoted in Figure 7.12. Note that it is almost impossible to determine the resultant pulls using Eqs (7.21) and (7.22) because the distribution of the contact force N and the friction coefficients are not known under real time CVT operation conditions. However, the resultant pulls are affected primarily by the contact force N, which is related to the thrust forces P_{in} and P_{out} by Eqs (7.8) and (7.9). Clearly, the two resultant pulls must be equal in magnitude and opposite in direction during fixed ratio CVT operations, i.e. $F_1 = F_2 = F_s$. Here, F_s is sometimes called the axial force in CVT related technical publications.

It is apparent from Figure 7.9 that the resultant projection of the block compression force and ring tension on the CVT center line O_1O_2 is equal to the resultant pull or the axial force, as represented by:

$$F_s = \begin{cases} [(K_2 - K_1) - C_2]\cos\beta & \text{when } i_{cvt} \geq 1 \text{ and } i_t < i_{tc} \\ [(K_2 - K_1) + C_1]\cos\beta & \text{when } i_{cvt} \geq 1 \text{ and } i_t \geq i_{tc} \\ [C_1 - (K_1 - K_2)]\cos\beta & \text{when } i_{cvt} < 1 \end{cases} \tag{7.23}$$

where angle β is as defined in Figure 7.5. In some CVT related literature, a so-called traction coefficient is defined as $\lambda = \left(\frac{T_{in}}{r_1}\right)/F_s$. This traction coefficient is physically a measure on the effectiveness of the block compression force and ring tension for torque transmission since the magnitude of the axial force is proportional to the resultant torque transmitting force by just a factor of $\cos\beta$, as shown in Eq. (7.23).

7.3.6 Relation between Thrusts on Input and Output Pulleys

The contact force is proportional to the thrust as defined by Eqs (7.8) and (7.9) respectively for the input and output pulleys. As mentioned previously, Eqs (7.12) and (7.13) cannot quantitatively define the ratio between the pulley thrusts, namely $\varepsilon = \frac{P_{in}}{P_{out}}$, because the friction coefficients during real time CVT operations are indeterminate. This ratio is critical for CVT design and control for the following reasons:

- For a given CVT ratio, the maximum input torque that can be transmitted depends on the output pulley thrust force P_{out}, as indicated by Eq. (7.16). Therefore, the output pulley thrust force can be controlled at different magnitudes for a specified CVT ratio for the transmission of input torque at different levels. Usually, the thrust pulley force is generated by a hydraulic piston and its magnitude is controlled through the hydraulic pressure in the piston chamber. As mentioned previously, the hydraulic pressure should be controlled at a level just high enough for the transmission of the input torque intended for the vehicle operation status.
- Now that the output thrust P_{out} is controlled as mentioned above, what should then be the input pulley thrust P_{in}? Clearly, the input pulley thrust P_{in} must generate sufficient contact force for the contact between input pulley and the metal blocks so that the input torque can be transmitted from the input pulley without slippage. Meanwhile, the contact force generated by P_{in} must guarantee that the dynamic equilibrium of the belt is maintained, i.e. $F_1 = F_2$. This means that the hydraulic pressure in the input piston chamber must be exactly equal to a certain value to maintain the belt dynamic equilibrium.
- It is theoretically possible to use two variable force solenoids, one for each movable pulley sheave, to control the hydraulic pressure in the input pulley piston chamber and in the output pulley chamber separately using feedback control. In practice, however, this control method needs active pressure control for both the input and output piston chambers with high accuracy to avoid control instability.
- In production CVTs, the hydraulic pressure in the piston chamber of the output pulley, p_{out}, is actively controlled at the calibrated value corresponding to the input torque and the CVT ratio, but the hydraulic pressure in the piston chamber of the input pulley, p_{in}, is not actively controlled. As detailed in the following section, the input piston chamber is isolated from the line pressure circuit and the CVT fluid is trapped inside during a fixed ratio operation. The thrust force on the output pulley P_{out}, that is proportional to p_{out}, generates the contact force and the friction forces between the belt and the output pulley, which then generate the resultant pull F_2 that would pull the belt toward the output pulley. However, the belt is tightly squeezed in the groove of the input pulley and cannot move toward the output side since the CVT fluid trapped in the input piston chamber is incompressible and does not allow the input movable sheave to move axially. Therefore, hydraulic pressure in the input piston chamber p_{in} will build up as a reaction, which leads to the generation of the input pulley thrust force P_{in} and thus to the contact force and friction forces between the input pulley and the belt. The resultant pull F_1 in Eq. (7.21) on the input side is then generated as a reaction to F_2. Since F_1 reacts to F_2, the two forces are always the same in magnitude, i.e. $F_1 = F_2$.
- As mentioned above, the hydraulic pressure p_{in} in the input piston chamber is generated as a reaction during fixed ratio CVT operation. The value of pressure p_{in} is inversely proportional to the input piston area. Knowing the ratio between the pulley thrusts under various CVT ratios and the torque ratios, the piston effective areas, A_{in} and A_{out}, can be optimized so that both hydraulic pressures p_{in} and p_{out} will be close to the line pressure or in the range conducive to CVT efficiency.

- During ratio changes, pressure p_{in} is controlled by the line pressure of the CVT control system so as to control the motion of the belt in the pulley grooves. For ratio control responsiveness, the pressure p_{in} before a shift is initiated should not differ too much from the line pressure. Knowing the pulley thrust ratio allows CVT control and calibration engineers to optimize piston chamber pressures and line pressure levels for all CVT operation conditions.

Analytical formulations on pulley thrusts for rubber belt drives was proposed decades ago by Gerber [5,6], Miloiu [7], and Worley [8,9]. The traction coefficient λ and the axial force F_s defined previously are used in the closed form formulations of the output pulley thrust by Miloiu and on the input pulley thrust by Worley, as represented respectively in the following equations:

$$\left.\begin{aligned} \frac{P_{out}}{F_s} &= \frac{\cot(\alpha+\gamma)(\phi_2-\psi)}{4}(1-\lambda)+\frac{\lambda\cos\alpha}{2\mu} \\ \psi &= \frac{\sin\alpha}{\mu}\ln\left(\frac{1+\lambda}{1-\lambda}\right) \\ \gamma &= \tan^{-1}\mu \end{aligned}\right\} \tag{7.24}$$

$$P_{in} = \frac{\cot(\alpha+\gamma)\phi_1}{4}\left(F_s+\frac{T_{in}}{r_1}\right) \tag{7.25}$$

where μ is the maximum friction coefficient between the block and the pulley groove surface and γ is the friction angle. To enhance the accuracy of these two equations for belt CVTs, Fujii et al. [19,20] introduced a so-called effective friction coefficient μ' defined as $\mu' = \frac{\pi}{Larger\ of\ (\phi_1,\ \phi_2\ )}$ to reflect the effect of the active contact arc in the pulley with the larger pitch radius, where ϕ_1 and ϕ_2 are the contact angles for the input and output pulleys as shown in Figure 7.5; α is the pulley groove angle; P_{out} is the output pulley thrust corresponding to the maximum transmitted torque T_{max} defined by Eq. (7.16) for a given CVT ratio with the maximum effective friction coefficient μ', i.e. $\mu_{tmax} = \mu'$. In the paper series [19–22], Fujii et al. further simplified Eq. (7.24) by linearizing the term $\ln\left(\frac{1+\lambda}{1-\lambda}\right)$ as $a\lambda$ (with $a = 2.29$); this simplification does not affect the accuracy of the original equation proposed by Miloiu since the traction coefficient λ is in the range 0.25 –0.45 for the practical torque ratio range between 0.5 and 0.9. With this simplification, the first equation in Eq. (7.24) is rewritten as:

$$\frac{P_{out}}{F_s} = \frac{\cot(\alpha+\gamma)\left(\phi_2-\frac{a\lambda\sin\alpha}{\mu'}\right)}{4}(1-\lambda)+\frac{\lambda\cos\alpha}{2\mu'} \tag{7.26}$$

By substituting the traction coefficient $\lambda = \left(\frac{T_{in}}{r_1}\right)/F_s$, Fujii et al. further transformed Eq. (7.26) into a quadratic equation in terms of the axial force F_s in the following form:

$$AF_s^2 - BF_s + C = 0 \tag{7.27}$$

$$\left.\begin{aligned} A &= \phi_2 \\ B &= \left(\frac{a\sin\alpha}{\mu'}+\phi_2\right)\frac{T_{in}}{r_1} - 4\tan(\alpha+\gamma)\left(\frac{\cos\alpha}{2\mu'}\frac{T_{in}}{r_1}-P_{out}\right) \\ C &= \frac{a\sin\alpha}{\mu'}\left(\frac{T_{in}}{r_1}\right)^2 \end{aligned}\right\} \tag{7.28}$$

In this quadratic equation, A, B, and C can be calculated if the CVT ratio and the input torque are given. As indicated in the paper series of Fujii et al., for the torque ratio range $0 \le i_t \le 1$ and CVT ratio range $0.5 \le i_{cvt} \le 2.5$, the solution to the axial force F_s from Eq. (7.27) should be $\left(B + \sqrt{B^2 - 4AC}\right)$. After the axial force F_s is solved, Eq. (7.25) is then used to determine the input pulley thrust P_{in}. The procedure to determine the thrusts and the thrust ratio follows three steps: (a) the output pulley thrust P_{out} is determined using Eq. (7.16) for a specified maximum input torque T_{max} at a given CVT ratio i_{cvt} with the maximum friction coefficient; (b) the input torque T_{in} is calculated as $(i_t T_{max})$ with $0 \le i_t \le 1$ and is then plugged into Eqs (7.28) and (7.27) for the determination of the axial force F_s; and (c) the axial force F_s is then plugged into Eq. (7. 25) to determine the input pulley thrust P_{in}, and the thrust ratio is then calculated as $\varepsilon = \frac{P_{in}}{P_{out}}$. In summary, the thrust ratio ε is a bi-variable function of the CVT ratio and the torque ratio, i.e. $\varepsilon = f(i_{cvt}, i_t)$, as determined by these three steps. For illustration purposes, the curve describing the relation between the thrust ratio ε and the CVT ratio i_{cvt} obtained by Fujii et al.'s modified formulation is shown in Figure 7.13 when the torque ratio i_t is equal to 0.77.

As far as engineering application is concerned, experimental data represented in the paper series of Fujii et al. [19–22] have shown good agreement between the thrust ratios measured in testing and the thrust ratios that are calculated from the closed form formulation. The following summarizes the conclusions made in the paper series by Fujii et al. on the thrust ratio based on experimental data and analytical analysis:

- The thrust ratio at a specific CVT ratio i_{cvt} does not depend on the maximum transmittable torque defined by Eq. (7.16) and is only a function of the torque ratio i_t. Generally, the thrust ratio is a bi-variable function of the CVT ratio and the torque ratio.
- The maximum transmittable input torque T_{max} is proportional to the output pulley thrust P_{out} for a specified CVT ratio. Note that T_{max} is the maximum input torque that is transmitted on the threshold of belt slippage when $\mu = \mu_{max}$.
- For any given CVT ratio, the thrust ratio is a near constant ε_0 specific to the CVT ratio and is independent of the torque ratio i_t if $i_t \ge 0.4$. This means that the thrust ratio only depends on the torque load when it is light. If the input torque is higher than $0.4T_{max}$, it will not affect the thrust ratio significantly.

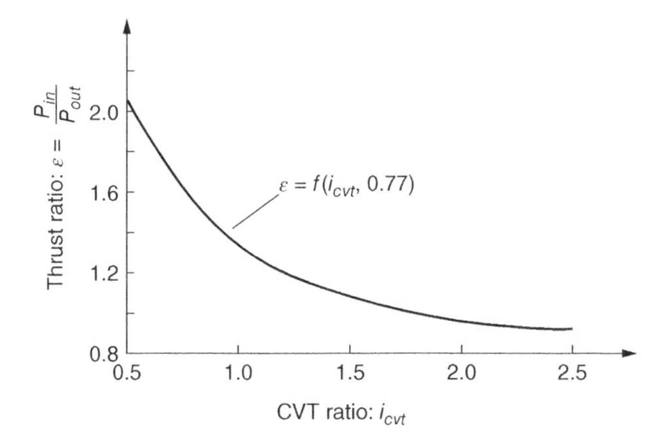

Figure 7.13 Typical thrust ratio plotted against CVT ratio.

- When $i_t \geq 0.4$, the near constant ε_0 for thrust ratio mentioned above decreases as the CVT ratio increases. It remains higher than 1.0 and almost becomes equal to 1.0 when the CVT ratio is 2.0. This means that the input pulley thrust is always higher than the output pulley thrust when $i_t \geq 0.4$.
- When $i_{cvt} \leq 1.0$, the thrust ratio is always higher than 1.0. This means that the input pulley thrust is always higher than the output pulley thrust for overdrive CVT operations.
- The input pulley angular velocity has insignificant effect on the thrust ratio when the CVT ratio is above 1.0. At low CVT ratios, i.e. $i_{cvt} < 0.5$, the thrust ratio at low input pulley angular velocity is higher than the thrust ratio at high input pulley angular velocity.

7.3.7 Ratio Changing Mechanism

As discussed previously, when the CVT runs with a fixed ratio, the belt must be at dynamic equilibrium and the two pulling forces in Figure 7.12 must be equal, i.e. $F_1 = F_2$. During a ratio changing process, this dynamic equilibrium must be broken so that the metal belt will move toward either the input or the output pulley as the metal blocks rise or fall in the pulley grooves. The motion of the belt assembly is complicated during ratio changes, but the motion of the belt mass center is constrained along the CVT center line O_1O_2 and is governed by the following equation:

$$M_B \frac{dV}{dt} = F_1 - F_2 \tag{7.29}$$

where M_B is the mass of the metal belt assembly and $\frac{dV}{dt}$ is the acceleration of the belt mass center in the direction that coincides with the CVT center line O_1O_2 as shown in Figure 7.12. During a ratio change, $\frac{dV}{dt} \neq 0$ and $F_1 \neq F_2$. In an upshift, the belt moves with its mass center toward the input pulley as the pitch radius r_1 increases and the pitch radius r_2 decreases, and in a downshift, the belt moves toward the output pulley as the pitch radius r_2 increases and the pitch radius r_1 decreases. It is therefore apparent that the two pulling forces F_1 and F_2 must observe respectively the following qualitative relations during CVT fixed ratio operation, upshift process and downshift process:

$$\left.\begin{array}{ll} F_1 = F_2 & \text{Fixed ratio operation} \\ F_1 > F_2 & \text{Upshift process} \\ F_1 < F_2 & \text{Downshift process} \end{array}\right\} \tag{7.30}$$

As shown in Figure 7.11 and represented by Eqs (7.21) and (7.22), the two pulling forces are related to the contact forces on the input and output pulleys. The contact forces are related to the pulley thrusts which depend on the piston chamber hydraulic pressures as illustrated in Figure 7.3. As mentioned previously, the hydraulic pressure in the output pulley piston chamber is actively controlled to generate the required thrust force, which also generates the pulling force F_2, and the hydraulic pressure in the input pulley piston chamber is generated as a reaction to resist the pulling of the metal belt by F_2.

7.4 CVT Control System Design and Operation Control

This section represents the CVT control system designs that implement the mechanisms of CVT torque transmission and ratio change analysed in the previous section. There are two basic design architectures for belt CVT control: one uses two variable force solenoids (VFS) or variable bleed solenoids (VBS) to control separately the pressures in the piston chambers of the movable sheaves, as shown in Figure 7.14, and is hereafter termed a **VBS based control system**. The other uses a servo mechanism to control the pressure in the piston chambers, as shown in Figure 7.15 and is termed a **servo mechanism control system**. Both control system architectures share the same circuit designs for the line pressure control and the torque converter clutch control shown in Figure 7.14.

The hydraulic circuits for the control of system line pressure and the operation of the torque converter clutch were presented in detail in Chapter 6. These circuits are basically the same for CVT control systems and will not be repeated in this section.

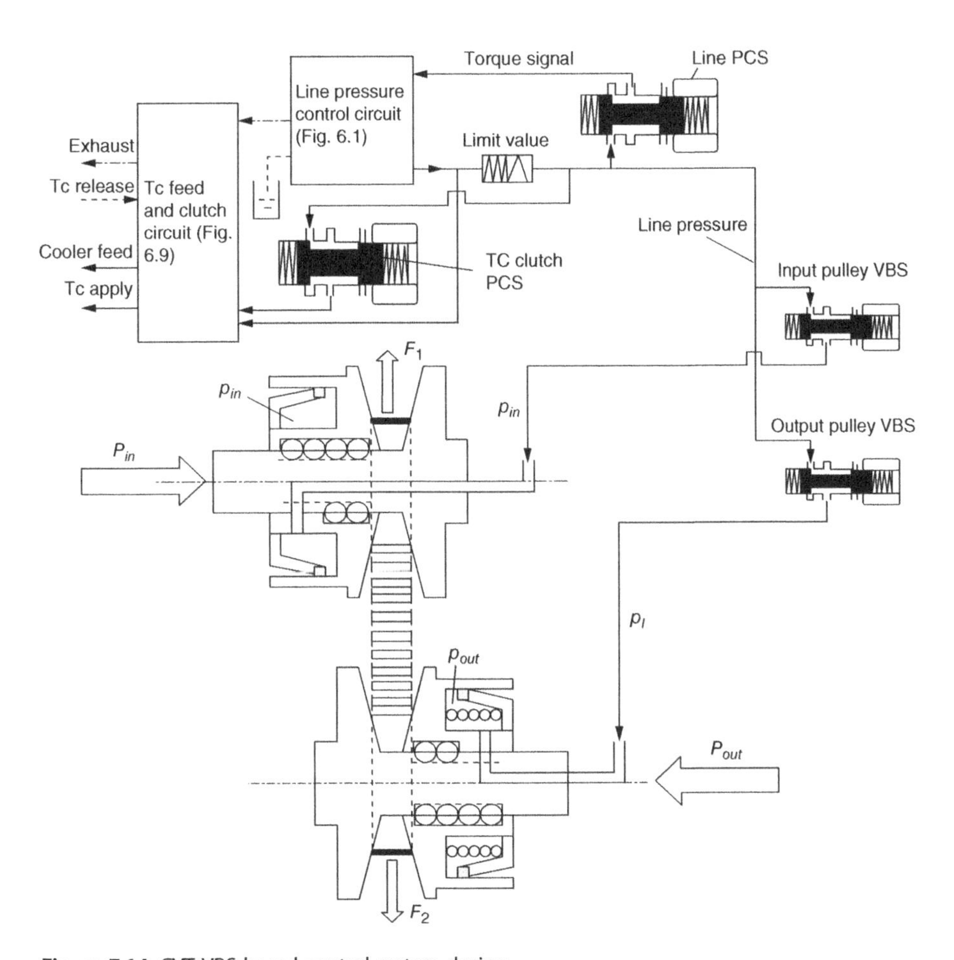

Figure 7.14 CVT VBS based control system design.

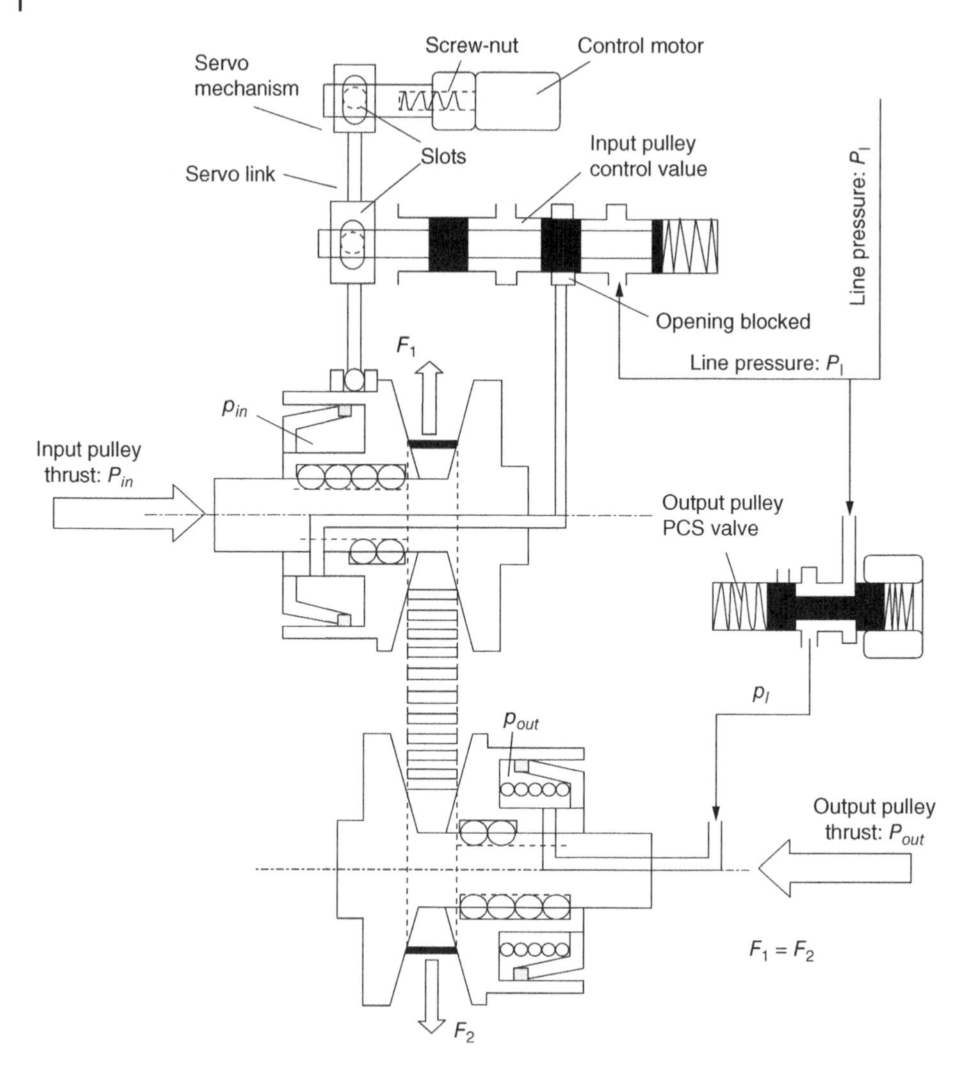

Figure 7.15 CVT servo mechanism control system design.

7.4.1 VBS Based Control System

As shown in Figure 7.14, the pressure in the piston chamber of the input pulley movable chamber p_{in} is controlled by the input pulley VBS, and the pressure in the piston chamber of the output pulley movable chamber p_{out} is controlled by the output pulley VBS. Note that the input pulley VBS has a position that fully blocks the line pressure CVT fluid from entering the piston chamber, so that the CVT fluid trapped in the piston chamber will be isolated from the line pressure circuit. The operations of this control system is described in the following.

Fixed ratio operation: When the CVT operates with a fixed ratio, the input pulley VBS is controlled to be at the position that fully blocks the valve out-port that is linked to the input piston chamber, meanwhile the output pulley VBS is controlled at the position that fully connects the line pressure circuit to the output piston chamber. Therefore, the hydraulic pressure in the output piston chamber, p_{out}, is equal to the line pressure p_l. The output pulley thrust P_{out} is actively generated by pressure p_{out} and thus produces the resultant pulling force F_2 that tends to pull the metal belt towards the output pulley, as formulated in Eqs (7.9) and (7.22). However, the CVT fluid trapped in the input piston chamber is incompressible and does not allow any axial displacement of the input pulley movable sheave. As a result, the metal belt is tightly pinched in the pulley grooves, creating the contact force between the input pulley and the metal blocks and the friction forces with it. Consequently, the input pulley thrust P_{in} is generated as a reaction. During fixed ratio operations, the following equations are observed:

$$\left.\begin{aligned} p_{out} &= p_l \\ P_{out} = A_{out}p_{out} &= \frac{\cos\alpha}{2\mu_{t2}r_1}T_{in} \\ \sum_{j=1}^{m} N_j = \frac{P_{out}}{(\cos\alpha \mp \mu_{r2}\sin\alpha)} &\approx \frac{P_{out}}{\cos\alpha} \end{aligned}\right\} \tag{7.31}$$

$$F_1 = F_2 \tag{7.32}$$

$$\left.\begin{aligned} \sum_{i=1}^{n} N_i = \frac{P_{in}}{(\cos\alpha \pm \mu_{r1}\sin\alpha)} &\approx \frac{P_{in}}{\cos\alpha} \\ P_{in} = A_{in}p_{in} &= \frac{\cos\alpha}{2\mu_{t1}r_1}T_{in} \\ p_{in} &= \frac{P_{in}}{A_{in}} \end{aligned}\right\} \tag{7.33}$$

where r_1 is the input pulley pitch radius which corresponds to a specific CVT ratio i_{cvt}; T_{in} is the input torque that needs be transmitted; and A_{in} and A_{out} are respectively the effective piston area of the input pulley and the output pulley. Note here that the sum of the contact forces on the output pulley $\sum_{j=1}^{m} N_j$ can be determined by Eq. (7.31) if the line pressure p_l is given, but the sum of contact forces on the input pulley $\sum_{i=1}^{n} N_i$ cannot be determined by Eq. (7.33) unless the thrust ratio is otherwise determined, for example, by the function $\varepsilon = f(i_{cvt}, i_t)$ pertaining to Eqs (7.24) and (7.25).

Upshift operation: When the CVT is in the process of an upshift, the input pulley VBS is controlled to be at the position that partially opens the line pressure circuit to the input piston chamber, as shown in Figure 7.14. Meanwhile, the output pulley VBS is controlled at the position that bleeds the CVT fluid in the output piston chamber to the exhaust circuit. In this setup, the pressure in the input piston chamber p_{in} increases, and the pressure in the output piston chamber p_{out} decreases simultaneously, making F_1 larger than F_2. This breaks the dynamic equilibrium of the metal belt before the upshift is initiated, and the belt starts to move toward the input pulley because $F_1 > F_2$. As the belt moves toward the input pulley, the CVT ratio decreases toward the target value, with input pulley pitch radius r_1 increasing and output pulley pitch radius r_2 decreasing. During the upshift process, both pressures p_{in} and p_{out} are controlled by the input pulley VBS and the output pulley VBS respectively based on the feedback on the real time CVT ratio, which is calculated via the input and output angular velocities measured by the respective speed sensors. As soon as the target CVT ratio is achieved, the input pulley VBS will be controlled at the position that fully blocks the valve out-port linked to the input piston chamber and the output pulley VBS will be controlled at the position that fully connects the line pressure circuit to the output piston chamber. The CVT then operates with the target ratio. During upshift operations, the following equations are observed:

$$\left.\begin{aligned} p_{in} &\approx p_l \\ P_{in} = A_{in}p_{in} &= \frac{\cos\alpha}{2\mu_{t1}r_1}T_{in} \\ \sum_{i=1}^{n} N_i &= \frac{P_{in}}{(\cos\alpha \pm \mu_{r1}\sin\alpha)} \approx \frac{P_{in}}{\cos\alpha} \end{aligned}\right\} \tag{7.34}$$

$$F_1 > F_2 \tag{7.35}$$

$$\left.\begin{aligned} p_{out} &< p_l \\ P_{out} = A_{out}p_{out} &= \frac{\cos\alpha}{2\mu_{t2}r_1}T_{in} \\ \sum_{j=1}^{m} N_j &= \frac{P_{out}}{(\cos\alpha \mp \mu_{r2}\sin\alpha)} \approx \frac{P_{out}}{\cos\alpha} \end{aligned}\right\} \tag{7.36}$$

Downshift operation: When the CVT is in the process of a downshift, the output pulley VBS remains at the position that links the line pressure circuit to the output piston chamber, as shown in Figure 7.14. Meanwhile, the input pulley VBS is controlled at the position that bleeds the CVT fluid in input piston chamber to the exhaust circuit. In this setup, the pressure in the input piston chamber p_{in} decreases and the pressure in the output piston chamber p_{out} is almost the same as the line pressure, making F_2 larger than F_1. This breaks the dynamic equilibrium of the metal belt before the downshift is initiated, and the belt starts to move toward the output pulley because $F_2 > F_1$. As the belt moves toward the output pulley, the CVT ratio increases toward the target value. During the downshift process, both pressures p_{in} and p_{out} are controlled by the input pulley VBS and the output pulley VBS respectively based on the feedback on the real time CVT ratio. As soon as the target CVT ratio is achieved, the input pulley VBS and the output pulley VBS will be controlled at the positions for fixed ratio CVT operation. The CVT then operates with the downshifted target ratio. During downshift operations, the following equations are observed:

$$\left.\begin{aligned} p_{out} &\approx p_l \\ P_{out} = A_{out}p_{out} &= \frac{\cos\alpha}{2\mu_{t2}r_1}T_{in} \\ \sum_{j=1}^{m} N_j &= \frac{P_{out}}{(\cos\alpha \mp \mu_{r2}\sin\alpha)} \approx \frac{P_{out}}{\cos\alpha} \end{aligned}\right\} \tag{7.37}$$

$$F_2 > F_1 \tag{7.38}$$

$$\left.\begin{aligned} p_{in} &< p_l \\ P_{in} = A_{in}p_{in} &= \frac{\cos\alpha}{2\mu_{t1}r_1}T_{in} \\ \sum_{i=1}^{n} N_i &= \frac{P_{in}}{(\cos\alpha \pm \mu_{r1}\sin\alpha)} \approx \frac{P_{in}}{\cos\alpha} \end{aligned}\right\} \tag{7.39}$$

7.4.2 Servo Mechanism Control System

Instead of using two variable bleed solenoids, the servo mechanism CVT control system uses two simpler valves for the control of input piston chamber pressure and the output piston chamber pressure. Figure 7.15 shows the structural layout of this control system. As shown in Figure 7.15, the output pulley pressure control valve (PCS) can be positioned to connect the line pressure with the output piston chamber or to bleed the CVT fluid in the output piston chamber to the exhaust circuit. The input pulley control valve has three positions: one fully blocks the port to the input piston chamber, the other connects the line pressure with the input piston chamber, and the third bleeds the CVT fluid in the input pulley piston chamber via the exhaust port. The valve body of the input pulley control valve is connected to the servo link by a pin-slot joint. The screw-nut converts the rotation of the control motor to a linear displacement at the top end of the servo link, also via a pin-slot joint.

Fixed ratio operation: The positions of the input pulley control valve and the output pulley PCS valve are shown in Figure 7.15 when the CVT operates with a fixed ratio. The output pulley PCS valve directly connects the line pressure circuit to the output piston chamber, thus the hydraulic pressure in the output pulley piston is the same as the line pressure, i.e. $p_{out} = p_l$. The input pulley control valve is controlled at the position that fully blocks the valve port to the input piston chamber. The output pulley thrust P_{out} is actively generated by pressure p_{out} and thus produces the resultant pulling force F_2 which tends to pull the metal belt toward the output pulley, as formulated in Eqs (7.9) and (7.22). However, the CVT fluid trapped in the input piston chamber is incompressible and does not allow the any axial displacement of the input pulley movable sheave. As a result, the metal belt is tightly pinched in the pulley grooves, creating the contact force between the input pulley and the metal blocks and the friction forces with it. Consequently, the input pulley thrust P_{in} is generated as a reaction. Note that there is no difference in fixed ratio operations between the servo mechanism control system and the VBS based control system discussed previously. During fixed ratio operations, the following equations are observed, which are identical to Eqs (7.31), (7.32) and (7.33) and are repeated here for reader's convenience:

$$\left.\begin{aligned} p_{out} &= p_l \\ P_{out} &= A_{out} p_{out} = \frac{\cos\alpha}{2\mu_{t2} r_1} T_{in} \\ \sum_{j=1}^{m} N_j &= \frac{P_{out}}{(\cos\alpha \mp \mu_{r2}\sin\alpha)} \approx \frac{P_{out}}{\cos\alpha} \end{aligned}\right\} \tag{7.40}$$

$$F_1 = F_2 \tag{7.41}$$

$$\left.\begin{aligned} \sum_{i=1}^{n} N_i &= \frac{P_{in}}{(\cos\alpha \pm \mu_{r1}\sin\alpha)} \approx \frac{P_{in}}{\cos\alpha} \\ P_{in} &= A_{in} p_{in} = \frac{\cos\alpha}{2\mu_{t1} r_1} T_{in} \\ p_{in} &= \frac{P_{in}}{A_{in}} \end{aligned}\right\} \tag{7.42}$$

As shown by Eq. (7.16), the maximum input torque T_{max} that can be transmitted for a given CVT ratio depends on the output thrust force P_{out}. Therefore, if the input torque T_{in} that needs to be transmitted under a specific CVT ratio i_{cvt} is given, then the output thrust force P'_{out} and the corresponding line pressure p'_l that are required for the CVT to transmit the input torque at the belt slippage threshold, i.e. when $\mu_{t2} = \mu_{max}$, are determined as follows:

$$\left.\begin{aligned} P'_{out} &= A_{out}p'_l = \frac{\cos\alpha}{2\mu_{max}r_1}T_{in} \\ p'_l &= \frac{P'_{out}}{A_{out}} \end{aligned}\right\} \tag{7.43}$$

where μ_{max} is the maximum friction coefficient between the metal belt and the pulley surface. If the line pressure p_l is controlled at p'_l determined by Eq. (7.43), the CVT can then barely transmit the input torque T_{in}; in other words, the torque ratio i_t would be 1.0. This would let the CVT operate under the slippage threshold. In CVT control practice, the line pressure p_l should be controlled at a higher value so that the torque ratio i_t is in the range of $0.7 \sim 0.9$, as recommended by Fujii et al. [19–22] based on experimental data. The line pressure p_l and the output pulley thrust force P_{out} are then determined by the following equations:

$$\left.\begin{aligned} p_l &= \frac{p'_l}{i_t} \\ P_{out} &= A_{out}p_l \end{aligned}\right\} \tag{7.44}$$

Upshift operation: When the CVT transmission control unit (TCU) decides to make an upshift, it will command the control motor to rotate and the motor rotation is converted to a linear displacement Δ_M at the top end of the servo link, as shown in Figure 7.16. Consequently, the valve body of the input pulley control valve moves to the left with a displacement Δ_V as the servo link pivots about the joint on the input pulley movable sheave. This opens up the valve port to the input piston chamber and increases the hydraulic pressure in it, p_{in}, toward the line pressure p_l. Meanwhile, the output pulley PCS is controlled at the position that connects the output piston chamber to the exhaust circuit as shown in the figure, decreasing the pressure in the output piston chamber, p_{out}. This breaks the dynamic equilibrium of the metal belt before the upshift is initiated, both the input and output pulley movable sheaves move to the right and the belt starts to move toward the input pulley because $F_1 > F_2$.

As the belt moves toward the input side, pitch radius r_1 increases and pitch radius r_2 decreases, lowering the value of the CVT ratio i_{cvt}. Meanwhile, as the input pulley movable sheave moves to the right side with displacement Δ_{s1}, it carries the low end of the servo link with the same displacement, as shown in Figure 7.17. As can be observed from the figure, the displacement Δ_{s1} causes the valve body to move to the right side via the servo link. This valve body movement is in the direction to close the valve port to the input piston chamber. Therefore, the control motor generates displacement Δ_M that causes the valve body to move to the left and consequently opens up the valve port to the input piston chamber, forcing the input moveable sheave to move to the right with displacement Δ_{s1}. But the displacement Δ_{s1} moves the valve body toward the closed position against the effect of displacement Δ_M.

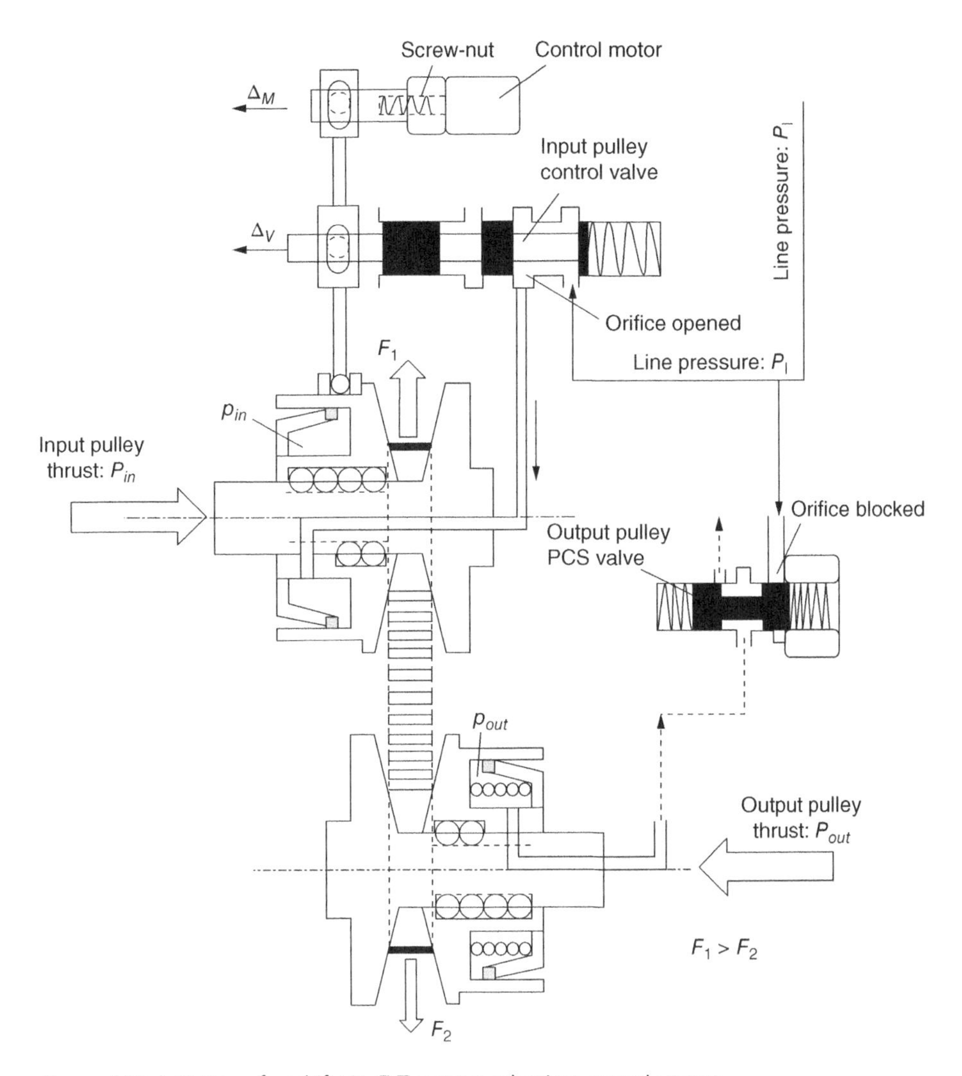

Figure 7.16 Initiation of upshifts in CVT servo mechanism control system.

The three displacements of the servo mechanism are linearly related, with Δ_M as the actuator. During an upshift, the motor rotation angle, i.e. displacement Δ_M, can be controlled by a PID as given by:

$$\left.\begin{aligned} \Delta i_{cvt} &= i^0_{cvt} - i^t_{cvt} \\ \Delta_M &= k_p \Delta i_{cvt} + k_i \int \Delta i_{cvt} dt + k_d \frac{d(\Delta i_{cvt})}{dt} \\ \Delta_V &= \frac{c}{b+c}\Delta_M - \frac{b}{b+c}\Delta_{s1} \end{aligned}\right\} \tag{7.45}$$

where b and c are the proportion factors of the servo link as shown in Figure 7.17, i^0_{cvt} and i^t_{cvt} are the target CVT ratio and the real time CVT ratio calculated based on the speeds of

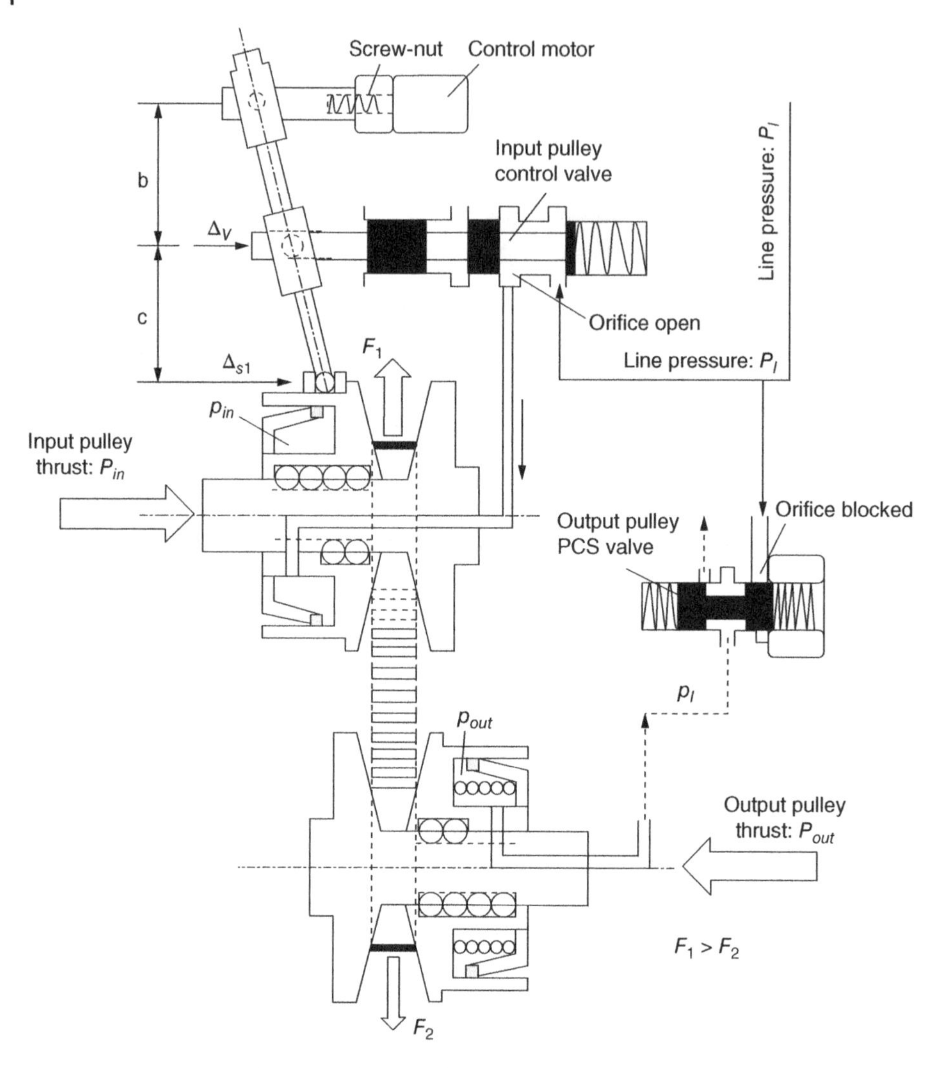

Figure 7.17 Operation of servo mechanism during upshifts.

the input and output pulleys measured by speed sensors, and k_p, k_i, and k_d are respectively the proportional, integral, and differential factors of the PID controller.

During the upshift process, the pressure in the exhaust circuit is also controlled so that the pressure in the output piston chamber will be ramped down slowly. In addition, the speeds of both the input and output pulleys, as well as the pressures in the input and output piston chambers, are measured by the respective sensors on a real time basis. The data from these sensors are processed to determine the instantaneous CVT ratio and the likelihood of belt slippage during the shift process. Once slippage is deemed likely, the controller will signal the engine controller via the control area network (CAN) to reduce the engine output torque by spark retarding or other means. The following equations are observed during an upshift process:

$$\left.\begin{aligned} p_{in} &= p_l \\ P_{in} = A_{in}p_{in} &= \frac{\cos\alpha}{2\mu_{t1}r_1}T_{in} \\ \sum_{i=1}^{n} N_i &= \frac{P_{in}}{(\cos\alpha \pm \mu_{r1}\sin\alpha)} \approx \frac{P_{in}}{\cos\alpha} \end{aligned}\right\} \tag{7.46}$$

$$F_1 > F_2 \tag{7.47}$$

$$\left.\begin{aligned} p_{out} &< p_l \\ P_{out} = A_{out}p_{out} &= \frac{\cos\alpha}{2\mu_{t2}r_1}T_{in} \\ \sum_{j=1}^{m} N_j &= \frac{P_{out}}{(\cos\alpha \mp \mu_{r2}\sin\alpha)} \approx \frac{P_{out}}{\cos\alpha} \end{aligned}\right\} \tag{7.48}$$

As can be observed from Eq. (7.46), if the line pressure p_l and the input torque T_{in} are known, the thrust force on the input pulley P_{in} and the friction coefficient μ_{t1} can be calculated by Eq. (7.46). The sum of contact forces on the input pulley $\sum_{i=1}^{n} N_i$ can also be calculated if the term $(\mu_{r1}\sin\alpha)$ is neglected. To initiate the upshift responsively, the effective piston area of the input pulley should be designed larger than that of the output pulley, that is, $A_{in} > A_{out}$, such that the hydraulic pressure in the input piston is lower than the line pressure when the CVT runs at the fixed ratio before the upshift is initiated. The ratio between the effective piston areas, i.e. $\frac{A_{in}}{A_{out}}$, should be optimized according to the ratio between the thrust forces shown in Figure 7.13 for a given CVT design. During an upshift, the value of the tangential friction coefficient μ_{t1} should be below the maximum friction coefficient μ_{max} to avoid belt slippage on the input pulley, i.e. $\mu_{t1} < \mu_{max}$. This inequality is readily satisfied because the pressure in the input piston chamber p_{in} is the line pressure p_l during the upshift process, which is higher than the pressure in the piston chamber before the upshift is initiated. To avoid slippage on the output pulley, the tangential friction coefficient μ_{t2} must also be lower than the maximum friction coefficient μ_{max}, i.e. $\mu_{t2} < \mu_{max}$. This means that the CVT fluid in the output piston chamber must be controlled to bleed through the opening in the output pulley PCS valve, as shown in Figure 7.17, so that pressure p_{out} drops gradually in small decrements. If necessary, the engine torque will be reduced during shifts to guarantee that there is no slippage on the output pulley, as mentioned previously.

As the input pulley movable sheave moves further to the right, the target CVT ratio i_{cvt}^0 will be gradually reached. Meanwhile, the opening in the valve port to the input piston chamber diminishes and eventually closes. The CVT fluid in the input piston chamber is then blocked off from the line pressure. As soon as the input pulley control valve fully blocks the valve port to the input piston chamber, the output pulley PCS valve will be controlled to move to the position that connects the line pressure with the output piston chamber, as shown in Figure 7.18. This completes the upshift process and the CVT then runs at the target ratio i_{cvt}^0.

Downshift operation: When the CVT transmission control unit (TCU) decides to make a downshift, it will signal the control motor to rotate in the direction opposite to that for upshifts. The displacement Δ_M at the top end of the servo link is then

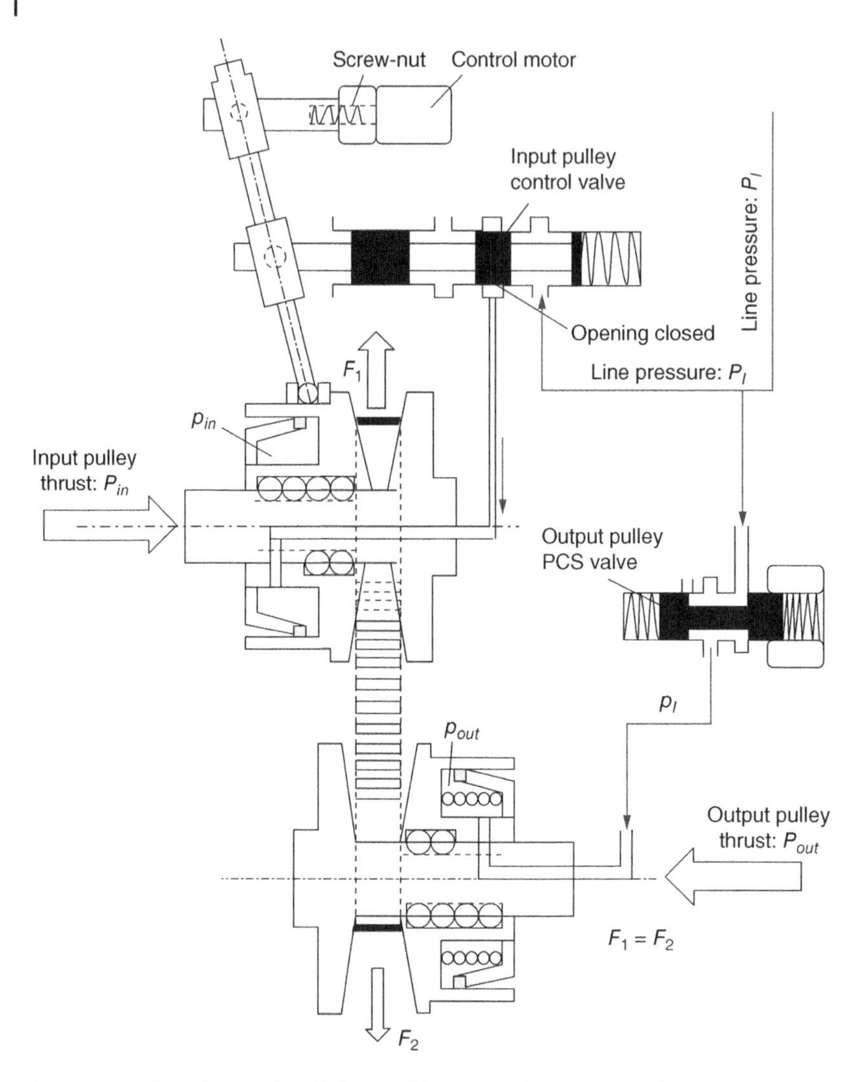

Figure 7.18 Completion of upshifts in CVT servo mechanism control system.

directed to the right, as shown in Figure 7.19. Consequently, the valve body of the input pulley control valve moves to the right with a displacement Δ_V. This connects the input piston chamber with the exhaust circuit, while still blocking the line pressure from the input piston chamber, as shown in Figure 7.19. The output pulley PCS valve is still at the same position as in fixed ratio operation. Therefore, the hydraulic pressure in the input piston chamber, p_{in}, decreases as CVT fluid bleeds through the valve opening to the exhaust circuit, and the hydraulic pressure in the output pulley p_{out} remains to be equal to the line pressure p_l. This breaks the dynamic equilibrium of the metal belt before the downshift is initiated, both the input and output pulley movable sheaves then move to the left and the belt starts to move toward the output pulley because $F_2 > F_1$.

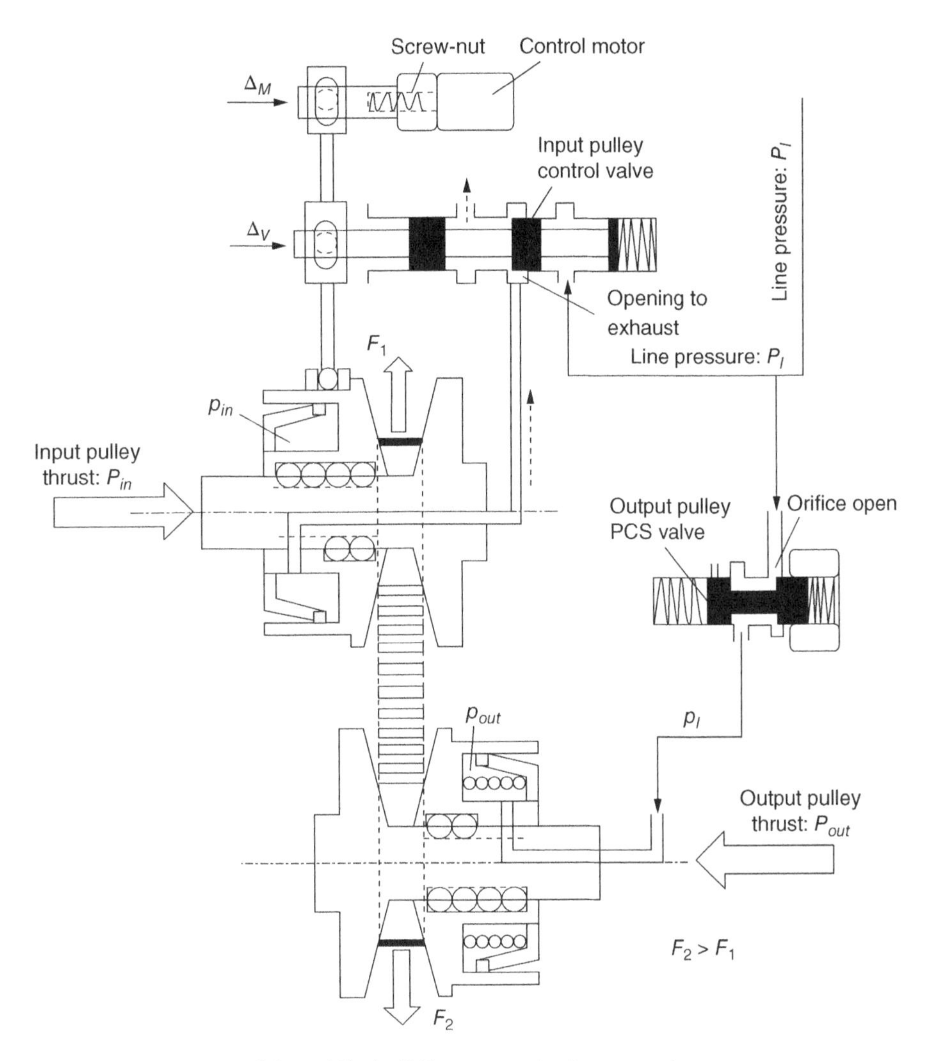

Figure 7.19 Initiation of downshifts in CVT servo mechanism control system.

As the belt moves toward the output side, pitch radius r_2 increases and pitch radius r_1 decreases, raising the value of the CVT ratio i_{cvt}. Meanwhile, as the input pulley movable sheave moves to the left side with displacement Δ_{s1}, it carries the low end of the servo link with the same displacement, as shown in Figure 7.20. As can be observed from the figure, the displacement Δ_{s1} causes the valve body to move to the left side via the servo link. This valve body movement is in the direction to close the valve port to the input piston chamber. Therefore, the control motor generates displacement Δ_M which causes the valve body move to the right and consequently links up the input piston chamber with the exhaust circuit via the valve port opening, forcing the input moveable sheave to move to the left with displacement Δ_{s1}. But the displacement Δ_{s1} moves the valve body toward the position that blocks the valve port to the input piston chamber, against the effect of displacement Δ_M.

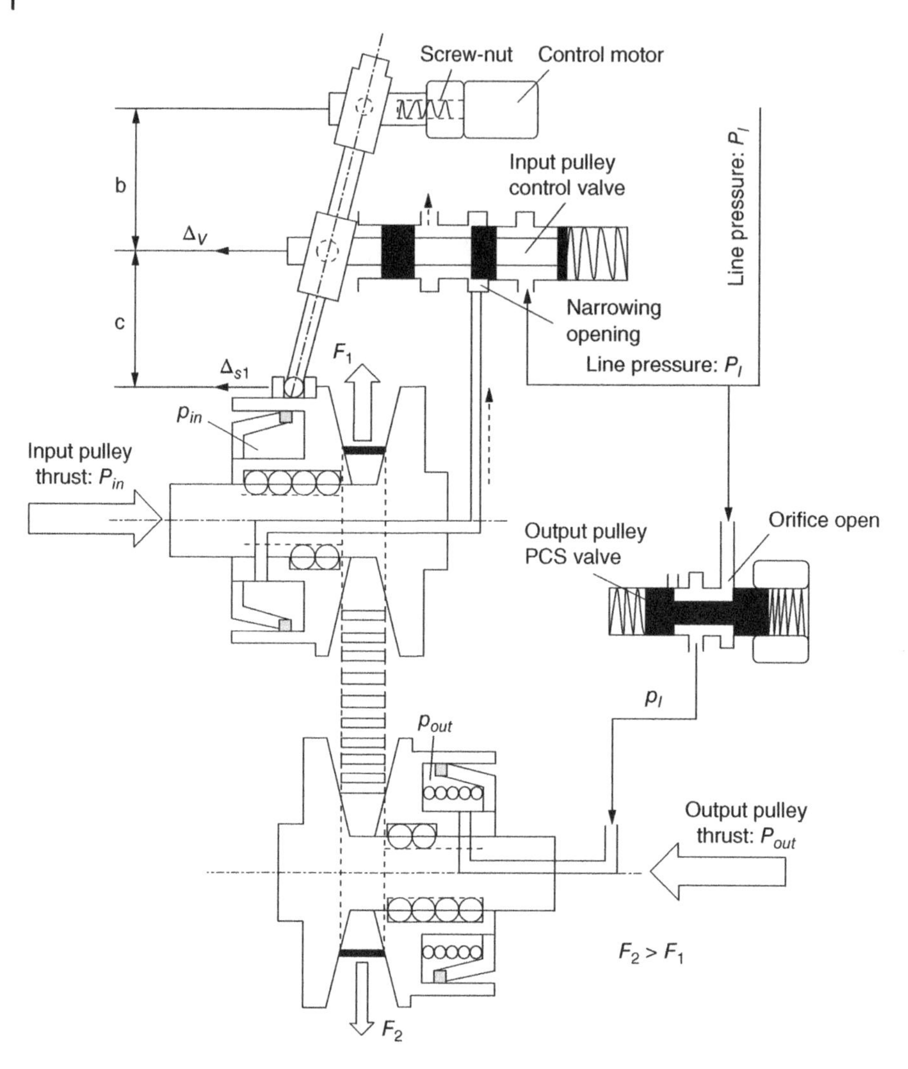

Figure 7.20 Operation of servo mechanism during downshifts.

For the downshift process, the relation between the three displacements of the servo mechanism and the PID controller for the motor rotation angle are similar to Eq. (7.45). During the downshift process, the pressure in the input piston chamber is controlled by the servo mechanism to decrease gradually. Similar to upshifts, the speeds of both the input and output pulleys, as well as the pressures in the input and output piston chambers, are measured by the respective sensors on a real time basis. The data from these sensors are processed to determine the instantaneous CVT ratio and the likelihood of belt slippage during the shift process. Once slippage is deemed likely, the controller will signal the engine controller via the CAN, to reduce the engine output torque by spark retarding or other means. The following equations are observed during a downshift process:

$$\left.\begin{aligned} p_{out} &= p_l \\ P_{out} = A_{out}p_{out} &= \frac{\cos\alpha}{2\mu_{t2}r_1}T_{in} \\ \sum_{j=1}^{m} N_j = \frac{P_{out}}{(\cos\alpha \mp \mu_{r2}\sin\alpha)} &\approx \frac{P_{out}}{\cos\alpha} \end{aligned}\right\} \tag{7.49}$$

$$F_2 > F_1 \tag{7.50}$$

$$\left.\begin{aligned} p_{in} &< p_l \\ P_{in} = A_{in}p_{in} &= \frac{\cos\alpha}{2\mu_{t1}r_1}T_{in} \\ \sum_{i=1}^{n} N_i = \frac{P_{in}}{(\cos\alpha \pm \mu_{r1}\sin\alpha)} &\approx \frac{P_{in}}{\cos\alpha} \end{aligned}\right\} \tag{7.51}$$

As can be observed from Eq. (7.49), if the line pressure p_l and the input torque T_{in} to be transmitted are known, the thrust force on the output pulley P_{out} and the friction coefficient μ_{t2} can be calculated by Eq. (7.49). The sum of contact forces on the output pulley $\sum_{j=1}^{m} N_j$ can also be calculated if the term $(\mu_{r2} \sin \alpha)$ is neglected. Note that the dynamic equilibrium of the CVT belt is broken as soon as the downshift is initiated regardless of the value of the input piston chamber pressure. This is because the input piston chamber pressure that keeps the belt in dynamic equilibrium when the CVT runs at the fixed ratio before the downshift is lowered instantaneously, as soon as the piston chamber is connected to the exhaust circuit via the valve opening. During a downshift, the value of the tangential friction coefficient μ_{t1} should be below the maximum friction coefficient μ_{max} to avoid belt slippage on the input pulley, i.e. $\mu_{t1} < \mu_{max}$. This inequality is satisfied by controlling the pressure in the input piston chamber p_{in} to be above what is required for input torque transmission. The slippage on the output pulley is readily avoided since the output piston chamber pressure p_{out} is the same as the line pressure p_l, which has been controlled to be high enough in the fixed ratio operation before the downshift. If necessary, the engine torque will be reduced during shifts to guarantee that there is no slippage on the input pulley, as mentioned previously.

As the input pulley movable sheave moves further to the left, the real time CVT ratio will gradually approach the target CVT ratio i_{cvt}^0. Meanwhile, the opening in the valve port to the input piston chamber diminishes and eventually closes due to the effect of the servo mechanism. The CVT fluid in the input piston chamber is then blocked from the line pressure. The downshift is thus completed as soon as the input pulley control valve fully blocks the valve port to the input piston chamber, as shown in Figure 7.21. The CVT then runs at the target ratio i_{cvt}^0.

7.4.3 Comparison of the Two Control System Designs

There is no fundamental difference in the operation principles of the two CVT control system designs, VBS based control system and servo mechanism control system, as detailed above. Both designs have the same hydraulic circuits and control methods for the system line pressure and for the torque converter clutch. In both designs, the

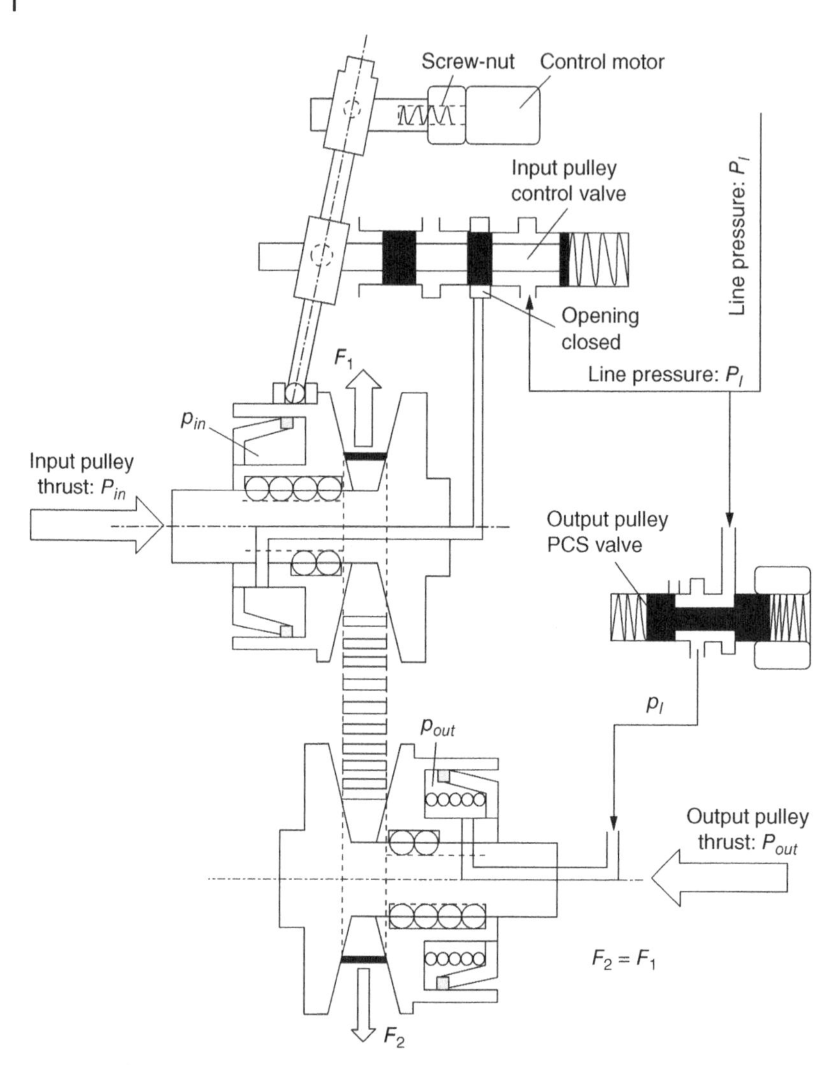

Figure 7.21 Completion of downshifts in CVT servo mechanism control.

input piston chamber is isolated from the line pressure circuit during fixed ratio operations, so that the input pulley thrust force is generated as a reaction to the action of the output pulley thrust force, which is controlled through control of the line pressure. The VBS based system has certain cost advantages in comparison, since it does not have the motor actuated servo mechanism. The servo mechanism system design features a negative feedback structurally on the control valve opening and thus possesses better stability for CVT shift operations. Most production CVTs today are controlled by the servo mechanism control system. With the availability of large volume multi-port variable force solenoids (VFS) or variable bleed solenoids (VBS), it can be stated that the VBS or VFS based control system can be used for production CVTs as an alternate to the servo mechanism system.

7.5 CVT Control Strategy and Calibration

CVT control features several technical aspects in general: torque converter clutch control, line pressure control, shift process control and ratio control strategy. There are two ratio control strategies: continuous ratio control for which the CVT was originally invented, and stepped ratio control which emulates conventional transmissions. In CVT equipped powertrains, the torque converter is primarily used as the vehicle launcher. Therefore the converter is only able to function as a torque multiplier when the vehicle is being launched and to function as a fluid damper when the vehicle speed is low, as shown in Figure 7.26. The torque converter clutch is applied under all other operation conditions. The hydraulic circuit and control process for the torque converter clutch were presented in Chapter 6 for conventional ATs. CVT torque converter clutch control is similar to that for conventional ATs. The CVT shift process control was detailed in Section 7.4. This current section will focus on CVT line pressure control and ratio control strategy.

7.5.1 Line Pressure Control

The CVT line pressure control is crucial for both fixed ratio and shift operations. As detailed in Section 7.4, during fixed operations, the line pressure is directly applied to the output piston chamber for the generation of active thrust force on the output pulley. The line pressure defines the torque ratio for CVT fixed ratio operations and must be controlled accurately so that the output thrust force is just high enough for transmitting the input torque required for various operation conditions. During ratio changes, the line pressure is applied in the input piston chamber for upshifts and in the output piston chamber for downshifts. It is apparent that the line pressure control determines the effectiveness and efficiency of CVT torque transmission and the CVT shift responsiveness. The basic hydraulic circuit for CVT line pressure control is similar to that for conventional ATs shown in Figure 6.1 and the control block diagram is shown in Figure 7.22.

As shown in the block diagram, the target line pressure value for a driver selected mode, normal, economy, or sport, is determined according to the major sensor signals on engine RPM, engine throttle, and speeds of the input and output pulleys, and is modified according to the auxiliary sensor signals on the CVT range, CVT fluid temperature, converter lock status, and brake system signals. The engine output torque T_e is interpolated from the engine map in terms of the engine RPM and the throttle opening. The

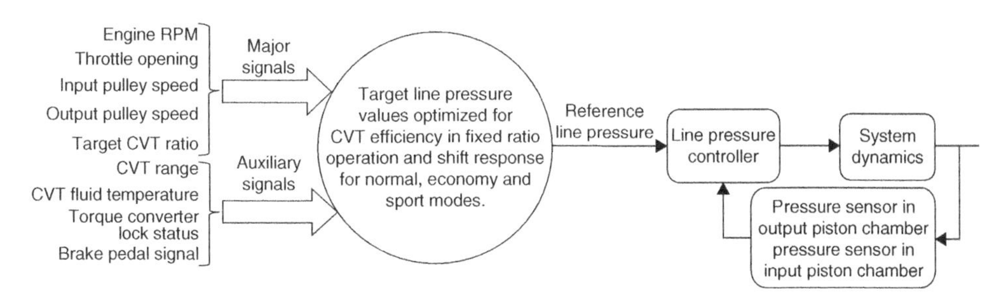

Figure 7.22 Block diagram of CVT line pressure control.

torque that needs to be transmitted by CVT, T_{in}, is the engine output torque minus the loads from auxiliary systems, such as the air conditioner and the power steering pump. The instantaneous CVT ratio i_{cvt} is determined by the speeds of the input and the output pulleys. The target CVT ratio i^0_{cvt} is selected by the CVT controller for the vehicle performance and fuel economy priorities. As shown in Eqs (7.43) and (7.44), the target line pressure p_l for fixed ratio operation is determined by:

$$p_l = \frac{\cos\alpha}{2\mu_{max}A_{out}r_1}\frac{T_{in}}{i_t} \tag{7.52}$$

where μ_{max} is the maximum friction coefficient between the belt and the pulley, A_{out} is the output piston effective area, as mentioned previously, r_1 is the input pulley pitch radius as defined by the CVT ratio i_{cvt}, and i_t is the torque ratio that can be selected in the range $0.7 \sim 0.9$. If the torque ratio is selected as 0.9, then the CVT transmits the input torque T_{in} with a 10% reservation on the capacity, i.e. it is 10% below the slippage threshold. It is apparent that the higher the torque ratio, the higher the power transmission efficiency. The target line pressure thus determined is used as the reference signal for the controller, which receives real time feedback on the pressures in the output piston chamber and the input piston chambers. The line pressure control strategy illustrated in Figure 7.22 is integrated into the CVT control system shown in Figure 7.23.

The command signals from the TCU in Figure 7.23 are shown in solid lines with arrows for servo mechanism control design and in dashed lines with arrows for VBS based control design. The two designs share similar basic control logic with the same sensors for the controller inputs. Interactions between the CVT, engine, and other systems are via the CAN. As explained in Section 7. 4, engine output torque reduction may be introduced during CVT ratio changes via the interaction between the CVT TCU and the engine ECU.

7.5.2 Continuous Ratio Control Strategy

As discussed in Chapter 1, CVTs are the ideal transmission that in theory matches the engine output perfectly. For the perfect match, the CVT ratio must be controlled to vary continuously as it was originally designed to do. The objective of continuous ratio control is to optimize the engine operation status for dynamic performance and fuel economy under real world road conditions. In the engine fuel map, there is a so-called optimal

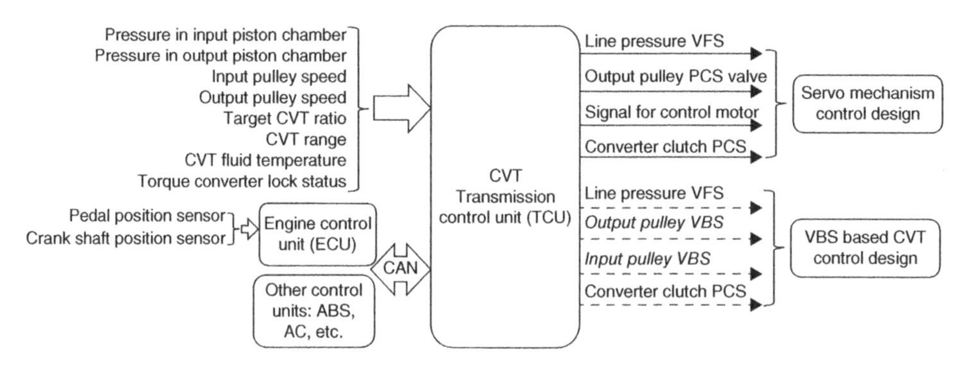

Figure 7.23 Block diagram of CVT control system.

operating line that defines the most fuel efficient engine operation status, as shown in Figure 7.24. By controlling the CVT ratio continuously, the engine will run along the optimal operating line, also called steady state operation line [24], for the optimized trade-off between performance and fuel economy. As mentioned in Section 7.1, this control strategy is realized by integrated engine–CVT control, in which the engine speed and the CVT ratio are controlled simultaneously [23,24].

The first step for continuous ratio control is to establish the optimal operating line or S-S operation line from the engine fuel map. For a typical engine, there is a theoretical optimal operating line that passes the unstable engine operation area in the fuel map and lacks sufficient torque reserves for transient maneuvers. The S-S line modifies the theoretical optimal operating line to avoid the unstable area for better vehicle drivability, as shown in Figure 7.24. The CVT ratio is thereby controlled continuously so that the engine runs along the S-S operation line which optimizes drivability without significant fuel economy penalty.

There are various control modes for transient operations along the S-S operation line, depending on how the driver's priorities are interpreted by the CVT controller. A transient operation defines how the engine–CVT joint operation will be shifted from a current point to the next point on the S-S operation line, for example, from point *A* to point *B*, as shown in Figure 7.24. The following gives details of two control modes – economy mode and performance mode [24] – for transient operations. Clearly the economy mode is selected with a priority on fuel economy and the other with a priority on power or performance.

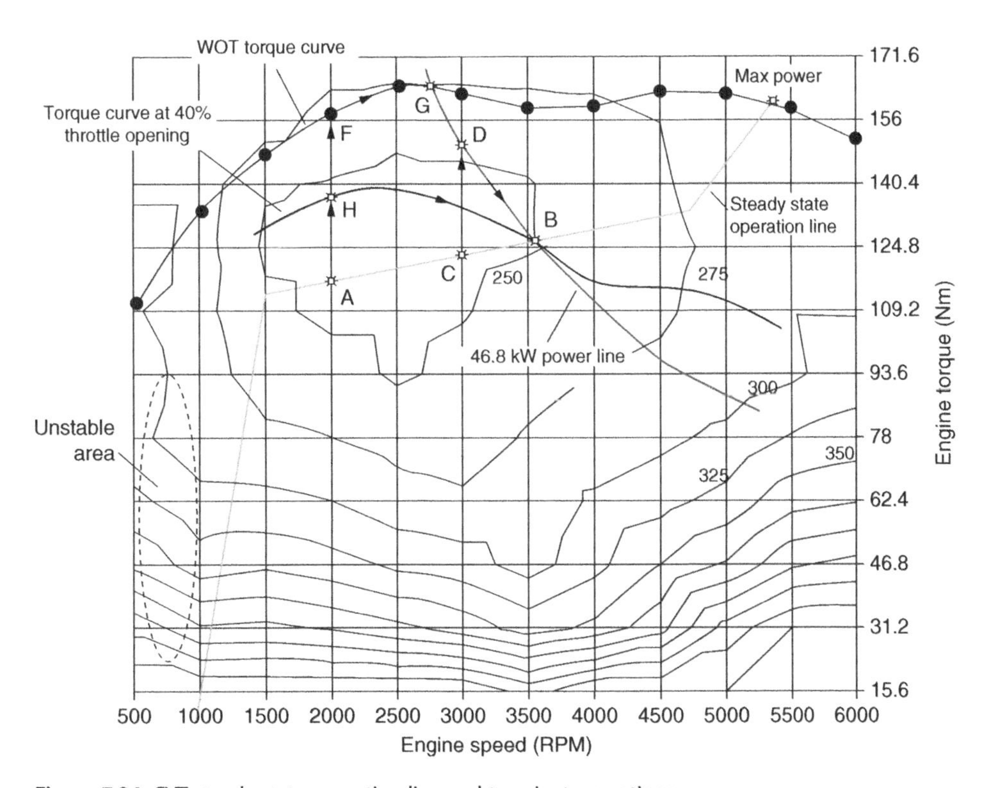

Figure 7.24 CVT steady state operation line and transient operations.

Economy mode: Suppose the vehicle is running at point A on the S-S operation line at a certain power level that can be determined based on road load or from the engine operation status. To accelerate the vehicle to a higher speed, the driver depresses the accelerator pedal and the pedal depression is interpreted by the CVT controller (TCU) as the driver's intention to increase the engine power to a higher level, at 46.8 kW for example. This elevated power level uniquely defines a point on the S-S operation line, i.e. point B, as shown in Figure 7.24. There is a unique torque curve that contains curve segment HB, corresponding to an engine throttle opening of 40%, shown as an example in the Fig. 7.24. This throttle opening percentage can be found in general by interpolating the engine output map. In the economy mode, the vehicle operation status will be transferred from point A to point B along the 40% throttle line. As soon as the driver depresses the accelerator pedal, the throttle opening is increased to be at 40% instantaneously by electronic throttle control. The CVT ratio and the engine operation are then controlled jointly so that the engine speed traces the 40% throttle torque curve, until steady-state vehicle operation status is reached at point B. For simplicity, the following assumptions are made in the formulation of the CVT powertrain dynamics during the transition from point A to point B:

- Engine transients are not considered. That means that the engine throttle is instantaneously controlled at point H and the engine also reaches steady state instantaneously. The engine RPM at point H is the same as at point A based on this assumption.
- Powertrain component inertias are not considered in the formulation.
- Powertrain efficiency η is considered to be constant.

Since A and B are both on the S-S operation line, dynamic equilibrium is achieved by the vehicle powertrain system and the following equations define the steady state at both points:

$$\left.\begin{aligned} \omega_e &= \frac{\pi(RPM)_e}{30} \\ V &= \frac{\omega_e}{i_{cvt} i_a} R_t \\ \left(fW + \frac{G}{100} W + C_c V^2\right) V &= \eta P_e \end{aligned}\right\} \tag{7.53}$$

where ω_e and $(RPM)_e$ are the engine angular velocity and RPM respectively. V is the vehicle speed, R_t is the vehicle tire radius, i_{cvt} and i_a are the CVT ratio and the final drive ratio, W is the vehicle weight, and P_e is the engine power at either point A or point B. It is apparent that the CVT ratio at target point B can be determined by Eq. (7.52), where G is the slope grade number and C_c is the vehicle air drag constant as represented by Eq. (1.14) in Chapter 1. During the transition from point A to point B along the 40% throttle line, the vehicle motion is governed by the following equation:

$$\left.\begin{aligned} \omega_e &= \frac{\pi(RPM)_e}{30} \\ V &= \frac{\omega_e}{i_{cvt} i_a} R_t \\ \frac{W}{g}\frac{dV}{dt} &= \frac{\eta T_e(\omega_e) i_{cvt} i_a}{R_t} - \left(fW + \frac{G}{100} W + C_c V^2\right) \end{aligned}\right\} \tag{7.54}$$

where $\frac{dV}{dt}$ is the vehicle acceleration and $T_e(\omega_e)$ is the engine output torque as a function of engine angular velocity. Note that the vehicle acceleration at both points A and B is equal to zero since the two points are on the S-S operation line. The vehicle acceleration at point H is not equal to zero, but can be determined by Eq. (7.54) using the engine torque found on the 40% throttle torque curve corresponding the engine RPM at point A.

Performance mode: This mode is used for the quickest transition from a current point to the target point on the S-S operation line, for example, from point C to point B, as shown in Figure 7.24. Similar to the economy mode, the pedal depression by the driver at point C is interpreted as an intention to increase the engine power to the level at point B, at which the 46.8 kW constant power line intersects the S-S operation line. In performance mode, the engine throttle opening is instantaneously increased to the percentage corresponding to point D on the 46.8 kW constant power line by electronic throttle control. The CVT and engine operations are then controlled jointly for the engine to run along the 46.8 kW constant power line from point D to the steady state point B. The performance mode can also be implemented for the transition from point A to point B. In this case, the engine throttle opening at point A will be increased instantaneously to WOT opening at point F, the CVT and the engine will then be jointly controlled so that the engine runs from point F to point G along the WOT torque curve and then from point G to point B along the 46.8 kW constant power line. Along the line of constant power P_e, i.e. line GB as shown in Figure 7.24, the following equations govern the motion of the vehicle powertrain system:

$$\left.\begin{aligned} \left(fW + \frac{G}{100}W + C_c V^2 + \frac{W}{g}\frac{dV}{dt}\right) V &= \eta P_e \\ V &= \frac{\omega_e}{i_{cvt} i_a} R_t \\ \omega_e &= \frac{\pi (RPM)_e}{30} \end{aligned}\right\} \tag{7.55}$$

As already mentioned, the engine and the CVT must be jointly controlled for the engine to run along the steady state operation line. The two control modes described above are used to make the transition from one power level to another. The CVT–engine joint control during the transition operations is based on the logic illustrated by the block diagram of Figure 7.25 [24]:

Figure 7.25 contains two control loops, one for the engine throttle and another for the engine speed. The value of θ_t is the driver accelerator depression, which is interpreted by

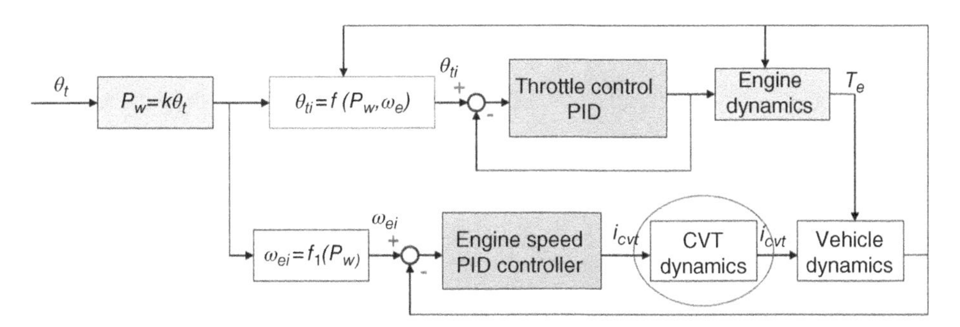

Figure 7.25 CVT-engine joint control block diagram.

the TCU as the intention to increase the engine power to the value P_w, while θ_{ti} is the engine throttle opening interpolated from the engine map according the selected transition mode. For example, if the economy mode is selected, θ_{ti} is 40% for the transition from point A to point B, as shown in Figure 7.24. If the performance mode is selected, θ_{ti} then needs to be interpolated from the engine map in terms of the target power and the engine speed. Then ω_{ei} is the engine angular velocity at the next steady state point when the transition is completed, for example, at point B, as shown in Figure 7.24, and i_{cvt} is the CVT ratio which is controlled by the methods discussed in Section 7.4, while i'_{cvt} is the time derivative of the CVT ratio.

7.5.3 Stepped Ratio Control Strategy

Continuous ratio control discussed above takes full advantage of CVTs and theoretically offers the optimal powertrain fuel efficiency. However, it has drawbacks in drivability in comparison to vehicles equipped with conventional ATs. For drivers who are accustomed to driving conventional AT vehicles, CVT vehicles with continuous ratio control give a sloppy or unresponsive driving feel. Therefore, most production CVT vehicles today are controlled by the stepped ratio strategy. As the name indicates, a stepped ratios strategy controls the CVT with multiple stepped gear ratios that emulate a conventional automatic transmission. Usually, the CVT is controlled with six to eight stepped gear ratios in this strategy, although more ratios are possible if necessary. Note that, unlike conventional ATs, each of the multiple gear ratios of the CVT controlled with this strategy may not always be a constant but may be controlled at different values to best fit the vehicle operation condition. For example, the engine speed and the vehicle speed are shown in Figure 7.26 for a CVT vehicle controlled with six gear ratios. As shown in the figure, the first and sixth gear ratios are not held constant. This is because both the low CVT ratio and the overdrive CVT ratio are achieved with the smallest contact angle, as defined in Figure 7.5 either in the input pulley or in the output pulley. For loading and strength considerations, it is beneficial to avoid running the CVT in these two extreme positions at high engine RPM and high vehicle speed.

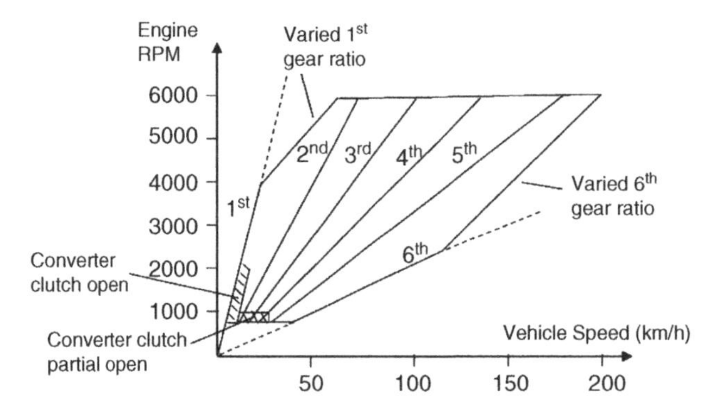

Figure 7.26 CVT stepped gear ratios and ratio variations.

A transmission shift schedule similar to Figure 6.17 must be established to implement a CVT stepped ratio control strategy. The attributes of conventional automatic transmission shift schedules, and guidelines to establish them, were presented in Section 6.4. The shift schedule for a CVT controlled with stepped ratios has similar attributes and can be established similarly. Since the number of gears and the gear ratio in each gear can be flexible, the CVT shift schedule naturally can be designed to better fit the engine output to vehicle operation conditions. Similar to the shift schedule of multiple ratio conventional ATs, a CVT shift schedule can also be designed with different shift modes, such as normal, economy, and sport. The driver can choose a mode for different priority at the convenience of pushing a button or turning a knob. As mentioned previously, the torque converter is primarily used as the vehicle launcher in CVT vehicles and is partially locked up to dampen powertrain transients at low vehicle speed, as shown in Figure 7.26. For efficiency considerations, the torque converter is locked up by the converter clutch and just behaves as a mechanical couple under all other operation conditions. Because of the CVT shifting mechanism discussed in Section 7.4, ratio changes can be completed smoothly without the torque converter functioning as a fluid damper in the process.

Similar to conventional ATs, the CVT is controlled to operate either in a fixed ratio or to implement a shift according to the shift schedule based on the inputs from the speed sensor and throttle position sensor. CVT control during fixed ratio operation and shift operations were detailed in Section 7.4.

7.5.4 CVT Control Calibration

Calibration for automatic transmission control was discussed in Section 6.5 on both component and system levels. In principle, CVT control calibration is similar to conventional ATs. As detailed in Sections 7.4 and 7.5, line pressure control is the most critical for both CVT fixed ratio operation and shifting operations. In fixed ratio operation, the line pressure is directly applied in the output piston chamber and should be controlled just above the value required to transmit the torque required for vehicle operation so that the CVT runs most efficiently. During ratio changes, the line pressure is applied on either the input piston chamber or the output piston chamber to actuate the axial motions of the movable sheaves of the input and output pulleys. For a CVT that is controlled with n gear ratios, there are n upshift threshold lines and n downshift threshold lines, as illustrated in Figure 7.27. Both the engine throttle opening and the engine output speed vary along a threshold line. The engine torque to be transmitted by CVT is determined by the engine throttle opening, which is the vertical axis in the shift schedule, as shown in Figure 7.27, and the engine output speed that can be determined from the CVT ratio and the vehicle speed, which is the horizontal axis of the shifting schedule. A huge number of line pressure values would be required to optimize the CVT operation in the whole drive range, but this cannot be implemented for practical CVT control. Therefore, a number of points are chosen along an upshift threshold line to define the CVT–engine operation status, as shown in the 1–2 upshift line in Figure 7.27. At each such point, the corresponding engine output torque, i.e. the CVT input torque that needs to be transmitted, is determined by interpolation from the engine map. If 10 points are chosen on the 1–2 upshift line, there will be 10 CVT input torque values, and these 10 torque values will then be used to determine the line pressure values according Eq. 7.52, as previously

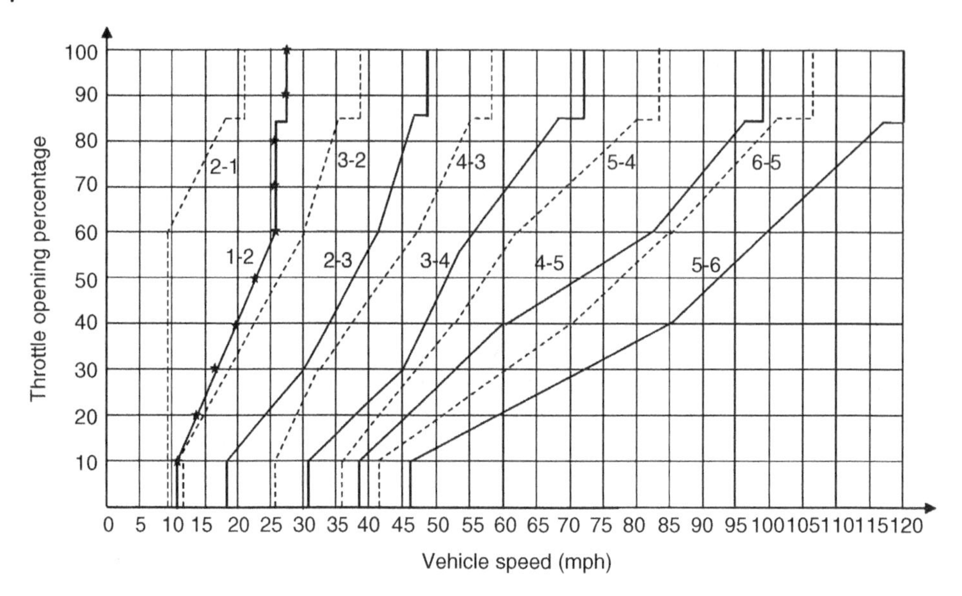

Figure 7.27 Shift lines for a CVT controlled with six stepped gear ratios.

described in line pressure control. A number of these line pressure values can then be used as the reference for line pressure control when the CVT runs with the first gear. For the whole drive range, there will be 50 different line pressure values if 10 points are chosen on each of the five upshift lines. A database can be built for the line pressure values for the CVT input torque in terms of throttle opening, gear ratio, and vehicle speed. The real time line pressure during CVT operation is interpolated from the database to best fit the vehicle operation condition. As a general guideline, the line pressure can be selected and calibrated in the following steps:

- Select a number of points along each upshift line. These points should reflect the most frequent vehicle operation conditions in daily traffic. Experiences and data on existing vehicles provide a good reference on the selection of these points.
- At each point mentioned above, the engine torque is determined from the engine map since throttle opening, vehicle speed, and CVT ratio are known at the point. The CVT input torque is then found by subtracting the loads on auxiliary systems, such as the load from the air conditioner and the power steering pump.
- The line pressure value is then determined by Eq. (7.52) for the particular point. This line pressure allows the CVT to transmit the input torque with a torque ratio of about 0.9. This line pressure is about 10% higher than is needed to transmit the required input torque with a 10% margin below the belt slippage threshold.
- A database is established for the line pressure values determined as above for all points along each upshift line and for all the five upshift lines shown in Figure 7.27.
- Before an upshift, the CVT will be controlled at line pressure determined at the point along the upshift line that is the closest to the engine–CVT operation status, or at a line pressure value determined by interpolating the points on the upshift line. After the upshift, the CVT will be controlled at the line pressure found at the next upshift line

corresponding to the particular engine throttle opening. The line pressure values before and after a downshift are determined similarly.

- The vehicle prototype is controlled on a testing rig to operate with the line pressure values determined above under driving conditions that emulate daily traffic; for example, the standard EPA drive ranges can be used to control the operation of the prototype vehicle on the testing rig. Data from these tests are used to validate line pressure control strategy, to modify control variables, and to adjust the line pressure values themselves for CVT operation optimization.
- Road tests are then performed to fine tune the line pressure values and related control variables under real world traffic conditions.

References

1 Continuously variable transmission – Wikipedia https://en.wikipedia.org/wiki/Continuously_variable_transmission
2 DAF 600 – Wikipedia: https://en.wikipedia.org/wiki/DAF_600.
3 Subaru Justy – Wikipedia: https://en.wikipedia.org/wiki/Subaru_Justy.
4 Srivastava, N. and Haque, I.: *A Review on Belt and Chain Continuously Variable Transmissions (CVT): Dynamics and Control*, Mechanism and Machine Theory, 44 (2009), pp. 19–41.
5 Gerbert, G.: *Belt Slip – A Unified Approach*, ASME Journal of Mechanical Design, 1996, Vol. 118, No. 3, pp. 432–438.
6 Gerbert, G.: *Metal V-belt Mechanics*, ASME Design Automation Conference, ASME Paper No. 84-DET-227, Boston, MA, 1984.
7 Miloiu, G.: *Die Druckkraft in Stufenlosen Getrieben II*, Antriebstechnik, Vol. 13, No. 1, 1969, pp. 450–467.
8 Worley, W.S. and Dolan, J.P.: *Closed-Form Approximations to the Solution of V-Belt Force and Slip Equations*, ASME Journal of Mechanical Design, 1985, Vol. 107, No. 2, pp. 292–300.
9 Worley, W.S.: *Designing Adjustable-Speed V-belt Drives for Farm Implements*, SAE Transaction, Vol. 63, 1955, pp. 321–323.
10 Srivastava, N. and Haque, I.: *On the Transient Dynamics of a Metal Pushing at High Speed*, International Journal of Vehicle Design, Vol. 37, No. 1, 2005, pp. 46–66.
11 Srivastava, N. and Haque, I.: *Transient Dynamics of Metal V-belt CVT: Effects of Pulley Flexibility and Friction Characteristics*, ASME Journal of Computation and Nonlinear Dynamics, Vol. 2, No. 1, 2007, pp. 86–97.
12 Srivastava, N. and Haque, I.: *Transient Dynamics of Metal V-belt CVT: Effects of Band Pack Slip and Friction Characteristics*, Mechanism and Machine Theory, Vol. 43, No. 4, 2008, pp. 459–479.
13 Carbone, G., Mangialardi, L., and Mantriota, G.: *Theoretical Model of Metal V-belt Drives During Ratio Changing Speed*, ASME Journal of Mechanical Design, Vol. 123, No. 1, 2000, pp. 111–117.
14 Carbone, G., Mangialardi, L., and Mantriota, G.: *Influence of Clearance between Plates in Metal Pushing V-belt Dynamics*, ASME Journal of Mechanical Design, Vol. 124, No. 3, 2002, pp. 543–557.

15 Carbone, G., Mangialardi, L., Mantriota, G.: *EHL Visco-plastic Friction Model in CVT Shifting Behaviour*, International Journal of Vehicle Design, Vol. 32, No. 3/4, 2003, pp. 333–357.
16 Micklem, J.D., Longmore, D.K., and Burrows, C.R.: *Modelling of the Steel Pushing V-belt Continuously Variable Transmission*, Proceedings of the Institute of Mechanical Engineers, Part C: Journal of Mechanical Engineering Science, Vol. 208, No. 1, 1994, pp. 13–27.
17 Pfeiffer, F. and Sedlmayr, M.: *Spatial Contact Mechanics of CVT Chain Drives*, Proceedings of ASME 2001 Design Engineering Technical Conference, Paper No. DETC2001/VIB-21511, Pittsburgh, USA, Vol. 6B, Sept. 2001, pp. 1789–1795.
18 Pfeiffer, F. and Sedlmayr, M.: *Force Reduction in CVT Chains*, International Journal of Vehicle Design, Vol. 32, No. 3/4, 2003, pp. 290–303.
19 Fujii, T., Kurokawa, T., and Kanehara, S.: *A Study of a Metal Pushing V-belt Type CVT – Part 1: Relation between Transmitted Torque and Pulley Thrust*, SAE Paper # 930666, 1993.
20 Fujii, T., Kurokawa, T., and Kanehara, S.: *A Study of a Metal Pushing V-belt Type CVT – Part 2: Compression Force between Metal Blocks and Ring Tension*, SAE Paper # 930667, 1993.
21 Kurokawa, T., Fujii, T., Kurokawa, T., and Kanehara, S.: *A Study of a Metal Pushing V-belt Type CVT – Part 3: What Forces Act on Metal Blocks*, SAE Paper # 950671, 1995.
22 Takash, K., Fujii, T., and Kanehara, S.: *A Study of a Metal Pushing V-belt Type CVT – Part 4: Forces Act on Metal Blocks When the Speed Ratio Is Changing*, SAE Paper # 940735, 1994.
23 Sakagachi, S., Kimula, E., and Yamamoto, K.: *Development of an Engine – CVT Integrated Control System*, SAE Paper # 1999–01-0754, 1999.
24 Vahabzadeh, H. and Linzell, M., *Modelling, Simulation and Control Implementation for a Split-Torque, Geared Neutral, Infinitely Variable Transmission*, SAE Paper No. 910409, 1991.
25 Tanaka, H. and Ishihara, T: *Electro-Hydraulic Digital Control of Cone-Roller Toroidal Traction Drive Automatic Power Transmission*, ASME Journal of Dynamic System and Measurement, Vol. 106, pp. 305–310.
26 Tanaka, H.: *Power Transmission of a Cone Roller Toroidal Traction Drive*, JSME Int. J., Series III, Vol. 32, No. 1, pp. 82–90, 1989.
27 Zhang, Y., Zhang, X., and Tobler, W.: *A Systematic Model for the Analysis of Contact, Side Slip and Traction of Toroidal Drives*, ASME Journal of Mechanical Design, Vol. 122, No. 4, pp. 523–528, 2000.
28 Zou, Z., Zhang, Y., Zhang, X., and Tobler, W.: *Modelling and Simulation of Traction Drive Dynamics and Control*, ASME Journal of Mechanical Design, Vol. 123, No. 4, pp. 556–561, 2001.

Problems

1 A belt CVT has the following design data: pulley center distance $E = 185$ mm; Belt length $L = 720$ mm; pulley groove angle: $11°$. The piston effective areas are 160 and 130 cm^2 for the primary pulley and secondary pulley respectively. The CVT ratio

range is [0.4, 2.3]. The static tangential friction coefficient is 0.1 for both pulleys. The final drive ratio is 6.12. The efficiency before the primary pulley is considered to be 1.0. The driveline efficiency is 0.82 after all extra engine loads and losses are considered. The CVT vehicle data and the engine fuel map are provided as shown.

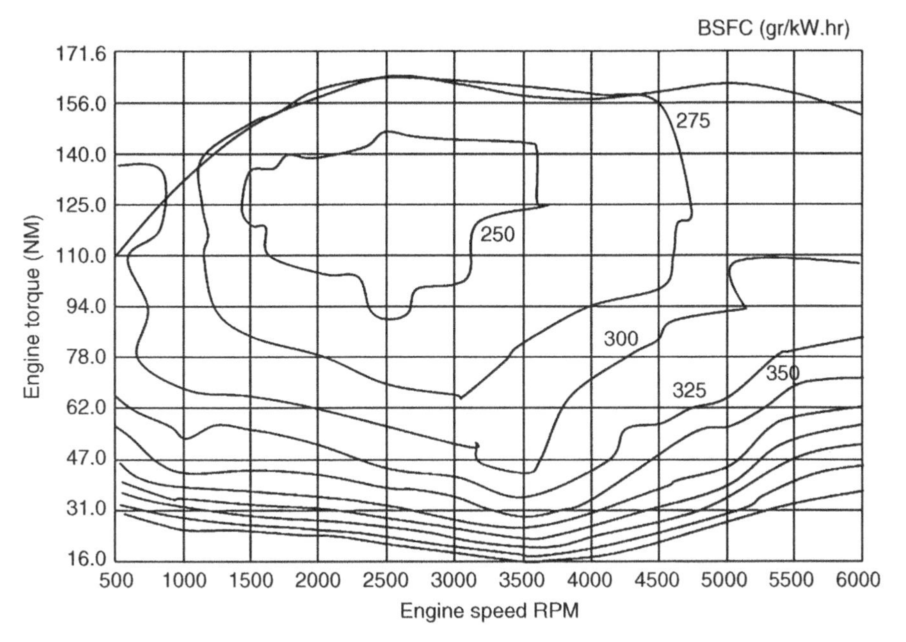

a) The CVT is required to be capable of transmitting a maximum input torque on the primary pulley of 160 Nm with the CVT ratio equal to 2.3 when the torque converter is locked. With a safety factor of 1.1, what should be the line pressure in Bar (note: 1 Bar = 10 N/cm^2) that is directly applied to the secondary piston chamber?

b) What should be the pressure in the primary piston chamber if the same safety factor of 1.1 is required for the input side when the CVT is operating under the condition in (a).

c) When the vehicle is cruising at a constant speed of 110 km/h on level ground, the engine torque is controlled at 78 Nm. What is the CVT ratio and the fuel economy in liter/100 km. Note: gas density is 750gram/liter. Show how data are taken from the fuel map.

Vehicle data:

Air drag coefficient: $C_d = 0.29$	Vehicle Mass = 1200 kg	Tire radius: $R = 0.28$ m	Rolling coefficient: 0.2
Final drive ratio: $i_a = 6.12$	Final drive efficiency: 0.97	Frontal projected area: $A = 2.15$ m^2	

2 A CVT is used for a vehicle with data shown below. The engine map, the steady state operating line, the constant power lines at 25 kW, 34 kW, and 45 kW, and the constant throttle line are shown in the figure. The torque converter is locked and all losses are neglected (i.e. $\eta = 1$). Points A, B, and C are on the steady-state operating line.

a) Determine the vehicle speed and CVT ratio at A, B, and C respectively.
b) At a point of time during the transition from A to B in economy mode, the vehicle speed sensor reads a vehicle speed of 108 km/h and the engine tachometer reads an engine speed of 2300 RPM. Determine the vehicle acceleration at this point.
c) The performance mode is used for the transition from C to B. Determine the vehicle acceleration at the beginning of the transition.
d) At a point of time during the transition from C to B in performance mode, the speed sensor reads a speed of 118 km/h and the CVT ratio is 1.1. Determine the vehicle acceleration at this point.

Vehicle data:

$C_d = 0.28$ (air drag coefficient)	$A = 2.15\ \text{m}^2$ (front projected area)	$m = 1250\ \text{kg}$ (vehicle mass),
$R_{tire} = 0.32\ \text{m}$ (tire radius)	$f = 0.02$ (rolling coefficient)	$i_a = 2.70$ (final drive ratio)

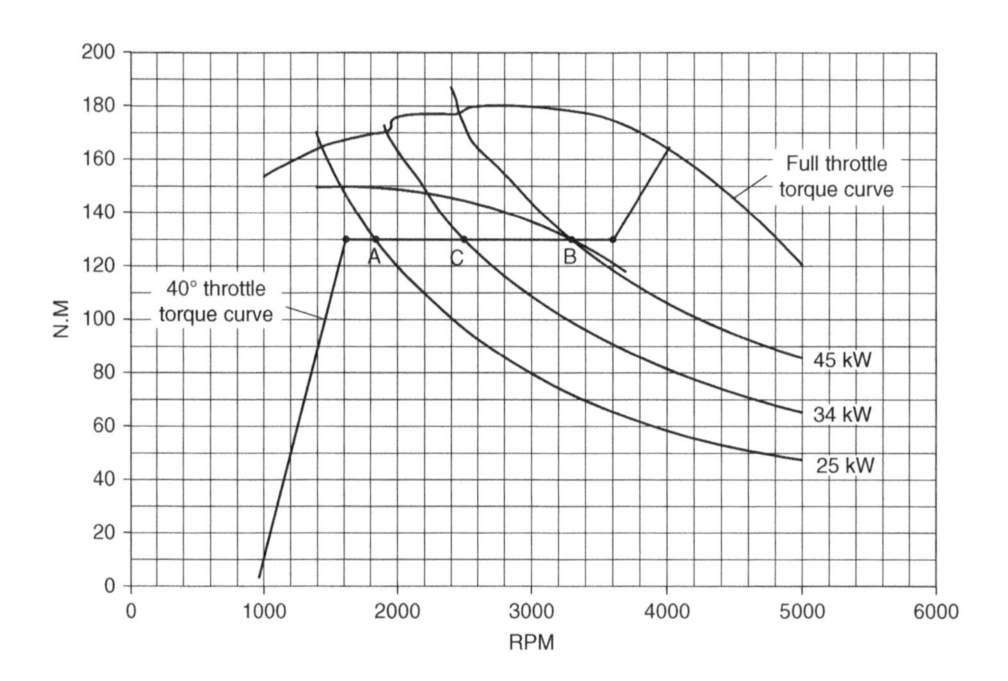

8

Dual Clutch Transmissions

8.1 Introduction

Dual clutch transmission (DCT) combines the advantages of manual transmission (MT) and conventional automatic transmission (AT). DCT vehicles feature the convenience and comfort of AT vehicles and a fuel economy comparable to that of MT vehicles. In addition, DCT is less costly to manufacture in comparison with AT since it shares similar structure and components with MT.

As illustrated in Section 8.2, a DCT is structurally a combination of two manual transmissions, one with even gears and the other with odd gears. While the transmission operates in the current gear, the next gear is already engaged, since the related clutch is open and the related gears freewheel. A gear shift is realized by releasing the currently applied clutch and applying the clutch for the next gear. Gear shifting of a dual clutch transmission is then in kinematics similar to that of clutch-to-clutch shift in a conventional AT. However, the dynamic characteristics are different between the two types of transmissions since the AT is equipped with a torque converter that dampens shift transients. The existing control technologies for conventional AT cannot be readily applied for DCT shift control. Moreover, more precise torque control is required to achieve launch and shift smoothness of DCT vehicles since there is no torque converter between the engine and the transmission input. Extensive research and testing has been conducted by engineers and researchers in the automotive industry on the launch and shift control of DCT vehicles. A model-based study was successfully conducted on DCT applications for medium duty trucks [1]. Shift dynamics and transmission control were investigated by model simulation for vehicles equipped with DCTs [2,3]. A gearshift control strategy for twin wet clutch transmission was developed and integrated with engine control to achieve synchronization during the transfer of engine torque from clutch to clutch through clutch slip control [4]. A system control approach, consisting of various control loops including engine speed control, clutch slip control, and transmission output torque control, was developed for gear shifts of twin clutch transmissions [5]. Torque based control, which synergizes the engine torque control and clutch torque control, was developed for AT vehicles to optimize shift qualities [6–8] and is applicable for DCT shift controls. To enhance the DCT torque control precision, a systematic model is proposed to analyze the dynamic characteristics of DCTs based on prototype vehicle testing [9], and a further study is focused on the clutch torque formulation and calibration for dry DCTs [10,11].

Automotive Power Transmission Systems, First Edition. Yi Zhang and Chris Mi.

Due to its advantages in fuel efficiency and cost, DCT products attracted extensive development interests in the automotive industry and DCT equipped passenger vehicles have been on the market for more than a decade. General descriptions of DCT development status can be found in the public domain [12–14]. Ford Motor Company developed DCTs for applications in both its subcompact Fiesta and compact Focus models around 2010. VW has applied DCTs across its passenger car product line for over a decade. Other OEMs have also occasionally applied DCTs for their passenger vehicle products, even though the applications are not widespread. It can be predicted that DCT vehicles will continue to take market share from vehicles equipped with other types of transmission.

Following this introduction section, the chapter will continue with Section 8.2 on the structural layouts of DCT systems and key components, including the basic DCT structure and kinematics. Current DCT production types will be presented also in the section. Section 8.3 will then concentrate on the dynamic modeling and analysis of DCT operations, including DCT vehicle launch and shifts. Section 8.4 will deal with the control system design and the control of ratio changing processes. Section 8.5 will present DCT clutch torque formulation for accurate clutch torque calculations during launch and shifts, using an electrically actuated dry DCT as the example in the case study. Since DCTs share with manual transmission in basic layouts and gear design, readers are referred to Chapter 2 and Chapter 3 for related topics.

8.2 DCT Layouts and Key Components

Dual clutch transmissions share similar components and basic structure with manual transmissions. A DCT can be considered in structure and in kinematics as a combination of two manual transmissions, with one providing the odd gears and the other the even gears. Two input shafts in a compact hollow-solid layout take up the engine output through the respective clutch. The structure of a typical DCT is illustrated in Figure 8.1. This DCT layout has been used in Ford subcompact and compact passenger cars. Some other DCT designs may also have the two input shafts in a parallel layout.

The transmission shown in Figure 8.1 has six forward speeds and one reverse, with odd gears on the solid shaft that is connected to the engine by clutch 1 (CL1) and even gears on the hollow shaft connected by clutch 2 (CL2). Reverse gear is achieved by using the sixth output gear as the idler. There are two final drive pinions, one on transfer shaft 1 and the other on transfer shaft 2, that drive the same ring gear, providing two different final drive ratios. Gear shifts only involve the engagement of the oncoming clutch and the release of the off-going clutch since the oncoming gear is preselected. For example, when the vehicle runs in first gear, clutch CL1 is engaged but clutch CL2 is open, and both the first gear and the second gear are engaged by the respective synchronizers which were discussed in Chapter 2. The first gears transmit power from the engine to the transmission output, while the second gears freewheel with the open clutch CL2. A 1–2 shift is then realized by disengaging clutch CL1 and engaging clutch CL2 in the 1–2 shift process. The operations in other gears and related shifts follow similar patterns.

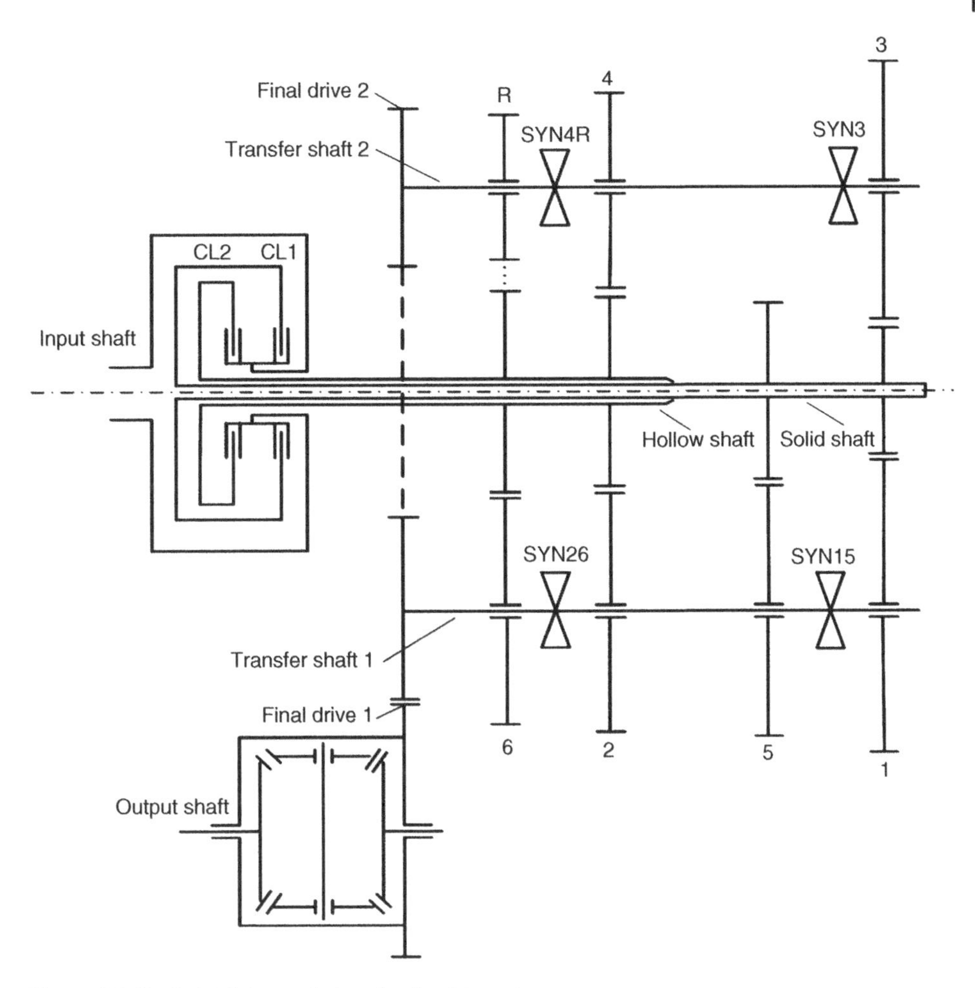

Figure 8.1 Dual clutch transmission structural layout.

8.2.1 Dry Dual Clutch Transmissions

Dual clutch transmissions are categorized as dry or wet according to the actuation of the dual clutch module. In a dry DCT, the two clutches in a compact pack are a single-disk type that is actuated by the respective release bearing via a diaphragm spring. For dry DCTs, the release bearings and the synchronizers can be controlled either electrically or hydraulically. In an electrically actuated dry DCT, the two clutches are controlled by electric motors through the respective control mechanisms, and the gear shifts are also controlled by electric motors through shifting drums that convert the motor's rotational motion to linear motion for the positioning of the synchronizers. The dual clutch module, the clutch control actuator, and the shift drums of the six-speed dry DCT shown in Figure 8.1 are illustrated in Figures 8.2, 8.3, and 8.4, respectively.

As can be observed in Figure 8.2, the dry clutch module contains two friction disks and two release bearings, one for each clutch. Torsional spring dampers are assembled with both friction disks to dampen the dynamic transients during vehicle launch and shifts.

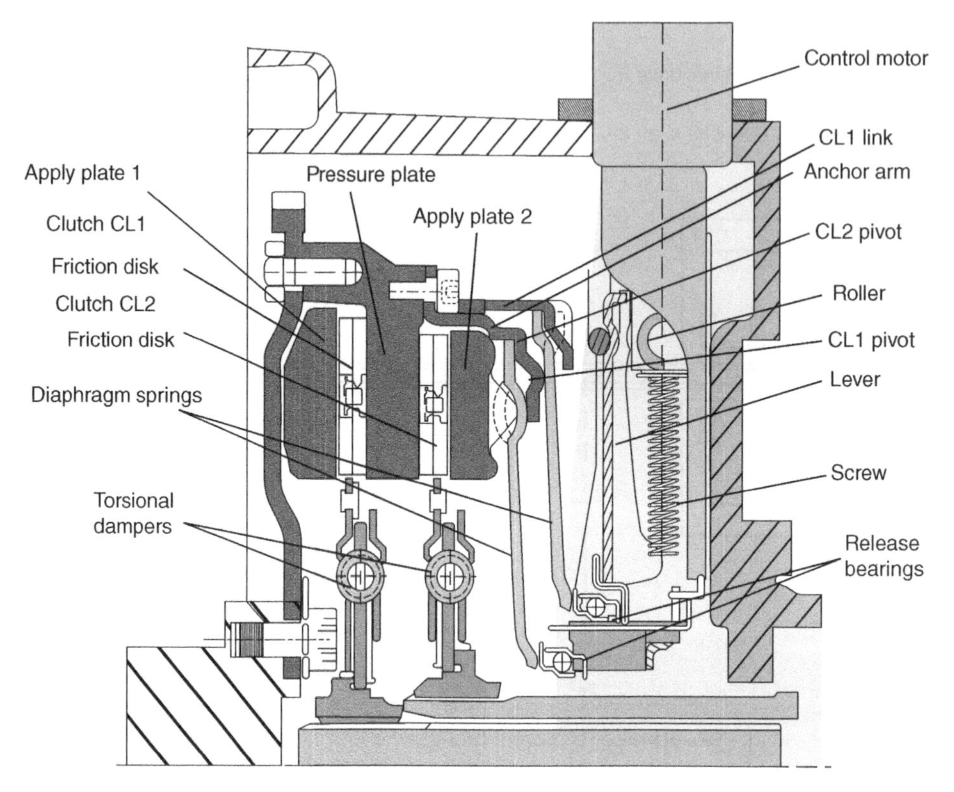

Figure 8.2 Section view of dry dual clutch module and clutch actuator.

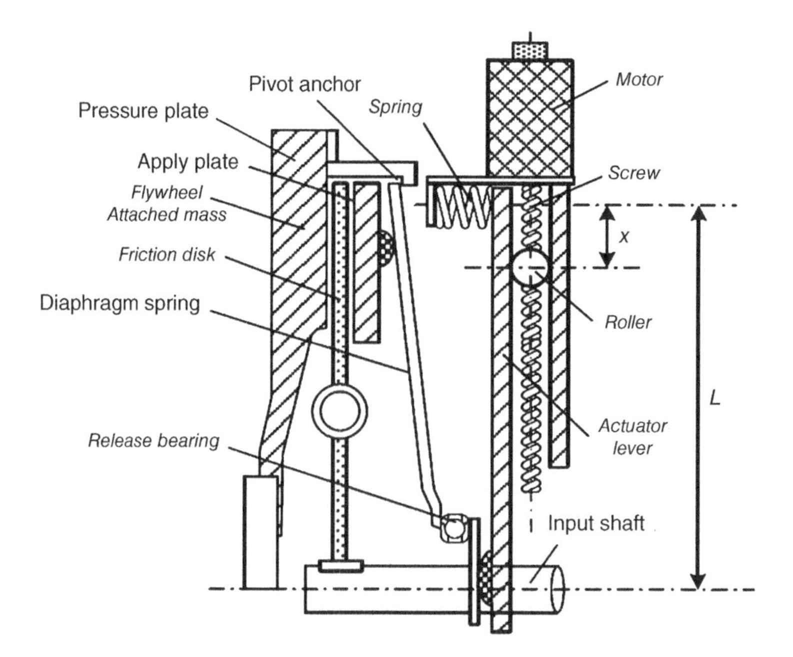

Figure 8.3 Dry DCT clutch control mechanism.

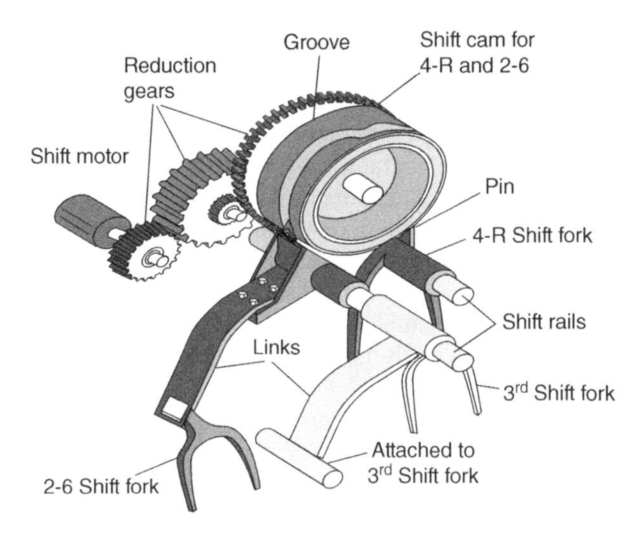

Figure 8.4 Dry DCT gear shifting cams.

In some other designs, a torsional spring damper may also be installed on the flywheel side to enhance the damping effects. The pressure plate assembly is attached to the engine flywheel and carries the two apply plates via flexible links (as shown in Figure 8.6c) to rotate with the flywheel as discussed in Chapter 2. The two apply plates are allowed to move slightly along the axial direction due to the flexible links. The pressure plate also carries the pivoting anchor on the anchor arm for each diaphragm spring, so the two diaphragm springs also always turn with the flywheel. The upper end of the CL1 diaphragm spring then contacts the CL1 link which is connected to the CL1 apply plate, as shown in Figure 8.2. Unlike the clutch in a manual transmission, which has a normally engaged clutch, clutches in a DCT are normally open, i.e. the clutch stays open unless engaged by the clamping force. As shown in Figure 8.2, the inner side of each diaphragm spring then contacts the respective release bearing which travels slightly on the input shaft when actuated by the control motor. Clamping force will be generated on the friction disk of a clutch if the respective release bearing is actuated by the motor to move leftward. For example, when release bearing 1 moves leftward, it will carry the inner side of the diaphragm spring of clutch CL1 and will displace the outer side of the CL1 diaphragm spring rightward, carrying with it the apply plate 1 to clamp the CL1 friction disk against the pressure plate. And the diaphragm spring also works as a lever that magnifies the force applied by the release bearing on its inner side. The friction torque generated by the clutch depends on the apply force on the release bearing, as will be detailed in later sections.

The core of the actuator for the dual clutch module is the mechanism that controls the travel of the release bearing along the input shaft. There are two such mechanisms for the dual clutch module. The one illustrated in Figure 8.3 is for clutch CL2 for the clutch module shown in Figure 8.2. The control mechanism for clutch CL1 is identical in structure and in operation principle.

As shown in Figure 8.3, the screw is turned with the control motor and displaces the roller along the screw axis. The roller displacement is linearly proportional to the motor rotation. The spring attached to the top of the actuator level is under compression at assembly. The force applied by the release bearing on the inner side of the diaphragm spring depends on the compressive spring force at the top of the actuator lever and the displacement of the roller along the screw axis. Therefore by controlling the rotation of the control motor, the roller displacement along the screw axis will be controlled to the amount that the force applied by the release bearing to the diaphragm inner side is what is needed to generate the required clutch torque. For the dual clutch module of a dry DCT, two identical control mechanisms are assembled in a compact package as illustrated in Figure 8.2. The quantitative formulation on the correlation of clutch torque and motor rotation will be discussed in Section 8.5.

The drums or cams for gear shifting are illustrated in Figure 8.4. A shift drum is, in kinematics, a cylindrical cam with screw grooves. Figure 8.4 illustrates the shifting mechanisms for the Ford six-speed DCT shown in Figure 8.1. There are two shifting cams, one for the 4–R and 2–6 synchronizers, and the other (not shown in Figure 8.4) for the 5–1 and 3rd synchronizers. During gear shifts, the shifting sleeve of the synchronizer involved is translated along the gear shaft by the shift fork which slides on the shift rail. A shift fork can be supported on two shift rails to make its axial motion more conducive. A round pin on the shift fork is positioned into the screw groove of the shifting cam, as shown in Figure 8.4. As the cam turns, the pin in the screw groove moves axially and the displacement of the pin is linearly proportional to the rotational angle of the shifting cam by the screw pitch. The rotation of the shifting cam is actuated by the shift motor through a set of reduction gears, which can be spur gears or worm gears, depending on the gear box layouts. By controlling the rotational angle of the shift motor, the shift fork – i.e. the shifting sleeve of the synchronizer – will be accurately controlled at the position required for the engagement or disengagement of the target gear.

Each of the shifting cams shown in Figure 8.4 is machined with two screw grooves, each of which locates a pin attached to a respective fork shaft. The pitch lines of the two screw grooves are designed with ramps that enable the fork shaft to move axially during gear engaging or disengaging and with flat slots that keep the fork shaft at the same position. The ramps and the flat slots are aligned such that the axial motions of the fork shafts do not interfere with one another.

For the DCT in Figure 8.1, the 4–R and the 2–6 synchronizers share the same shifting cam. The two pitch lines for the two cam groves are shown in Figure 8.5. The pitch line of a cam groove consists of ramps and flat sections. Ramps are for the screw profiles that enable fork shaft motions for gear engaging or disengaging. Flat sections are for the flat grooves that keep the fork shaft (i.e. the respective gear) at the neutral position. As shown in Figure 8.5, there is no overlapping between the ramps for the 4–R synchronizer and for the 2–6 synchronizer. Therefore the engaging and disengaging of the 4th and reverse gears will not interfere with that of the 2nd and 6th gears.

The clutch actuation and gear shifting of dry DCTs can also be realized hydraulically. A production DCT of this type is shown in Figure 8.6. This dry DCT has been used in VW compact to midsize passenger cars. In this DCT, the actuation of the dual clutches is controlled by two hydraulic pistons via actuation levers. The gear engaging and disengaging motions are also realized by hydraulic pistons. In this type

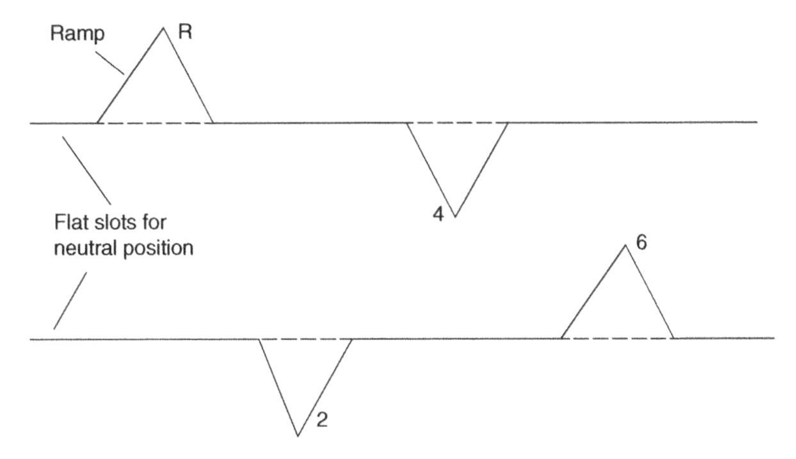

Figure 8.5 Pitch lines for shifting cam grooves.

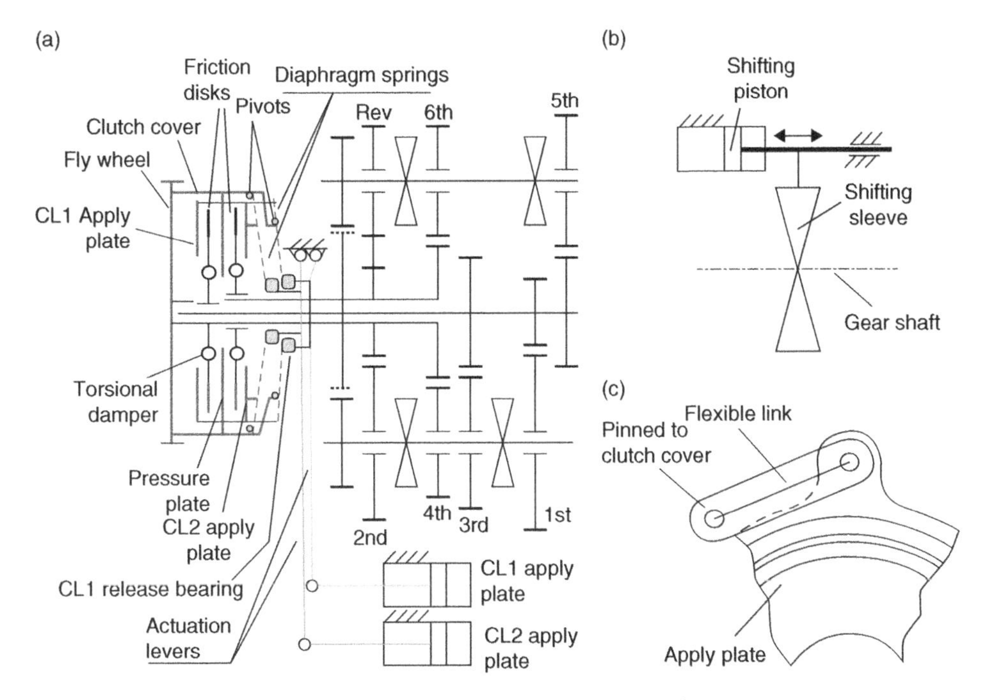

Figure 8.6 Dry DCT with hydraulically actuated clutches and gear shifting.

of DCT, a hydraulic booster is usually used to compensate the fluctuations of hydraulic pressure during clutch actuation and shifts caused by the fairly large piston displacements.

The actuation of the dual clutches in dry DCTs can also be realized by hydraulic pistons that push the release bearings. The two pistons with ring-shaped effective piston areas, one for each release bearing, are in a compact nested package that is installed on the hollow shaft and is fixed to the transmission housing. These two pistons are linked by hydraulic hoses to two slave pistons with smaller effective areas. The slave pistons are located away from the clutch well, resulting in more packaging room and better heat dissipation for the dual clutch module. The slave pistons can be controlled by electric motors via screw mechanisms that convert the rotations to linear displacements. The piston force on the release bearing is the force on the slave piston multiplied by the ratio between the respective piston effective areas. This clutch actuator design is also conducive to reducing vehicle launch and shift harshness. Gear shifting in this design is also realized by hydraulic pistons, as shown in Figure 8.6b.

8.2.2 Wet Dual Clutch Transmissions

The gear shaft layouts in wet dual clutch transmissions are similar to those in dry DCTs, as shown in Figure 8.7. The dual clutches in wet DCTs are multiple disk clutches similar to those in conventional automatic transmissions. The two clutches are nested in a compact module as illustrated in Figure 8.7a. Transmission shifts are realized by engaging the oncoming clutch and releasing the off-going clutch synchronously as in clutch to clutch shifts of conventional ATs. Gears are engaged or disengaged by hydraulic pistons that position the synchronizer sleeves according to the target gear as illustrated in Figure 8.7b. As mentioned previously, the next target gear is engaged already while the DCT operates in the current gear. Gear engaging or disengaging is therefore not related to the transmission shifting process and shifting quality only depends on clutch control. A hydraulic system with the pump powered by engine or electric motor provides the pressure required for the control of clutches and the gear shifting pistons.

The actuation of shifting pistons is controlled by shifting solenoids that are actually hydraulic switches routing the ATF flow paths according to the shift logic. The actuation

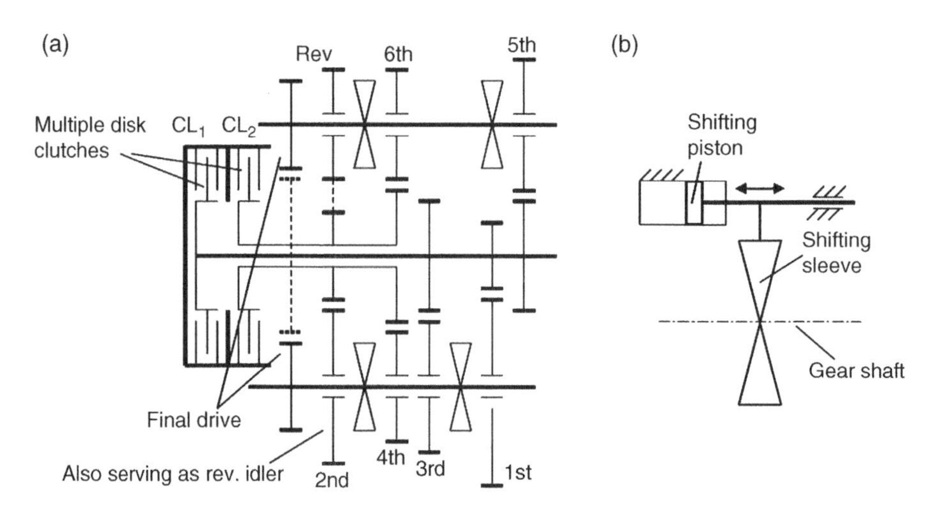

Figure 8.7 Wet DCT layout and gear shifting pistons.

of the dual clutches is realized by variable force solenoids, one for each clutch, that accurately control the hydraulic pressure in the oncoming clutch and off-going clutch, as detailed in Chapter 6.

Both dry and wet DCTs have been successfully applied for production vehicles. Dry DCTs have cost advantage over wet DCTs because of the simplicity in the control system for clutch actuation and gear shifts. Wet DCTs intrinsically have higher torque capacity because multiple disk clutches are used instead of single disk clutches. In the following sections, dry DCTs will be the focus of discussion since the shift dynamics and control of wet DCTs are in line with that of conventional ATs.

8.3 Modeling of DCT Vehicle Dynamics

The dynamic model for the transmission in Figure 8.1 just described is shown in Figure 8.8. The input and output shafts are modeled as spring-dampers. Gear shafts are modeled as lumped masses. The four synchronizers are modeled as switches that route power flows since gear engaging or disengaging does not occur during shifts and thus does not affect shift quality. As indicated in Figure 8.8, the mass moments of inertia of the lumped masses are denoted as follows: engine output assembly (I_e), clutch input side (I_1), hollow shaft (I_2), solid shaft (I_3), transfer shaft 1 (I_4), transfer shaft 2 (I_5), output shaft (I_6). In similar fashion, ω_e, ω_1, ω_2, ω_3, ω_4, ω_4, and ω_6, denote the respective angular velocities. The stiffness and damping coefficient for the input shaft and output shaft are denoted by k_i, k_o, k_i and c_i, c_o, respectively. The vehicle equivalent mass moment of inertia on the output shaft is denoted by I.

8.3.1 Equations of Motion during Launch and Shifts

Separate systems of equations are required to govern the DCT power train system dynamics because of the different power flow paths and clutch statuses. The equation of motion for vehicle launch, 1–2 upshift and 5–4 downshift are presented in the following. For other operation modes, the equations of motion can be derived according to the power flow path in a similar fashion.

Launch: In the launch operation, both the first and second gears are engaged. The clutch torque in clutch CL1 is gradually increased by its actuator until it is fully closed while clutch CL2 is open. The system of equations of motion is as follows:

$$T_e - T_{in} = I_e \cdot \dot{\omega}_e \tag{8.1}$$

$$T_{in} - T_{CL1} = I_1 \cdot \dot{\omega}_1 \tag{8.2}$$

$$T_{CL1} - \frac{T_a}{i_{a1} i_1} = I_{eq}^1 \cdot \dot{\omega}_3 \tag{8.3}$$

$$T_a - T_o = I_6 \cdot \dot{\omega}_6 \tag{8.4}$$

$$T_o - T_{Load} = I \cdot \dot{\omega}_w \tag{8.5}$$

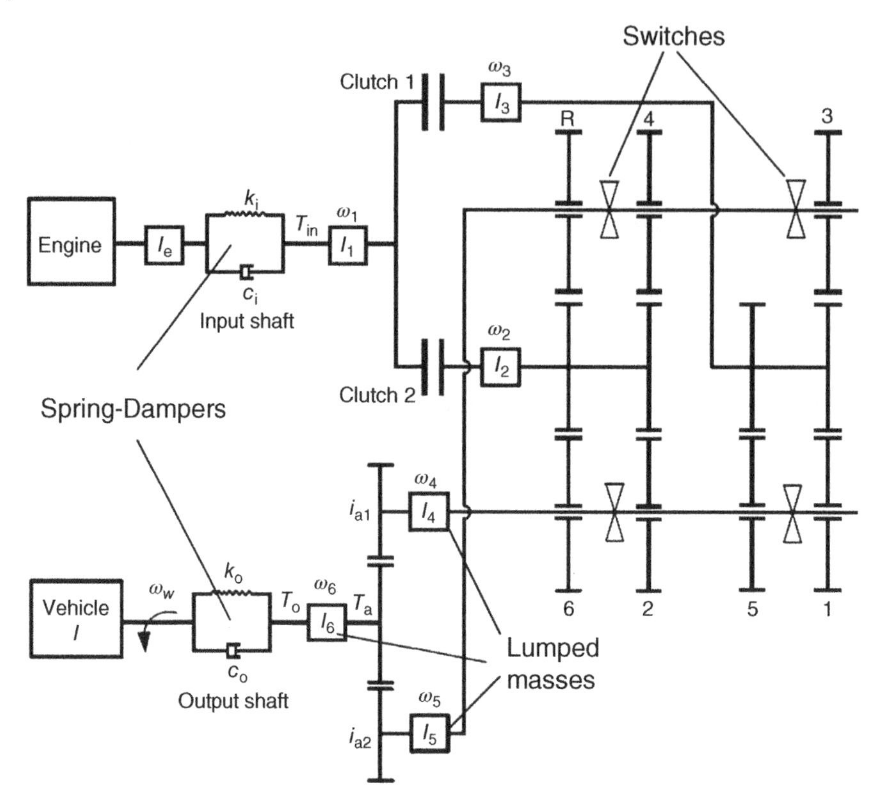

Figure 8.8 DCT dynamic model structure.

where i_1 is first gear ratio, i_{a1} is final drive ratio which is shared by first, second, fifth, and sixth gears, T_a is the final drive output torque, I_{eq}^1 is the equivalent mass moment of inertia on the solid shaft in first gear for the lumped masses, including the transfer shaft 1, assembly of the solid shaft, and all other components rotating accordingly in first gear. Equations (8.4) and (8.5) can be combined to drop out T_a so as to form a system of four state variable equations. There are four independent state variables – $\dot{\omega_e}$, $\dot{\omega_1}$, $\dot{\omega_3}$, and $\dot{\omega_w}$ – since ω_3 and ω_6 are related by $\omega_3 = i_{a1} i_1 \omega_6$. The input shaft torque T_{in}, output shaft torque T_o, and road load torque T_{Load} are expressed by the following equations:

$$T_{in} = k_i(\theta_e - \theta_1) + c_i(\omega_e - \omega_1) \tag{8.6}$$

$$T_o = k_o(\theta_6 - \theta_w) + c_o(\omega_6 - \omega_w) \tag{8.7}$$

$$T_{Load} = (f \cdot W + R_A + R_G) \cdot r \tag{8.8}$$

where ω_w and θ_w are the angular velocity and angle of rotation of the wheel, f is rolling resistance coefficient, r is tire radius, and R_A and R_G are the air and grade resistance respectively, as formulated in Chapter 1. The engine torque T_e in Eq. (8.1) is a

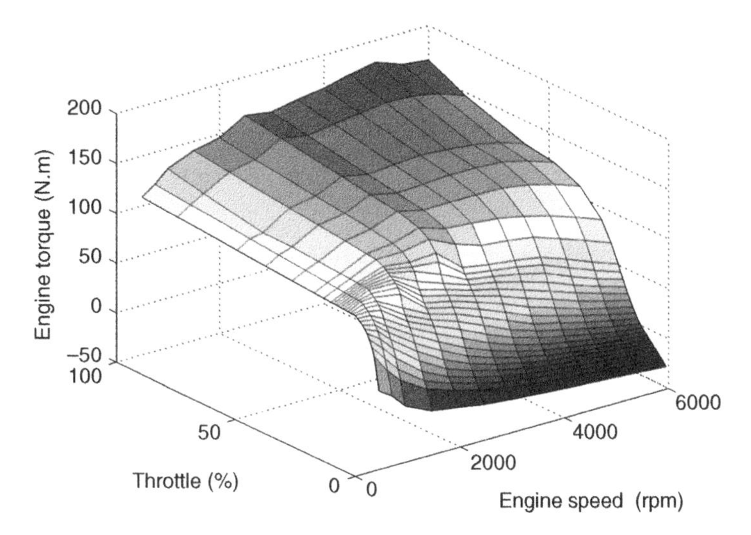

Figure 8.9 Engine torque output in terms of throttle opening and speed.

function of engine velocity ω_e and throttle angle ϕ_T, as shown in the engine map Figure 8.9.

Clutch torque T_{CL1} is a control variable during the vehicle launch process. When fully engaged, the vehicle runs in first gear and the clutch then mechanically links the engine output shaft to the transmission input. The clutch status is determined by the angular velocities of the clutch input and output sides with clutch torque determined by the following equation:

$$T_{CL1} = \begin{cases} T_{in} - I_1\dot{\omega}_1 & \text{if clutch is engaged} \\ P_1 C_1 \mu_1 & \text{if clutch slips} \end{cases} \tag{8.9}$$

where P_1 is the clamping force on clutch pressure plate which is controlled by the clutch actuator, C_1 is a constant related to clutch dimension, and μ_1 is the clutch friction coefficient. The clutch actuator for the DCT used in the example vehicle uses a screw-lever mechanism shown in Figure 8.3. The structure and kinematics of the clutch actuator design have been previously described.

The initial condition for the system of Eqs (8.1–8.5) is given as: $t = 0$; $\omega_e = \omega_e^0$, $\omega_1 = \omega_6 = \omega_w = 0$; $\theta_e = \theta_1 = \theta_6 = \theta_w = 0$. Here, ω_e^0 is the engine angular velocity at which the vehicle is launched. When the vehicle runs in first gear, angular velocities ω_1 and ω_3 are the same. Eqs (8.2) and (8.3) are then combined into one equation. The remaining equations then form the equation system for first gear operation.

Upshifts: Similar to conventional automatic transmissions, DCT shifts involve two stages: the torque phase when gear ratio remains unchanged and the inertia phase when the gear ratio is gradually changed to the value of the next target gear. Due to the difference in clutch status, two sets of equations are required to describe

the system dynamics for the two stages in a shift process. The system equations for 1–2 shift are presented in the following, which can be easily extended to other shifts.
In the torque phase we have:

$$T_e - T_{CL1} - T_{CL2} = I_e \cdot \dot{\omega}_e + I_1 \dot{\omega}_1 \tag{8.10}$$

$$T_{in} = T_{CL1} + T_{CL2} + I_1 \dot{\omega}_1 \tag{8.11}$$

$$\begin{aligned} T_{CL1} \cdot i_1 + T_{CL2} \cdot i_2 = & \left(I_2 \cdot i_2^2 + I_3 \cdot i_1^2 + I_4\right) \frac{\dot{\omega}_1}{i_1} \\ & + {}^{T_{Load}}\!/_{i_{a1}} + \frac{I_6}{i_1 \cdot i_{a1}^2} \dot{\omega}_1 + \frac{I}{i_{a1}} \dot{\omega}_w \end{aligned} \tag{8.12}$$

$$T_o = I \dot{\omega}_w + T_{Load} \tag{8.13}$$

where ω_6 is equal to $\omega_1/(i_1 i_a)$ since the first gear ratio has not changed yet. The state variable equation system is formed by the four equations Eqs (10–13). The variables that need to be solved from the equation system above are: $\dot{\omega}_e$, $\dot{\omega}_1$, $\dot{\omega}_w$, and T_{CL1}. Clutch torque T_{CL2} is the control variable that is gradually ramped up following a pre-determined profile as shown in Figure 8.13. Equations (8.6–8.8) have given us expressions for T_{in}, T_o, and T_{Load}. The initial condition for the torque phase of 1–2 shift corresponds to the status of the vehicle operation in first gear when the shift starts.

As soon as clutch CL1 starts to slip, the torque phase finishes and the inertia phase starts. In real time control, the threshold point can be detected by speed sensors that measure the angular velocities of the transmission input and output. In simulation, this point is determined by comparing the clutch torque T_{CL1} solved from the system of Eqs (8.10–8.13) and the clutch torque capacity T_{CL1}^c which depends on clutch design and control parameters. As soon as the inertia phase starts, the system dynamics is then governed by another set of equations presented as follows.

During the inertia phase, the angular velocities ω_1 and ω_6 are no longer related by the gear ratios i_{a1} and i_1. Angular velocities $\dot{\omega}_e$, $\dot{\omega}_1$, $\dot{\omega}_6$, and $\dot{\omega}_w$ are independent variables to be solved from the equation system governing the system dynamics. Both clutch torques, T_{CL1} and T_{CL2}, are control variables during the inertia phase as shown in Figure 8.13. The system equations of motion for the inertia phase are expressed as follows:

$$T_e - T_{CL1} - T_{CL2} = I_e \cdot \dot{\omega}_e + I_1 \cdot \dot{\omega}_1 \tag{8.14}$$

$$T_{in} = T_{CL1} + T_{CL2} + I_1 \dot{\omega}_1 \tag{8.15}$$

$$\begin{aligned} T_{CL1} \cdot i_1 + T_{CL2} \cdot i_2 = & \left(I_2 \cdot i_2^2 + I_3 \cdot i_1^2 + I_4\right) i_{a1} \dot{\omega}_6 \\ & + {}^{T_o}\!/_{i_{a1}} + I_6 {}^{\dot{\omega}_6}\!/_{i_{a1}} \end{aligned} \tag{8.16}$$

$$T_o = I \dot{\omega}_w + T_{Load} \tag{8.17}$$

The initial condition for the solution to the system of Eqs (8.14–8.17) corresponds to the system status at the end of the torque phase. The inertia phase ends when ω_1 is equal to $i_{a1} i_2 \omega_6$. This completes the process of 1–2 shift and the vehicle enters operation in second gear.

Downshifts: In a power-on downshift, the torque of the off-going clutch must be decreased rapidly to slip the clutch before applying pressure in the oncoming clutch in order to avoid clutch tie-up and backward power circulation. Therefore, the inertia phase must come before the torque phase in a power-on down shift. A 5–4 shift is used as example for the system of equations for down shifts. This example can be extended to all other downshifts.

First, in the inertia phase, to start the 5–4 power-on downshift, clutch torque T_{CL1} must be decreased rapidly to produce slippage in clutch CL1 before clutch CL2 is pressurized. The slip threshold of clutch CL1 is reached when $T_e \geq T_{CL1}^C$, where T_{CL1}^C is the torque capacity of clutch CL1 which is controlled to decrease rapidly by the control motor as shown in Figure 8.3. The inertia phase starts as soon as the off-going clutch CL1 starts slipping. During the inertia phase of a 5–4 power-on downshift, both the off-going clutch CL1 and the oncoming clutch CL2 slip, with the oncoming clutch torque T_{CL2} ramped up rapidly and the off-going clutch torque T_{CL1} ramping down even more rapidly. Both clutch torques, T_{CL1} and T_{CL2}, are control variables in the inertia phase. The vehicle powertrain dynamics is governed by the following:

$$T_e - T_{CL1} - T_{CL2} = I_e \cdot \dot{\omega}_e + I_1 \cdot \dot{\omega}_1 \tag{8.18}$$

$$T_{in} = T_{CL1} + T_{CL2} + I_1 \dot{\omega}_1 \tag{8.19}$$

$$T_{CL1} i_5 i_{a1} + T_{CL2} i_4 i_{a2} = T_o + I_6 \dot{\omega}_6 + \left(I_4 + I_3 i_5^2\right) i_{a1}^2 \dot{\omega}_6 + \left(I_5 + I_2 i_4^2\right) i_{a2}^2 \dot{\omega}_6 \tag{8.20}$$

$$T_o = I \dot{\omega}_w + T_{Load} \tag{8.21}$$

Since both clutches slip, the angular accelerations $\dot{\omega}_4$ and $\dot{\omega}_5$ are not related to $\dot{\omega}_1$ by the ratios of the fourth and fifth gears. Therefore, the equation of motion for the gear shafts is written on the transmission output shaft whose mass moment of inertia is labeled as I_6. In the system of state variable equations above, $\dot{\omega}_e$, $\dot{\omega}_1$, $\dot{\omega}_6$, and $\dot{\omega}_w$ are independent state variables to be solved. During the downshift inertia phase, both the oncoming and off-going clutches slip. Both clutch torques T_{CL1} and T_{CL2} are control variables. T_{CL1} is ramped down rapidly and T_{CL2} is ramped up rapidly by the controller as shown in Figure 8.14. The initial condition for the equation system corresponds to the system dynamic status in fifth gear operation when the 5–4 downshift starts. The inertia phase ends when $\omega_1 \geq \omega_2 = i_{a2} i_4 \omega_4$. For control purpose, ω_1 is set to be equal to 1.01 ω_2, allowing a small amount of engine flare.

Torque phase: During the torque phase of 5–4 down shift, the oncoming clutch CL2 is applied already, but the off-going clutch CL1 is still slipping under pressure and transferring friction torque to the transmission input. Both final drives are involved in the power flow. The equations of motion are presented as follows for this operation status:

$$T_e - T_{CL1} - T_{CL2} = I_e \cdot \dot{\omega}_e + I_1 \cdot \dot{\omega}_1 \tag{8.22}$$

$$T_{in} = T_{CL1} + T_{CL2} + I_1 \dot{\omega}_1 \tag{8.23}$$

$$T_{CL1} i_5 i_{a1} + T_{CL2} i_4 i_{a2} = T_o + I_6 \dot{\omega}_1 / (8.i_{a1} i_4) + \left(I_4 + I_3 i_5^2\right) i_{a1}^2 \dot{\omega}_1 / (i_{a1} i_4) + \left(I_5 + I_2 i_4^2\right) i_{a2}^2 \dot{\omega}_1 / (i_{a1} i_4) \tag{8.24}$$

$$T_o = I \dot{\omega}_w + T_{Load} \tag{8.25}$$

where, i_{a1} is the final drive ratio for the first, second, fifth, and sixth gears, i_{a2} is the final drive ratio for the third, and fourth gears, $\dot{\omega}_6$ has been replaced by $\dot{\omega}_1 / (i_{a1} i_4)$ in

Eq. (8.24) above. There are four state variables – $\dot{\omega}_e$, $\dot{\omega}_1$, $\dot{\omega}_w$, and the oncoming clutch torque T_{CL2}. It is noted that the off-going clutch torque T_{CL1} is a control variable and the oncoming clutch torque T_{CL2} is now a state variable. The initial condition for the system above corresponds to the system status at the end of the inertia phase. For power-on 5–4 downshift, off-going clutch torque T_{CL1} must be brought to zero as soon as clutch CL1 starts to slip. The 5–4 down shift is completed when ω_1 is equal to ω_2 and the off-going clutch torque T_{CL1} becomes equal to zero.

In summary, the system of equations for all other fixed ratio operations and shifts can be derived similarly according to the respective gear mesh paths. For the launching operation, T_{CL1} is the control variable. For power-on upshifts from odd gears, oncoming clutch torque T_{CL2} is a control variable in both the torque phase and the inertia phase, and off-going clutch torque T_{CL1} is a state variable determined by system dynamics in the torque phase and a control variable in the inertia phase. For power-on upshifts from even gears, the roles of clutch CL1 and clutch CL2 are just reversed, and so are the two clutch torques. For power-on downshifts from even gears, the off-going clutch torque T_{CL2} is a control variable in both the inertia phase and the torque phase, and oncoming clutch torque T_{CL1} is a control variable in the inertia phase and a state variable in the torque phase. The roles of the two clutches can be defined similarly for power-on downshifts from odd gears.

Note that the state variable equation systems for the various DCT operations statuses represented above are derived for the dynamic model structure shown in Figure 8.8. For DCTs that are equipped with dual mass flywheels, the system dynamics can be modeled in a structure shown in Figure 8.10, which is a minor modification of the model shown in Figure 8.8. The system equations for dual mass DCTs can be derived for all operation modes following similar steps.

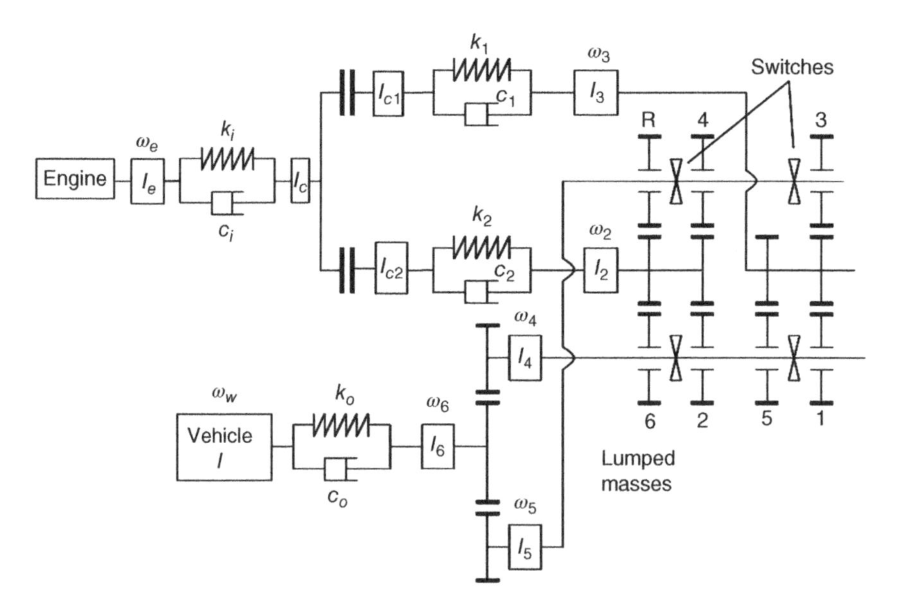

Figure 8.10 Dynamic model structure for DCTs with dual mass flywheels.

The dynamic model presented in this section was applied for a test vehicle with data shown in the following table. The comparison between model simulation and testing is presented in the next section for this test vehicle.

Example vehicle data	
Vehicle mass (m)	1400 kg
Transmission gear ratios (i)	$i_1 = 3.917$ $i_2 = 2.429$ $i_3 = 1.436$ $i_4 = 1.021$ $i_5 = 0.848$ $i_6 = 0.667$ $i_r = 3.292$
Final drive gear ratio (ia)	$i_{a1} = 3.762$ $i_{a2} = 4.158$
Tire radius (r)	0.2975 m
Air drag coefficient (C_d)	0.328
Frontal area (A)	2.12 m^2

8.4 DCT Clutch Control

Similar to conventional ATs, DCT control concerns with shift schedule control and shift process control. The DCT shift schedule balances vehicle performance, fuel economy, and drivability. The establishment of a DCT shift schedule is similar to that for conventional ATs discussed in Chapter 6. Shift decisions are made by the transmission control unit (TCU) primarily based on signals from the speed sensor and throttle sensor. Other sensors, such as ATF temperature sensor and brake pedal sensor, provide supplemental information for the TCU to make adjustments in shift decision making. The shift schedule for the example DCT is shown in Figure 8.11.

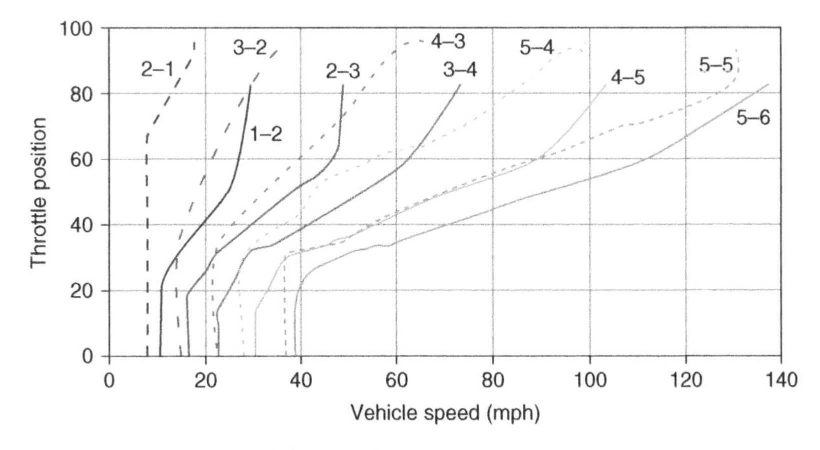

Figure 8.11 Typical DCT shift schedule.

Unlike vehicles equipped with conventional automatic transmissions, which use a one-way clutch to avoid backward torque transfer and a torque converter to dampen dynamic impacts, dual clutch transmission vehicles rely only on clutch torque control for responsiveness and smoothness during launch and shifts. It is critical to control the clutch torque magnitude and the timing of actuation for both the oncoming and off-going clutches. In addition, engine torque control during shifts must be implemented to interact with clutch torque control. The main objective of clutch control is to realize power-on shifts without torque interruption and engine flare so as to achieve optimized drivability and comfort similar to that of conventional ATs.

Launch; Launch control aims at acceleration responsiveness and smoothness. Excessive torque in clutch CL1 at the beginning of launch causes vehicle jerk, resulting in substandard passenger feel. On the other side, inadequate clutch torque is not enough to launch the vehicle quickly and the vehicle will be perceived as lacking in power. The launch clutch torque profile in normal road traffic mainly depends on the acceleration pedal depression which is controlled by the electronic throttle control (ETC). In such operations, the engine torque increases with respect to its RPM and is known by interpolating the engine map. The clutch torque T_{CL1} is quickly increased by the control motor to catch up the engine torque. The transmission input shaft RPM traces the engine RPM while the vehicle is being launched. The launching process is considered to be completed when the transmission input speed is just a few RPMs below that of the engine. The vehicle then operates in first gear afterward. Under stop and go traffic conditions, DCT clutch control can be technically challenging because the DCT vehicle must possess the "creeping" capability as a requirement for ATs. Excessive clutch slippage in traffic jams leads to overheating in the dual clutch module. This issue can be addressed by using heat resistant materials for clutch design, and in real time traffic, the DCT controller may temporally cancel "creeping" and activate the ABS system to hold the vehicle in position in uphill road conditions.

Example torque and RPM profiles of launch clutch CL1 are illustrated in Figure 8.12. Experimental data and simulation results by the dynamic modeling described in the previous section are compared in Figure 8.12. In general, a high level of agreement between simulation and test is observed for engine and transmission input speeds. Some discrepancy on the engine RPM is observed at the beginning of the launch due to the engine RPM sensitivity and modeling accuracy. Note that the measured output torque is obtained by a torque sensor installed on the half shaft.

Upshifts: Clutch tie-up and engine flare are the two key issues for DCT upshift control similar to clutch to clutch shift in conventional ATs. For example, if clutch CL1 is released too late in a 1–2 upshift, then the two clutches will be tied up, yielding backward power recirculation. On the other hand, if the clutch CL1 is released too early, the engine speed will flare up in the absence of vehicle load. In order to make sure that the engine torque is transferred to the driving wheels smoothly and continuously, the torque of dual clutches must be controlled with high accuracy in magnitude and timing.

Different control strategies are used for clutch torque control during torque phase and inertia phase. Without loss of generality, a 1–2 upshift is used as an example. During the torque phase, the pressure in clutch CL1 is gradually reduced to decrease the torque capacity (T_{CL1}^{C}) until clutch CL1 starts to slip as shown in Figure 8.13. The threshold of slippage is reached when the off-going clutch torque capacity T_{CL1}^{C} is equal to the engine torque. Meanwhile the oncoming clutch CL2 is gradually applied as soon as

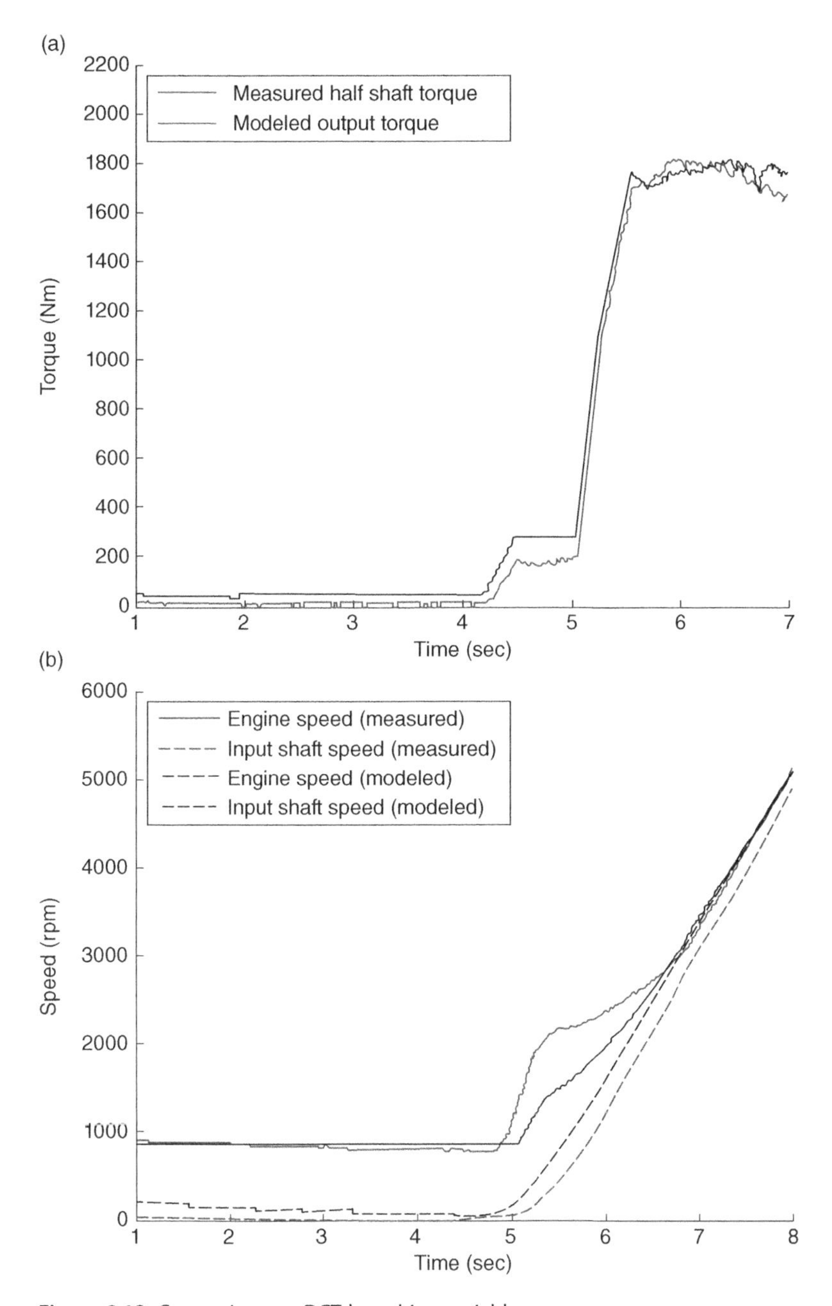

Figure 8.12 Comparison on DCT launching variables.

the upshift decision is made by the DCT controller. This will not cause any backward torque transfer since the input side of clutch CL2 turns faster than its output side. This feed-forward control is based on calibration data and is adopted until the end of the torque phase when clutch CL1 starts to slip. As the shift enters the inertia phase, the torque

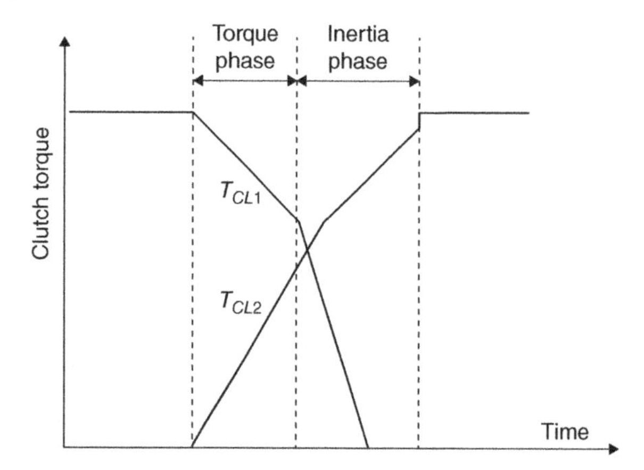

Figure 8.13 Typical clutch torque profiles during upshift.

in the off-going clutch CL1 is quickly reduced to zero as shown in Figure 8.13. Meanwhile, the gear ratio starts to change and the torque in the oncoming clutch CL2 is ramped up with feedback on the difference between a pre-designed gear ratio change function and the actual speed ratio, as formulated in the following:

$$\begin{cases} \Delta i = i_{designed} - i_{actual} \\ T_{CL2} = K_p \Delta i + K_i \int \Delta i \cdot dt + K_d \dfrac{d\Delta i}{dt} \\ 0 \le T_{CL2} \le T_{CL2}^{Max} \\ i_2 \le i_{designed} \le i_1 \end{cases} \tag{8.26}$$

where $i_{designed}$ is a time function that bridges the difference between i_1 and i_2, i_{actual} is calculated as $\frac{\omega_1}{\omega_2}$. K_p, and K_i and K_d, are the PID gains. T_{CL2} is the torque in clutch CL2 limited by the torque capacity T_{CL2}^{Max}. Note that the clutch torque control will be initially implemented by the PID controller, and the real time clutch torque profiles will be finalized through the calibration of test vehicles.

Downshifts: Downshift to give acceleration at high vehicle speed is to increase transmission output torque. Quick response is the top priority in this case since the driver anticipates the shift occurring. Unlike upshifts, a downshift occurs rapidly with the inertia phase coming first and torque phase after. For example, to realize a 5–4 down shift, the torque in clutch CL1 must be reduced quickly to the point of clutch slip before clutch CL2 is applied. The engine speed flares above the speed of the oncoming clutch output side as soon as clutch CL1 starts to slip. The inertia phase starts as soon as clutch CL1 slips and the transmission ratio begins to change from the lower value of the high gear toward the higher value of the target low gear. Both clutches slip during the inertia phase until the low gear ratio is achieved. The shift then enters the torque phase when there may still be some residual torque in the off-going clutch CL1. The downshift is completed when the residual torque in CL1 is brought to zero quickly in the torque phase. The oncoming clutch torque T_{CL2} is then further increased to securely engage the target gear. The torque profiles in the 5–4 power-on downshift example are shown in

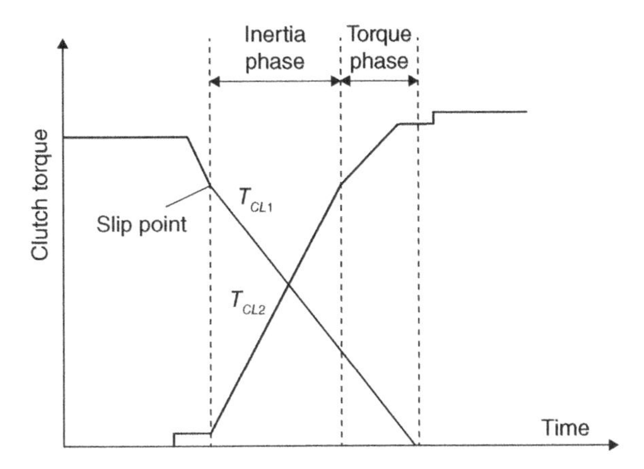

Figure 8.14 Typical clutch torque profiles during downshifts.

Figure 8.14, which are typical for other power-on downshifts. Note here that it is technically possible that the off-going clutch torque can be brought down to zero in the inertia phase of a downshift before the target low gear ratio is achieved. If this happens, there will no torque phase as in typical shifts.

Engine torque control during shifts: As discussed previously in Chapters 5 and 6, the reduction of transmission input torque, which is the same as the engine torque for DCT vehicles because there is no torque converter, lowers the oncoming clutch torque required to complete the shift and also lowers the transmission output torque overshoot that may occur in the inertia phase. The effects of engine torque reduction for DCT upshifts and downshifts are analysed in the following.

During DCT upshifts, the engine speed is higher than that of the input shaft for the target gear, as illustrated for the 1–2 upshift in Figure 8.15. Engine torque reduction must not occur until the end of the torque phase, otherwise a deep torque hole would result, creating power interruption as felt by the driver. As the shift enters the inertia phase when clutch CL1 starts to slip, the engine speed must be reduced quickly to be equal to the speed of the second gear input shaft, i.e. the output side of clutch CL2. This is only possible if the engine torque is smaller than T_{CL2} during most of the inertia phase, as shown in Figure 8.15. Because the time delay in engine torque reduction by spark retarding, the engine torque will still be above the oncoming clutch torque T_{CL2} soon after the turning point from the torque phase to the inertia phase. Engine deceleration occurs as soon as the oncoming clutch torque T_{CL2} is higher than the engine torque and continues toward the end of the upshift. The timing and duration of engine torque reduction by spark retarding can be precisely calibrated for various shifts to optimize shift response and smoothness. The optimized result is obtained if the recovered engine torque is just equal to the oncoming clutch torque T_{CL2} at the time when the engine speed is equal to the speed of the second gear input shaft. The oncoming clutch torque T_{CL2} is then further increased by a step to secure the clutch engagement. To achieve the seamless coupling of the engine and the second gear input shaft by the oncoming clutch CL2, it is crucial to establish an accurate correlation between the oncoming clutch torque T_{CL2} and the clutch control variable, as detailed in the next section.

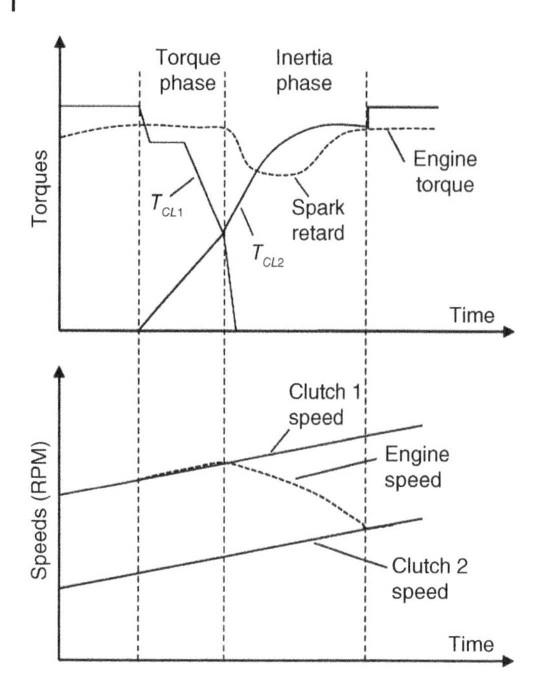

Figure 8.15 Engine torque reduction by spark retarding for DCT upshifts.

The torque profiles, the engine speed, and the speeds of the two input shafts are illustrated in Figure 8.16, for a 1–2 shift of the example DCT vehicle in Section 8.3. The plot at the top shows the profiles of the off-going clutch torque T_{CL1} and the oncoming clutch torque T_{CL2}. These profiles follow the pattern shown in Figure 8.13 and are implemented in the test vehicle. The simulation results are obtained by the model formulations presented previously in Section 8.3. It can be observed that decent agreements on the engine speed and the speed of the two input shafts are obtained between model simulation and measurements on the test vehicle. Discrepancy exists because of inaccuracy in clutch torque calculations and in the mass moments of inertia of DCT components.

During DCT downshifts, the engine speed is lower than that of the input shaft for the target gear, as illustrated for the 5–4 downshift in Figure 8.17. When the DCT TCU commands the 5–4 downshift, the torque capacity of the off-going clutch is decreased quickly to the point of slippage, at which the engine torque is equal to the off-going clutch torque T_{CL1}. The oncoming clutch torque T_{CL2} is immediately ramped up as soon as the off-going clutch CL1 starts slipping, as shown in Figure 8.17. During the inertia phase, the engine torque is higher than the sum of the off-going clutch torque and the oncoming clutch torque, and the engine speed is therefore quickly accelerated to catch up the speed of the input shaft of the target gear. At the end of the inertia phase, the engine speed increases to be above the speed of the target gear input shaft, while there may still be some residual torque in the off-going clutch CL1. The 5–4 downshift shift now enters the torque phase, with the engine speed slightly higher than the speed of the output side of clutch CL2. The engine torque is then reduced in the torque phase by spark retarding so that the oncoming clutch torque T_{CL2} becomes higher than the engine torque. The engine speed is then decelerated to be equal to the speed of the fifth gear input gear. Through accurate calibration, it is possible to reach the point in the torque phase at

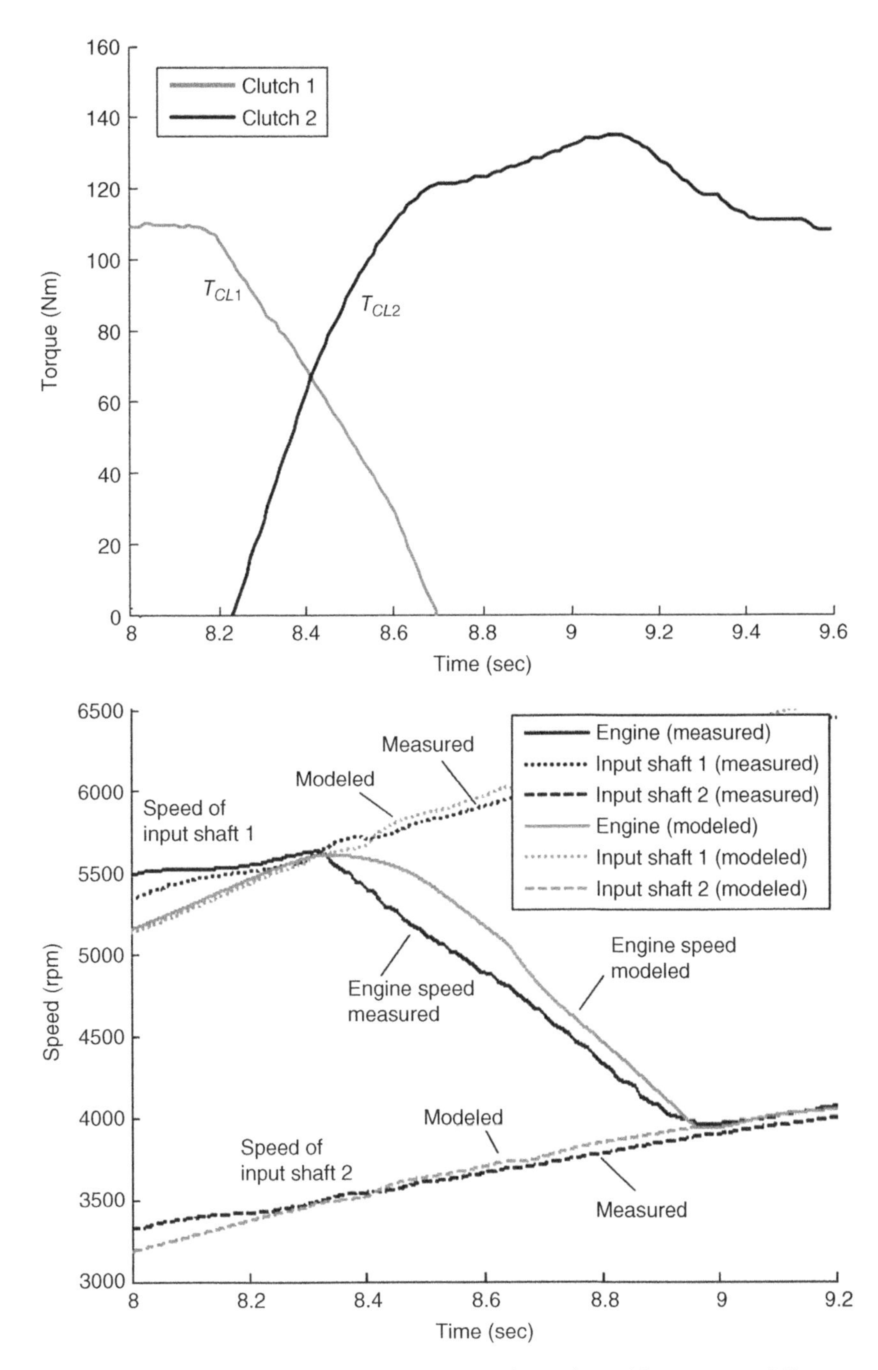

Figure 8.16 Simulation and test results torque and speed variables in 1–2 upshift.

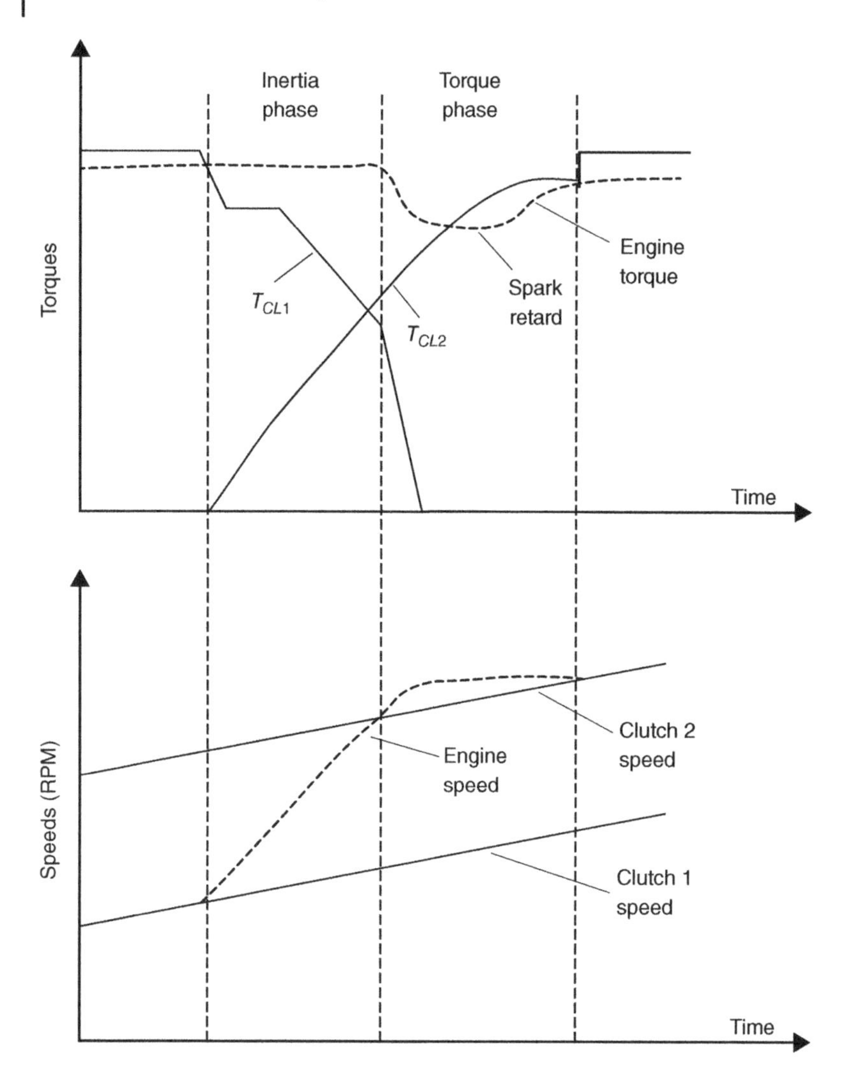

Figure 8.17 Engine torque reduction by spark retard for DCT downshifts.

which the engine speed is equal to the target gear input shaft speed, and the engine torque is equal to the oncoming clutch torque. For secure clutch engagement, the oncoming clutch is further increased by a step after the 5–4 downshift is completed. Note that the torque phase in power-on downshifts may not be clearly defined because the off-going clutch torque may be ramped down to zero even before the target gear ratio is achieved. If this happens, the downshift is completed very quickly and there could be some shift harshness because the engine and the input shaft of the target gear cannot be coupled seamlessly. However, power-on downshifts are usually triggered by the drivers for high power and is anticipated by the driver. Acceleration is the priority here and some amount of shift harshness should not cause too much an unpleasant discomfort.

The torque profiles, the engine speed, and the speeds of the two input shafts are illustrated in Figure 8.18, for a 5–4 power-on downshift for the example DCT vehicle.

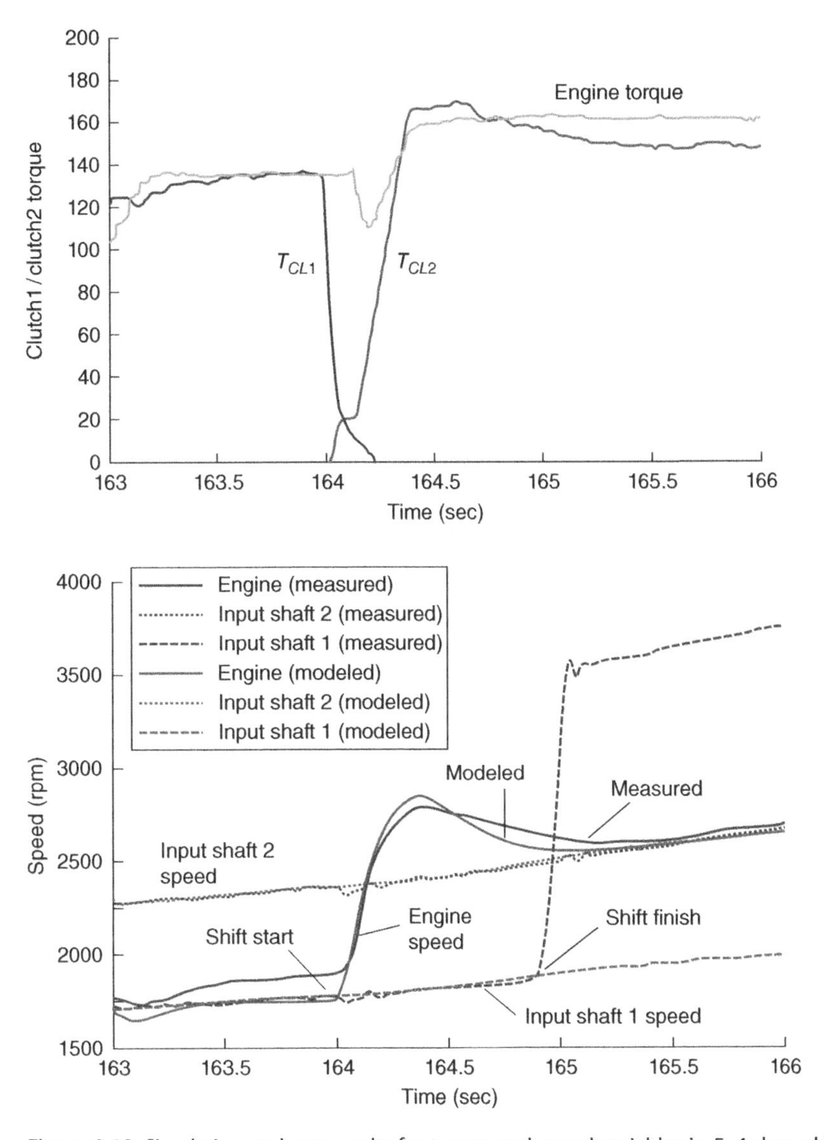

Figure 8.18 Simulation and test results for torque and speed variables in 5–4 downshift.

The plot at the top shows the profiles of the off-going clutch torque T_{CL1} and the oncoming clutch torque T_{CL2}, as well as the engine torque with spark retarding during the torque phase. It can be observed from Figure 8.18 that the engine speed is increased to be above the speed of the target gear input shaft (input shaft 2) during the inertia phase. This is because the off-going clutch CL1 is slipping with T_{CL1} being sharply reduced and the oncoming clutch CL2 has just being pressurized. Toward the end of the 5–4 downshift, the oncoming clutch torque T_{CL2} is ramped up to be higher than the reduced engine torque, decelerating the engine speed to be equal to the speed of the target gear input shaft. The engine is then coupled with the fourth gear input shaft by the oncoming clutch

CL2. For the 5–4 power-on downshift, it can also be observed from Figure 8.18 that good agreements on the engine speed and the speed of the two input shafts are obtained between model simulation and measurements on the test vehicle.

8.5 Clutch Torque Formulation

As mentioned previously, a DCT differs from a conventional AT in that the latter has a torque converter between the engine output and transmission input, despite the similarity in clutch-to-clutch shift characteristics. The presence of the torque converter cushions the powertrain dynamic transients and is therefore conducive to smoothness during vehicle launch and shifts. Without the cushion effect of the torque converter, clutch torque control requires higher precision to achieve launch and shift quality comparable to ATs. This requires the accurate correlation between the real time clutch torque and the clutch control variable. The contents of this section concentrate on the clutch torque formulation and calibration for dry dual clutch transmissions, using the Ford six-speed DCT shown in Figure 8.1 as the case study example. Firstly, the theoretical or nominal torque capacity of each of the dual clutches with given design parameters is correlated to the clutch control variable, which is the angle of rotation of the control motor as shown in Figures 8.2 and 8.3. The nominal clutch torque is formulated in terms of the clutch design parameters based on the assumption that the friction power is constant over the friction disk face, as detailed in Chapter 2. Secondly, an algorithm based on powertrain dynamics is established for the calculation of clutch torque in the launching clutch during vehicle launch and in both clutches during shifts. This algorithm uses wheel speed sensor data as the input and is capable of accurately calculating the clutch torque while both clutches are slipping on a real time basis. Thirdly, experimental data will be presented to validate the clutch torque formulation mentioned above.

8.5.1 Correlation on Clutch Torque and Control Variable

In Section 2.4, the torque capacity of disk type clutches was formulated by Eq. (2.8) based on the assumption of uniform disk face wear in terms of the clutch design parameters, disk friction coefficient, and clutch clamping force. The assumption of uniform wear is equivalent to even distribution of friction power over the friction disk face. Typically, each of the dual clutches in dry dual clutch transmissions has a single friction disk and the nominal clutch torque capacity is expressed by the following equation:

$$T_{CL} = fF\frac{D+d}{2} \tag{8.27}$$

where D and d are the friction disk outer and inner diameters, which measure the radial dimension of the dual clutch module, f is the friction coefficient between the friction disk and the pressure plate, and F is the clamping force on the friction disk. Eq. (8.27) applies to both clutch CL1 and CL2, and for a given dual clutch module design, the clutch torque depends on the clamping force F and the friction coefficient f, which varies with clutch temperature and clutch slip rate.

The clamping force F on the friction disk is related to the force applied on the release bearing through the diaphragm spring, which functions as the pressure plate lever, as shown in Figure 8.3. However, due to the deflection of the diaphragm spring and the existence of backlash in the clutch actuation mechanism, there are nonlinear characteristics between the clutch torque and the actuator control parameter. To account for this nonlinearity, tests need to be performed to measure the force applied to the release bearing (i.e. the engagement load) versus the release bearing travel. Based on test data, the engagement load is correlated to the release bearing travel, also termed engagement travel, as shown in Figure 8.19 for the six-speed DCT shown in Figure 8.1. Each of the two curves in Figure 8.19 corresponds to the respective release bearing. It is observed from Figure 8.19 that the two curves correlating engagement load and engagement travel differ from each other despite their similarity in shape. It is important to note from Figure 8.19 that there is an engagement load for each clutch even when the release bearing travel or engagement travel is zero. This is because that the diaphragm spring is fairly stiff and it takes a fairly large force to create a measureable deflection along the input shaft. In general, each of the two curves can be divided into two linear segments as can be observed from Figure 8.19, indicating typical characteristics of diaphragm spring stiffness.

As shown in Figure 8.19, there are substantial forces (denoted as F_0) on the release bearing of both clutches when the engagement travels are zero due to high rigidity of the diaphragm springs that also serve as for the pressure plate levers as shown in Figure 8.3. Because of this, two separate functions must be used to correlate the release bearing force F_b with the roller displacement. As shown in Figure 8.20a, the release bearing force F_b and the spring force F_s are related as follows before F_b becomes equal to F_0,

$$F_b = \frac{x_{roller}}{L - x_{roller}} F_s \tag{8.28}$$

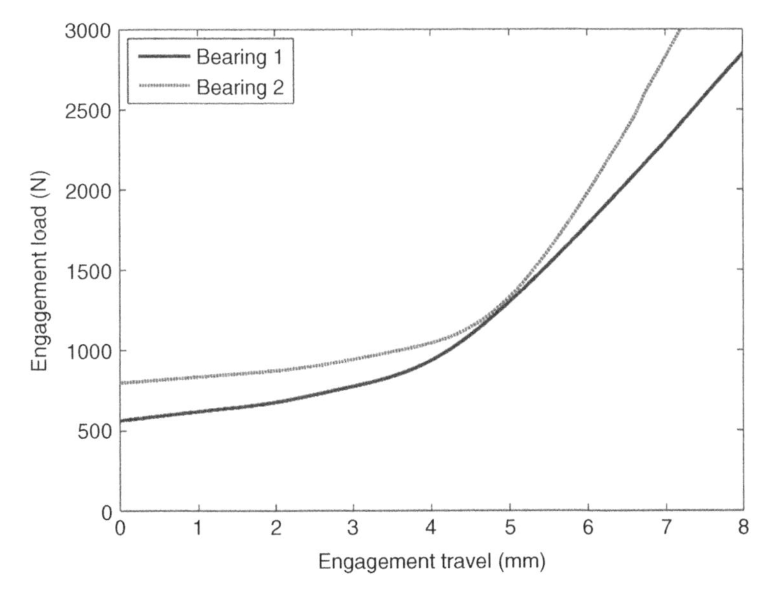

Figure 8.19 Relationship between release bearing travel and engagement load.

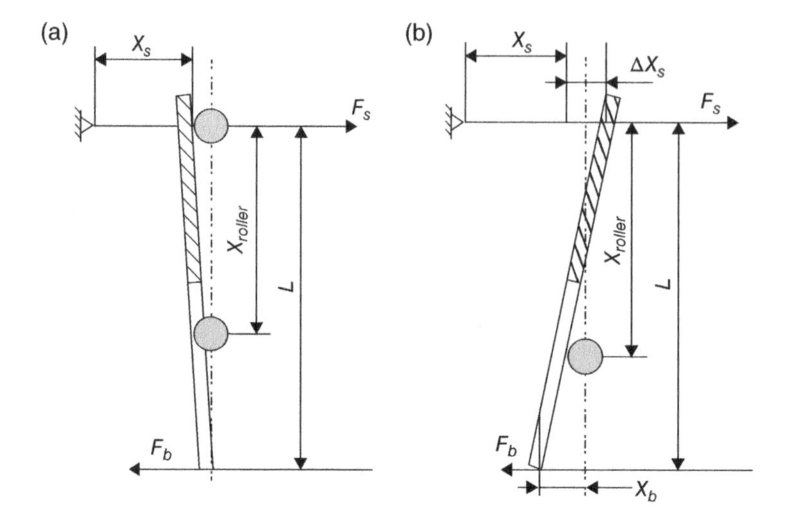

Figure 8.20 Engagement load before and after bearing travel.

where x_{roller} is the roller travel on the screw as shown in Figures 8.3 and 8.20, L is the total effective length of actuator lever, and F_s is the spring force with an initial value of F_{s0} that is caused by spring compression at assembly. The spring displacement is very small when $F_b \leq F_0$ since the release bearing displacement is near zero and the spring force remains almost the same as the initial value, i.e. $F_s = F_{s0}$ if $F_b \leq F_0$. At the threshold when $F_b = F_0$, the displacement of roller x_p can be solved from Eq. (8.28) as:

$$x_p = \frac{F_0}{F_{s0} + F_0} L \tag{8.29}$$

Therefore, when $x_{roller} \leq x_p$, the release bearing force is represented in terms of roller displacement by Eq. (8.28). After the release bearing begins to travel, a separate function is required to correlate the release bearing force or the engagement load and the roller displacement since the spring compression is affected by the bearing travel. The release bearing travel and its effect on the spring compression is shown in Figure 8.20b, and the extra amount of spring compression caused by the bearing travel is determined as:

$$\Delta x_s = \frac{x_b}{L - x_{roller}} x_{roller} \tag{8.30}$$

where Δx_s is the increment of spring deflection and x_b is the engagement travel of the release bearing. Due to this increment, the spring force with bearing travel x_b is expressed as:

$$F_s = F_{s0} - k \frac{x_{roller}}{L - x_{roller}} x_b \tag{8.31}$$

where k is the spring stiffness. The equilibrium of the actuator lever requires that the sum of moments about the contact point between the roller and the actuator lever is zero, leading to the following relation between the engagement load F_b and the spring force F_s:

$$F_s x_{roller} = F_b(L - x_{roller}) \tag{8.32}$$

Combining Eqs (8.28–32), the release bearing force or the engagement force F_b can be represented in terms of the roller displacement respectively for the two clutches as:

$$F_b = \begin{cases} F_b = \dfrac{x_{roller}}{L - x_{roller}} F_{s0} & x_{roller} \le x_p = x_p = \dfrac{F_0}{F_{s0} + F_0} L \\ \left(F_{s0} - k \dfrac{x_{roller}}{L - x_{roller}} x_b\right) \dfrac{x_{roller}}{L - x_{roller}} & x_{roller} > x_p = x_p = \dfrac{F_0}{F_{s0} + F_0} L \end{cases} \tag{8.33}$$

where the release bearing travel x_b is related to the engagement F_b by the respective curve in Figure 8.18.

8.5.2 Case Study on Clutch Torque and Control Variable Correlation

As indicated in Eq. (8.27), the clutch torque is a function of the clamping force on the pressure plate, the friction coefficient, and the clutch dimensions. The main design parameters of the two clutches used in the example DCT are:

Main clutch parameters

Parameters	Clutch 1	Clutch 2
Clutch outer diameter	$D_1 = 232.5$ mm	$D_2 = 225$ mm
Clutch inner diameter	$d_1 = 157$ mm	$d_2 = 157$ mm
Diaphragm lever ratio	$i_{ratio1} = 3.6$	$i_{ratio2} = 4.2$
Friction coefficient	$f_1 = 0.35$	$f_2 = 0.35$

According to Eq. (8.27), the nominal clutch torque in clutch 1 and clutch 2 can be calculated respectively as:

$$\begin{cases} T_{CL1} = \dfrac{f_1 (F_{b1} i_{ratio1})(D_1 + d_1)}{2(1000)} = 0.2454 F_{b1} \quad (\text{Nm}) \\ T_{CL2} = \dfrac{f_2 (F_{b2} i_{ratio2})(D_2 + d_2)}{2(1000)} = 0.2808 F_{b2} \quad (\text{Nm}) \end{cases} \tag{8.34}$$

where F_{b1} and F_{b1} are the engagement forces for clutch 1 and clutch 2 respectively. The spring constants are selected to be 150 N/mm for both actuators and the length of the actuator lever is $L = 100$ mm. The roller displacements at which release bearings begin to move are $x_{p1} = 25$ mm and $x_{p2} = 30$ mm. The initial spring forces are determined by Eq. (8.29) as $F_{s1} = 1689$ N and $F_{s1} = 1860$ N.

Using Eqs (8.33) and (8.34), the clutch torque and the roller displacement before the release bearing starts to travel is related respectively by the following equations,

$$\begin{cases} T_{CL1} = 0.2454 F_{b1} = (0.2454) \dfrac{x_{roller1}}{L - x_{roller1}} F_{s1} = \dfrac{414.48 x_{roller1}}{100 - x_{roller1}} & x_{roller1} \le 25 \\ T_{CL2} = 0.2808 F_{b2} = (0.2808) \dfrac{x_{roller2}}{L - x_{roller2}} F_{s2} = \dfrac{522.29 x_{roller2}}{100 - x_{roller2}} & x_{roller2} \le 30 \end{cases} \tag{8.35}$$

After the bearings start to move, the relationship between engagement travel x_b and the engagement load F_b can be obtained from Figure 8.19, which means that F_b is a function of x_b, i.e. $F_b = f(x_b)$. When the engagement travel is smaller than 4 mm, it is accurate enough to fit the function $f(x_b)$ by the following linear function as:

$$F_{b1} = 99.5\,x_{b1} + 563 \qquad x_{b1} \le 4\,\text{mm} \tag{8.36}$$

Solving x_{b1} from the equation above in terms of F_{b1} and plugging it into the second equation in Eq. (8.33), one can further represent the engagement load F_{b1} in terms of the roller displacement $x_{roller1}$ and finally express the torque of clutch CL1 using Eq. (8.34) in the following form:

$$T_{CL1} = \frac{41241\beta_1 + 20724\,\beta_1^2}{99.5 + 150\,\beta_1^2} \qquad x_{roller1} > 25 \tag{8.37}$$

where, $\beta_1 = \dfrac{x_{roller1}}{L - x_{roller1}}$. Similarly, the clutch torque in clutch CL2 can be expressed as a function $x_{roller2}$ by the following equation, with $\beta_2 = \dfrac{x_{roller2}}{L - x_{roller2}}$.

$$T_{CL2} = \frac{19978\beta_2 + 33570\,\beta_2^2}{38.25 + 150\,\beta_2^2} \qquad x_{roller2} > 30 \tag{8.38}$$

Eqs (8.37) and (8.38) allow the direct calculation of clutch torque in terms of the roller displacement, which is linearly related to the rotational angle of the control motor. These two equations correlate the nominal clutch torques in terms of the control variables respectively. The clutch torques represented by Eqs (8.37) and (8.38) can also be plotted against the roller displacements for convenience, as shown in Figure 8.21.

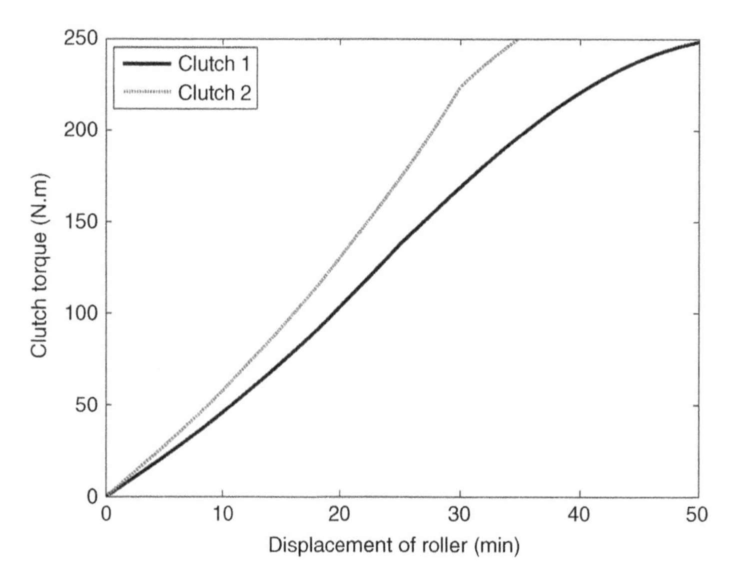

Figure 8.21 Clutch torques plotted against roller displacements.

8.5.3 Algorithm for Clutch Torque Calculation under Real Time Conditions

As detailed in the previous section, the clutch torque is a function of the clamping force on the pressure plate, and the friction coefficient for a given clutch. The nominal clutch torque is calculated by Eqs (8.37) or (8.38) in terms of roller displacements. However, this calculation must be calibrated for real world applications since the clutch friction coefficient is temperature dependent. In this section, an algorithm is presented for the accurate calculation of the clutch torque based on powertrain dynamics.

The system of equations of motion has been derived in Section 8.3 for vehicle launch and shift operations. These equations can be transformed to express the clutch torques, T_{CL1} and T_{CL2}, in terms of vehicle road load and inertia. For launch operation, only clutch CL1 is transmitting engine torque to the transmission input and T_{CL1} can be solved by combining Eqs (8.1–8.5) in the following form:

$$T_{CL1} = \left(I_{eq}^{1} i_{a1} i_1 + \frac{I + I_6}{i_{a1} i_1}\right)\dot{\omega}_w + \frac{T_{Load}}{i_{a1} i_1} \tag{8.39}$$

This equation is derived by considering a solid connection between the final drive output and the vehicle inertia, i.e. the spring–damper for the output shaft in the dynamic model structure shown in Figure 8.8 is dropped. During vehicle launch, the speed sensor measures the wheel speed at specified sampling time steps and the wheel acceleration $\dot{\omega}_w$ is obtained via a numerical algorithm that differentiates the wheel speed versus time. Therefore, the clutch torque during launch can be calculated by Eq. (8.39) if the road load can be estimated accurately.

During shift operations, both clutches participate in transmitting engine torque to the transmission input. For the 1–2 shift, the system of equations of motion for the vehicle has been expressed by Eqs (8.10–13) for the torque phase and by Eqs (8.14–17) for the inertia phase. For the whole 1–2 shift, the two clutch torques can be correlated to the road load and the inertia terms as:

$$i_1 T_{CL1} + i_2 T_{CL2} = \left[\left(I_2 i_2^2 + I_3 i_3^2 + I_4 + \right) i_{a1} + \frac{I_6 + I}{i_{a1}}\right]\dot{\omega}_w + \frac{T_{Load}}{i_{a1}} \tag{8.40}$$

When one of the two clutches are open, i.e. when one of the two clutch torques is zero, the other can be determined by Eq. (8.40) as previously explained. Therefore, for vehicle launching and fixed gear operations, the clutch torque is uniquely defined by Eq. (8.39) in terms of the road load and the inertia term. However, during shift operations, both clutches slip under pressure, and both clutch torques, T_{CL1} and T_{CL1}, exist and participate in transmitting the engine torque to the transmission input. An additional condition is needed besides Eq. (8.40) in order to solve T_{CL1} and T_{CL1} during shift operations.

As formulated by Eqs (8.28–38), the clutch torque is a function of clutch design parameters, the friction coefficient and the control variable. The proportion between the two clutch torques should be independent of the friction coefficient since the temperature effect is the same for both clutches. Therefore, torques in clutch CL1 and clutch CL2 are proportionate in the following ratio:

$$K = \frac{T_{CL1}}{T_{CL2}} = \frac{K_1}{K_2} \tag{8.41}$$

where K_1 and K_2 are the factors for the respective clutch that depend on clutch dimension, actuator parameters, and the roller position as detailed previously. For the example DCT in the chapter, K_1 and K_2 are defined in Eqs (8.35–8.38). For a given clutch module design, the clutch torque ratio K only depends on the two control variables: $x_{roller1}$ and $x_{roller2}$. Note again that this ratio is not affected by the clutch temperature since it affects the clutch disk friction coefficients equally. Combining Eqs (8.40) and (8.41) leads to the determination of the two clutch torques T_{CL1} and T_{CL1} in terms of $\dot{\omega}_w$ and T_{Load} as:

$$\begin{cases} T_{CL2} = \dfrac{\left[\left(I_2\, i_2^2 + I_3\, i_3^2 + I_4 + \right) i_{a1} + \dfrac{I_6 + I}{i_{a1}}\right] \dot{\omega}_w + \dfrac{T_{Load}}{i_{a1}}}{i_2 + i_1 K} \\ T_{CL1} = K T_{CL2} \end{cases} \tag{8.42}$$

8.5.4 Case Study for the Clutch Torque Algorithm

The algorithm derived above has been applied for the test vehicle whose data are provided at the end of Section 8.3. Vehicle acceleration, wheel speed, and roller positions of the two clutch controller are measured by the respective sensors and recorded during test runs on a proving ground. The test vehicle is driven on flat track with rolling resistance coefficient well established. A torque sensor is mounted on the half shaft to measure the drive train output torque. The measured half shaft torque is converted by the related gear ratios to be the equivalent torque on the input shaft. This equivalent torque is compared with the clutch torque calculated by the algorithm.

In the launch operation, the DCT runs in first gear and the clutch torque T_{CL1}. is calculated directly by Eq. (8.39). The vehicle acceleration or wheel speed from the test is the only model input. The comparison of clutch torque is shown in Figure 8.22. The clutch torque calculated from the torque algorithm closely agrees with the clutch torque obtained from measurement.

During a shift, the measured half shaft torque cannot be converted to be the equivalent torque values on the respective input shaft since the proportion of the measured clutch

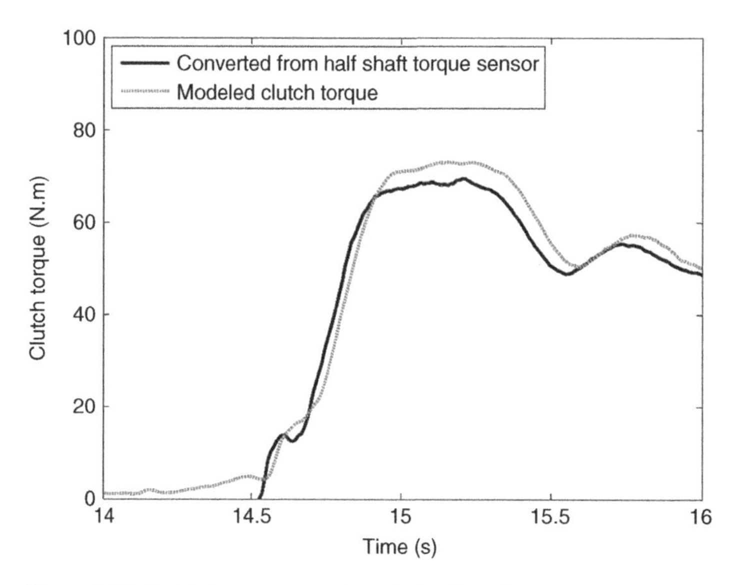

Figure 8.22 Clutch torque comparison during launch.

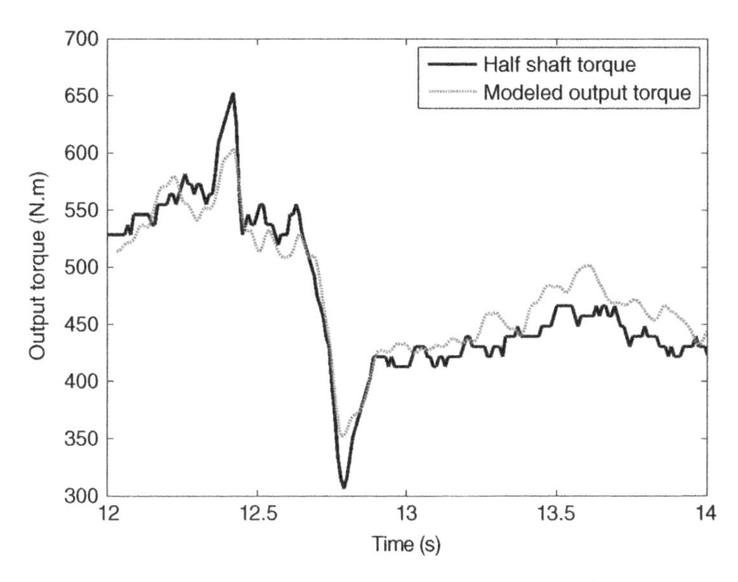

Figure 8.23 Clutch torque comparison during the 1–2 upshift.

torques is not known. Therefore, the resultant half shaft torque, that is the sum of the two clutch torques multiplied by the respective gear ratios, is compared with the measured half shaft torque. As shown in Figure 8.23, the output torque during 1–2 upshift is in close agreement with the test data, indicating the effectiveness and accuracy of the torque calculation algorithm.

The clutch torque when the vehicle operates in a fixed gear can also be calculated by the algorithm represented in this section. The results in fourth gear operation are shown in Figure 8.24. The torque in clutch CL1 equals zero because only clutch CL2 is now transferring the engine torque to drive the vehicle. As shown in the figure, the torque in clutch CL2 is almost the same as the test results, which reconfirms the accuracy of the proposed algorithm.

In summary, this section is focused on the dual clutch torque formulation and calibration. The clutch torque formulation is proposed based on a constant energy conversion rate over the friction disk face. The correlation on clutch torque and the parameters of the clutch actuator has been established in terms of roller position, and the related design parameters. For calibration purposes, a clutch torque calculation algorithm has been proposed based on DCT powertrain dynamics. This algorithm uses vehicle wheel speed obtained from a speed sensor as the input.

The algorithm has several advantages: (a) it enables the determination of clutch torque without using the friction coefficient of the friction disk which varies as a function of temperature; (b) it provides an effective way to calibrate the clutch torque against the design and control variables of the clutch and its actuator; (c) it provides a reliable correlation between clutch torque and clutch control variable during real time operation for adaptive transmission control. Finally, the analytical formulation and algorithm for clutch torque calculation are validated against proving ground test data and good agreement is achieved between analytical and test data.

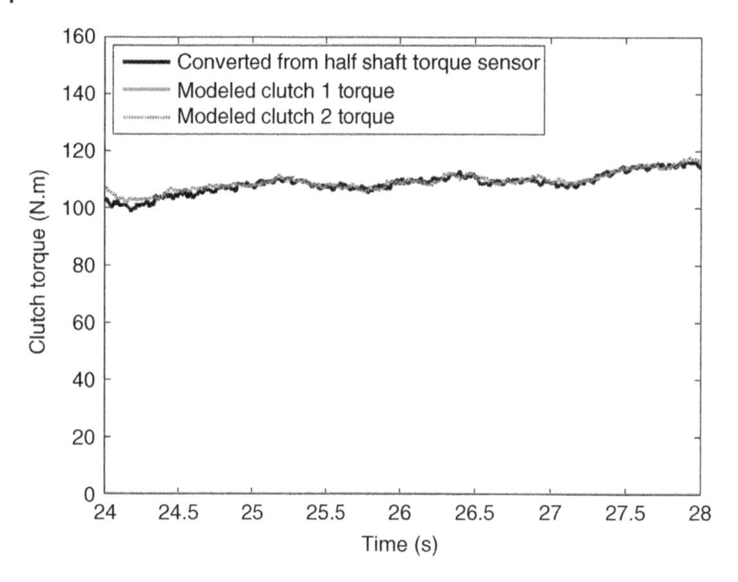

Figure 8.24 Clutch torque comparison during operation in the fourth gear.

References

1 Song, X., Liu, J. and Smedley, D.: *Simulation Study of Dual Clutch Transmission for Medium Duty Truck Applications*, SAE Paper No. 2005-01-3590.

2 Kulkarni, M., Shim, T., and Zhang, Y.: *Shift Dynamics and Control of Dual-Clutch Transmissions*, Mechanisms and Machine Theory, 2007, 42(2), pp. 168–182.

3 Zhang, Y., Chen, X., Zhang, X., Jiang, H., and Tobler, W.: *Dynamic Modeling and Simulation of a Dual-Clutch Automated Lay-shaft Transmissions*, ASME Journal of Mechanical Design, 2005, Vol. 127, No. 2, pp. 302–307.

4 Goetz, M., Levesley, M., and Crolla, D.: *Integrated Powertrain Control of Gearshifts on Twin Clutch Transmissions*, Proceedings of the Transmission and Driveline Systems Symposium 2004, SAE Paper No. 2004-01-1637.

5 Goetz, M., Levesley, M., and Crolla, D.: *Dynamics and Control of Gearshift on Twin Clutch Transmissions*, Proc. Inst. Mech. Engr, Part D (J. Automobile Engr), 2005, 219(8), pp. 951–963.

6 Livshiz, M., Kao, M., and Will, A.: *Validation and Calibration Process of Powertrain Model for Engine Torque Control Development*, SAE Paper No. 2004-01-0902.

7 Walamatsu, H., Ohashi, T., Asatsuke, S., and Saitou, Y.: *Honda's 5-speed All Clutch to Clutch Automatic Transmission*, Proceedings of the Transmission and Driveline Systems Symposium 2002, SAE Paper No. 2002-01-0932.

8 Minowa, T., Ochi, T., and Kuroiwa, H.: *Smooth Gear Shift Control Technology for Clutch to Clutch Shifting*, Proceedings of the Transmission and Driveline Systems Symposium 1999, SAE Paper No. 1999-01-1054.

9 Liu, Y., Qin, D., Jiang, H., and Zhang, Y.: *A Systematic Model for Dynamics and Control of Dual Clutch Transmissions*, Journal of Mechanical Design, ASME Transaction, 2009, Vol. 131, No. 6, pp. 061012-1-061012-7.

10 Liu, Y., Qin, D., Jiang, H., Liu, C., and Zhang, Y.: *Clutch Torque Formulation and Calibration for Dry Dual Clutch Transmissions,* Mechanism and Machine Theory, 2014, Vol. 75, pp. 41–53.
11 Liu, Y., Qin, D., Jiang, H., and Zhang, Y.: *Shift Control Strategy and Experimental Validation for Dry Dual Clutch Transmissions,* Mechanism and Machine Theory, 2011, Vol. 46, pp. 218–277.
12 Matthes, B.: *Dual Clutch Transmissions – Lessons Learned and Future Potential,* Proceedings of the Transmission and Driveline Systems Symposium 2005, SAE Paper No. 2005-01-1021.
13 Wheals, J., Turner, A., Ramasy, K., and O'Neil, A.: *Double Clutch Transmission (DCT) Using Multiplexed Linear Actuation Technology and Dry Clutches for High Efficiency and Low Cost,* SAE Paper No. 2007-01-1096.
14 Razzacki, S. and Hottenstein, J.: *Synchronizer Design nd Development for Dual Clutch Transmission (DCT),* SAE Paper No. 2007-01-0114.

Problems

1 For the dry DCT with hydraulically actuated clutches and gear shifting shown in Figure 8.6 for a dual mass flywheel DCT, construct the hydraulic system circuit for the actuation of the dual clutches and the gear shifters. The hydraulic system should use one VFS for line pressure control, two VFS valves for clutch actuation (one for each clutch) and shift solenoids for gear shifters. The hydraulic system should also include a hydraulic booster in the circuit.

2 For the dynamic model structure shown in Figure 8.10 for a dual mass flywheel DCT, formulate the state variable equations. In the formulation, only state variables should appear on the left side of the equations. In addition, represent the initial conditions for launch, 1–2 shift torque phase, and inertia phase respectively.

9

Electric Powertrains

An electric vehicle (EV) is similar to an engine-powered vehicle except that the engine-powertrain is now replaced by an electric machine, and the onboard fuel (gasoline or diesel) is replaced by an electric energy storage device, such as a battery pack.

9.1 Basics of Electric Vehicles

Figure 9.1 shows the basic structure of an EV. In this configuration, the battery stores energy in a chemical form. The most popular is a lithium-ion battery at the present time. The battery is charged from an electric outlet with an electric charger either carried on board or installed at the charging station. Typically, a low power charger is carried on board (3.3 kW, 6.6 kW) and fast chargers are installed at charge stations. The inverter converts the battery's DC voltage to a multi-phase AC to drive the electric machine. The inverter can change the amplitude and frequency of the power flow into the motor so that the torque, speed, and direction of the motor are controlled to drive the vehicle in the desired operation mode. During braking of the vehicle, the battery is charged by regenerative energy recovered from the kinetic energy of the vehicle which is converted from mechanical energy to electric energy by the electric motor, acting as a generator. Due to the characteristics of the electric machines, the mechanical transmission is usually simpler for an EV than in a conventional vehicle. Many EVs use a single speed gear reduction to satisfy all the driving needs, while some others use two-speed automated transmissions for the purpose similar to the CVT in a conventional vehicle. EVs generally do not need a multiple-speed automatic transmission as used in conventional vehicles. There may be a DC-DC converter between the battery and the inverter in some cases, in order to match the voltage of the battery to the inverter/motor voltage.

Figure 9.2 shows a typical electric drivetrain which consists of an electric motor, a power electronics converter, and a gearbox.

9.2 Current Status and Trends for EVs

EVs have many advantages and challenges. Electricity is more efficient than the combustion process in a car. Well-to-wheel studies show that, even if the electricity is generated

Automotive Power Transmission Systems, First Edition. Yi Zhang and Chris Mi.

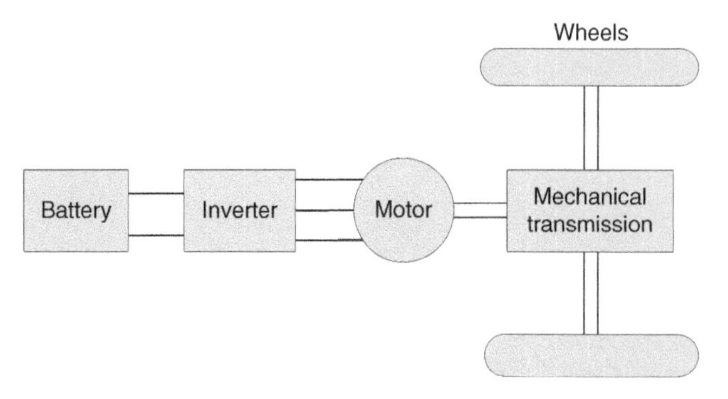

Figure 9.1 Basic structure of an EV.

Figure 9.2 Electric system of a passenger car, which includes a PM motor, a power electronics inverter, and a gearbox.

from petroleum, the equivalent miles that can be driven by 1 gallon (3.8 liter) of gasoline is 108 miles (173 km) in an electric car, compared to 33 miles (53 km) in an internal combustion engine (ICE) car [6–8]. In a simpler comparison, it costs 2 cents per mile to use electricity (at USD 0.12 per kWh) but 10 cents per mile to use gasoline (at USD 3.30 per gallon) for a compact car. Since electricity can be generated through renewable sources, such as hydroelectric, wind, solar, and biomass, EVs can be cleaner and more sustainable than combustion-based vehicles. On the other hand, the current electricity grid has extra capacity available at night when usage of electricity is off-peak. It is therefore ideal to charge EVs at night when the grid has extra energy capacity available. This is particularly so when differentiated pricing structures are available where the price at night is significantly cheaper. The batteries on board an electric vehicle can also potentially be used for electric grid support, such as peak shaving, frequency regulation, backup power, and other purposes.

However, there are many challenges for EVs. High cost, limited driving range, and long charging time are the main shortcomings for battery-powered EVs. Electric energy storage is currently the bottleneck for mass penetration of EVs due to the unsatisfactory energy density, power density, lifespan, and cost.

EVs were invented in 1834, that is, about 60 years earlier than gasoline-powered cars, which were invented in 1895. By 1900, there were 4200 automobiles sold in the USA, of which 40% were electric cars (http://sites.google.com/site/petroleumhistoryresources/Home/cantankerous-combustion). In the USA, there were a number of electric car companies in the 1920s, with two of them dominating the EV markets – Baker of Cleveland and Woods of Chicago. Both car companies offered hybrid electric cars. However, the hybrid cars were more expensive than gasoline cars and sold poorly. Electric vehicles and hybrid electric vehivles (HEVs) had faded away by 1930 and the electric car companies all failed. There were many reasons leading to the disappearance of EVs.

It was not until the Arab oil embargo in 1973 that the soaring price of gasoline sparked new interest in EVs. The US Congress introduced the Electric and Hybrid Vehicle Research, Development, and Demonstration Act in 1976 recommending the use of EVs as a means of reducing oil dependency and air pollution. In 1990, the California Air Resource Board (CARB), in consideration of the smog affecting Southern California, passed the zero emission vehicle (ZEV) mandate, which required 2% of vehicles sold in California to have no emissions by 1998 and 10% by 2003. California car sales is approximately10% of the total car sales in the United States. Major car manufacturers were afraid that they might lose the California car market without a ZEV. Hence, every major automaker developed EVs and HEVs. Fuel cell vehicles were also developed during this period. Many EVs were made, such as GM's EV1, Ford's Ranger pickup EV, Honda's EV Plus, Nissan's Altra EV, and Toyota's RAV4 EV.

In 1993, the US Department of Energy set up the Partnership for Next Generation Vehicle (PNGV) program to stimulate the development of EVs and HEVs. The partnership was a cooperative research program between the US government and major auto corporations, aimed at enhancing vehicle efficiency dramatically. Under this program, three US car companies demonstrated the feasibility of a variety of new automotive technologies, including an HEV that could achieve 70 MPG. This program was cancelled in 2001 and was transitioned to the Freedom CAR (Cooperative Automotive Research), which was responsible for the HEV, plug-in HEV (PHEV), and battery research programs under the US Department of Energy.

Unfortunately, the EV program faded away by 2000, with thousands of EV programs terminated by the auto companies. This is due partially to the fact that consumer acceptance was not overwhelming, and partially to the fact that the CARB relaxed its ZEV mandate.

From 1997 to 2009, the US market was mostly focused on hybrid cars. Many models were available to the consumer, with the best-selling car being the Toyota Prius, which sold nearly 4 million units worldwide from 1997 to 2016.

EVs reemerged in the USA and other parts of the world around 2009 when several policies and incentive programs were established by the US government partially to stimulate the economy after the 2008 economic crisis. In particular, Tesla Motors was established in 2003 and produced its first battery-powered pure electric sports car, Roadster, in 2008, and sold to more than 30 countries. A total of 2450 Roadsters were sold in 2008. This was the first EV that had a range of more than 200 miles per charge, and was also the first EV to use a lithium-ion battery pack. As of September 2016, Tesla has sold more than 160,000 battery electric cars worldwide, including Model S (145,000 units), Model X (16,000 units). The latest model, Model 3, has already taken orders of more than 500,000 with an expected delivery of 2018.

GM's Chevrolet Volt is the all-time best-selling plug-in electric car in the world with over 117,000 units sold from 2012 to September 2016. The Volt is not a pure EV, rather, it is a plug-in hybrid car with the pure-electric driving capability using the battery energy for 40 miles. The Chevrolet Bolt, a purely battery-powered EV from GM, was available in late 2016. It has a driving range of 238 mile per charge and is the only all-electric car under $50,000 that is capable of more than 200 miles per charge as of October 2016. The starting price in the US is $37,495.

Nissan produced its Leaf EV in 2010. The Leaf has a range of 84 miles (with a 24 kWh battery pack) to 107 miles per charge (with a 30 kWh battery pack). It sold more than 230,000 units from 2010 to July 2016, ranking it the best-selling all-electric car in history.

Ford released its Focus EV in 2011, with a 23 kWh liquid-cooled lithium-ion battery pack which delivers a range of 76 miles on a single charge. The production of the Ford Focus is limited, and sales have been in the hundreds to less than 2000 a year.

The Fiat 500e is a pure-electric vehicle produced by Fiat Chrysler Automotive Group (FCA). It is powered by an 83 kW, 199 Nm permanent-magnet motor, and a 24 kWh liquid-cooled/heated li-ion battery pack which delivers a range of 80 miles (130 km), and up to 100 mi (160 km) in city driving, according to FCA.

Recently, Toyota announced its all-new Prius Prime; unlike the Prius, which has a hybrid powertrain, the Prius Prime is similar to the Chevy Volt. Toyota said that the car will offer 640 miles of range per charge/refuel and will achieve 133 MPG equivalent (MPGe), the highest MPGe of all electric and plug-in electric cars, better than the Tesla Model X (289 miles range and 89 MPGe) and Chevy Bolt (238 miles range and 119 MPGe) by a large margin.

The US was the largest EV market in 201, but in 2015, China took over the first place. In 2015 and 2016, China sold more than 350,000 electric vehicles per year (including passenger, light-duty trucks, delivery vehicles, and buses) and approximately 300,000 low-speed electric cars. The estimated highway-capable electric vehicle sales will reach 450,000 in 2017, predicted by many. Government incentive plays a crucial role in the growth of EV manufacturing and sales. In 2015, even though the reported sales topped 300,000, there were extensive frauds, faking the EV sales to obtain the government subsidy, according to China's National Development and Reform Commission (NDRC).

9.3 Output Characteristic of Electric Machines

Electric machines have many unique characteristics that are especially suited for automotive powertrain applications. There are many kinds of electric machines, such as brushed DC machines, induction machines (IM), synchronous machines (SM), permanent magnet machines (PM), and switched reluctance machines (SRM). However, due to the space, weight, cost, and other limitations, EV powertrain applications are dominated by just three: induction machines, brushless DC machines, and permanent magnet synchronous machines. The SRM and other machines have been proposed for EV applications but have not become popular up to now.

When using electric motors for EV powertrain applications, there are a few possible configurations. Today's electric motors, combined with inverters and associated controllers, have a wide speed range of operations, which covers a normal range for constant

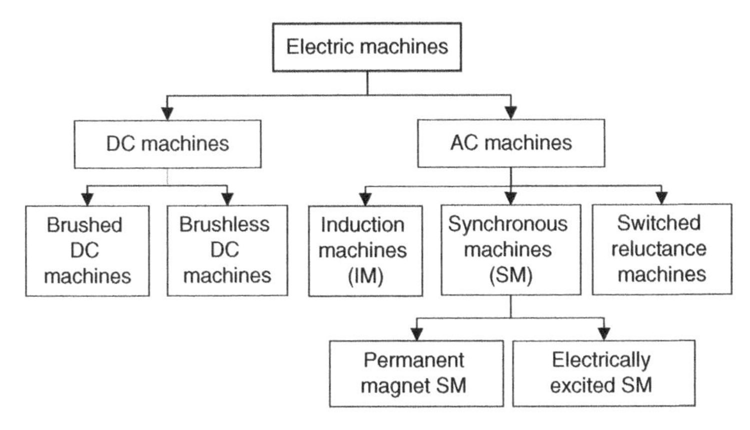

Figure 9.3 Classification of electric machines.

torque operations, and an extended speed range for constant power operations, which makes the design of the powertrain much easier. The powertrain motor needs to be able to provide the required torque and speed for all driving conditions of an EV. Hence, the size of the motor will be fairly large, usually rated at 100 kW or more for highway-capable passenger cars. Traditional automatic transmissions or continuously variable transmissions (CVT) used in conventional cars are no longer required for EVs. However, a two-speed automatic transmission may be beneficial in reducing vehicle energy consumption.

Electric motors are extensively discussed in various textbooks and many technical publications. In this chapter, we will first briefly discuss the principles of DC motors, induction motors, and permanent magnet machines, and then focus on a few unique aspects of electric motors that are specific to traction applications.

Depending on the type of current supplied and the principle of operation, electric machines can be classified to various categories as shown in Figure 9.3. Typically, electric machines consist of direct current (DC) and alternative current (AC) machines. Within DC machines, there are brushed and brushless DC machines, then AC machines are further classified into induction, synchronous, and switched reluctant machines, based on their principle of operations. For synchronous machines, the magnetic field can be created either by permanent magnets (hence, PM synchronous machines) or by a coil (hence, electrically excited synchronous machines).

Traditionally, electric machines, including brushed DC machines, induction machines, and electrically excited synchronous machines, are widely used in industrial and residential applications with a fixed speed or a limited range of operating speeds for a given power supply. With the advances in power electronics, microcomputers, and control, variable-speed drives have become more popular today.

9.4 DC Machines

Due to the availability of variable-speed drives, DC machines are no longer popular in industrial and residential applications, due to their bulky size and high maintenance cost. However, DC machines are often used to illustrate the principle of electric machines because of their simplicity and controllability. In particular, many variable-speed AC machines, such as induction machines and PM synchronous machines, can be controlled to behave in a similar way to DC machines.

9.4.1 Principle of DC Machines

Figure 9.4 shows the basic principle and operation of a DC machine. The DC machine includes a stator, a rotor with commutators, and a pair of brushes. In the given space, we fix two permanent magnets on the stator, which will generate a magnetic field between the two magnets. A rectangular coil with its two sides perpendicular to the magnetic fields are placed inside the space between the two magnets. To increase the magnetic field strength as well as the mechanical strength, we put the steel rotor inside the two magnets and attach the coil to the surface of the rotor.

The stator consists of the magnet, a housing that holds the magnet, bearings, and connectors. The rotor consists of the winding, a steel cylinder that is usually made of laminated silicon steel which has slots to fill the windings. The rotor that contains the winding is also called the armature. The laminated steel is used to reduce the magnetic losses due to the time-varying magnetic field.

The coil or armature is powered by a DC source. Assuming the DC power has a voltage of V, the coil has a resistance of R as shown in Figure 9.5, and both the magnets and coil are stationary, then the current will form in the coil, as expressed as (here, the inductance is neglected due to the DC voltage):

$$I = V/R \tag{9.1}$$

Now, since the current is under the magnetic field, a few things will happen sequentially. First, a force will be generated on the two sides of the coil which can be expressed as:

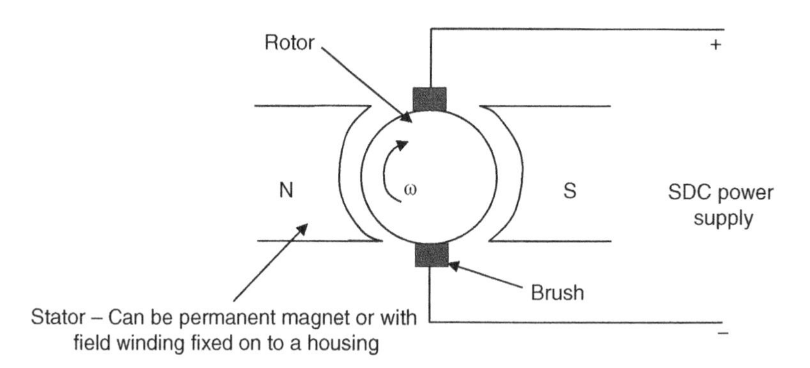

Figure 9.4 Principle of DC machines (cross-sectional view).

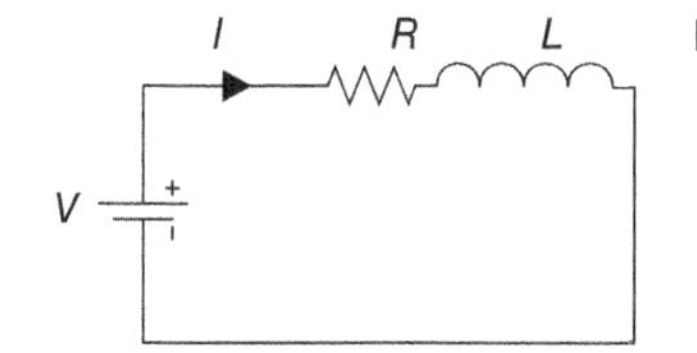

Figure 9.5 Equivalent circuit when stationary.

$$f = BIl \tag{9.2}$$

where B is the magnetic field generated by the magnets, and l is the side length of the coil under the magnetic field. The force of the coils will generate a torque T since it is wound on the surface of the rotor. Assume there are N conductors inside the magnet, and the diameter of the rotor is D, the torque is (Figure 9.4 has two conductors, with each side of the coil is considered a conductor):

$$T = f \times \frac{D}{2} N \tag{9.3}$$

This torque will move the coil to rotate around the center of the rotor. Assume the angular speed of the coil is ω, the inertia of the rotor is J, and the torque (including friction of the rotor and shaft, and any load that is connected to the shaft) is T_L, then we have,

$$T - T_L = J\frac{d\omega}{dt} \tag{9.4}$$

The linear speed of the coil (the surface of the rotor) is $u = \omega D/2$, where D is the diameter of the rotor.

From the Lorentz law, the rotating coil will generate a back electromotive force (emf), or an internal voltage by each conductor, which can be expressed as

$$e = Blu \tag{9.5}$$

The total voltage on the coil, assume it has N conductors, will be

$$E = BluN = Bl\frac{\omega DN}{2} \tag{9.6}$$

Now, the current in the coil will be different from the one when the rotor is in a standstill condition. The new current, as shown in Figure 9.6, can be expressed as:

$$I = \frac{V - E}{R} \tag{9.7}$$

The total flux under each magnet can be expressed as:

$$\Phi = B \times l \times \frac{\pi D}{2} \tag{9.8}$$

Therefore, we can replace the above equations as:

$$E = Bl\frac{\omega DN}{2} = \frac{2\Phi}{\pi l D} l \frac{\omega DN}{2} = \frac{N}{\pi}\Phi\omega = k\Phi\omega \tag{9.9}$$

$$T = f * \frac{D}{2} N = BIl\frac{D}{2} N = \frac{2\Phi}{\pi l D} lI \frac{D}{2} N = \frac{N}{\pi}\Phi I = = k\Phi I \tag{9.10}$$

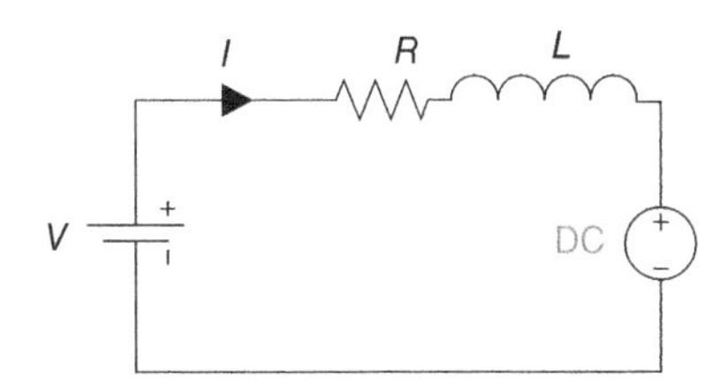

Figure 9.6 Equivalent circuit when rotating.

$$V = E + IR = k\Phi\omega + IR = k\Phi\omega + \frac{T}{k\Phi}R \tag{9.11}$$

where $k = N/\pi$.

The above expressions are derived from Figure 9.4 where only one pair of poles is shown. DC machines can be made of one pair or multiple pairs of poles. Also, the number of conductors are assumed to be all in series in the above derivations. In DC motor designs, the conductors can be designed to be connected in series or parallel depending on the number of poles. Therefore, for a more generic formula, we assume that the DC motor has p pairs of poles, and the N conductors have $2a$ branches in parallel. Each parallel branch will only get $I/2a$ current. Therefore, the following equations are derived. Only the constant k changes with the introduction of parallel circuits $2a$ and number of pole pairs, p:

$$\Phi = B \times l \times \frac{\pi D}{2p}$$

$$E = Bl\frac{\omega DN}{2}\frac{1}{2a} = \frac{2p\Phi}{\pi lD}l\frac{\omega DN}{2}\frac{1}{2a} = \frac{pN}{2\pi a}\Phi\omega = k\Phi\omega \tag{9.12}$$

$$T = f^*\frac{D}{2}N = Bl\frac{I}{2a}\frac{D}{2}N = \frac{2p\Phi}{\pi lD}l\frac{I}{2a}\frac{D}{2}N = \frac{pN}{2\pi a}\Phi I = = k\Phi I$$

where $k = \frac{pN}{2\pi a}$

In the steady state, the voltage Eq. (9.11) still holds for multiple pairs of poles. Plugging Eqs (9.13) and (9.14) to (9.11), the torque–speed relationship can be expressed as:

$$T = k\Phi\frac{V}{R} - \frac{(k\Phi)^2}{R}\omega \tag{9.13}$$

The torque–speed relationship can be plotted as shown in Figure 9.7. In Figure 9.7, when there is no load connected to the motor shaft, the speed of the motor is $\omega_o = V/k\Phi$. When the motor shaft speed is zero – in other words, when the motor shaft is locked – the back emf is also 0, therefore, the shaft torque is $T_{locked} = k\Phi(V/R)$.

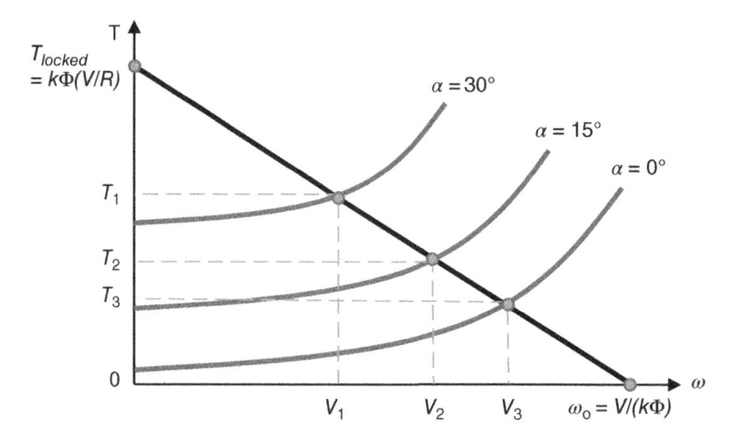

Figure 9.7 Torque–speed relationship of DC machines and the operating points when driving a typical vehicle.

In Figure 9.7, we also plotted the torque–speed of the traction force of a vehicle for three different grades. The vehicle resistance torque can be express as

$$T_{vehicle} = gFr = g\left(F_{roll} + F_{AD} + F_{grade} + ma\right)r \tag{9.14}$$

where $T_{vehicle}$ is the vehicle resistive torque on the motor shaft, r is the radius of the wheel, g is the gear ratio from the wheel to the motor shaft (which includes the final drive and the transmission), F_{roll} is the rolling resistance, F_{AD} is the aerodynamic resistance, F_{grade} is the hill climbing resistance, m is the mass of the vehicle, and a is the acceleration of the vehicle.

Three vehicle resistance torque curves are plotted in Figure 9.7, for grades of 0°, 15°, and 30° as examples. The cross points of the motor curve with the vehicle resistive torque curves indicate the operating points of the motor. The difference between the motor torque and the resistive torque indicates how much acceleration force is available at a certain vehicle speed.

Equation (9.16) can also be rewritten as a speed–torque relationship:

$$\omega = \frac{V}{k\Phi} - \frac{R}{(k\Phi)^2}T \tag{9.15}$$

Figure 9.8 shows the speed–torque relation expressed by Eq. (9.15). The motor speed will decrease as the torque increases, with a slope of $R/(k\Phi)^2$. The load torque in Figures 9.7 and 9.8 represents a typical vehicle resistance force. When the motor (or vehicle) starts, the vehicle speed is zero, so the back emf is zero. The motor produces a current of V/R (neglecting the inductances), so a large torque is produced, shown as the cross point of motor torque with the vertical axis in Figure 9.7. This torque is to overcome the vehicle resistance and provide a large acceleration needed to accelerate the vehicle. Depending on the slope of the road, the acceleration can be calculated. As the vehicle speed increases, the emf will increase, hence the stator current will decrease, which results in the decrease of stator current and less torque. At a certain speed, depending on the slope of the road, the motor torque is equal to the vehicle resistive torque and a maximum speed is reached, shown as the cross points of the motor torque curve and the vehicle resistive curve in Figure 9.8.

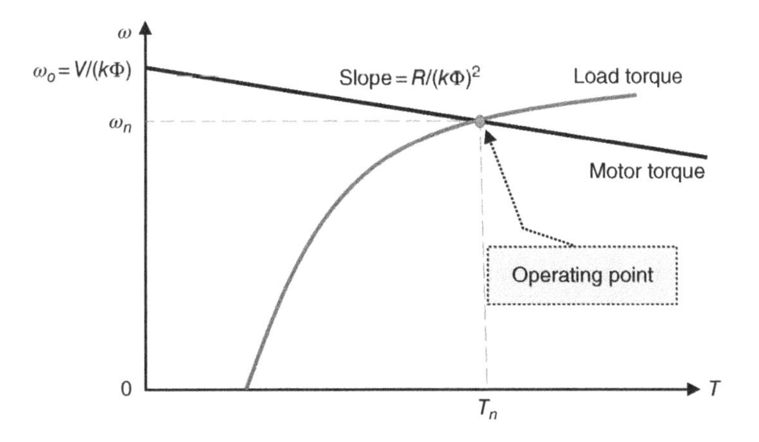

Figure 9.8 Motor speed–torque relationship and as a function of load torque.

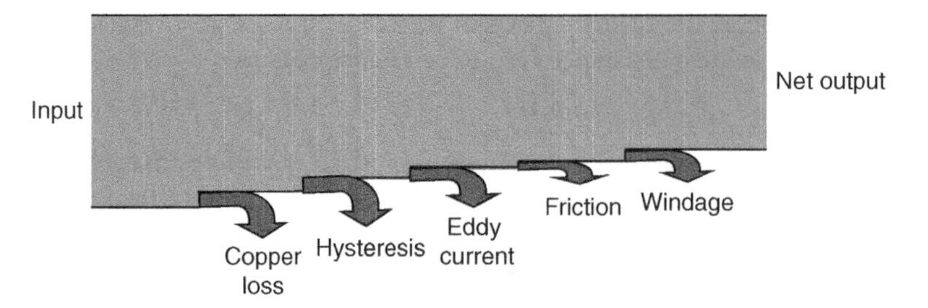

Figure 9.9 Power flow in a DC machine.

The motor will have a number of losses before it transfers the electric power to the shaft as mechanical power. Figure 9.9 shows the power flow and types of losses in a typical electric machine. First, when current flows in the stator windings, it creates conduction losses in the winding, also known as copper losses. Second, the magnetic field will generate eddy currents and hysteresis losses in the rotor material, which is typically composed of laminated silicon steel. Third, when the rotor spins, it introduces frictional losses in the bearing and windage loss due to wind resistance acting on the rotor.

The power balance equations are:

$$\begin{aligned} P_{em} &= P_1 - p_{cu1} - p_{iron} \\ P_2 &= P_{mec} - p_{fw} - p_{ad} \end{aligned} \tag{9.16}$$

P_1 is the input power from the voltage supply; P_{em} is the electromagnetic power transferred from the stator to the rotor; P_{mec} is the total mechanical power on the rotor shaft; P_2 is the output power to the load connected to the shaft; p_{cu1} is the copper loss of the rotor winding; p_{iron} is the iron loss of the stator core; p_{fw} is the frictional and windage loss; and p_{ad} is the stray load loss.

The efficiency is the ratio of output mechanical power and the input electrical power.

$$\eta = \frac{P_2}{P_1} = \frac{P_2}{P_2 + p_{cu1} + p_{iron} + p_{fw} + p_{ad}} \tag{9.17}$$

The typical efficiency of an electric machine ranges from 50% for very small motors to 99% for large motors. The typical efficiency for EV powertrain motors – which may be rated at tens of kilowatts to over 100 kW – is about 85–95%. The power loss in the machine needs to be dissipated to the surrounding air or cooling medium through air, liquid, or natural convection.

9.4.2 Excitation Types of DC Machines

The magnetic field is the key to generating torque in an electric machine. In DC motors, the magnetic field can be generated through excitation by an external voltage source, in parallel or series with the main DC supply, or from a separate DC power source, or using permanent magnets. The four different types of excitation of DC machines are shown in Figure 9.10.

Figure 9.10a shows the equivalent circuit of a parallel-excited or shunt DC motor, in which the field winding is in parallel with the armature winding and shares the

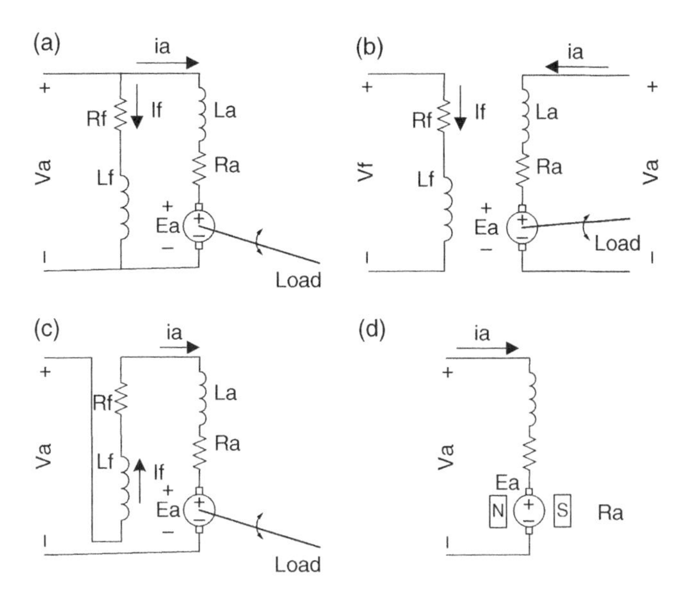

Figure 9.10 Excitation of DC motors: (a) parallel-excited or shunt DC machine; (b) separately excited DC machines; (c) Series-excited DC machines; (d) Permanent magnet DC machines.

same DC source. The field winding usually has a large number of turns and a large resistance. So the field winding current is typically small, in the order of less 5% of the armature current. A variable resistor can be connected in series with the field winding to adjust the magnetic field to adjust the speed of the motor.

Figure 9.10b shows the equivalent circuit of a separately excited DC motor. In this arrangement, the field winding has its own supply voltage, separate from the main power source. It is convenient to adjust the field winding voltage to adjust the motor speed.

Figure 9.10c shows the equivalent circuit of a series-wound DC motor. In this setup, the field winding is in series with the armature winding. Hence the field winding and the armature winding share the same current. To reduce losses in the field winding, the field winding usually has a very small number of turns and a small resistance. It is difficult to adjust the motor speed since the current is the same for both field and armature windings.

Figure 9.10d shows the equivalent circuit of a permanent magnet DC motor. The field is generated by the magnets, hence, it is not possible to adjust the magnetic flux. However, since there is no electrical loss for the generation of the magnetic field, the efficiency of permanent magnet DC motors are typically 3–5% higher than electrically excited DC motors. Flux weakening is possible in PM machines, in which case an electric current in the stator is controlled to generate a magnetic field that is in the opposite direction of the stator magnetic field, so as to reduce the total air-gap magnetic flux.

9.4.3 Speed Control of DC Machines

From Eq. (9.18), we can see that, for a given load torque, the DC motor's speed is related to three important parameters: the total flux ϕ, the stator resistance R, and the supply voltage. Hence, we can change the speed of a DC motor in three different ways, namely, adjusting the terminal voltage; adjusting the magnetic field flux, and adjusting the stator resistance.

9.4.3.1 Adjust Terminal Voltage

From Eq. (9.15), we can see that, by adjusting the terminal voltage, the steady-state motor speed will change for a given load torque characteristics, as shown in Figure 9.11. With the reduction of terminal voltage, the operating speed will decrease. The control of terminal voltage can be realized via power electronics circuits, namely, a buck converter, a half-bridge converter, or a full bridge converter.

In Figure 9.12, the motor is controlled by a buck converter. The DC supply, *Va*, is connected via a semiconductor switch, which opens and closes at a very high frequency (in the kilohertz range). When the switch closes, current flows from the power supply *Va* to the motor terminal, where the armature inductance will limit the rate of increase of current into the stator. When the switch opens, the current in the stator will continue to flow through the diode, which is called a freewheeling diode. In the steady state, the average voltage at the motor terminal is determined by how much the switch closes during one period, which is also defined as the duty ratio:

$$V_o = V_a\frac{t_{on}}{T_s} + 0\frac{t_{off}}{T_s} = DV_a \tag{9.18}$$

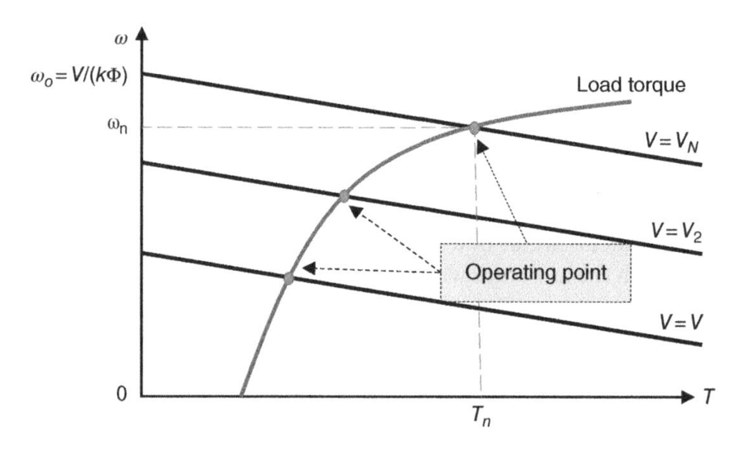

Figure 9.11 Motor speed–torque control by adjusting terminal voltage.

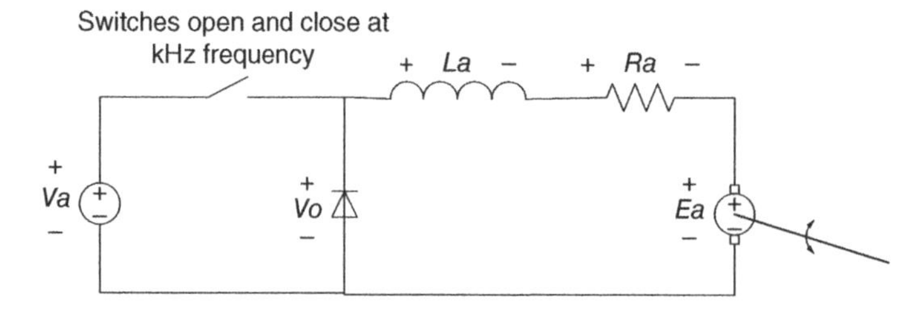

Figure 9.12 Voltage control of DC motors via a buck converter.

where t_{on} is the time of the switch is turned on, T_s is the switching cycle, t_{off} is the time the switch is turned off during one period, and D is the duty ratio of the switch. It is apparent that when the switch is open all the time, there will be no current flowing to the motor terminal, hence the motor speed will be zero. When the switch is closed all the time, the motor will have the maximum current and the speed will reach its maximum.

In the circuit in Figure 9.12, the motor can only rotate in one direction because the range of voltage applied to the motor is from 0 (when $D = 0$) to V_a (when $D = 1$). Another disadvantage of the circuit is that the motor cannot send power back to the source due to the unidirectional power flow of the power switch.

Figure 9.13 shows a motor controlled by a half bridge converter, also referred to as a two-quadrant chopper. The half bridge converter can control the motor in one direction and also realize energy feedback to the source. In Figure 9.13, switch S2 is open all the time during motoring while S1 is controlled to open and close at a certain frequency. During motoring, when switch S1 is closed, current flows to the motor through S1. When S1 is open, motor current continues to flow through D2 for freewheeling. The motor speed is proportional to the duty ratio of S1. During regenerative braking, S1 is kept open all the time, while S2 is controlled. When S2 is closed, the motor terminal forms a short circuit through S2. Due to the back emf and inductance of the motor, the current will flow out of the motor through S2 with a certain rate limited by the inductance. Once the current reaches a certain amount, controlled by the duty ratio of S2, S2 opens, the current that originally flows out of the motor through S2 is now forced to flow through D1 to the source, hence the energy is fed back to the supply.

The two-quadrant chopper can control the motor in motoring and regenerative braking but can only rotate in one direction. To overcome the issues of a buck converter and two-quadrant chopper, a full bridge converter can be used, as shown in Figure 9.14. In Figure 9.14, when S1 and S4 close, S2 and S3 will open. When S1 and S4 open, S2 and S3 will close. So, when S1 and S4 close all the time, the motor gets a positive voltage and rotates in the positive direction at its maximum speed. When S2 and S3 close all the time, the motor gets a negative voltage and rotates in the negative direction with its maximum speed. If S1/S4 and S3/S4 both opens and closes 50% of the time, the average voltage at the motor is zero, hence the motor will not rotate.

Each switch is also paralleled with a freewheeling diode which is to conduct current in the opposite direction to the switch. Hence, the full bridge can operate the motor in all four conditions: positive direction motoring (forward driving); positive direction

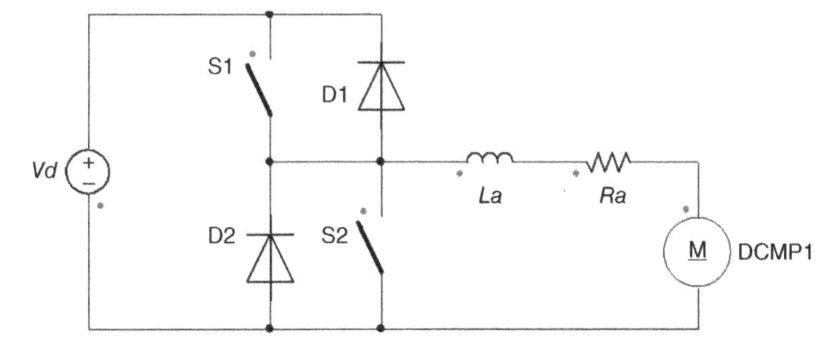

Figure 9.13 Voltage control of DC motors via a half bridge converter.

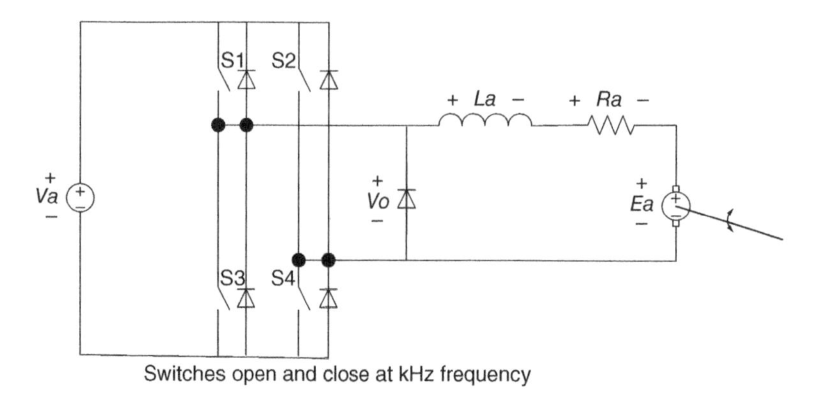

Figure 9.14 Voltage control of DC motors via a full bridge converter.

generating (driving while braking); negative direction motoring (vehicle backing up), negative direction generating (braking when reversing the vehicle).

Controlling the terminal voltage, whether through a buck or two-quadrant chopper or a full bridge converter, the motor terminal voltage will reach its maximum value when the duty ratio reaches 1. Hence, the motor speed will reach its maximum value. Therefore, the motor speed can be only controlled below its rated speed when controlling the armature voltage.

9.4.3.2 Adjust Magnetic Flux

The second method of controlling the motor speed is to control the magnetic flux. Depending on the type of excitation, the flux can be controlled by adjusting the voltage of the excitation (separately excited machines), or the resistance in the excitation branch (parallel excited machines), or by flux weakening (PM machines). When the flux is adjusted, the speed–torque will change so the speed is adjusted, as shown in Figure 9.15. When controlling the field winding resistance in a DC machine, the increase of field resistance will result in the decrease of magnetic flux. Hence the speed of the motor will increase for a given load torque. So typically, DC motor speed can be controlled above the rated speed by adjusting the magnetic flux. Field weakening has the same effect, that is, the motor speed can be increased by reducing the total flux.

9.4.3.3 Adjust Stator Resistance

The last method, adjusting the armature resistance, although it is not popular, can also be used. From Eq. (9.15), when an external resistance is connected in series with the armature winding, the slope of the speed–torque will change, hence the speed will change for a given load torque, as shown in Figure 9.16. The motor speed can be controlled to go down, as shown in Figure 9.16. Note too that, since the resistance will share the same armature current, it will consume a lot of power. So the efficiency of the motor for this method will be compromised. Hence, this method is no longer prevalent in the motor speed control, partially due to the advance of power electronics technology.

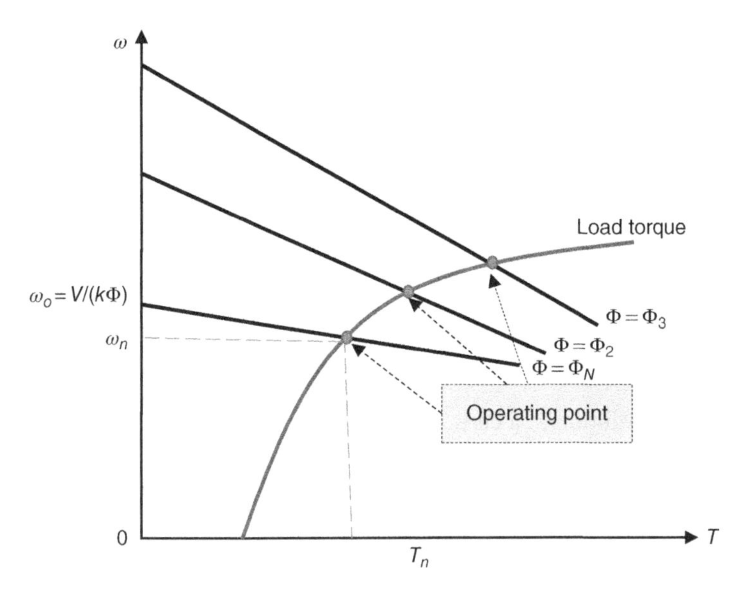

Figure 9.15 Motor speed–torque control by adjusting magnetic flux.

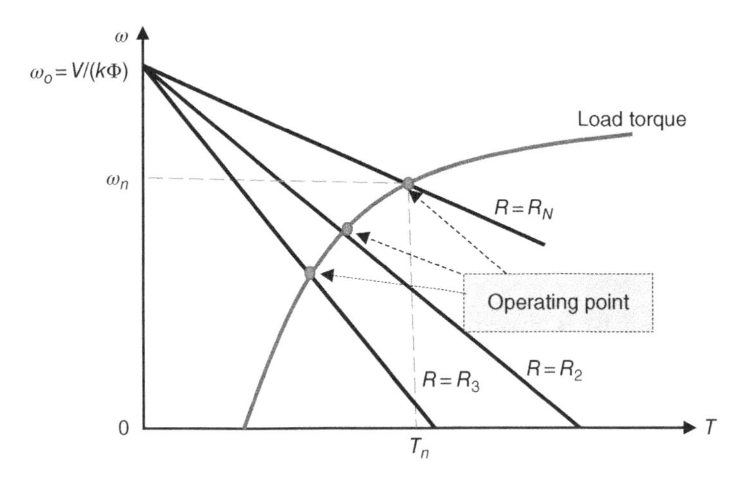

Figure 9.16 Motor speed–torque control by armature resistance.

9.5 Induction Machines

Induction motors are a popular choice for traction applications due to their robust construction, low cost, wide field weakening range, and high reliability. Especially for EVs, PHEVs, and HEVs that require a high-power motor, induction motors can provide more reliable operation than other types of electric motors [21–37]. However, when compared to PM motors, induction motors have lower efficiency and less torque density.

One typical induction motor used for traction applications is the squirrel cage induction motor. An inverter is used to control the motor so that the desired torque can be

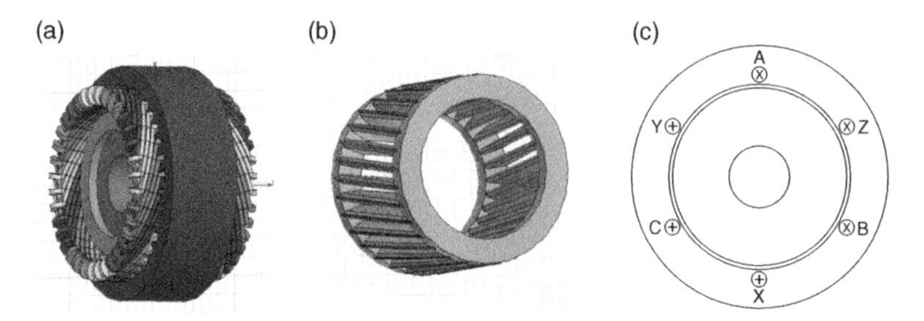

Figure 9.17 An induction motor: (a) rotor and stator assembly; (b) rotor squirrel cage; (c) cross-sectional view of an ideal induction motor with six conductors on the stator.

delivered for a given driving condition at a certain speed. Advanced control methodologies, such as vector control, direct torque control, and field-oriented control, are popular in induction motor control for traction applications.

9.5.1 Principle of Induction Motors

The basic structure of an induction machine is shown in Figure 9.17. The two main parts of an induction motor are the stator (which houses the winding) and the rotor (which houses the squirrel cage). Both stator and rotor are made out of laminated silicon steel with a thickness of 0.35, 0.5, or 0.65 mm. The laminated steel sheets are first stamped with slots and are then stacked together to form the stator and rotor, respectively. Windings are put inside the stator slots while the rotor is cast in aluminum.

There are some additional components to make up the whole machine: the housing that encloses and supports the whole machine, the shaft that transfers torque, the bearing, an optional position sensor, and a cooling mechanism (such as a fan or liquid cooling tubes).

In Figure 9.17c, AX is phase a, BY is phase b, and CZ is phase c. The direction of the phase currents is for a particular moment $\omega t = 60$ electric degrees; "+" indicates positive and "–" indicates negative. It can be seen that conductor AZB forms one group and XCY forms another group. Together they create a magnetic field at 30° NW–SE. The direction of the field will change as the current changes over time.

The stator windings shown in Figure 9.17c are supplied with a three-phase AC sinusoidal current. Assuming that the amplitude of the currents is I_m amperes, and the angular frequency of the current is ω radians per second, then the three-phase currents can be expressed as:

$$\begin{aligned} i_a &= I_m \cos(\omega t) \\ i_b &= I_m \cos(\omega t - 2\pi/3) \\ i_c &= I_m \cos(\omega t - 4\pi/3) \end{aligned} \tag{9.19}$$

Since the currents of each of the three phases are functions of time, the direction of current as shown in Figure 9.17c will change with time. If we mark the direction of the current at any given time, we can see the magnetic field generated by the stator current with its peak changing position over time.

Mathematically, we can derive this magnetic field. Each of the three phase currents will generate a magnetic field. Since the three windings are located 120° from each other in space along the inside surface of the stator, the field generated by each phase can be written as follows, assuming the spatial magnetic field distribution in the air gap due to winding currents is sinusoidal by design:

$$\begin{aligned} B_a &= Ki_a(t)\cos(\theta) \\ B_b &= Ki_b(t)\cos(\theta-120^\circ) \\ B_c &= Ki_c(t)\cos(\theta-240^\circ) \end{aligned} \tag{9.20}$$

where K is a constant. Using Equations (9.1) and (9.2), considering that $\cos(\omega t)\cos(\theta) = [\cos(\omega t-\theta) + \cos(\omega t+\theta)]/2$ and $\cos(\omega t+\theta) + \cos(\omega t+\theta-240^\circ) + \cos(\omega t+\theta-480^\circ) = 0$, we get

$$\begin{aligned} B_{gap} &= Ki_a(\omega t)\cos(\theta) + Ki_b(\omega t-120^\circ)\cos(\theta-120^\circ) + Ki_c(\omega t-240^\circ)\cos(\theta-240^\circ) \\ &= \frac{3}{2}KI_m\cos(\omega t-\theta) + \frac{1}{2}KI_m[\cos(\omega t+\theta) + \cos(\omega t+\theta-240^\circ) \\ &\quad + \cos(\omega t+\theta-480^\circ)] \\ &= B_m\cos(\omega t-\theta) \end{aligned} \tag{9.21}$$

Equation (9.21) shows that the magnetic field is a traveling wave along the inner surface of the stator. In other words, the total magnetic field is a sinusoidal field with its peak rotating at an angular speed ω rad/s.

Since $\omega = 2\pi f$, the rotational speed of the field will be the same as the supply frequency: f revolutions per second or $n_S = 60f$ revolutions per minute (RPM). Noting that the above derivation is based on one pair of poles, a more general equation for the field speed (or synchronous speed) of an induction machine can be given as

$$n_S = \frac{60f}{p} \text{ and } \omega_S = \frac{2\pi n_S}{60} = \frac{2\pi f}{p} = \frac{\omega}{p} \tag{9.22}$$

where p is the number of pairs of poles. Figure 9.18 shows the arrangement of a four-pole squirrel-cage induction motor with flux distribution.

Assuming initially that the rotor is stationary, an electromotive force (emf) will be induced inside the rotor bars of the squirrel cage. A current is therefore formed inside the rotor bars through the end rings. Similarly, since the field is rotating, this current will generate a force on the rotor bars (the rotor bar current is inside the stator magnetic field). If the force (or torque) is sufficiently large, the rotor will start to rotate.

The maximum speed of the rotor will be less than the synchronous speed because, if the rotor reaches the synchronous speed, there will be no relative movement between the rotor bars and the stator field, hence no emf or force will be generated. The difference between the rotor speed and the synchronous speed is defined as slip s, that is, $s = (n_S - n_m)/n_S = (\omega_S - \omega_m)/\omega_S$, where n_m and ω_m are the rotor speeds in RPM and radians per second, respectively. Typical slips of induction motors are within 1–3%.

9.5.2 Equivalent Circuit of Induction Motors

We can represent the induction motor by two separate circuits, one for the stator and one for the rotor. Since the three phases are symmetrical, we only need to analyze one

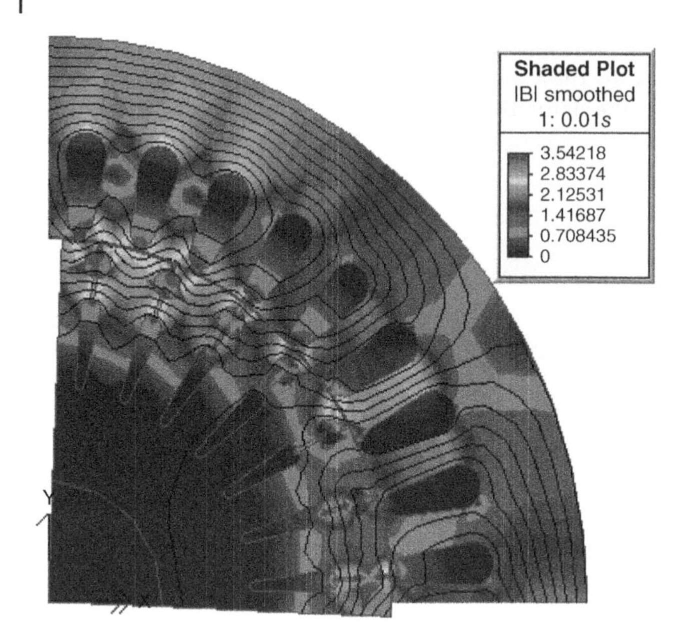

Figure 9.18 Flux distributions of a four-pole induction motor during transient finite element analysis.

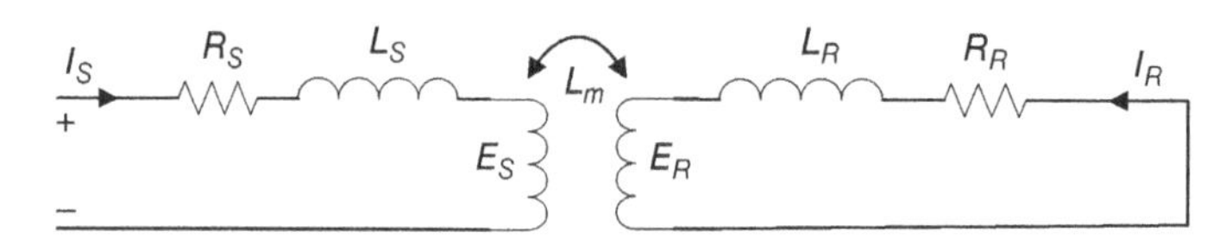

Figure 9.19 Stator and rotor circuits of an induction machine.

phase as shown in Figure 9.19. We use phasors for the analysis of the AC circuit. Here we have defined the direction of current flow using the transformer convention. It is worth noting that the rotor and the stator quantities will have different frequencies except when the rotor is stationary.

The voltage equation of the primary and the secondary circuit can be written as:

$$V_S = I_S R_S + j\omega L_S I_S + E_S \tag{9.23}$$

$$0 = I_R R_R + j\omega_R L_R I_R + E_R \tag{9.24}$$

where V is the phase voltage, I the phase current, R the phase resistance, and L the leakage inductance of the winding. The subscripts S and R represent the stator and rotor respectively.

Since the field is rotating at synchronous speed ω_S and the rotor is rotating at speed ω_m, the speed of the magnetic field relative to the rotor bar is $\omega_S - \omega_m = s\omega_S = s\omega/p$, and $\omega_R = ps\omega_S = s\omega$ is the frequency of the rotor current.

If we multiply both sides equation (9.24) by k and then divide by s, we get

$$0 = \left(k^2\frac{R_R}{s}\right)(I_R/k) + j\left(k^2\frac{\omega_R}{s}L_R\right)(I_R/k) + \frac{kE_R}{s} \tag{9.25}$$

The rotor has AC quantities a slipping frequency $\omega_R = s\omega$. By using the following, $R'_R = k^2 R_R$, $X'_R = k^2\omega L_R = k^2 X_R$, $I'_R = I_R/k$, $E'_R = kE_R/s$, we have

$$0 = \frac{R'_R}{s} I'_R + jX'_R I'_R + E'_R \tag{9.26}$$

We will choose k such that $E_S = E'_R$.We can then redraw the equivalent circuit of the induction motor as shown in Figure 9.20a. Here we neglected the magnetic loss in the stator core. If we include the magnetic loss, then the equivalent circuit can be illustrated as in Figure 9.20b.

In the equivalent circuit in Figure 9.20, for a given voltage supply the current of the circuit can be written as:

$$I_S = \frac{V_S}{R_S + j\omega L_S + (R_m + jX_m)||\left(R'_R/s + j\,X'_R\right)} \tag{9.27}$$

To simplify the analysis, we can neglect $R_m + j\omega X_m$. Under this assumption, the electromagnetic power transferred from the stator to the rotor is:

$$P_{em} = mI_S^2 \frac{R'_R}{s} = \frac{m\,V_S^2}{\left(R_S + R'_R/s\right)^2 + \left(X_S + X'_R\right)^2} \frac{R'_R}{s} \tag{9.28}$$

Noting that electromagnetic power or rotor power has two parts, namely, the loss of the rotor winding and the power transferred to its shaft, Eq. (9.28) can be rewritten as:

$$P_{em} = \frac{m\,V_S^2}{\left(R_S + R'_R/s\right)^2 + \left(X_S + X'_R\right)^2} \left[R'_R + \frac{(1-s)}{s} R'\right]_R \tag{9.29}$$

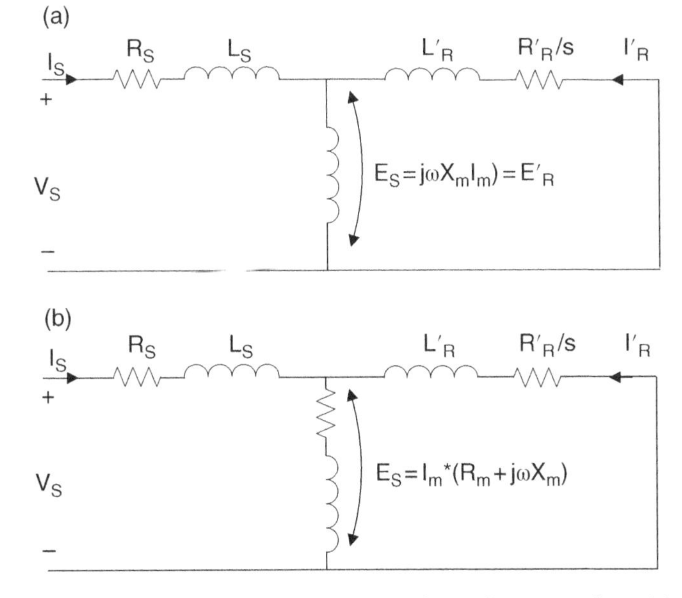

Figure 9.20 Modified equivalent circuit of an induction machine: (a) neglecting iron loss; (b) considering iron loss.

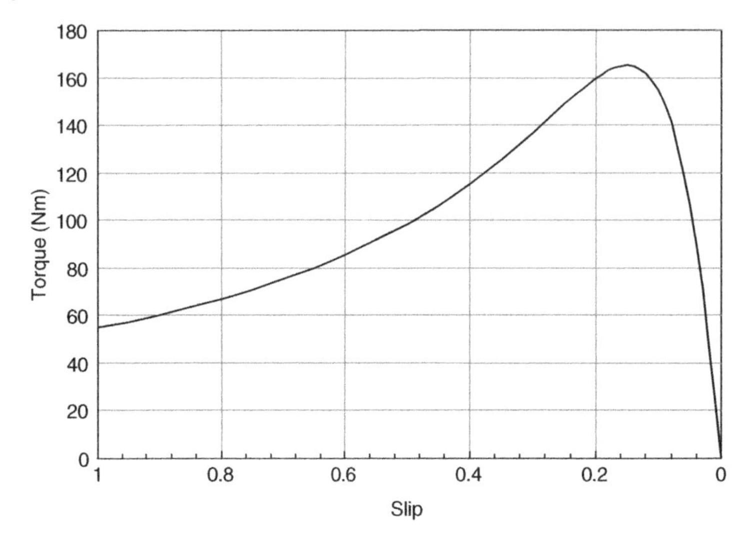

Figure 9.21 Torque–speed characteristics of an induction motor for a constant frequency and constant voltage supply.

The first term represents the rotor copper loss and the second term the mechanical power on the shaft. The electromagnetic torque of the motor can be written as:

$$\begin{aligned} T_{em} &= \frac{m}{\omega_S} \frac{V_S^2}{\left(R_S + R_R'/s\right)^2 + \left(X_S + X_R'\right)^2} \frac{1}{s} R_R' \\ &= \frac{m}{\omega_m} \frac{V_S^2}{\left(R_S + R_R'/s\right)^2 + \left(X_S + X_R'\right)^2} \frac{(1-s)}{s} R_R' \end{aligned} \tag{9.30}$$

We can plot torque T_{em} as a function of slip s from Eq. (9.30) and obtain the torque–speed characteristics of an induction motor as shown in Figure 9.21.

9.5.3 Speed Control of Induction Machine

The speed of an induction motor, in RPM, can be expressed as:

$$n = (1-s)n_S = (1-s)\frac{60f}{p} \tag{9.31}$$

Hence, we will have three approaches to change the speed of an induction motor: change the number of poles, change the frequency, and change the slip:

1) **Change number of poles**: The stator winding is designed such that, by changing the winding configuration, the number of poles will change. For example, some induction motors are designed as 4/6, 6/8, or 4/8 pole capable. While changing the number of poles has been used in controlling the induction motor speed in the past, it is used less and less today due to the complexity of the stator winding configuration and low efficiency.

2) **Change frequency of the supply voltage**: This is the most popular method for controlling induction motor speed in modern drive systems, including traction drives. This will be discussed in more detail in the next section.
3) **Change slip**: Since the electromagnetic torque of an induction motor is closely related to slip as shown in Eq. (9.31), there are a few ways to change the slip to control induction motor speed:
   a) **Change the magnitude of the supply voltage**: As shown in Figure 9.22, as the voltage is changed, the speed of the motor is also changed. However, this method provides limited variable speed range since the torque is proportional to the square of voltage.
   b) **Change stator resistance or stator leakage inductance**: This can be done by connecting a resistor or inductor in series with the stator winding.
   c) **Change rotor resistance or rotor leakage inductance**: This is only applicable to wound-rotor induction motors.
   d) **Apply an external voltage to the rotor winding**: This voltage has the same frequency as the rotor back emf or rotor current. Modern, doubly fed wind power generators belong to this group. This method is only applicable to wound-rotor induction motors.

When an external resistance is in series with the stator or rotor winding, there is loss associated with this resistor. Hence the system efficiency is compromised. When an external inductor is in series with the stator or rotor, the power factor is compromised. Hence, adding resistance or inductance is no longer a popular method in modern electric drive systems.

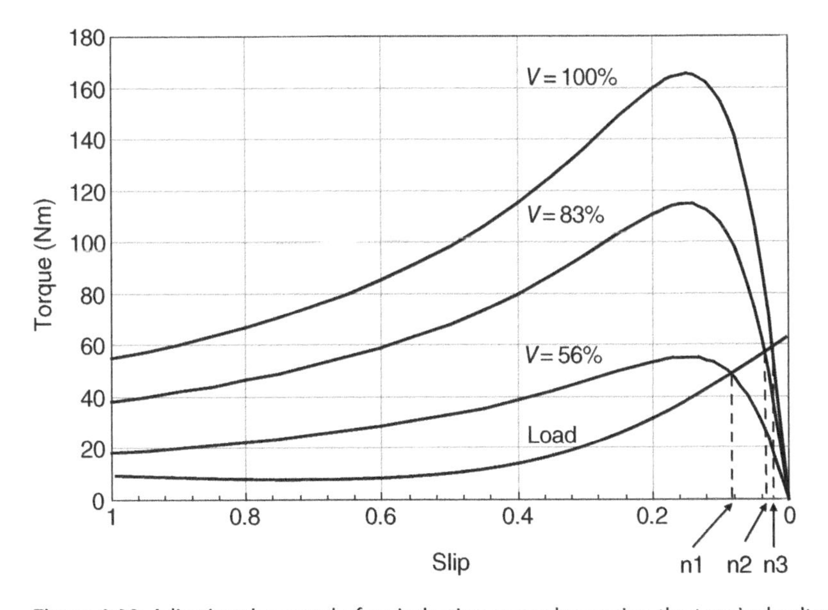

Figure 9.22 Adjusting the speed of an induction motor by varying the terminal voltage.

9.5.4 Variable Frequency, Variable Voltage Control of Induction Motors

Varying the frequency of power supply is by far the most effective and most popular method of adjusting the speed of an induction motor. If we neglect the stator resistance, leakage inductance, and the magnetic loss, the stator voltage equation can be written as:

$$V_S = E_S = k_S \omega \phi = k_S 2\pi f \Phi \tag{9.32}$$

where k_S is the machine constant and Φ is the total flux. Hence, when changing the frequency, the stator voltage should also be changed proportionally in order to maintain a relatively constant flux so that the stator and rotor core do not get saturated, while the output torque can be maintained constant:

$$\frac{V_S}{f} = \text{constant} \tag{9.33}$$

When the frequency and voltage are adjusted, the torque–speed characteristics are as shown in Figure 9.23. Although the above expression is generally true, three observations can be made:

1) For low-frequency operations, the voltage drop across the stator resistance and inductance are no longer negligible, so the stator voltage has to be increased to compensate.
2) The motor speed corresponding to the rated frequency and rated voltage is called the rated speed or base speed.
3) When the stator voltage reaches its rated supply (maximum), in order to further increase frequency (or speed), the stator flux must be reduced in order to satisfy Eq. (9.33). This is called the flux weakening operation. The ratio of the maximum speed to the rated base speed of the motor is defined as the adjustable speed range or X. Modern induction motors can achieve up to $X = 5$ adjustable speed range.

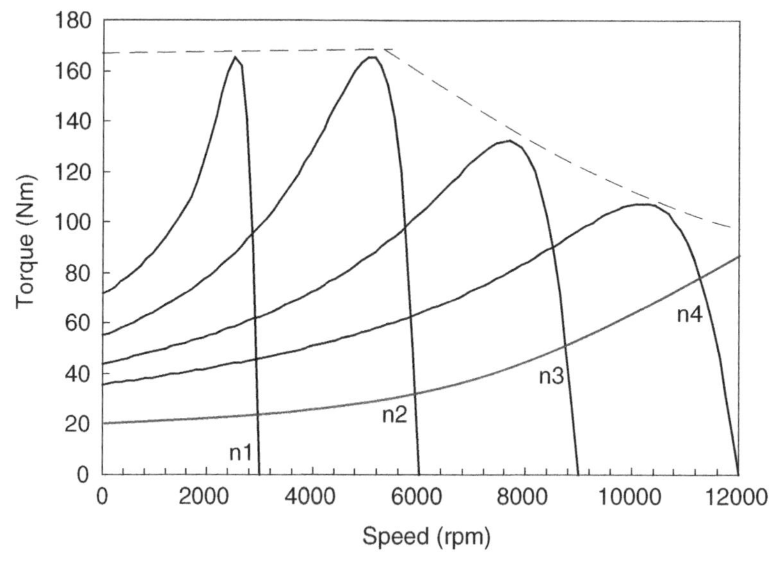

Figure 9.23 Adjusting induction motor speed using variable frequency supply. In this example, the rated speed is 6000 RPM, and the maximum speed is 12,000 RPM. The adjustable speed range $X = 2$.

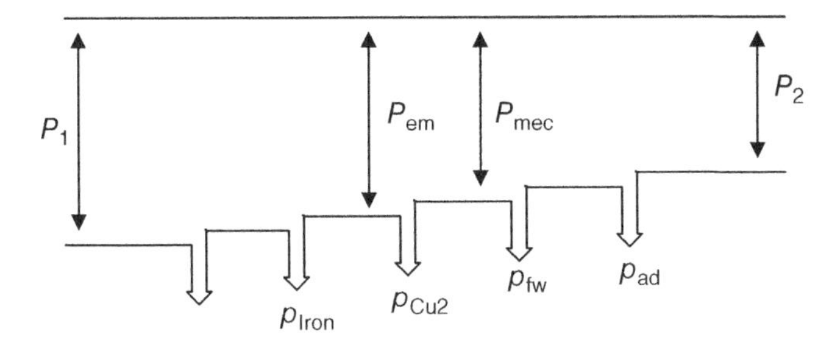

Figure 9.24 Losses in an induction motor.

9.5.5 Efficiency and Losses of Induction Machine

The losses in an induction machine are shown in Figure 9.24. The losses include: (1) copper loss in the stator winding; (2) magnetic loss in the stator iron (or core loss or iron loss); (3) copper loss in the rotor winding; (4) windage loss due to the rotation of the rotor and frictional loss in the bearing; and (5) additional losses that cannot be accounted for by the above components, also called additional loss or stray load loss.

The power balance equations are:

$$\begin{aligned} P_{em} &= P_1 - p_{cu1} - p_{iron} \\ P_{mec} &= P_{em} - p_{cu2} \\ P_2 &= P_{mec} - p_{fw} - p_{ad} \end{aligned} \tag{9.34}$$

P_1 is the input power from the voltage supply; P_{em} is the electromagnetic power transferred from the stator to the rotor; P_{mec} is the total mechanical power on the rotor shaft; P_2 is the output power to the load connected to the shaft; p_{cu1} is the copper loss of the stator winding; p_{cu2} is the copper loss of the rotor; p_{iron} is the iron loss of the stator core; p_{fw} is the frictional and windage loss; and p_{ad} is the stray load loss.

The efficiency can be expressed as:

$$\eta = \frac{P_2}{P_1} = \frac{P_2}{P_2 + p_{cu1} + p_{iron} + p_{cu2} + p_{fw} + p_{ad}} \tag{9.35}$$

One aspect of traction motors for modern HEVs is high-speed operation. Traditionally, laminated silicon steel sheets were designed for use at low frequencies (50 or 60 Hz), and today's traction drives typically operate at about 6000–15,000 RPM. With four-pole designs, the operating frequency is 500 Hz. Some traction motors operate at frequencies as high as 800–1200 Hz. Since eddy current loss and hysteresis loss are proportional to frequency or the square of the frequency, the core loss will be significant at high frequencies. In order to keep the core loss within a reasonable range, the magnetic flux in the iron has to be relatively lower than that used in low-speed motors, and the thickness of the silicon steel sheets may have to be reduced as well.

The second aspect is that the inverter-operated induction motor will contain harmonics in its voltage and current. These harmonics will introduce additional losses in the winding and stator and rotor core. As is well known, the eddy current loss can be doubled in many induction motors due to the pulse width modulated (PWM) supply. These additional losses may cause excessive temperature rise which must be considered during the design and analysis of induction motors.

9.5.6 Field-Oriented Control of Induction Machine

With field-oriented control, an induction machine can perform somewhat like a DC machine. This section explains the theory and implementation of the field-oriented control of an induction machine [42].

When expressed in phasors, the voltage equation for a three-phase induction machine with three symmetrical stator windings is given as:

$$V_S = R_S i_S + p\lambda_S \tag{9.36}$$

$$V_R = R_R i_R + p\lambda_R \tag{9.37}$$

where p is the differential operand d/dt, and V, I, and λ are phasors of voltage, current, and flux linkage respectively. Subscript S relates to stator quantities and R refers to rotor quantities. Equations (9.36) and (9.37) are expressed in stator and rotor coordinates respectively. Therefore, stator frame S is stationary and rotor frame R is rotational (rotor quantities are at rotor frequency or slip frequency).

Suppose there is a frame B, and the angle between the stator and this frame B is δ, therefore the angle between the rotor and this frame is $(\delta - \theta)$. Multiplying Eq. (9.36) by $e^{-j\delta}$ and Equation 37 by $e^{-j(\delta-\theta)}$, we get:

$$\begin{aligned} V_S \cdot e^{-j\delta} &= R_S i_S \cdot e^{-j\delta} + p\lambda_S \cdot e^{-j\delta} \\ V_R \cdot e^{-j(\delta-\theta)} &= R_R i_R \cdot e^{-j(\delta-\theta)} + p\lambda_R \cdot e^{-j(\delta-\theta)} \end{aligned} \tag{9.38}$$

Let:

$$\begin{aligned} V_S^{(B)} &= V_S \times e^{-j\delta}, \quad V_R^{(B)} = V_R \times e^{-j(\delta-\theta)} \\ i_S^{(B)} &= i_S \times e^{-j\delta}, \quad i_R^{(B)} = i_R \times e^{-j(\delta-\theta)} \\ \lambda_S^{(B)} &= \lambda_S \times e^{-j\delta}, \quad \lambda_R^{(B)} = \lambda_R \times e^{-j(\delta-\theta)} \end{aligned} \tag{9.39}$$

By employing the equation:

$$p\left(\lambda_S^{(B)}\right) = p\left(\lambda_S e^{-j\delta}\right) = -j \cdot \lambda_S (p\delta) e^{-j\delta} + p\lambda_S e^{-j\delta} \tag{9.40}$$

Or:

$$p\lambda_S e^{-j\delta} = p\left(\lambda_S^{(B)}\right) + j\lambda_S (p\delta) e^{-j\delta} \tag{9.41}$$

$$p\lambda_R e^{-j(\delta-\theta)} = p\left(\lambda_R^{(B)}\right) + j\lambda_R e^{-j\delta} p(\delta - \theta) \tag{9.42}$$

Equations (9.41) and (9.42) can then be transferred to a general frame B, where all space phasors are expressed in frame B with the superscript (B) as:

$$V_S^{(B)} = R_S i_S^{(B)} + p\lambda_S^{(B)} + j \cdot \lambda_S^{(B)} p\delta \tag{9.43}$$

$$V_R^{(B)} = R_R i_R^{(B)} + p\lambda_R^{(B)} + j \cdot \lambda_R^{(B)} p(\delta - \theta) \tag{9.44}$$

The superscript (B) will be omitted further in this section for convenience. When expressed in phasors, the flux linkage can be expressed as:

$$\lambda_S = (L_m + L_{1\sigma}) i_S + L_m i_R \tag{9.45}$$

$$\lambda_R = L_m i_S + (L_m + L_{2\sigma}) i_R \tag{9.46}$$

where L_m is the stator inductance and $L_{1\sigma}$ and $L_{2\sigma}$ are the stator and rotor leakage inductance respectively.

Note that although the phasors are in a different frame, the stator flux and rotor flux are rotating at the same speed.

For squirrel cage induction machines, the rotor current i_R is not accessible. Therefore, a fictitious rotor magnetizing current i_{mr} is defined such that the rotor flux can be expressed in terms of this fictitious rotor magnetizing current and stator inductance in the same way as in Eq. (9.45):

$$\lambda_R = i_{mr} \cdot L_m \tag{9.47}$$

The rotor current can then be expressed as a function of magnetizing current and stator current from Eq. (9.46):

$$i_R = \frac{i_{mr} - i_S}{1 + \sigma} \tag{9.48}$$

where:

$$\sigma = L_{2\sigma}/L_m \tag{9.49}$$

Substituting Eqs (9.47) and (9.48) into Eq. (9.44) and considering that V_R is normally set to 0 for squirrel cage induction motors, the rotor equation can be rewritten as:

$$0 = i_{mr} - i_S + T_r p i_{mr} + jT_r\, i_{mr} p(\delta - \theta) \tag{9.50}$$

where T_r is the rotor time constant which can be expressed as:

$$T_r = L_m(1 + \sigma)/R_R \tag{9.51}$$

As stated above, the rotor magnetizing current is a fictitious current. The magnitude of this current can be observed through the following approach. If the rotor equation is written in the stator frame then $\delta = 0$, $p\theta$ is equal to the speed of the rotor ω, and Eq. (9.50) has the following form:

$$0 = i_{mr} - i_S + T_r p i_{mr} - jT_r\, i_{mr}\omega \tag{9.52}$$

Since this equation is written in the stator frame, we can find the α and β components of phasors i_S and i_{mr}:

$$\begin{aligned} i_S &= i_{S\alpha} + j i_{S\beta} \\ i_{mr} &= i_{mr\alpha} + j i_{mr\beta} \end{aligned} \tag{9.53}$$

Therefore Eq. (9.52) becomes:

$$\begin{aligned} \frac{di_{mr\alpha}}{dt} &= \frac{1}{T_r}(i_{s\alpha} - i_{mr\alpha}) - i_{mr\beta}\omega \\ \frac{di_{mr\beta}}{dt} &= \frac{1}{T_r}(i_{s\beta} - i_{mr\beta}) + i_{mr\alpha}\omega \end{aligned} \tag{9.54}$$

Stator current can be easily transferred from the *abc* system to the $\alpha\beta$ system. Eq. (9.54) can be implemented discretely in the time domain, therefore, $i_{mr\alpha}$ and $i_{mr\beta}$ can be observed. Once this has been done, i_{mr} and δ_r can finally be calculated:

$$i_{mr} = \sqrt{i_{mr\alpha}^2 + i_{mr\beta}^2},\ \cos(\delta_r) = i_{mr\alpha}/i_{mr},\ \sin(\delta_r) = i_{mr\beta}/i_{mr} \tag{9.55}$$

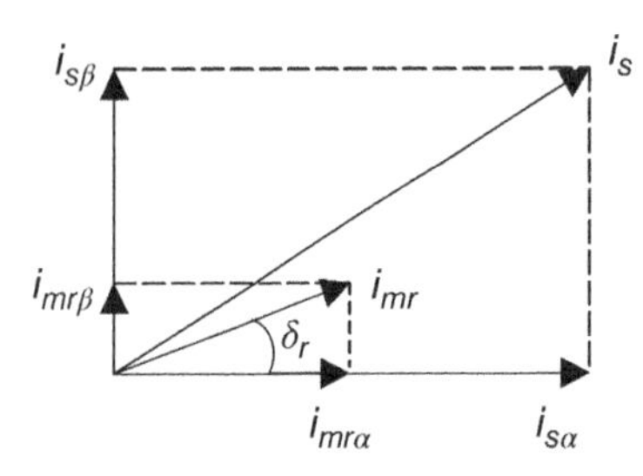

Figure 9.25 Stator and rotor current in α, β coordinates.

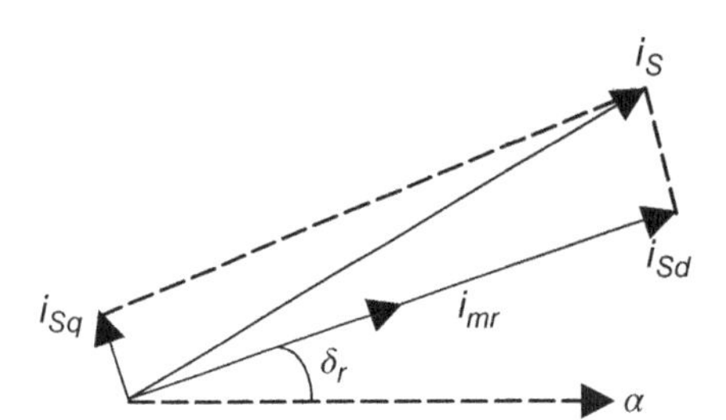

Figure 9.26 Stator current in d, q and α, β coordinates.

where δ_r is the angle between the fictitious current i_{mr} and the stator current $i_{S\alpha}$ as shown in Figure 9.25.

If the frame is chosen such that B is aligned with λ_R as shown in Figure 9.26, i_{mr} will only have real components. Therefore this rotor equation can then be decomposed into its direct and quadrature components as:

$$\begin{aligned} i_{sd} &= i_{S\alpha}\cos\delta_r + i_{s\beta}\sin\delta_r \\ i_{sq} &= -i_{S\alpha}\sin\delta_r + i_{s\beta}\cos\delta_r \end{aligned} \tag{9.56}$$

From Eq. (9.45), when i_s is decomposed to d, q components, the equation can be written as:

$$i_{mr} - i_{sd} + T_r p i_{mr} = 0 \tag{9.57}$$

$$-i_{sq} + T_r\, i_{mr} p(\delta - \theta) = 0 \tag{9.58}$$

From Eq. (9.57) it can be seen that i_{mr} is only related to i_{sd}. Therefore, i_{mr} can be controlled by controlling i_{sd}.

The torque in the machine is:

$$T_q = \frac{3}{2}\cdot\frac{p}{2}\cdot(\lambda_S \times i_S) \tag{9.59}$$

which has to be balanced with the load and acceleration torque:

$$T_q = T_L + J \cdot p\omega \tag{9.60}$$

where T_q is the developed torque, T_L is the load torque, and ω is the angular speed of the motor. If θ is the angle between the stator and the rotor, then $\omega = p\theta$.

It can also be proved that i_{sq} is directly related to motor torque as follows. By substituting i_S and i_R into the torque in Eq. (9.59), the torque can be derived:

$$T_q = \frac{3}{2}\cdot\frac{p}{2}\cdot\frac{L_S}{1+\sigma} i_{mr}\, i_{sq} \tag{9.61}$$

Magnetizing current i_{mr} can be controlled by controlling the real component of stator current, and torque control are achieved by controlling the imaginary component of the stator current.

For ease of implementing control, we will introduce the per-unit system. A per-unit system is essentially a system of dimensionless parameters occurring in a set of wholly or partially dimensionless equations. This kind of system can extensively simplify the phenomena of problems. The parameters of the machines fall in a reasonably narrow numerical range when expressed in a per-unit system related to their ratings and therefore this is extremely useful in simulating machine systems and implementing the control of electric machine by digital computers. Generally, rated power and frequency can be chosen respectively as the base values of power and frequency for normalization, whereas the peak values of rated phase current and phase voltage may be chosen respectively as the base values of current and voltage. Derived base values of impedance, inductance, and flux leakage are as follows (with subscript B indicating the variable as base value):

$$\begin{aligned} Z_B &= V_B / I_B \\ L_B &= Z_B / \omega_B \\ \lambda_B &= L_B I_B \end{aligned} \tag{9.62}$$

Normalized torque can be expressed as:

$$T_{qB} = \frac{3}{2}\frac{p}{2}\lambda_B I_B = \frac{3}{2}\frac{p}{2} L_B I_B^2 \tag{9.63}$$

The torque equation can then be normalized. Dividing Eq. (9.61) by Eq. (9.63), we get:

$$T_q^* = \frac{1}{1+\sigma}\frac{L_S}{L_B}\frac{i_{mr}}{I_B}\frac{i_{sq}}{I_B} = \frac{L_S^*}{1+\sigma} i_{mr}^* i_{sq}^* \tag{9.64}$$

where superscript $*$ donates the normalized value. For convenience, superscript $*$ will be omitted in further derivations. To implement the control strategy, a technique has to be developed to identify the magnitude of the magnetizing current i_{mr} and the angle δ_r.

There are two ways to implement the flux observer of Eq. (9.54). One way is to take the Laplace transform of Eq. (9.49) and apply a bilinear transformation to convert the Laplace transform to the z transform. The inverse z transform can be used to obtain $i_{mr\alpha}$ and $i_{mr\beta}$ in the discrete time domain. An alternative method is to discretize Eq. (9.49) directly in the time domain. Assuming that the sample time is T_s, then the following equation can be obtained from Eq. (9.54):

$$\begin{aligned} \frac{(i_{mr\alpha_i} - i_{mr\alpha_{i-1}})}{T_s} &= \frac{1}{T_r}\left(\frac{i_{s\alpha_i} + i_{s\alpha_{i-1}}}{2} - \frac{i_{mr\alpha_i} + i_{mr\alpha_{i-1}}}{2}\right) - \frac{i_{mr\beta_i} + i_{mr\beta_{i-1}}}{2}\omega \\ \frac{(i_{mr\beta_i} - i_{mr\beta_{i-1}})}{T_s} &= \frac{1}{T_r}\left(\frac{i_{s\beta_i} + i_{s\beta_{i-1}}}{2} - \frac{i_{mr\beta_i} + i_{mr\beta_{i-1}}}{2}\right) - \frac{i_{mr\alpha_i} + i_{mr\alpha_{i-1}}}{2}\omega \end{aligned} \tag{9.65}$$

Therefore $i_{mr\alpha}$ and $i_{mr\beta}$ can be derived from Eq. (9.60):

$$\begin{aligned} i_{mr\alpha_i} &= \frac{1-\kappa}{1+\kappa} i_{mr\alpha_{i-1}} + \frac{\kappa}{1+\kappa}\left(i_{s\alpha_i} + i_{s\alpha_{i-1}}\right) - T_r \frac{\kappa}{1+\kappa}\left(i_{mr\beta_i} + i_{mr\beta_{i-1}}\right)\omega \\ i_{mr\beta_i} &= \frac{1-\kappa}{1+\kappa} i_{mr\beta_{i-1}} + \frac{\kappa}{1+\kappa}\left(i_{s\beta_i} + i_{s\beta_{i-1}}\right) + T_r \frac{\kappa}{1+\kappa}\left(i_{mr\alpha_i} + i_{mr\alpha_{i-1}}\right)\omega \end{aligned} \tag{9.66}$$

where κ is the ratio of sampling time to rotor constant:

$$\kappa = T_s/2T_r \tag{9.67}$$

The time variables can also be made dimensionless by multiplying both sides of the equations by ω_B. Therefore both T_s and T_r are expressed in per-unit values in Eqs (9.66) and (9.67.) A block diagram of the flux observer is shown in Figure 9.27. The flux observer takes the phase currents and speed as input and calculates i_{mr}, $\cos\alpha$, and $\sin\alpha$.

It has been shown in the previous sections that it is possible to control the magnetizing component and torque component of the stator current separately. A PI controller is one way to implement control. The numerical expression for a PI controller is:

$$V_o = K_{PI}\left(T_{PI}\varepsilon + \int \varepsilon dt\right) \tag{9.68}$$

where V_o is the output of the PI controller and ε is the error signal of input V_i (here V_i can be the measured current or torque of the motor, and V_o can be the PWM signal). In order to get the time domain discrete expression, we differentiate Eq. (9.68):

$$\frac{dV_o}{dt} = K_{PI}\left(T_{PI}\frac{d\varepsilon}{dt} + \varepsilon\right) \tag{9.69}$$

Further implementation is straightforward:

$$\frac{V_{oi} - V_{oi-1}}{T_S} = K_{PI}\, T_{PI}\frac{\varepsilon_i - \varepsilon_{i-1}}{T_S} + K_{PI}\frac{\varepsilon_i + \varepsilon_{i-1}}{2} \tag{9.70}$$

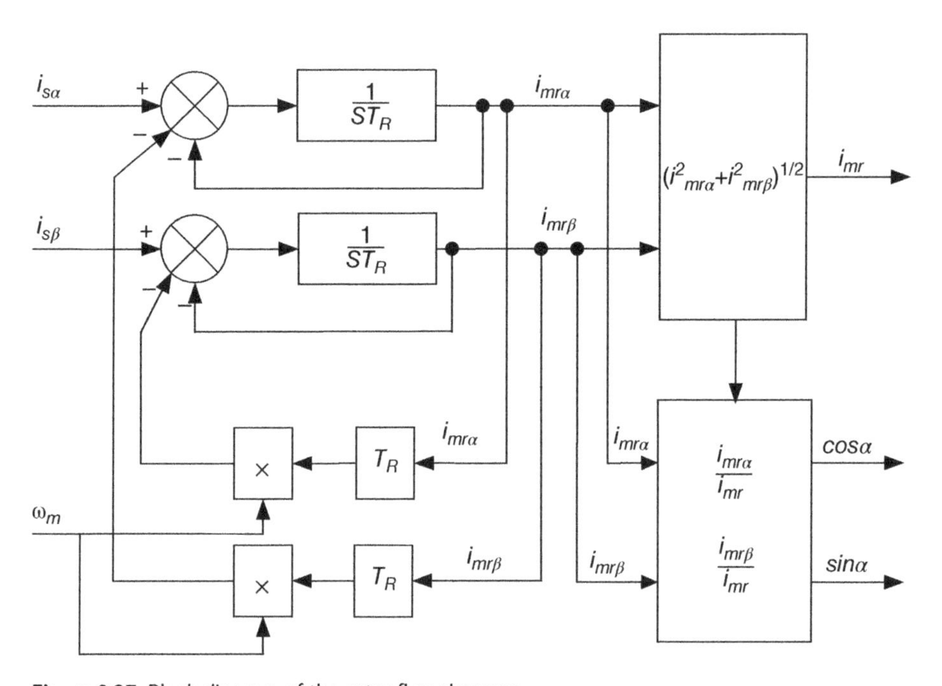

Figure 9.27 Block diagram of the rotor flux observer.

$$V_{oi} = V_{oi-1} + K_1(\varepsilon_i - K_2\,\varepsilon_{i-1}) \tag{9.71}$$

where K_1 and K_2 can be expressed as:

$$\begin{aligned} K_1 &= (1 + T_S/2T_r)K_{PI}\,T_{PI} \\ K_2 &= \frac{(1 - T_S/2T_r)}{(1 + T_S/2T_r)} \end{aligned} \tag{9.72}$$

For example, if the gain is chosen as $K_{PI} = 50$, $T_{PI} = 0.02$ seconds, sampling time $T_r = 0.02$ seconds, and $T_s = 0.67$ ms, then the constants K_1 and K_2 are $K_1 = 1.0168$, $K_2 = 0.9671$.

The purpose of field-oriented control is to control an induction machine in such a way that it behaves like a DC motor. A block diagram is shown in Figure 9.28 and the flowchart is shown in Figure 9.29. An incremental encoder is used to measure the speed of the motor. As shown in Eq. (9.54), the magnetizing current does not change instantaneously with i_{sd} as it does in a DC motor. Rather, the magnetizing current lags a time constant T_r corresponding to the change of i_{sd}.

In this setup, the flux observer uses the speed signal of an incremental encoder and the current measurement through two external current sensors. Only the currents of two phases are needed to perform the coordinate transformations due to symmetry.

9.6 Permanent Magnet Motor Drives

PM motors are the most popular choices for EV and HEV powertrain applications due to their high efficiency, compact size, high torque at low speeds, and ease of control for regenerative braking [43–90]. The PM motor in an HEV powertrain is operated either as a motor during normal driving or as a generator during regenerative braking and power splitting, as required by the vehicle operations and control strategies. PM motors with higher power densities are also now increasingly the choice for aircraft, marine, naval, and space applications.

The most commercially used PM material in traction drive motors is neodymium-ferrite-boron (Nd–Fe–B). This material has a very low Curie temperature and high-temperature sensitivity. It is often necessary to increase the size of magnets to avoid demagnetization at high temperatures and high currents. On the other hand, it is advantageous to use as little PM material as possible in order to reduce the cost without sacrificing the performance of the machine.

9.6.1 Basic Configuration of PM Motors

When PMs are used to generate the magnetic field in an electric machine, it becomes a PM motor. Both DC and AC motors can be made with PMs. Only PM synchronous motors and PM brushless DC motors are chosen for modern traction drives. We will primarily explain the operation of PM synchronous motors in this book.

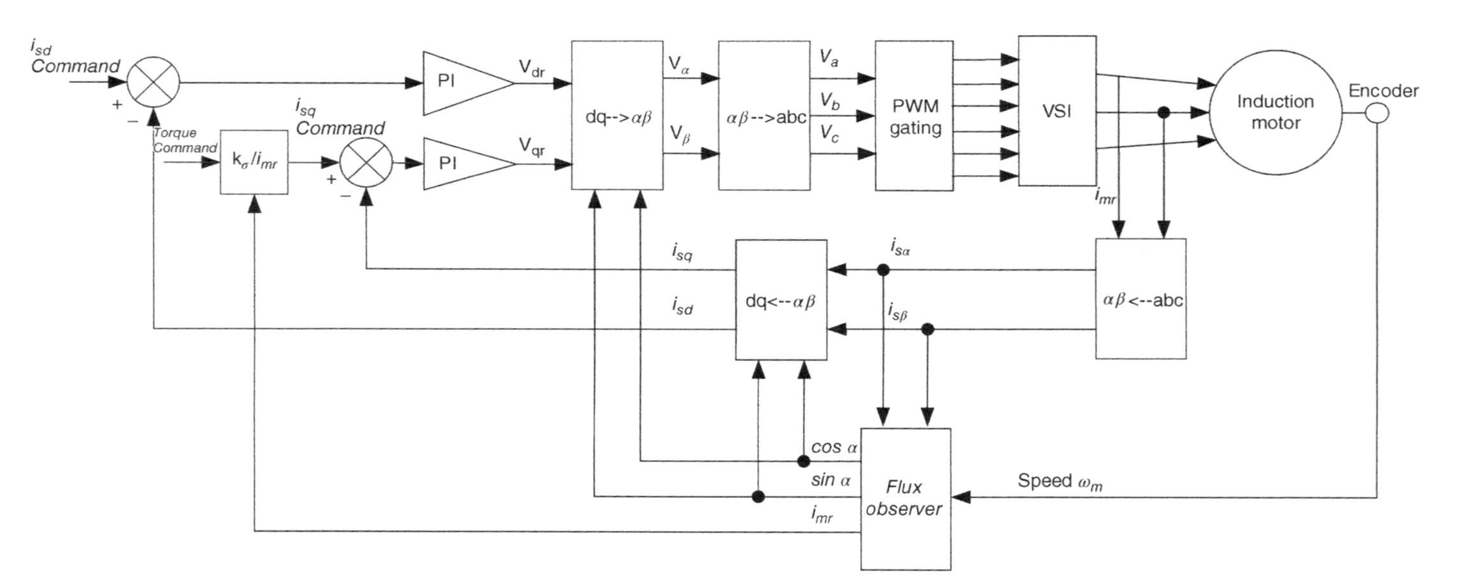

Figure 9.28 Field-oriented control of an induction machine.

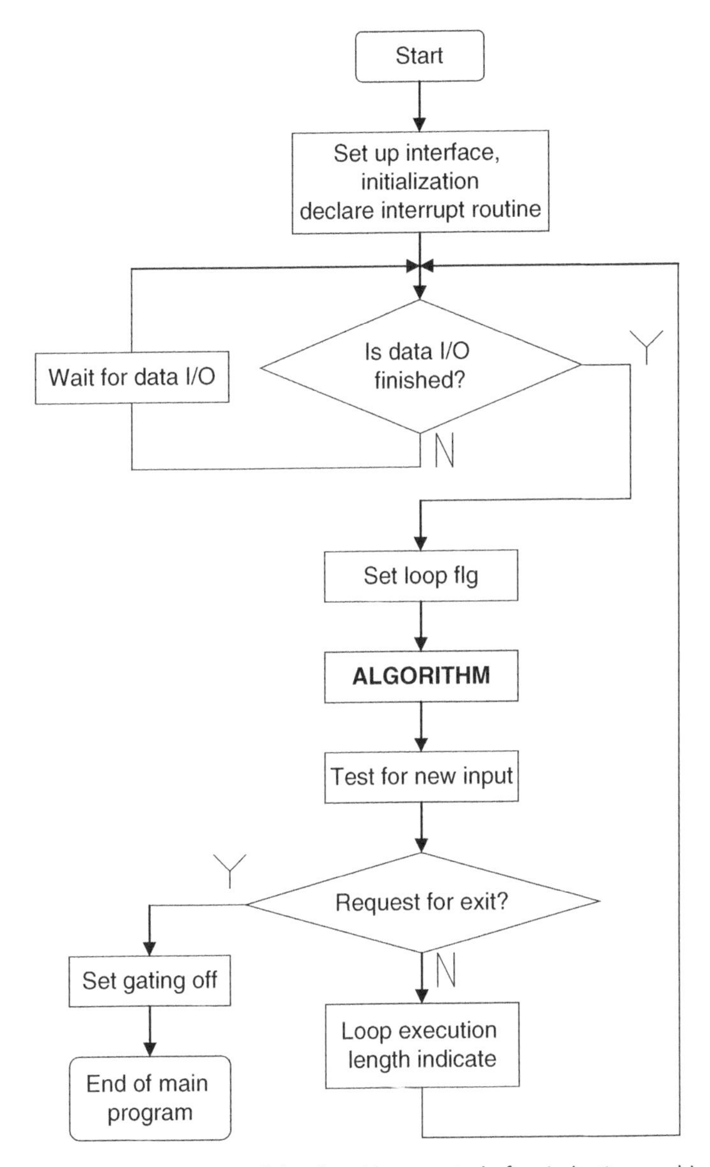

Figure 9.29 Flowchart of the closed-loop control of an induction machine.

A PM synchronous motor contains a rotor and a stator, with the stator similar to that of an induction motor, and the rotor contains the PMs. From the section on induction motors, we know that the three-phase winding, with three-phase symmetrical AC supply, will generate a rotating magnetic field. To generate a constant average torque, the rotor must follow the stator field and rotate at the same synchronous speed. This is also why these machines are called PM synchronous motors.

There are different ways to place the magnets on the rotor, as shown in Figure 9.30. If the magnets are glued to the surface of the rotor, it is called a surface-mounted PM motor (SPM). If the magnets are inserted inside the rotor in the pre-cut slots, then it is called an interior permanent magnet motor (IPM).

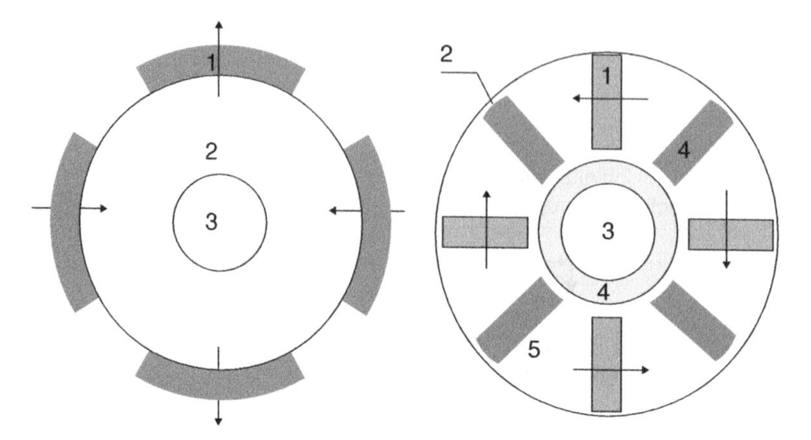

Figure 9.30 Surface-mounted magnets and interior magnets: left, SPM motor; right, IPM motor. 1 – magnet; 2 – iron core; 3 – shaft; 4 – non-magnetic material; 5 – non-magnetic material.

For an SPM motor, the rotor can be a solid piece of steel since the rotor iron core itself is not close to the air gap, hence the eddy current loss and hysteresis loss due to slot/tooth harmonics can be neglected. For the IPM motor, the rotor needs to be made out of laminated silicon steel since the tooth/slot harmonics will generate eddy current and hysteresis losses.

Due to the large air gap as well as the fact that the magnets have a permeability similar to that of air, SPM motors have similar direct-axis reactance x_d and quadrature-axis reactance x_q. On the other hand, IPM motors have different x_d and x_q. This difference will generate a so-called reluctance torque. It is worth pointing out that although there is a reluctance torque component, it does not necessarily mean that an IPM motor will have a higher torque rating than an SPM motor for the same size and same amount of magnetic material used. This is because, in IPM motors, in order to keep the integrity of the rotor laminations, there are so-called "magnetic bridges" that will have leakage magnetic flux. So for the same amount of magnet material used, an SPM motor will always have higher total flux. There are many different configurations for IPM motors as shown in Figure 9.31. The exploded view of a PM synchronous motor is shown in Figure 9.32.

9.6.2 Basic Principle and Operation of PM Motors

The no-load magnetic field of PM machines is shown in Figure 9.33. When the rotor is driven by an external source (such as an engine), the rotating magnetic field will generate a three-phase voltage in the three-phase windings. This is the generator mode operation of the PM machine.

When operated as a motor, the three-phase windings, similar to those of an induction motor, are supplied with either a trapezoidal form of current (brushless DC) or sinusoidal current (synchronous AC). These currents generate a magnetic field that is rotating at the same speed as the rotor or synchronous speed. By adjusting the frequency of the stator current, the speed of the rotor or the synchronous speed can be adjusted accordingly.

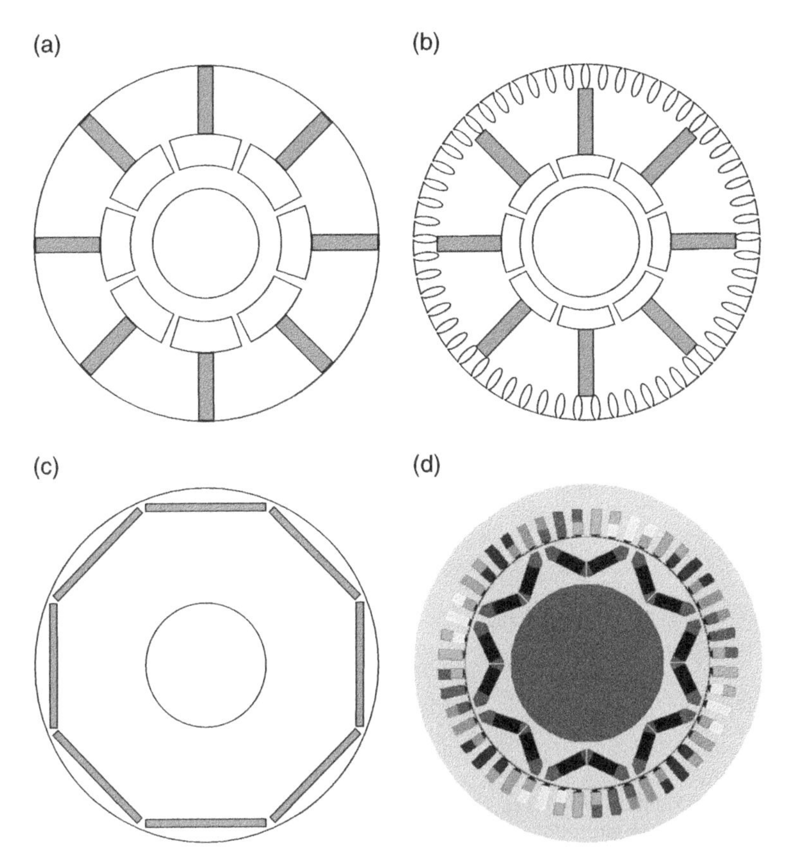

Figure 9.31 Four commonly used IPM rotor configurations: (a) circumferential-type magnets suitable for brushless DC or synchronous motor; (b) circumferential-type magnets for the line-start synchronous motor; (c) rectangular slots IPM motor; (d) V-type slots IPM motor.

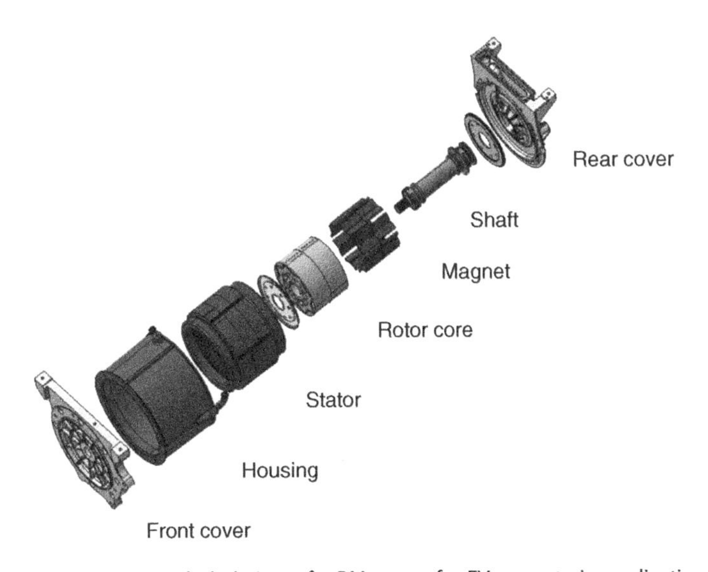

Figure 9.32 Exploded view of a PM motor for EV powertrain applications.

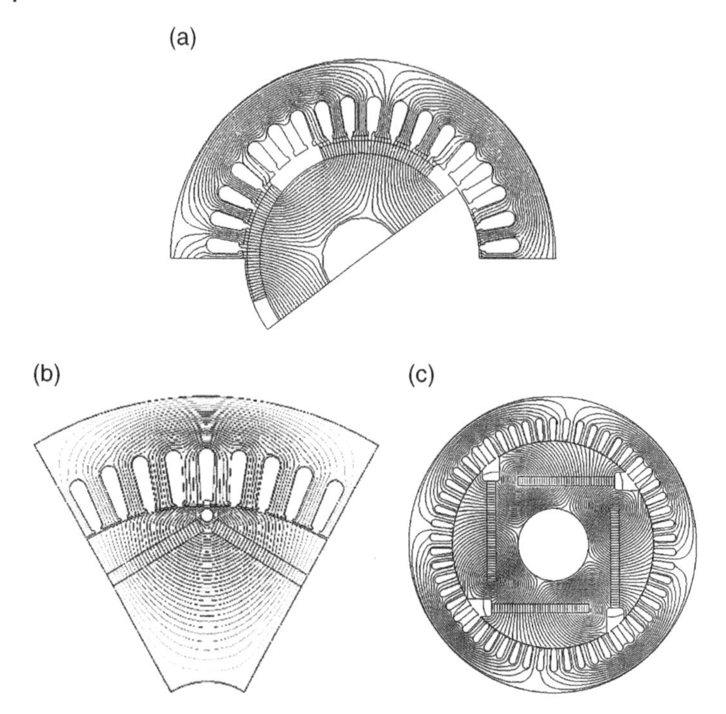

Figure 9.33 Magnetic field distribution of PM machines at no-load conditions (the stator current is zero): (a) a four-pole SPM motor; (b) an eight-pole symmetrical IPM motor; (c) a four-pole unsymmetrical IPM configuration.

The torque is the attraction between the rotor magnetic field and the stator magnetic field in the circumferential direction. Hence, under no-load conditions, the rotor and the stator field are almost lined up. When the angle between the rotor field and the stator field reaches 90 electric degrees, the maximum torque is reached in SPM motors. For IPM motors, the maximum torque occurs at an angle slightly larger than 90°, due to the existence of reluctant torque.

Figure 9.34 illustrates how a PM motor operates in different modes. The stator winding generates a rotating field that attracts the rotor magnets. If the two fields are lined up, the attraction between the two magnetic fields is in the radial direction, hence there is no electromagnetic torque. When the stator field is leading the rotor field, the stator will attract the rotor magnets. The machine then operates as a motor. When the stator field is lagging the rotor field, the machine becomes a generator.

At no load, the rotor magnetic field will generate a back emf E_o in the stator windings. When a voltage with the same frequency is applied to the stator windings, then a current will be generated and the voltage equation can be written as;

$$V = E_o + IR + jIX \tag{9.73}$$

where R is the stator resistance and X is the synchronous impedance. The phasor diagram is shown in Figure 9.35 when neglecting the stator resistance. From the diagram, the term jIX can be further decomposed into two components: jI_dX_d and jI_qx_q. In fact, in

Figure 9.34 Operation of a PM synchronous machine: (a) no load; (b) operating as generator; (c) operating as motor.

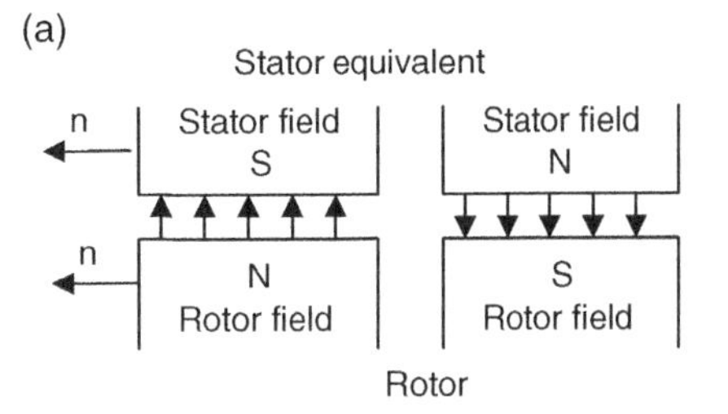

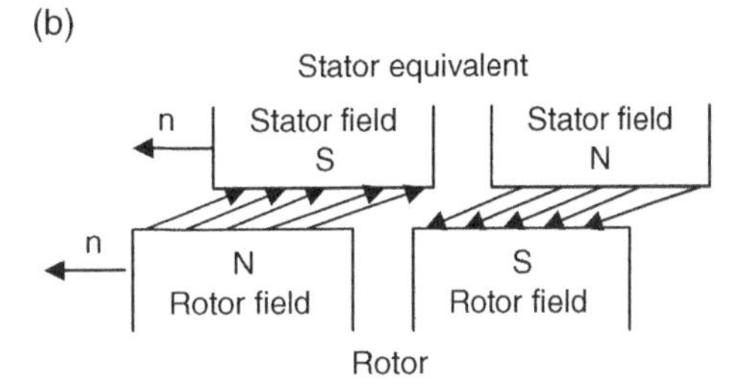

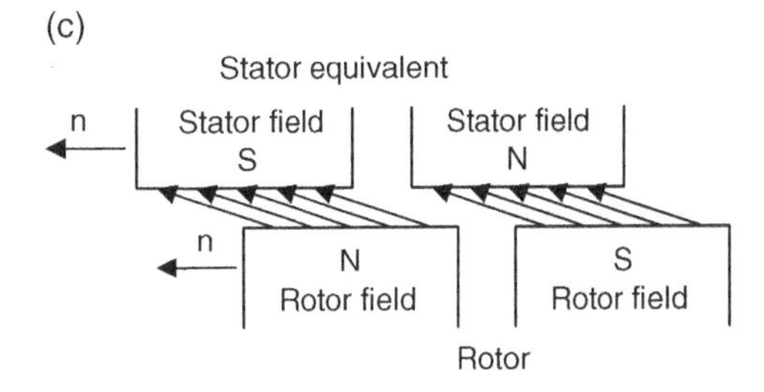

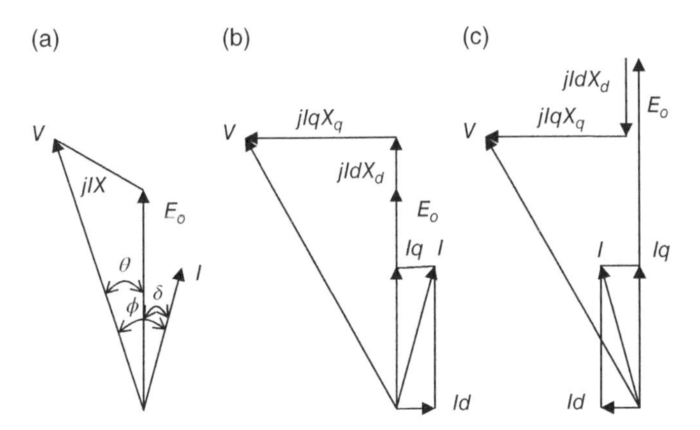

Figure 9.35 Phasor diagram of PM synchronous motors: (a) SPM; (b) IPM; (c) flux weakening mode of IPM.

IPM motors, the d axis and q axis will have different reactances. By using Figure 9.35, Eq. (9.73) can be rewritten for IPM motors as:

$$V = E_o + IR + jI_d X_d + jI_q X_q \tag{9.74}$$

The real power can be calculated, since from Figure 9.35, $\phi = \delta + \theta$:

$$\begin{aligned} P_1 &= mIV\cos\phi = mIE_o\cos\delta = mV(I\cos\delta\cos\theta - I\sin\delta\sin\theta) \\ &= mV\left(I_q\cos\theta - I_d\sin\theta\right) \end{aligned} \tag{9.75}$$

where ϕ is the power factor angle (the angle between the voltage and current), θ is the angle between the voltage and back emf, and δ is the inner power angle (the angle between the back emf and the voltage). From Figure 9.35:

$$\begin{aligned} I_q X_q &= V\sin\theta \\ I_d X_d &= V\cos\theta - E_o \end{aligned} \tag{9.76}$$

Therefore, the power of PM motors can be expressed as:

$$P = \frac{mE_oV}{X_d}\sin\theta + \frac{mV^2}{2}\left(\frac{1}{X_q} - \frac{1}{X_d}\right)\sin(2\theta) \tag{9.77}$$

The torque can be derived by dividing Eq. (9.77) by the rotor speed as shown in Figure 9.36, where the torque–speed characteristics of a typical PM motor are shown. For SPM motors, since $X_d = X_q$, the second term of Eq. (9.77) is zero. For IPM motors, the q axis has less reluctance due to the existence of soft iron in its path, and the d axis has magnets in its path which has larger reluctance. Therefore X_q is much larger than X_d.

On the other hand, from Eq. (9.75), and neglecting losses, we can see that:

$$\begin{aligned} T &= \frac{mIE_o\cos\delta}{\omega/p} = \frac{mIk\omega\phi}{\omega/p}\cos\delta = mpkI\phi\cos\delta \\ T_{\max} &= mpkI\phi = \text{constant} \end{aligned} \tag{9.78}$$

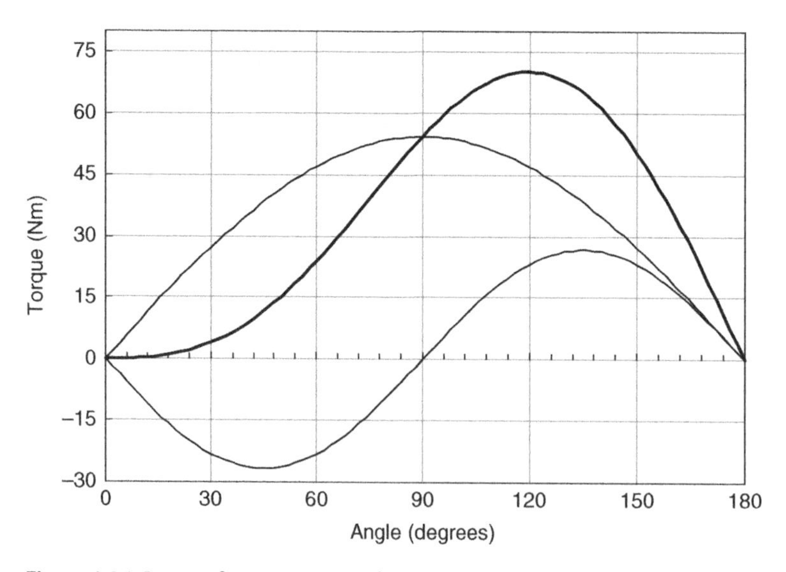

Figure 9.36 Power of IPM motor as a function of inner power angle.

Therefore, when inner power angle $\delta = 0$, for a given stator current, the torque of the motor reaches its maximum. In this condition, the stator current is in phase with the back emf E_o, and:

$$\begin{aligned} V^2 &= E_o^2 + \left(I_q X_q\right)^2 = (k\omega\phi)^2 + \left(I\omega L_q\right)^2 \\ \frac{V}{\omega} &= \sqrt{(k\phi)^2 + \left(IL_q\right)^2} = \text{constant} \end{aligned} \tag{9.79}$$

Hence, the stator voltage must be proportional to frequency to satisfy Eq. (9.79) and maintain maximum torque output at the same time. This operation is also called constant torque operation. It can also be seen from Eq. (9.77) that for a given θ, the power is inversely proportional to the frequency, since V, X_d, and E_o are all proportional to frequency ω. This is similar to the V/f control of induction motors.

When stator voltage reaches its maximum, Eq. (9.79) can no longer be maintained. As ω increases, V becomes constant, and a current in the d axis direction must be supplied, as shown in Figure 9.35c. The relationship between the voltage and frequency can be expressed as

$$\begin{aligned} V^2 &= (E_o - I_d X_d)^2 + \left(I_q X_q\right)^2 = (k\omega\phi - I_d\omega L_d)^2 + \left(I_q\omega L_q\right)^2 \\ \frac{V}{\omega} &= \sqrt{(k\phi - I_d L_d)^2 + \left(I_q L_q\right)^2} \end{aligned} \tag{9.80}$$

This operation is also called the flux weakening operation region because the d axis current generates a magnetic flux in the opposite direction to the PM field. Note that, due to constraints such as the current limit of the inverter, the q axis current may have to be decreased from its rated value so that the total current from the inverter is kept the same. Additional losses at higher speeds may make it necessary to further reduce the torque output. It can also be seen from Eq. (9.77) that for a given θ, the first term is constant since V is constant, and both X_d and E_o are proportional to frequency ω. In theory, the torque is inversely proportional to frequency in this operation, so the power is constant. Hence this mode is also referred to as the constant power operation range.

The torque–speed characteristics can be plotted as shown in Figure 9.37.

The efficiency of PM motors is typically higher than induction motors since they do not need excitation for the magnetic field, while it is needed for induction and DC motors. The losses in a PM machine are shown in Figure 9.38. The losses include: (1) copper loss in the stator winding; (2) magnetic loss in the stator iron (or core loss or iron loss); (3) magnetic loss in the rotor magnet as well as losses of the rotor steel; (4) windage loss due to the rotation of the rotor and frictional loss in the bearing; and (5) additional losses that cannot be accounted for by the above components, also called additional loss or stray load loss.

The power balance equations are:

$$\begin{aligned} P_{em} &= P_1 - p_{cu1} - p_{iron} \\ P_{mec} &= P_{em} - p_{mag} \\ P_2 &= P_{mec} - p_{fw} - p_{ad} \end{aligned} \tag{9.81}$$

where P_1 is the input power from the voltage supply, P_{em} is the electromagnetic power transferred from the stator to the rotor, P_{mec} is the total mechanical power of the rotor shaft, P_2 is the output power to the load connected to the shaft, p_{cu1} is the copper loss of

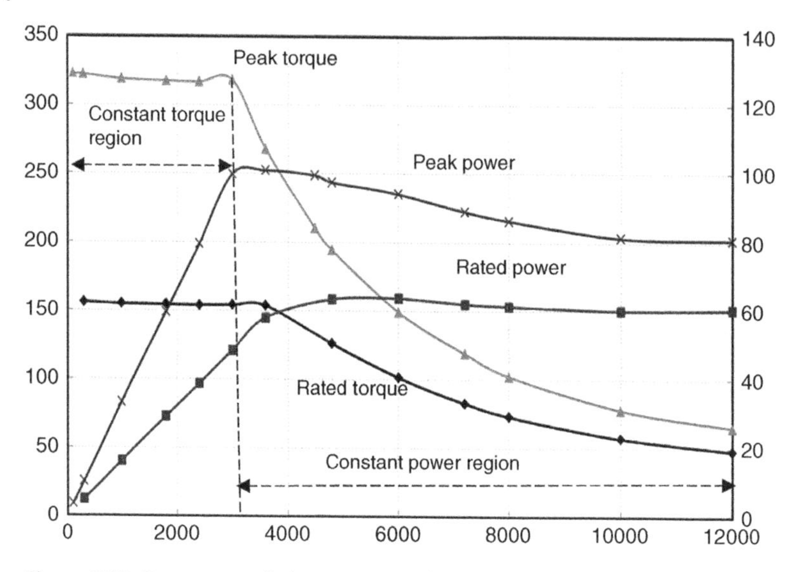

Figure 9.37 Torque–speed characteristics of a typical PM motor.

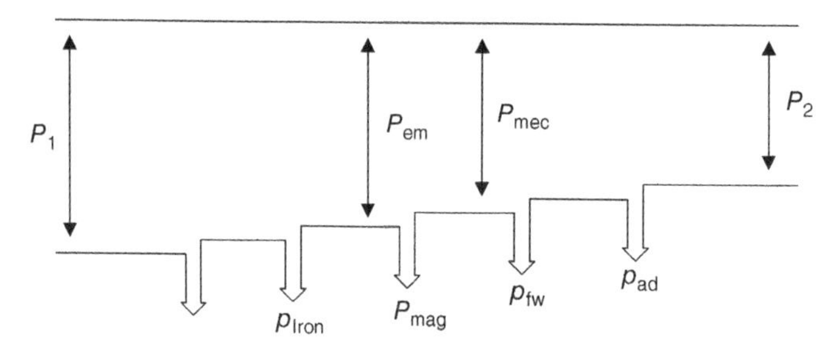

Figure 9.38 Losses in PM motor.

the stator winding, p_{mag} is the loss of the rotor magnet and rotor steel, p_{iron} is the iron loss of the stator core, p_{fw} is the frictional and windage loss, and p_{ad} is the stray load loss.

The efficiency can be expressed as:

$$\eta = \frac{P_2}{P_1} = \frac{P_2}{P_2 + p_{cu1} + p_{iron} + p_{mag} + p_{fw} + p_{ad}} \tag{9.82}$$

The typical efficiency of a PM motor is plotted in Figure 9.39.

9.7 Switched Reluctance Motors

Both switched reluctance motors and synchronous reluctance motors have attracted attention in traction applications due to their simple structure, not needing a squirrel cage or magnets on the rotor, very little loss on the rotor, and ease of control [93–114].

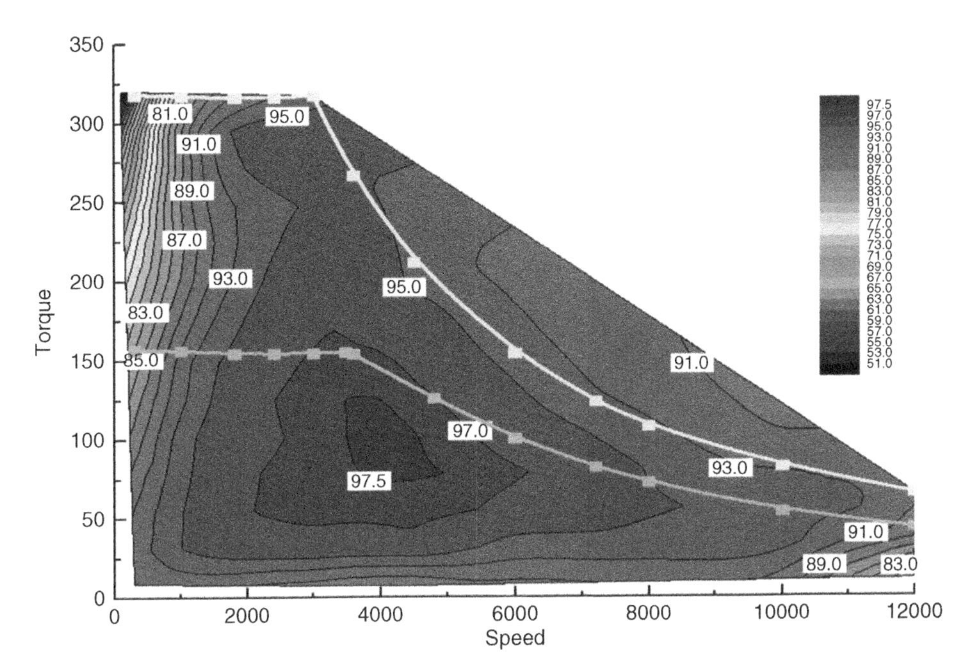

Figure 9.39 Efficiency map of a typical EV motor.

Although they have many advantages, PM motors and induction motors both have their own limitations. For example, PM motors face the possibility of demagnetization at extremely high temperature, limited speed range, and difficulty in protecting the powertrain during a fault condition. Induction motors have limited torque capability at low speeds, lower torque density and lower efficiency, noise due to stator/rotor slot combinations, and so on.

From the previous section we have seen that the torque of a synchronous motor has two terms, one related to E_o and X_d, which is induced by the rotor PM field, and one related to V, X_d, and X_q, which is induced by the difference in reactance of the d axis and q axis. In other words, even if the magnets are removed, an IPM motor can still generate torque with a sinusoidal supply due to the existence of salience of the rotor. This is called a synchronous reluctance motor. The stator and the rotor of a synchronous reluctance motor have the same number of poles.

Switched reluctance or synchronous reluctance motors do not use magnets or a squirrel cage. They simply use the difference in d axis and q axis reactance to produce reluctant torque. Therefore, they are similar to a synchronous motor without excitation and are therefore known as a switched reluctance motor. Hence only the second term of Eq. (9.77) exists. The torque of a switched reluctance motor with sinusoidal supply is:

$$T = \frac{mV^2}{2\omega_R}\left(\frac{1}{X_q} - \frac{1}{X_d}\right)\sin(2\theta) \tag{9.83}$$

In order to increase the torque of a switched reluctance motor, the q axis and d axis reactance is designed to have a large difference. A cross-section of a synchronous motor is shown in Figure 9.40.

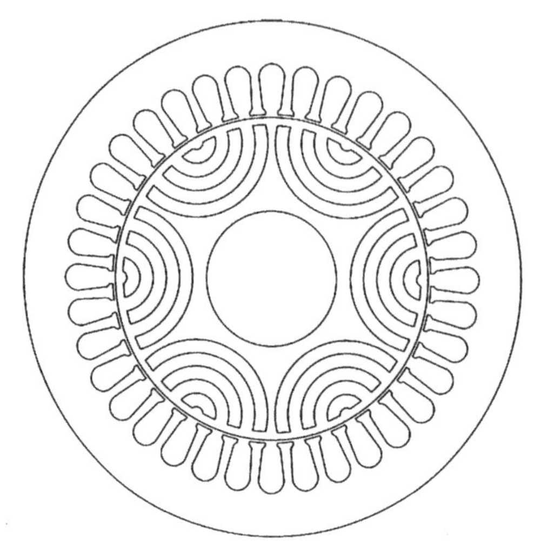

Figure 9.40 synchronous reluctance motor.

Switched reluctance motors are similar to synchronous motors but will have different numbers of poles on the stator and the rotor. Figure 9.41 shows the cross-section of a switched reluctance motor and its control circuit.

9.8 EV Transmissions

As discussed earlier in this chapter, EVs usually do not need a multiple-speed transmission due to the wide speed range of electric motors. However, in order to reduce the size and cost of the EV drivetrain, EV motors are often designed to operate at a relatively high speed. Hence, a speed-reduction gearbox is necessary for EVs. In some applications, a two-speed transmission may be employed to increase the overall efficiency of the EV system.

A reverse gear is not needed in EV powertrains since the motor can be controlled to turn in both directions.

9.8.1 Single-Speed EV Transmission

A single-speed transmission is the most popular choice for EVs. A single-speed can be made with multi-stage gears, or with a planetary-gear. Figures 9.42 and 9.43 shows the two configurations, respectively.

In Figure 9.42, a two-stage gear is used to realize the high ratio transmission. The speed relationship is;

$$\omega_{out} = \frac{N_1 N_3}{N_2 N_4} \omega_{motor} \tag{9.84}$$

where N_1, N_2, N_3, and N_4 are the number of teeth of each of the gears.

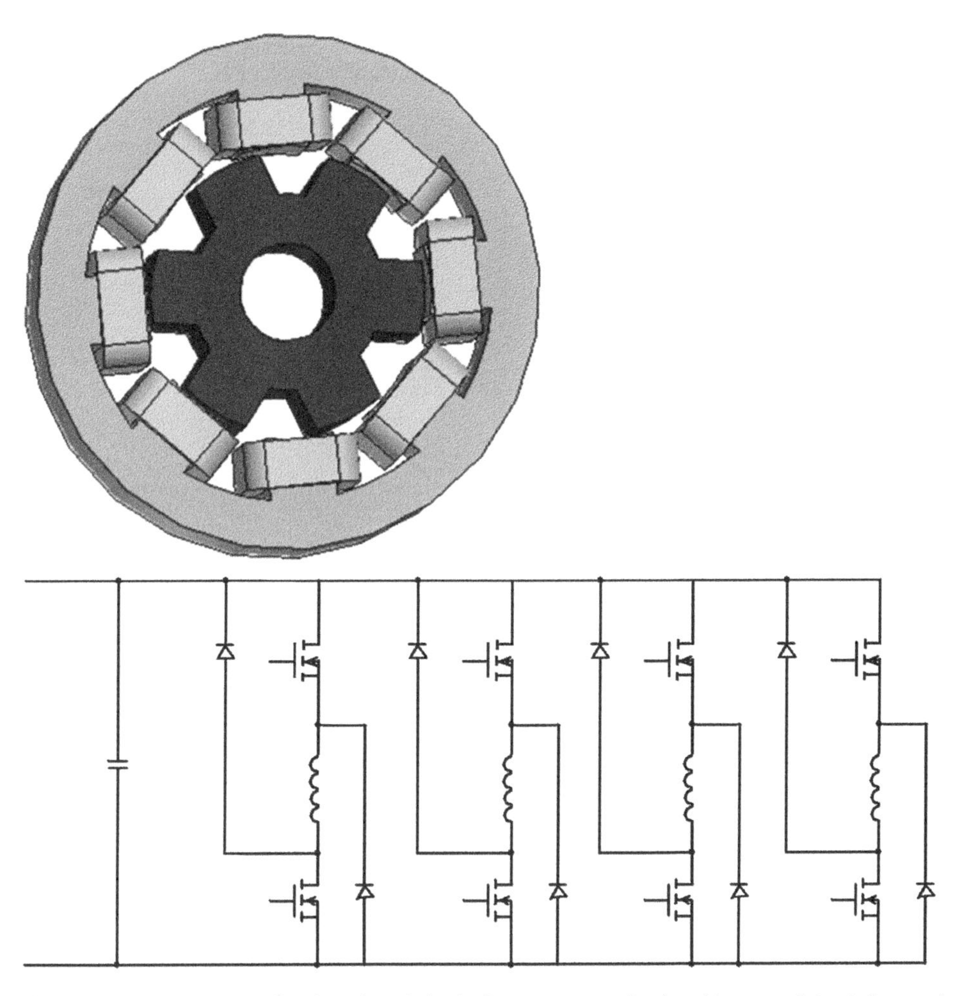

Figure 9.41 Cross-section of a 6/8-pole switched reluctance motor (top) and its control circuit (bottom).

Figure 9.42 A multi-stage gear based single speed transmission for EVs.

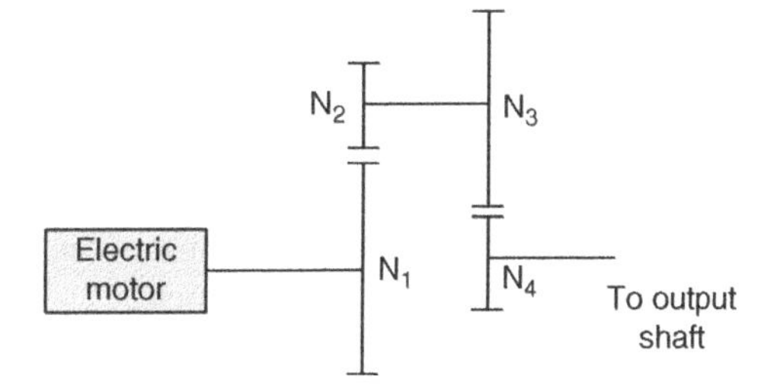

In Figure 9.43, the sun gear is connected to the electric motor, the planetary carrier is fixed to the case, and the output shaft is connected to the ring gear. The speed relationship of the gear train can be expressed as:

$$\omega_c = \frac{N_s}{N_r + N_s}\omega_s + \frac{N_r}{N_r + N_s}\omega_r \tag{9.85}$$

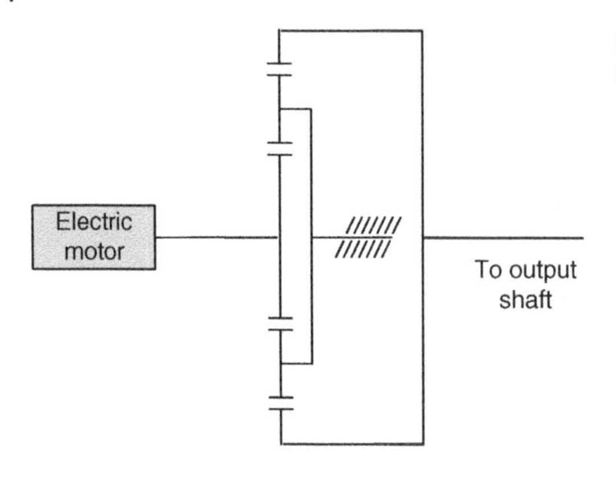

Figure 9.43 A planetary gear based single-speed transmission for EVs.

Since the carrier is grounded, the actual speed relation between the motor and the output shaft is:

$$\omega_{out} = \frac{N_r}{N_r}\omega_{motor} \tag{9.86}$$

For example, if the ring gear has 72 teeth and the sun gear has 28 teeth, then the speed ratio from the motor to the output shaft is 72/28 = 2.57, which is a very large gear ratio.

It is also possible to fix the sun gear or ring gear in the planetary based transmission. In these cases, the speed ratio can be expressed as follows. When the ring gear is fixed and the sun gear is used as the output:

$$\omega_s = \left(1 + \frac{N_r}{N_r}\right)\omega_c \tag{9.87}$$

When the sun gear is fixed and the ring gear is used as the output:

$$\omega_r = \left(1 + \frac{N_s}{N_r}\right)\omega_c \tag{9.88}$$

9.8.2 Multiple Ratio EV Transmissions

Two-speed automatic transmissions have been proposed for electric powertrain applications. Even though an electric motor can provide a large speed range to satisfy the operational needs of typical cars without the need of a multiple-speed transmission, there are imperfections in this arrangement.

With a single ratio reduction box, the requirement for the motor increases. The motor will need to provide a large stall torque (Torque generated at zero and very low speeds) and a large speed range at the same time. The efficiency of the motor will be compromised due to the wide operation range. Therefore, with a properly designed multiple-speed transmission, the system will provide many advantages. Figure 9.44 shows a two-speed transmission with automatic shifting.

First, the stall torque requirement is reduced due to the large gear ratio available. This will potentially reduce the size of the motor, which can result in the reduction of size,

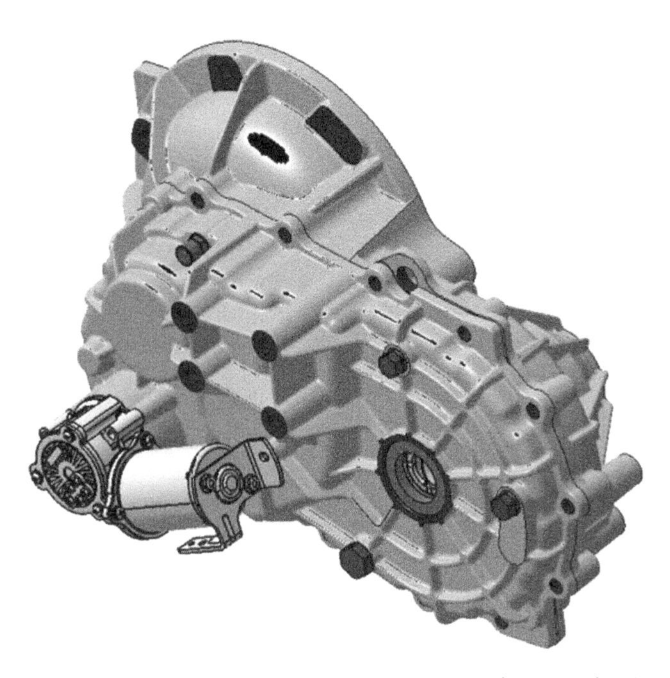

Figure 9.44 Single-ratio speed reduction gearbox for EV applications.

weight, and cost of the motor. It also reduces the requirements on the inverter (less current will be needed).

Second, the motor can be controlled to operate in its more efficient region by changing the gear ratios like the ones used for an internal combustion engine.

Last, the top speed of the motor can be reduced due to a lower gear ratio being available. This will reduce the bearing requirement, losses in the steel, operation frequencies, and cost of the motor.

Studies have shown that a two-speed transmission can fulfill the above purposes and at the same time provide energy savings of at least 5–10% while improving the acceleration, gradability, and top speed of the vehicle. However, one designing an EV powertrain with a multiple speed transmission, it needs to consider that the added cost should be reasonable or minimal. The total mass of the powertrain should remain the same. The gear shifting must be simple and smooth and no torque interruption should occur during gear shifting.

Traditional CVTs generally do not meet the above requirements because they are usually bulky, expensive, and inefficient. A special design of the transmission is therefore needed. For example, there are two methods for realizing the two-speed transmission for EV: an automatic gearbox based two-speed transmission and a planet gear based two-speed transmission.

9.8.2.1 Automatic Gearbox Based Two-Speed Transmission

Figure 9.45 shows the principle and internal structure of a typical two-speed EV transmission based on an automatic gearbox. There are three parallel shafts in the gearbox:

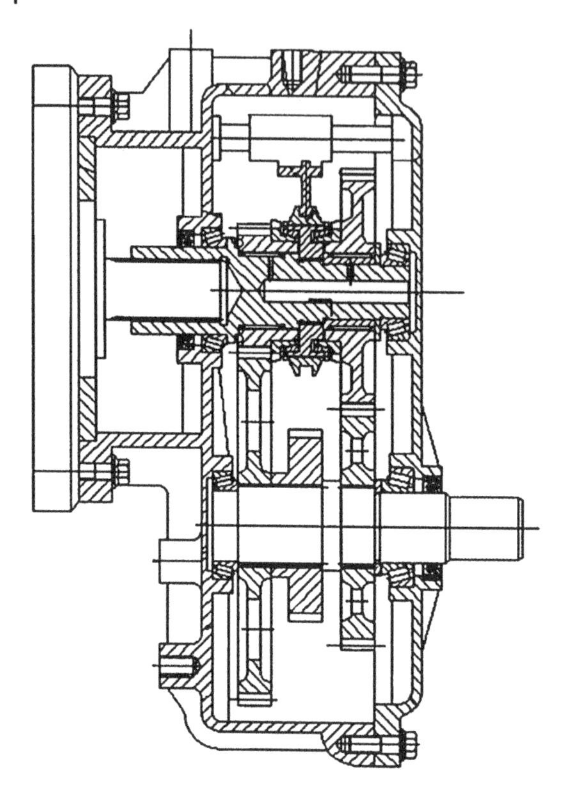

Figure 9.45 Structure of a two-speed EV transmission based on automatic gearbox.

input, output, and grooved shafts. For the input shaft, there are two gear wheels, labeled as the first and second wheel, and a synchronizer installed on it. The two gear wheels have different sizes and gear ratios that are used to provide different speed variable ratios to the output shaft. The synchronizer can be connected to the input shaft and be spinning at the same speed, which is used to realize the speed shift between the two gear wheels.

The synchronizer is an important intermediate device in this gearbox, and it is flexible to move along the input shaft, resulting in three working positions: first gear, second gear, and neutral gear. As shown in Figure 9.45, when the synchronizer moves to the left, it can connect the first gear wheel with the input shaft through a spline. Then, the transmission ratio is determined by the first gear wheel. Similarly, when the synchronizer moves to the right, it can connect the second gear wheel with the input shaft through another spline. Then, the transmission ratio is determined by the second gear wheel. Considering the different size of the gear wheels, the transmission ratio is therefore regulated. When the synchronizer is arranged in the middle, it has no connection with the gears, which is named as the neutral gear position. In this case, there is no direct power transmission between the input and output shafts.

The grooved shaft is in parallel with the input shaft and placed above it as shown in Figure 9.45. The spin of this groove shaft is driven by a motor, and its main function is to adjust the position of the synchronizer to achieve the shifting of the transmission. In the working process, the groove shaft is connected to a selector rod, which contains a shifting fork that allows the synchronizer to rotate. The shifting fork can move to the left and right to connect different gear wheels. When it is maintained at the direction around the grooved shaft, the neutral gear position is realized.

The output shaft is also in parallel with the input shaft, and two gear wheels, named as the third and fourth gear wheels, are connected to it. The third and fourth gear wheels can engage with the first and second gear wheels on the input shaft, respectively, and the selection of engagement is achieved by the synchronizer. There is also a measurement device on the output shaft to measure the rotational speed. The measured signals are then processed to acquire the vehicle speed, and the vehicle controller can react to optimize the driving status based on these data.

This proposed transmission has two-speed shifting. The first speed aims at the normal driving status. In this situation, the two gear wheels with smaller reduction ratio can provide higher speed. The second speed is more suitable for acceleration and gradability. In this situation, the reduction ratio of the gear wheels is higher, and larger torque can be provided for heavy-duty working scenarios. The reduction ratios are optimized based on the output property of the motor and the power requirement of the vehicle. Hence, the lacking of power in the heavy-duty mode and the waste of power in the normal driving mode are both solved. The switching of the two working modes is determined by an automatic controller, and the process is optimized based on the real-time status of the vehicle, which contributes to improving the performance of the electric vehicle. In addition, the shifting is realized by a low-power motor instead of by direct involvement of the driver, which can avoid any manual mistake and provide effectiveness and convenience.

This proposed gearbox-based two-speed transmission has three main advantages. First, the shifting performance and lift-time are significantly improved. There are two shifting steps in this system, and the neutral position is between them. The speed control of the rotating motor can be achieved in the neutral status to make its speed consistent with the following gear wheel. The synchronizing process is only performed when their speeds are exactly the same. Therefore, the shifting process is more effective and smoother. The wear and tear on the synchronizer can be alleviated and its lift-time can be extended.

Second, the power provided by the motor can perfectly match the power required by the driving profile. The reduction ratio of the gear wheels is automatically selected by the vehicle controller, and the reaction speed of the controller can be really fast. The shifting process can be much smoother since the speed of the synchronizer can be regulated at the neutral position. All these processes do not need the attention of the driver, which also makes this design friendly for the general public.

Third, the structure of this design is simple, which can also reduce the system cost and weight. Since this transmission system needs to be installed on the vehicle side, it is important to reduce its cost and weight for practical applications. The good news is that this design uses a minimum number of components to realize the speed shifting function without sacrificing any technical performance of the vehicle. Compared to existing transmission systems for electric vehicles, the proposed system has a significant cost–performance advantage. In addition, since this design considers the characteristic and output performance of the motor, the size is also reduced, which is suitable for the onboard installation.

9.8.2.2 Planet Gear based Two-Speed Transmission

Figure 9.46 shows the simplified structure of a planet gear based two-speed transmission system, which consists of four parts: input shaft, planet gear wheel system, synchronizer, and output shaft. Similar to the previous gearbox based transmission system, the input shaft is directly connected to the motor as the power input, and the output shaft is connected to the vehicle wheels as the power output. This system can provide two speed-reduction ratios, and a synchronizer is utilized to achieve the speed shifting as well. The difference from the previous system is that the planet gear wheel system is used to replace the previous automatic gearbox system, and the outside case of the system also contributes to the transmission process. The transmission ratio depends on the connection of the planet gear wheel system with the synchronizer. The unique function of the planet gear wheel system is explained in details as follows.

The planet gear wheel system consists of four parts: sun wheel, planet wheel, planet carrier, and ring gear. The sun wheel is in the center of the gear system, and it is connected to the input shaft, which is also the center of rotation. The planet carrier is connected to the output shaft, which determines the output rotation. The ring gear is at the outer edge of the planet wheels, and there is a spline connected to it. Note that the ring gear can rotate, and its speed depends on different working modes. The planet wheels are arranged inside of the ring gear, and usually there are three planet wheels, which have engagements with both the sun wheel and the ring gear. When the system is working, the planet wheels are not only self-rotating but also move around the sun wheel.

The structure of the synchronizer in this planet wheel system is similar to that in the gearbox system. The synchronizer is connected to the grooved shaft through a spline, and this connection can make the ring gear rotate together with the synchronizer. The synchronizer can move along the direction that is in parallel with the input shaft, resulting in three positions: first shifting position, second shifting position, and neutral position. A selector rod and shifting fork are used to fix the synchronizer in different positions. For example, in Figure 9.46, when the synchronizer moves to the left, the shifting fork is connected to the outer side case, which is defined as the first shifting position.

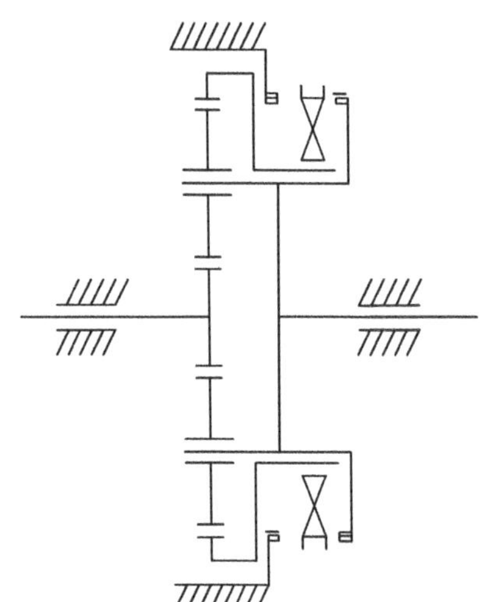

Figure 9.46 Structure of a two-speed EV transmission based on planet-gear.

When the synchronizer moves to the right, the shifting fork is connected to the planet carrier, which is also the output shaft. Therefore, the reduction ratio is different from the previous case, and it is defined as the second shifting position. When the synchronizer moves to the middle, it has no connection with either the input or the output shaft, which is defined as the neutral position.

The working principle of this planet gear based transmission is similar to the previous gearbox based system. The two shifting positions can provide different reduction ratios in the transmission. The first position is used for the normal driving status of the electric vehicle because it can provide a relatively high speed. The second position is used for the acceleration and gradability scenarios because it can provide a large rotation torque. The switch of these two shifting positions is controlled by an automatic controller, which can optimize the power output of the motor and improve the driving performance. In addition, in the neutral shifting position, the speed of the synchronizer can be controlled by a motor to smooth the shifting switch transient.

The planet gear based transmission also has three advantages over the gearbox based system. First, the size and weight of the transmission system are relatively small. Since all planet gear wheels are placed inside the ring gear, the internal space is fully used and the system structure is more compact. Therefore, the required size and usage of metal can be significantly reduced.

Second, the transmission performance is improved. The input and output shafts are on the same line. Considering the symmetrical structure of the planet gear wheels, the distribution of multiple planet wheels can balance the force on the sun gear and the bearings, which can help increase the transmission efficiency. Also, because there are more engagements of gears, the system can rotate more smoothly and be more robust to external shock and vibration.

Third, the range of transmission ratio can be increased. The planet gear system can realize the composition and decomposition of motion. With a proper design of dimensions, the planet gear system can achieve a high transmission ratio with a limited number of gear wheels. Especially in electric vehicle applications, the high transmission ratio can contribute to increasing the maximum achievable torque of the motor, which can further improve the acceleration and gradability performance.

9.9 Conclusions

Electric motors and associated controllers are one of the key enabling technologies for electric, hybrid electric, and plug-in hybrid electric vehicles. Various types of electric motors and drive systems are available for the powertrain of electric vehicles. Traction motors and drives experience very harsh environmental conditions, such as a wide temperature range (–30 to 60 °C), severe vibration and shock, high electromagnetic noise, size and weight constraints, and stringent safety and reliability requirements. As a result, there are many unique aspects in the design, development, analysis, manufacturing, and research of electric motors and drives for traction applications which are all important aspects but cannot all be covered in this chapter. Readers could consult the references below for further reading. For example, more in-depth studies about synchronous reluctance motor design and optimization can be found in [114] and studies of the uncontrolled generation in PM drive motors are covered in [140, 141].

Bibliography

1 Ehsani, M., Rahman, K.M., and Toliyat, H.A. (1997) Propulsion system design of electric and hybrid vehicles. *IEEE Transactions on Industrial Electronics*, 44, 19–27.

2 Muta, K., Yamazaki, M., and Tokieda, J. (2004) Development of new-generation hybrid system THS II – drastic improvement of power performance and fuel economy. SAE World Congress, March 8–11, No. 2004-01-0064.

3 Chan, C.C., Jiang, K.T., Xia, J.Z., et al. (1996) Novel permanent magnet motor drive for electric vehicles. *IEEE Transactions on Industrial Electronics*, 43 (2), 331–339.

4 Ehsani, M., Gao, Y., and Gay, S. (2003) Characterization of electric motor drives for traction applications. Industrial Electronics Society, IECON'03, November 2–6, Vol. 1, pp. 891–896.

5 Rahman, Z., Butler, K.L., and Ehsani, M. (2000) Effect of extended-speed, constant-power operation of electric drives on the design and performance of EV propulsion system. SAE Future Car Congress, April, No. 2001-01-0699.

6 Rahman, K.M. and Ehsani, M. (1996) Performance analysis of electric motor drives for electric and hybrid electric vehicle application. *Power Electronics in Transportation*, pp. 49–56.

7 Rahman, K.M., Fahimi, B., Suresh, G., et al. (2000) Advantages of switched reluctance motor applications to EV and HEV: design and control issues. *IEEE Transactions on Industry Applications*, 36 (1), 111–121.

8 Honda, Y., Nakamura, T., Higaki, T., and Takeda, Y. (1997) Motor design considerations and test results of an interior permanent magnet synchronous motor for electric vehicles. Proceedings of the IEEE Industry Applications Society Annual Meeting, October 5–9, pp. 75–82.

9 Kamiya, M. (2005) Development of traction drive motors for the Toyota hybrid system. International Power Electronics Conference, April 4–8.

10 Miller, J.M., Gale, A.R., McCleer, P.J., et al. (1998) Starter/alternator for hybrid electric vehicle: comparison of induction and variable reluctance machines and drives. Proceedings of the IEEE 1998 Industry Applications Society Annual Meeting, October 12–15, pp. 513–523.

11 Wang, T., Zheng, P., and Cheng, S. (2005) Design characteristics of the induction motor used for hybrid electric vehicle. *IEEE Transactions on Magnetics*, 41 (1), 505–508.

12 Harson, A., Mellor, P.H., and Howe, D. (1995) Design considerations for induction machines for electric vehicle drives. Proceedings of the IEE International Conference on Electrical Machines and Drives, September 11–13, pp. 16–20.

13 Krishnan, R. (1996) Review of flux-weakening in high performance vector controlled induction motor drives. Proceedings of the IEEE International Symposium on Industrial Electronics, June 17–20, pp. 917–922.

14 Miller, T.J.E. (1993) *Switched Reluctance Motors and their Control*, Magna Physics Publishing and Clarendon Press, Oxford.

15 (a) Miller, T.J.E. (2001) *Electronic Control of Switched Reluctance Machines*, Reed, New York, pp. 227–245; (b) Williamson, S.S., Emadi, A., and Rajashekara, K. (2007) Comprehensive efficiency modeling of electric traction motor drives for hybrid electric vehicle propulsion applications. *IEEE Transactions on Vehicular Technology*, 56 (4), 1561–1572.

16 West, J.G.W. (1994) DC, induction, reluctance and PM motors for electric vehicles. *Power Engineering Journal*, 8 (2), 77–88.
17 Welchko, B.A. and Nagashima, J.M. (2003) The influence of topology selection on the design of EV/HEV propulsion systems. *Power Electronics Letters*, 1 (2), 36–40.
18 Zhu, Z.Q. and Howe, D. (2007) Electrical machines and drives for electric, hybrid, and fuel cell vehicles. *Proceedings of the IEEE*, 95 (4), 746–765.
19 Chan, C.C. (2002) The state of the art of electric and hybrid vehicles. *Proceedings of the IEEE*, 90 (2), 247–275.
20 Jahns, T.M. and Blasko, V. (2001) Recent advances in power electronics technology for industrial and traction machine drives. *Proceedings of the IEEE*, 89 (6), 963–975.
21 Akin, B., Ozturk, S.B., Toliyat, H.A., and Rayner, M. (2009) DSP-based sensorless electric motor fault-diagnosis tools for electric and hybrid electric vehicle powertrain applications. *IEEE Transactions on Vehicular Technology*, 58 (6), 2679–2688.
22 Kou, B., Li, L., Cheng, S., and Meng, F. (2005) Operating control of efficiently generating induction motor for driving hybrid electric vehicles. *IEEE Transactions on Magnetics*, 41 (1), 488–491.
23 Zeraoulia, M., Benbouzid, M.E.H., and Diallo, D. (2006) Electric motor drive selection issues for HEV propulsion systems: a comparative study. *IEEE Transactions on Vehicular Technology*, 55 (6), 1756–1764.
24 Wang, T., Zheng, P., Zhang, Q., and Cheng, S. (2005) Design characteristics of the induction motor used for hybrid electric vehicles. *IEEE Transactions on Magnetics*, 41 (1), 505–508.
25 Li, W., Cao, J., and Zhang, X. (2010) Electrothermal analysis of induction motor with compound cage rotor used for PHEV. *IEEE Transactions on Industrial Electronics*, 57 (2), 660–668.
26 Asano, K., Okada, S., and Iwamam, N. (1992) Vibration suppression of induction-motor-driven hybrid vehicle using wheel torque observer. *IEEE Transactions on Industry Applications*, 28 (2), 441–447.
27 Diallo, D., Benbouzid, M.E.H., and Makouf, A. (2004) A fault-tolerant control architecture for induction motor drives in automotive applications. *IEEE Transactions on Vehicular Technology*, 53 (6), 1847–1855.
28 Benbouzid, M.E.H., Diallo, D., and Zeraoulia, M. (2007) Advanced fault-tolerant control of induction-motor drives for EV/HEV traction applications: from conventional to modern and intelligent control techniques. *IEEE Transactions on Vehicular Technology*, 56 (2), 519–528.
29 Proca, A.B., Keyhani, A., and Miller, J.M. (2003) Sensorless sliding-mode control of induction motors using operating condition dependent models. *IEEE Transactions on Energy Conversion*, 18 (2), 205–212.
30 Salmasi, F.R., Najafabadi, T.A., and Maralani, P.J. (2010) An adaptive flux observer with online estimation of DC-link voltage and rotor resistance for VSI-based induction motors. *IEEE Transactions on Power Electronics*, 25 (5), 1310–1319.
31 Proca, A.B., Keyhani, A., and Miller, J. (2002) Sensorless sliding-mode control of induction motors using operating condition dependent models. *Power Engineering Review*, 22 (7), 50–50.
32 Sudhoff, S.D., Corzine, K.A., Glover, S.F., et al. (1998) DC link stabilized field oriented control of electric propulsion systems. *IEEE Transactions on Energy Conversion*, 13 (1), 27–33.

33 Khoucha, F., Lagoun, S.M., Marouani, K., et al. (2010) Hybrid cascaded H-bridge multilevel-inverter induction-motor-drive direct torque control for automotive applications. *IEEE Transactions on Industrial Electronics*, 57 (3), 892–899.

34 McCleer, P.J., Miller, J.M., Gale, A.R., et al. (2001) Nonlinear model and momentary performance capability of a cage rotor induction machine used as an automotive combined starter-alternator. *IEEE Transactions on Industry Applications*, 37 (3), 840–846.

35 Degner, M.W., Guerrero, J.M., and Briz, F. (2006) Slip-gain estimation in field-orientation-controlled induction machines using the system transient response. *IEEE Transactions on Industry Applications*, 42 (3), 702–711.

36 Kim, J., Jung, J., and Nam, K. (2004) Dual-inverter control strategy for high-speed operation of EV induction motors. *IEEE Transactions on Industrial Electronics*, 51 (2), 312–320.

37 Neacsu, D.O. and Rajashekara, K. (2001) Comparative analysis of torque-controlled IM drives with applications in electric and hybrid vehicles. *IEEE Transactions on Power Electronics*, 16 (2), 240–247.

38 Liu, R., Mi, C., and Gao, W. (2008) Modeling of iron losses of electrical machines and transformers fed by PWM inverters. *IEEE Transactions on Magnetics*, 44 (8), 2021–2028.

39 Ding, X. and Mi, C. (2011) Impact of inverter on losses and thermal characteristics of induction motors. *International Journal on Power Electronics*, in press.

40 Mi, C., Slemon, G.R., and Bonert, R. (2003) Modeling of iron losses of permanent magnet synchronous motors. *IEEE Transactions on Industry Applications*, 39 (3), 734–742.

41 Khluabwannarat, P., Thammarat, C., Tadsuan, S., and Bunjongjit, S. (2007) An analysis of iron loss supplied by sinusoidal, square wave, bipolar PWM inverter and unipolar PWM inverter. International Power Engineering Conference, IPEC 2007, December 3–6, pp. 1185–1190.

42 Mi, C., Shen, J., and Natarajan, N. (2002) Field-oriented control of induction motors. IEEE Workshop on Power Electronics in Transportation (WPET'02), September.

43 Parker, R.J. and Studders, R.J. (1962) *Permanent Magnets and their Application*, John Wiley & Sons, Inc., New York.

44 Merrill, F.W. (1955) Permanent magnet excited synchronous motor. *AIEE Transactions*, 74, 1754–1760.

45 Binns, K.J., Jabbar, M.A., and Parry, G.E. (1979) Choice of parameters in hybrid permanent magnet synchronous motor. *Proceedings of the IEE*, 126 (8), 741–744.

46 Honsinger, V.P. (1980) Performance of polyphase permanent magnet machines. *IEEE Transactions on Power Apparatus and Systems*, 99 (4), 1510–1518.

47 Rahman, M.A. and Little, T.A. (1984) Dynamic performance analysis of permanent magnet synchronous motors. *IEEE Transactions on Power Apparatus and Systems*, 103 (6), 1277–1282.

48 Rahman, M.A., Little, T.A., and Slemon, G.R. (1985) Analytical models for interior-type permanent magnet synchronous motors. *IEEE Transactions on Magnetics*, 21 (5), 1741–1743.

49 Zhou, P., Rahman, M.A., and Jabbar, M.A. (1994) Field and circuit analysis of permanent magnet machines. *IEEE Transactions on Magnetics*, 30 (4), 1350–1359.

50 Vaez, S., John, V.I., and Rahman, M.A. (1999) An online loss minimization controller for interior permanent magnet motor drives. *IEEE Transactions on Energy Conversion*, 14 (4), 1435–1440.

51 Uddin, M.N., Radwan, T.S., and Rahman, M.A. (2002) Performance of interior permanent magnet motor drive over wide speed range. *IEEE Transactions on Energy Conversion*, 17 (1), 79–84.
52 Rahman, M.A., Vilathgamuwa, M., Uddin, M.N., and Tseng, K.J. (2003) Non-linear control of interior permanent magnet synchronous motor. *IEEE Transactions on Industry Applications*, 39 (2), 408–416.
53 Jahns, T.M. (1984) Torque production in permanent magnet synchronous motor drives with rectangular current excitation. *IEEE Transactions on Industry Applications*, 20 (4), 803–813.
54 Morimoto, S., Sanada, M., and Takeda, Y. (1994) Wide-speed operation of interior permanent magnet synchronous motors with high performance current regulator. *IEEE Transactions on Industry Applications*, 30 (4), 920–926.
55 Jahns, T.M. (1987) Flux-weakening regime operation of an interior permanent-magnet synchronous motor drive. *IEEE Transactions on Industry Applications*, 23 (4), 681–689.
56 Morimoto, S., Sanada, M., and Takeda, Y. (1996) Inverter-driven synchronous motors for constant power. *IEEE Industry Applications Magazine*, pp. 18–24.
57 Zhu, Z.Q., Chen, Y.S., and Howe, D. (2000) Online optimal field weakening control of permanent magnet brushless ac drives. *IEEE Transactions on Industry Applications*, 36 (6), 1661–1668.
58 Zhu, Z.Q., Shen, J.X., and Howe, D. (2006) Flux-weakening characteristics of trapezoidal back-EMF machines in brushless DC and AC modes. Proceedings of the International Power Electronics and Motion Control, August 13–16, pp. 908–912.
59 Shi, Y.F., Zhu, Z.Q., and Howe, D. (2006) Torque–speed characteristics of interior-magnet machines in brushless AC and DC modes, with particular reference to their flux-weakening performance. Proceedings of the International Power Electronics and Motion Control, August 13–16, pp. 1847–1851.
60 Safi, S.K., Acarnley, P.P., and Jack, A.G. (1995) Analysis and simulation of the high-speed torque performance of brushless DC motor drives. *IEE Proceedings – Electric Power Applications*, 142 (3), 191–200.
61 Bose, B.K. (1988) A microcomputer-based control and simulation of an advanced IPM synchronous machine drive system for electric vehicle propulsion. *IEEE Transactions on Industrial Electronics*, 35 (4), 547–559.
62 Bose, B.K. (1988) A high-performance inverter-fed drive system of an interior permanent magnet synchronous machine. *IEEE Transactions on Industry Applications*, 24 (6), 987–997.
63 Lovelace, E.C., Jahns, T.M., Kirtley, J.L. Jr., and Lang, J.H. (1998) An interior PM starter/alternator for automotive applications. Proceedings of the International Conference on Electrical Machines, December, pp. 1802–1808.
64 Lipo, T.A. (1991) Synchronous reluctance machines – a viable alternative for AC drives? *Electric Machines and Power Systems*, 19, 659–671.
65 Soong, W.L., Staton, D.A., and Miller, T.J.E. (1995) Design of a new axially-laminated interior permanent magnet motor. *IEEE Transactions on Industry Applications*, 31 (2), 358–367.
66 Soong, W.L. and Ertugrul, N. (2002) Field-weakening performance of interior permanent-magnet motors. *IEEE Transactions on Industry Applications*, 38 (5), 1251–1258.
67 Chaaban, F.B., Birch, T.S., Howe, D., and Mellor, P.H. (1991) Topologies for a permanent magnet generator/speed sensor for the ABS on railway freight vehicles. Proceedings of the

IEE International Conference on Electrical Machines and Drives, September 11–13, pp. 31–35.
68 Liao, Y., Liang, F., and Lipo, T.A. (1995) A novel permanent magnet machine with doubly salient structure. *IEEE Transactions on Industry Applications*, 3 (5), 1069–1078.
69 Chan, C.C., Jiang, J.Z., Chen, G.H., et al. (1994) A novel polyphase multipole square-wave permanent magnet motor drive for electric vehicles. *IEEE Transactions on Industry Applications*, 30 (5), 1258–1266.
70 Wang, J.B., Xia, Z.P., and Howe, D. (2005) Three-phase modular permanent magnet brushless machine for torque boosting on a downsized ICE vehicle. *IEEE Transactions on Vehicular Technology*, 54 (3), 809–816.
71 Russenschuck, S. (1990) Mathematical optimization techniques for the design of permanent magnet machines based on numerical field calculation. *IEEE Transactions on Magnetics*, 26 (2), 638–641.
72 Russenschuck, S. (1992) Application of Lagrange multiplier estimation to the design optimization of permanent magnet synchronous machines. *IEEE Transactions on Magnetics*, 28 (2), 1525–1528.
73 Rasmussen, K.F., Davies, J.H., Miller, T.J.E., et al. (2000) Analytical and numerical computation of air-gap magnetic fields in brushless motors with surface permanent magnets. *IEEE Transactions on Industry Applications*, 36 (6), 1547–1554.
74 Boules, N. (1990) Design optimization of permanent magnet DC motors. *IEEE Transactions on Industry Applications*, 26 (4), 786–792.
75 Proca, A.B., Keyhani, A., El-Antably, A., et al. (2003) Analytical model for permanent magnet motors with surface mounted magnets. *IEEE Transactions on Energy Conversion*, 18 (3), 386–391.
76 Pavlic, D., Garg, V.K., Repp, J.R., and Weiss, J.A. (1988) Finite element technique for calculating the magnet sizes and inductance of permanent magnet machines. *IEEE Transactions on Energy Conversion*, 3 (1), 116–122.
77 Miller, T.J.E., McGilp, M., and Wearing, A. (1999) Motor design optimization using SPEED CAD software – practical electromagnetic design synthesis. IEE Seminar, Ref. no. 1999/014, pp. 1–5.
78 ANSYS (2005) http://www.ansoft.com (accessed April 2005).
79 Kenjo, T. and Nagamori, S. (1985) *Permanent Magnet and Brushless DC Motors*, Clarendon Press, Oxford.
80 Miller, T.J.E. (1989) *Permanent Magnet and Reluctance Motor Drives*, Oxford Science Publications, Oxford.
81 Slemon, G.R. and Liu, X. (1992) Modeling and design optimization of permanent magnet motors. *Electric Machines and Power Systems*, 20, 71–92.
82 Bose, B.K. (1997) *Power Electronics and Variable Frequency Drives – Technology and Applications*, IEEE Press, Piscataway, NJ.
83 Balagurov, V.A., Galtieev, F.F., and Larionov, A.N. (1964) *Permanent Magnet Electrical Machines*, Energia, Moscow (in Russian, and translation in Chinese).
84 Gieras, J.F. and Wing, M. (2002) *Permanent Magnet Motor Technology: Design and Applications*, 2nd edn, Marcel Dekker, New York.
85 Mi, C., Filippa, M., Liu, W., and Ma, R. (2004) Analytical method for predicting the air-gap flux of interior-type permanent magnet machines. *IEEE Transactions on Magnetics*, 40 (1), 50–58.

86 Cho, D.H., Jung, H.K., and Sim, D.J. (1999) Multiobjective optimal design of interior permanent magnet synchronous motors considering improved core loss formula. *IEEE Transactions on Energy Conversion*, 14 (4), 1347–1352.
87 Borghi, C.A., Casadei, D., Cristofolini, A., et al. (1999) Application of multi-objective minimization technique for reducing the torque ripple in permanent magnet motors. *IEEE Transactions on Magnetics*, 35 (5), 4238–4246.
88 Upadhyay, P.R., Rajagopal, K.R., and Singh, B.P. (2004) Effect of armature reaction on the performance of axial field permanent magnet brushless DC motor using FE method. *IEEE Transactions on Magnetics*, 40 (4), 2023–2025.
89 Li, Y., Zou, J., and Lu, Y. (2003) Optimum design of magnet shape in permanent magnet synchronous motors. *IEEE Transactions on Magnetics*, 39 (6), 3523–3526.
90 Fujishima, Y., Wakao, S., Kondo, M., and Terauchi, N. (2004) An optimal design of interior permanent magnet synchronous motor for the next generation commuter train. *IEEE Transactions on Applied Superconductivity*, 14 (2), 1902–1905.
91 Mi, C. (2006) Analytical design of permanent magnet traction drives. *IEEE Transactions on Magnetics*, 42 (7), 1861–1866.
92 Ding, X. and Mi, C. Modeling of eddy current loss and temperature of the magnets in permanent magnet machines. *Journal of Circuits, Systems, and Computers*, submitted.
93 Fahimi, B., Emadi, A., and Sepe, R. (2004) A switched reluctance machine-based starter/alternator for more-electric cars. *IEEE Transactions on Energy Conversion*, 19 (1), 116–124.
94 Rahman, K.M. and Schulz, S.E. (2002) Design of high-efficiency and high-torque-density switched reluctance motor for vehicle propulsion. *IEEE Transactions on Industry Applications*, 38 (6), 1500–1507.
95 Ramamurthy, S.S. and Balda, J.C. (2001) Sizing a switched reluctance, motor for electric vehicles. *IEEE Transactions on Industry Applications*, 37 (5), 1256–1263.
96 Mecrow, B.C. (1996) New winding configurations for doubly salient reluctance machines. *IEEE Transactions on Industry Applications*, 32 (6), 1348–1356.
97 Mecrow, B.C., Finch, J.W., El-Kharashi, E.A., and Jack, A.G. (2002) Switched reluctance motors with segmental rotors. *IEE Proceedings – Electric Power Applications*, 149 (4), 245–254.
98 Krishnamurthy, M., Edrington, C.S., Emadi, A., et al. (2006) Making the case for applications of switched reluctance motor technology in automotive products. *IEEE Transactions on Power Electronics*, 21 (3), 659–675.
99 Rahman, K.M., Fahimi, B., Suresh, G., et al. (2000) Advantages of switched reluctance motor applications to EV and HEV: design and control issues. *IEEE Transactions on Industry Applications*, 36 (1), 111–121.
100 Long, S.A., Schofield, N., Howe, D., et al. (2003) Design of a switched reluctance machine for extended speed operation. Proceedings of IEEE International Electric Machines and Drives Conference, June 1–4, pp. 235–240.
101 Schofield, N. and Long, S.A. (2005) Generator operation of a switched reluctance starter/generator at extended speeds. Proceedings of IEEE Conference on Vehicle Power and Propulsion, September 7–9, pp. 453–460.
102 Edrington, C.S., Krishnamurthy, M., and Fahimi, B. (2005) Bipolar switched reluctance machines: a novel solution for automotive applications. *IEEE Transactions on Vehicular Technology*, 54 (3), 795–808.

103 Dixon, S. and Fahimi, B. (2003) Enhancement of output electric power in switched reluctance generators. Proceedings of IEEE International Electric Machines and Drives Conference, June 1–4, pp. 849–856.

104 Cameron, D.H., Lang, J.H., and Umans, S.D. (1992) The origin and reduction of acoustic noise in doubly salient variable-reluctance motors. *IEEE Transactions on Industry Applications*, 26 (6), 1250–1255.

105 Colby, R.S., Mottier, F., and Miller, T.J.E. (1996) Vibration modes and acoustic noise in a four-phase switched reluctance motor. *IEEE Transactions on Industry Applications*, 32 (6), 1357–1364.

106 Long, S.A., Zhu, Z.Q., and Howe, D. (2001) Vibration behaviour of stators of switched-reluctance machines. *IEE Proceedings – Electric Power Applications*, 148 (3), 257–264.

107 Long, S.A., Zhu, Z.Q., and Howe, D. (2002) Influence of load on noise and vibration of voltage and current controlled switched reluctance machines. Proceedings of IEE International Conference on Power Electronics, Machines, and Drives, April 16–18, pp. 534–539.

108 Blaabjerg, F., Pedersen, J.K., Neilsen, P., et al. (1994) Investigation and reduction of acoustic noise from switched reluctance drives in current and voltage control. Proceedings of International Conference on Electrical Machines, December, pp. 589–594.

109 Wu, C.Y. and Pollock, C. (1993) Time domain analysis of vibration and acoustic noise in the switched reluctance drive. Proceedings of International Conference on Electrical Machines and Drives, October, pp. 558–563.

110 Gabsi, M., Camus, F., Loyau, T., and Barbry, J.L. (1999) Noise reduction of switched reluctance machine. Proceedings of IEEE International Electric Machines and Drives Conference, May 12–19, pp. 263–265.

111 Gabsi, M., Camus, F., and Besbes, M. (1999) Computation and measurement of magnetically induced vibrations of switched reluctance machine. *IEE Proceedings – Electric Power Applications*, 146 (5), 463–470.

112 Wu, C.Y. and Pollock, C. (1995) Analysis and reduction of acoustic noise in the switched reluctance drive. *IEEE Transactions on Industry Applications*, 31 (6), 91–98.

113 Long, S.A., Zhu, Z.Q., and Howe, D. (2005) Effectiveness of active noise and vibration cancellation for switched reluctance machines operating under alternative control strategies. *IEEE Transactions on Energy Conversion*, 20 (4), 792–801.

114 Matsuo, T. and Lipo, T.A. (2004) Rotor design optimization of synchronous reluctance machine. *IEEE Transactions on Energy Conversion*, 9 (2), 359–365.

115 Li, Y. and Mi, C. (2007) Doubly salient permanent magnet machines with skewed rotor and six-state communication control. *IEEE Transactions on Magnetics*, 43 (9), 3623–3629.

116 Liao, Y., Liang, F., and Lipo, T.A. (1992) A novel permanent magnet motor with doubly salient structure. Industry Applications Society Annual Meeting, October, pp. 308–314.

117 Liao, Y. and Lipo, T.A. (1993) Sizing and optimal design of doubly salient permanent magnet motors. Sixth International Conference on Electrical Machines and Drives, September, pp. 452–456.

118 Liao, Y., Liang, F., and Lipo, T.A. (1995) A novel permanent magnet motor with doubly salient structure. *IEEE Transactions on Industry Applications*, 31 (5), 1069–1078.

119 Cheng, M., Chau, K.T., and Chan, C.C. (2000) Non-linear varying-network magnetic circuit analysis for DSPM motors. *IEEE Transactions on Magnetics*, 36 (1), 339–348.

120 Cheng, M., Chau, K.T., and Chan, C.C. (2001) Static characteristics of a new doubly salient permanent magnet motor. *IEEE Transactions on Energy Conversion*, 16 (1), 20–25.
121 Cheng, M., Chau, K.T., Chan, C.C., and Sun, Q. (2003) Control and operation of a new 8/6-pole doubly salient permanent-magnet motor drive. *IEEE Transactions on Industry Applications*, 39 (5), 1363–1371.
122 Cheng, M., Chau, K.T., and Chan, C.C. (2001) Design and analysis of a new doubly salient permanent magnet motor. *IEEE Transactions on Magnetics*, 37 (4), 3012–3020.
123 Chau, K.T., Sun, Q., Fan, Y., and Cheng, M. (2005) Torque ripple minimization of doubly salient permanent-magnet motors. *IEEE Transactions on Energy Conversion*, 20 (2), 352–358.
124 Deodhar, R.P., Staton, D.A., and Miller, T.J.E. (1996) Prediction of cogging torque using the flux-MMF diagram technique. *IEEE Transactions on Industry Applications*, 32 (6), 569–576.
125 Ding, X., Bhattacharyal, M., and Mi, C. (2010) Simplified thermal model of PM motors in hybrid vehicle applications taking into account eddy current loss in magnets. *Journal of Asia Electric Vehicles*, 8 (1), 1–7.
126 Mellor, P.H., Roberts, D., and Turner, D.R. (1991) Lumped parameter thermal model for electrical machines of TEFC design. *IEE Proceedings – Electric Power Applications*, 138 (5), 205–218.
127 Taylor, G.I. (1935) Distribution of velocity and temperature between concentric cylinders. *Proceedings of the Royal Society*, 159 (Pt A), 546–578.
128 Gazley, C. (1958) Heat transfer characteristics of rotational and axial flow between concentric cylinder. *Transactions of the American Society of Mechanical Engineers*, 80, 79–89.
129 Boglietti, A., Cavagnino, A., Lazzari, M., and Pastorelli, A. (2002) A simplified thermal model for variable speed self cooled industrial induction motor. Industry Applications Conference, 37th IAS Annual Meeting, October 13–17, Vol. 2, pp. 723–730.
130 Tang, S.C., Keim, T.A., and Perreault, D.J. (2005) Thermal modeling of Lundell alternators. *IEEE Transactions on Energy Conversion*, 20 (1), 25–36.
131 Sooriyakumar, G., Perryman, R., and Dodds, S.J. (2007) Analytical thermal modelling for permanent magnet synchronous motors. Universities Power Engineering 42nd International Conference, UPEC 2007, September 4–6, pp. 192–196.
132 Staton, D., Boglietti, A., and Cavagnino, A. (2005) Solving the more difficult aspects of electric motor thermal analysis in small and medium size industrial induction motors. *IEEE Transactions on Energy Conversion*, 20 (3), 620–628.
133 Guo, Y.G., Zhu, J.G., and Wu, W. (2005) Thermal analysis of soft magnetic composite motors using a hybrid model with distributed heat sources. *IEEE Transactions on Magnetics*, 41 (6), 2124–2128.
134 Funieru, B. and Binder, A. (2008) Thermal design of a permanent magnet motor used for gearless railway traction. Industrial Electronics, IECON 2008, 34th Annual Conference, November 10–13, pp. 2061–2066.
135 Staton, D.A. and Cavagnino, A. (2008) Convection heat transfer and flow calculations suitable for electric machines thermal models. *IEEE Transactions on Industrial Electronics*, 55 (10), 3509–3516.
136 Cassat, A., Espanet, C., and Wavre, N. (2003) BLDC motor stator and rotor iron losses and thermal behavior based on lumped schemes and 3-D FEM analysis. *IEEE Transactions on Industry Applications*, 39 (5), 1314–1322.

137 Kim, W.-G., Lee, J.-I., Kim, K.-W., et al. (2006) The temperature rise characteristic analysis technique of the traction motor for EV application. Strategic Technology, 1st International Forum, October 18–20, pp. 443–446.

138 Chowdhury, S.K. (2005) A distributed parameter thermal model for induction motors. International Conference on Power Electronics and Drives Systems, PEDS 2005, November 28–December 1, Vol. 1, pp. 739–744.

139 Hsu, J.S., Nelson, S.C., Jallouk, P.A., et al. (2005) Report on Toyota Prius Motor Thermal Management. http://www.ornl.gov/~webworks/cppr/y2001/rpt/122586.pdf (accessed February 2011).

140 Jahns, T.M. and Caliskan, V. (1999) Uncontrolled generator operation of interior PM synchronous machines following high-speed inverter shutdown. *IEEE Transactions on Industry Applications*, 35 (6), 1347–1357.

141 Liaw, C.Z., Soong, W.L., Welchko, B.A., and Ertugrul, N. (2005) Uncontrolled generation in interior permanent-magnet machines. *IEEE Transactions on Industry Applications*, 41 (4), 945–954.

10

Hybrid Powertrains

A hybrid electric vehicle (HEV) is the combination of a conventional internal combustion engine (ICE) powered vehicle and an electric vehicle. It uses both an ICE and one or more electric machines for the propulsion of a vehicle. The two power devices, the ICE and the electric motor, can be connected in series or in parallel from the power flow point of view. When the ICE and motor are connected in series, the HEV is a series hybrid in which only the electric motor is providing mechanical power to the wheels. When the ICE and the electric motor are connected in parallel, the HEV is a parallel hybrid in which both the electric motor and the ICE can deliver mechanical power to the wheels, separately or together.

In an HEV, the ICE is the main power converter that provides all the energy for the vehicle. The electric motor increases the system efficiency and reduces fuel consumption by recovering kinetic energy during regenerating braking; and optimizes the operation of the ICE during normal driving by adjusting the engine torque and speed. The ICE provides the vehicle with an extended driving range, thereby overcoming the disadvantages of a pure EV.

In a plug-in HEV (PHEV), in addition to the liquid fuel available on the vehicle, there is also electricity stored in the battery which can be recharged from the electric grid. Therefore, fuel usage can be further reduced.

In a series HEV or PHEV, the ICE drives a generator (referred to as the I/G set). The ICE converts energy in the liquid fuel to mechanical energy and the generator converts the mechanical energy of the engine output to electricity. An electric motor will propel the vehicle using electricity generated by the I/G set. This electric motor is also used to capture the kinetic energy during braking. There will be a battery between the generator and the electric motor to buffer the electric energy between the I/G set and the motor.

In a parallel HEV or PHEV, both the ICE and the electric motor are coupled to the final drive shaft through a mechanical coupling mechanism, such as a clutch, gears, belts, or pulleys. This parallel configuration allows both the ICE and the electric motor to drive the vehicle either in combined mode, or separately. The electric motor is also used for regenerative braking and for capturing the excess energy from the ICE during coasting.

HEVs and PHEVs can have either the series–parallel or a more complex configuration which usually contains more than one electric machine. These configurations can generally further improve the performance and fuel economy of the vehicle with added component cost.

Automotive Power Transmission Systems, First Edition. Yi Zhang and Chris Mi.

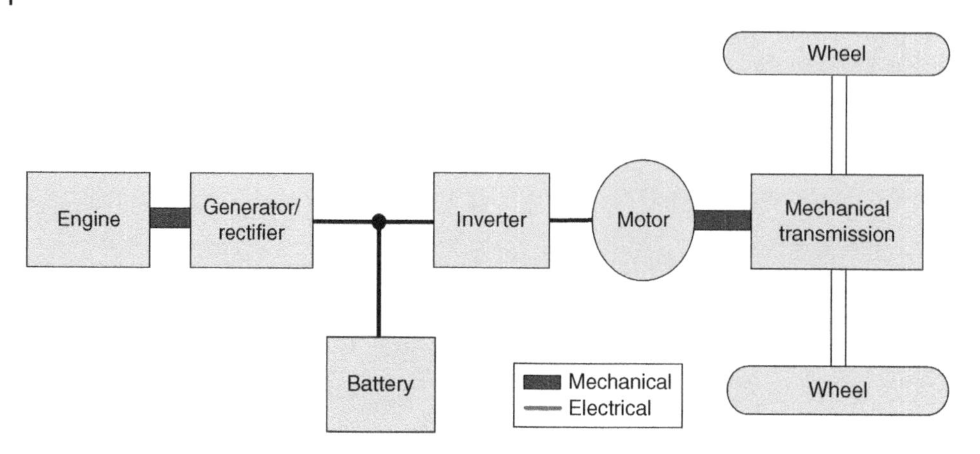

Figure 10.1 Architecture of a series HEV.

10.1 Series HEVs

Figure 10.1 shows the configuration of a series HEV. In this HEV, the ICE is the main energy converter that converts the original energy in gasoline to mechanical power. The mechanical output of the ICE is then converted to electricity using a generator. The electric motor moves the final drive using either electricity generated by the generator or electricity stored in the battery. The electric motor can receive electricity directly from the engine, or from the battery, or both. Since the engine is decoupled from the wheels, the engine speed can be controlled independently of vehicle speed. This not only simplifies the control of the engine, but most importantly, can allow operation of the engine at its optimum speed to achieve the best fuel economy. It also provides flexibility in locating the engine on the vehicle. There is no need for the traditional mechanical transmission in a series HEV. Based on the vehicle operating conditions, the propulsion components on a series HEV can operate with different combinations. Neglecting all the losses in the system, We have the following general equations:

$$P_g = \omega_g T_g$$
$$P_e = \omega_e T_e$$
$$P_m = \omega_m T_m$$
$$T_{fd} = T_m i_T$$
$$T_{shaft} = T_{fd} i_{fd}$$

where T stands for torque, P stands for power, and i stands for gear ratio. Subscript g means the generator, e means the engine, m means the motor, fd means the final drive, *shaft* means the driving shaft, B means battery, i_{fd} is the gear ratio of the final drive and i_T is the gear ratio of the transmission.

- Battery alone: When the battery has sufficient energy, and the vehicle power demand is low, the I/G set is turned off, and the vehicle is powered by the battery alone.

$$T_g = T_e = 0$$
$$P_{fd} = P_m = \omega_m T_m = P_B$$

- Combined power: At high power demands, the I/G set is turned on and the battery also supplies power to the electric motor.

$$P_g = P_e$$
$$P_m = \omega_m T_m = P_B + P_g = P_B + \omega_g T_g$$

- Engine alone: At highway cruising and at moderately high power demands, the I/G set is turned on. The battery is neither charged nor discharged. This is mostly due to the fact that the battery's state of charge (SOC) is already at a high level but the power demand of the vehicle prevents the engine from turning off, or it may not be efficient to turn the engine off.

$$P_g = P_e = P_m = \omega_m T_m$$

- Power split: When the I/G is turned on, the vehicle power demand is below the I/G optimum power, and the battery SOC is low, then a portion of the I/G power is used to charge the battery.

$$P_g = P_e$$
$$P_B = P_g - P_m$$

- Stationary charging: The battery is charged from the I/G power without the vehicle being driven.

$$T_{fd} = T_{shaft} = 0$$
$$P_B = P_g = P_e$$

- Regenerative braking: The electric motor is operated as a generator to convert the vehicle's kinetic energy into electric energy and charge to the battery.

$$T_g = T_e = 0$$
$$T_{fd} = T_{shaft} / i_{fd}$$
$$T_m = T_{fd} / i_T$$
$$P_B = P_m = \omega_m T_m$$

A series HEV can be configured in the same way that conventional vehicles are configured, that is, the electric motor in place of the engine as shown in Figure 10.2. Other choices are also available, such as in-wheel hub motors. In this case, as shown in Figure 10.2, there are four electric motors with each one installed inside each wheel. Due to the elimination of transmission and final drive, the efficiency of the vehicle system can be significantly increased. The vehicle will also have all-wheel drive (AWD) capability. However, controlling the four electric motors independently is a challenge.

10.2 Parallel HEVs

Figure 10.3 shows the configuration of a parallel hybrid. In this configuration, the ICE and the electric motor are coupled to the final drive through a mechanism, such as

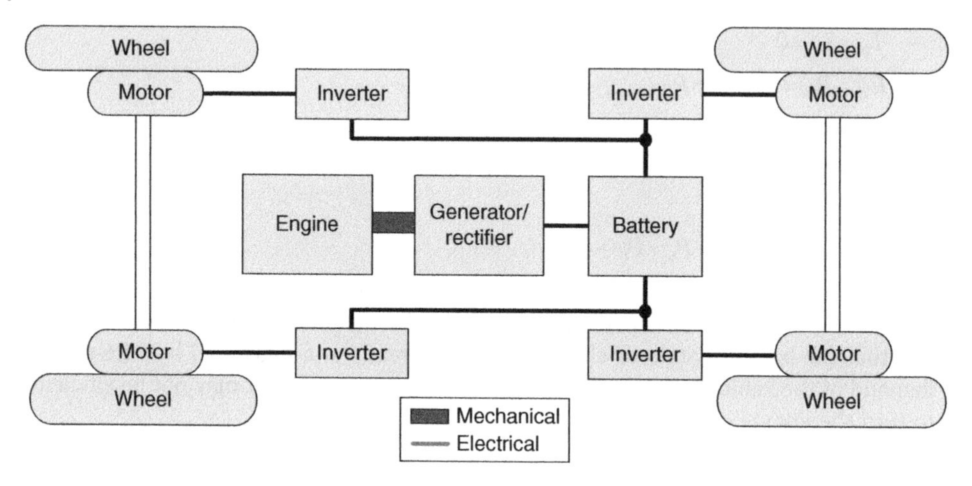

Figure 10.2 Hub motor configuration of a series HEV.

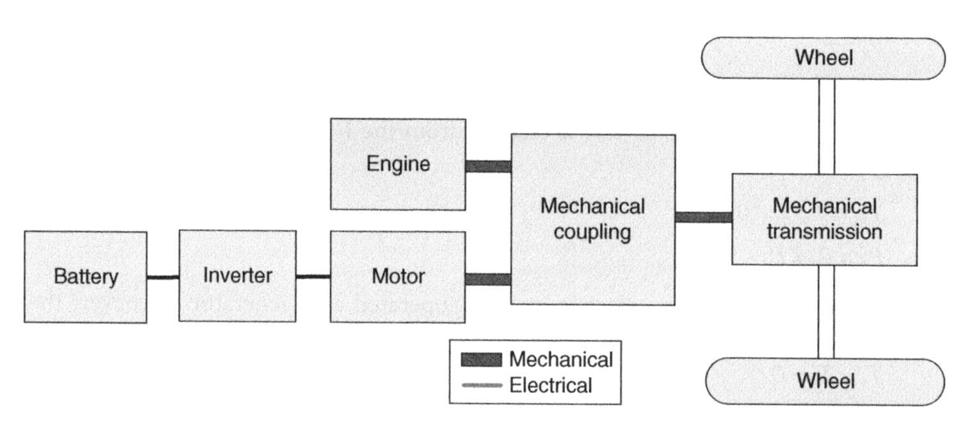

Figure 10.3 Architecture of a parallel HEV.

clutches, belts, pulleys, and gears. Both the ICE and the motor can deliver power to the final drive, either in combined mode, or separately. The electric motor can be used as a generator to recover the kinetic energy during braking or absorbing a portion of power from the ICE. The parallel hybrid needs only two propulsion devices – the ICE and the electric motor, which can be used in the following mode:

- Motor alone mode: When the battery has sufficient energy, and the vehicle power demand is low, then the engine is turned off, and the vehicle is powered by the motor and battery only.

$$T_e = 0$$

$$P_m = \omega_m T_m = P_B$$

- Combined power mode: At high power demands, the engine is turned on and the motor also supplies power to the wheels.

$$T_{fd} = (T_m i_T + T_e i_e)$$
$$P_B = P_m = \omega_m T_m$$
$$P_{shaft} = P_m + P_e$$

- Engine alone mode: At highway cruising and moderately high power demands, the engine provides all the power needed to drive the vehicle. The motor remains idle. This is mostly due to the fact that the battery SOC is already at a high level but the power demand of the vehicle prevents the engine turning off, or it may not be efficient to turn the engine off.

$$T_m = 0$$
$$P_B = P_m = \omega_m T_m = 0$$

- Power split mode: When the engine is on, but the vehicle power demand is low and the battery SOC is also low, then a portion of the engine power is converted to electricity by the motor to charge the battery.

$$T_{fd} = (T_e i_e - T_m i_T)$$
$$P_{fd} = P_e - P_m$$
$$P_B = P_m = \omega_m T_m$$

- Stationary charging mode: The battery is charged by running the motor as a generator and driven by the engine, without the vehicle being driven.

$$T_{fd} = 0$$
$$P_m = P_e = P_B = \omega_m T_m = \omega_e T_e$$

- Regenerative braking mode: The electric motor is operated as a generator to convert the vehicle's kinetic energy to electric energy and stored in the battery. Note that while in the regenerative mode, it is in principle possible to run the engine as well, and provide additional current to charge the battery more quickly (while the propulsion motor is in generator mode) and command its torque accordingly, i.e. to match the total battery power input. In this case, the engine and motor controllers have to be properly coordinated.

$$P_e = 0$$
$$T_{fd} = T_{shaft} / i_{fd}$$
$$T_m = T_{fd} / i_T$$
$$P_B = P_m = \omega_m T_m$$

The Honda Civic hybrid is a typical parallel hybrid. It has an electric motor mounted between the ICE and the conventional transmission (CVT), as shown in Figure 10.4. The electric motor either provides assistance to the engine in high power demand or splits the engine power during low power demand.

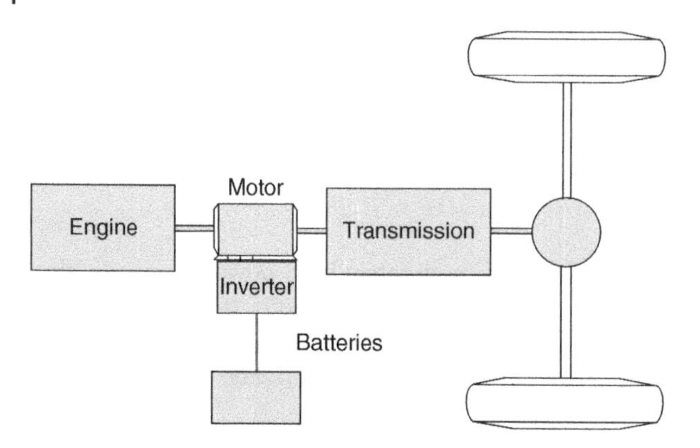

Figure 10.4 The powertrain layout of the Honda Civic hybrid.

10.3 Series–Parallel HEVs

The series–parallel HEV shown in Figure 10.5 incorporates the features of both a series and a parallel HEV. Therefore, it can be operated as a series or parallel HEV. In comparison to a series HEV, the series–parallel HEV adds a mechanical link between the engine and the final drive, so the engine can drive the wheels directly. When compared to a parallel HEV, the series–parallel HEV adds a second electric machine that serves primarily as a generator.

Because a series–parallel HEV can operate in both parallel and series modes, the fuel efficiency and driveability can be optimized based on the vehicle's operating condition. The increased degree of freedom in control makes the series–parallel HEV a popular choice. However, due to the increased components and complexity, the series–parallel HEV is generally more expensive than a series or parallel HEV. The detailed analysis of the operation of series–parallel or complex hybrid will highly depend on the configuration of the powertrain and will be discussed in detail in the later sections.

Toyota Prius is a typical series–parallel hybrid. Toyota produced the world's first mass-marketed modern HEV in 1997, the Prius, as shown in Figure 10.6. The worldwide sales of Prius exceeded 1 million units in 2009. The Prius hybrid powertrain uses a planetary

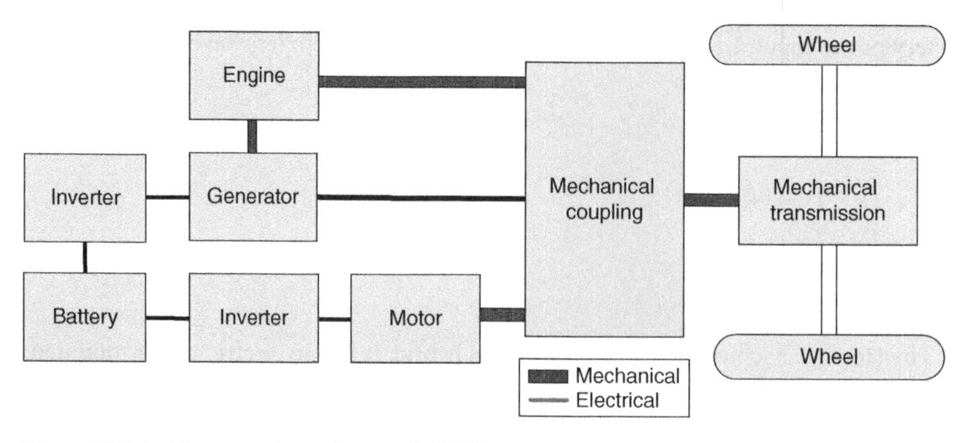

Figure 10.5 Architectures of a series–parallel HEV.

Figure 10.6 Toyota Prius (2010 model).

gear set to realize continuous variable transmission (CVT). Therefore, the conventional transmission is not needed in this system. As shown in Figure 10.7, the engine is connected to the carrier of the planetary gear, while the generator is connected to the sun gear. The ring gear is coupled to the final drive, as is the electric motor. The planetary gear set also acts as a power/torque split device. During normal operations, the ring gear speed is determined by the vehicle speed, while the generator speed can be controlled such that the engine speed is in its optimum efficiency range.

The 6.5 Ah, 21 kW nickel-metal-hydride battery pack is charged by the generator during coasting and by the propulsion motor (in generation mode) during regenerative braking. The engine is shut off during low-speed driving.

The same technology has been used in the Camry hybrid, the Highlander hybrid, and the Lexus brand hybrids. However, the Highlander and the Lexus hybrids add a third motor at the rear wheel. The drive performance, such as for acceleration and braking, can thus be further improved.

The Escape hybrid by the Ford Motor Company, shown in Figure 10.8, is the first hybrid in the sports utility vehicle category. The Escape hybrid adopted the same planetary gear concept as the Toyota system.

The Toyota Prius and the Ford Escape use similar powertrain transmission as shown in Figure 10.9. This consists of an engine, two electric machines, and a planetary gear train in the transmission. The engine is connected to the carrier, the electric motor MG2 is

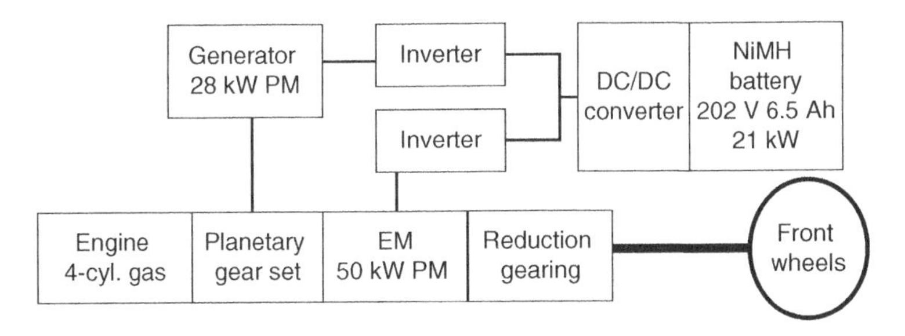

Figure 10.7 Powertrain layout of the Toyota Prius (PM – permanent magnet; EM – electric machine).

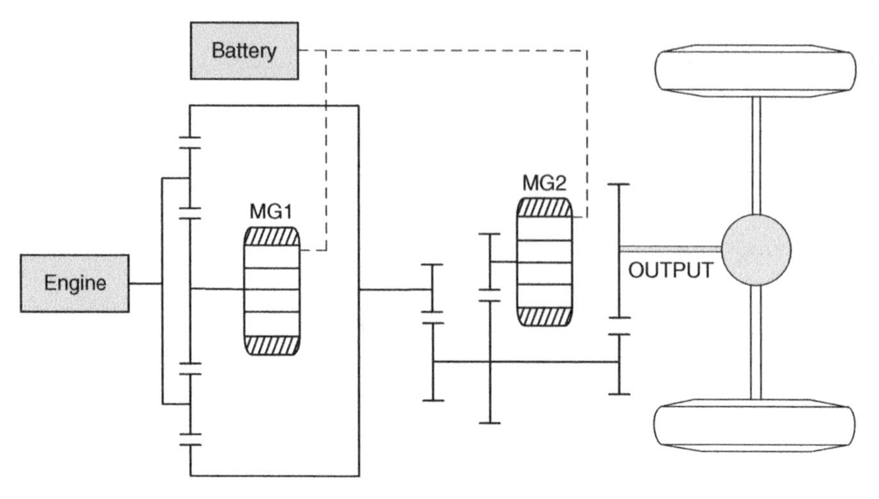

Figure 10.8 Ford Escape hybrid SUV.

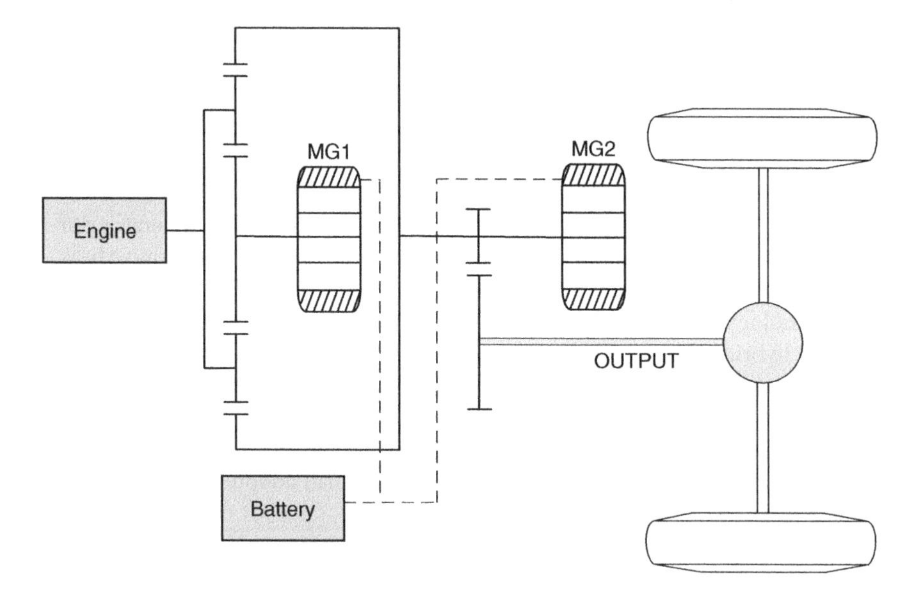

Figure 10.9 Toyota Prius Transmission.

connected to the ring gear as well as to the final drive, and generator MG1 is connected to the sun gear. Hence, the speed and torque relationships are:

$$\omega_e = \frac{N_s}{N_r + N_s}\omega_g + \frac{N_r}{N_r + N_s}\omega_r$$

$$T_r = T_e\frac{N_r}{N_r + N_s}$$

$$T_g = T_s = T_e\frac{N_s}{N_r + N_s}$$

$$T_{shaft} = (T_m + T_r) * i_1$$

where ω_e, ω_m and ω_g are the speed of the engine, motor and generator, respectively, ω_r is the ring gear speed, $\omega_m = \omega_r$; $\omega_s = \omega_g$; $i_1 = N_2/N_1$ is the gear final drive ratio, and N_1 and N_2 the gear teeth numbers of the final drive.

Since there is no clutch, the planetary gear is always running whenever the vehicle is moving. It can be seen from the above equation and the diagram of the powertrain that the speed of the motor MG2 is directly proportional to the linear speed of the vehicle through the radius of the front tires and the final drive ratio. The ring gear speed and the motor speed are identical for the Prius hybrid. There is a gear ratio between the ring gear and the motor for the Mercury Mariner hybrid vehicle.

There are four different operation modes:

Mode 0: Launch and backing up – the motor is powered by the battery. The vehicle is driven by the motor only.

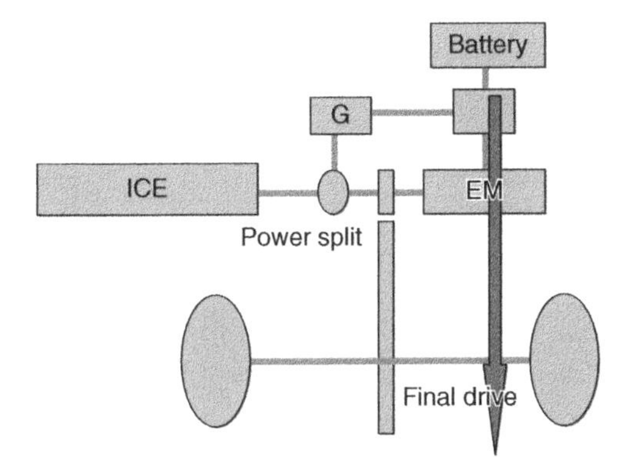

$$T_e = 0$$

$$T_g = 0$$

$$T_{shaft} = i_1 T_m$$

Mode 1: Cruising, or e-CVT mode 1.

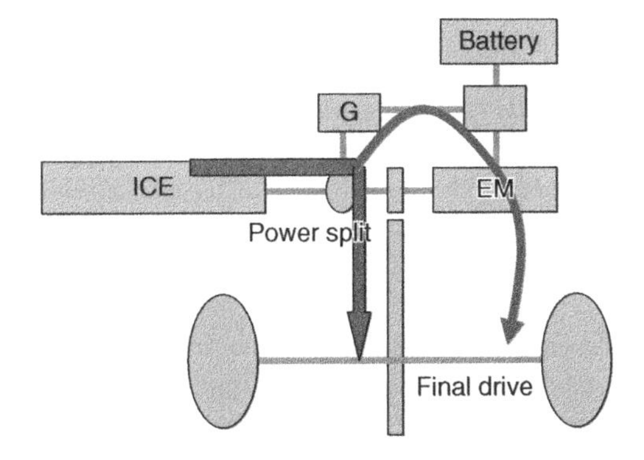

$$T_r = T_e \frac{N_r}{N_r + N_s}$$

$$T_g = T_s = T_e \frac{N_s}{N_r + N_s}$$

$$P_m = \omega_m T_m = P_g = \omega_g T_s$$

$$P_B = 0$$

$$T_{out} = i_1 (T_r + T_m)$$

Mode 2: Sudden acceleration, e-CVT mode 2.

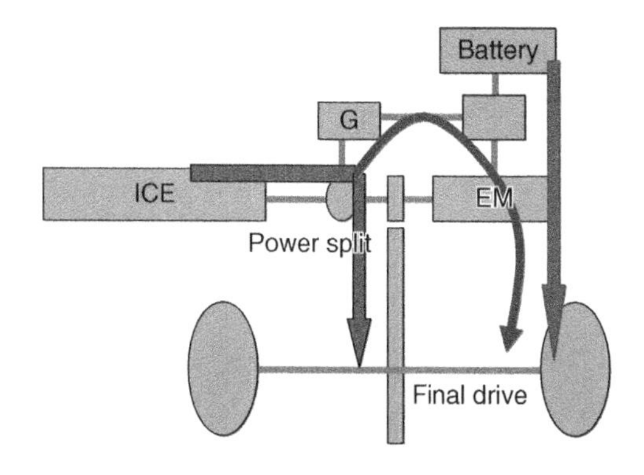

$$T_r = \frac{N_r}{N_r + N_s} T_e$$

$$T_g = T_s = T_e \frac{N_s}{N_r + N_s}$$

$$T_{out} = i_1 (T_r + T_m)$$

$$P_m = \omega_m T_m = P_g + P_B$$

Mode 3: Regenerative braking: MG2 is operating as a generator to produce electricity to charge the battery and at the same time to provide braking torque to the final drive. This operation is the reverse of launch and backing up operation.

During normal operation (e-CVT or acceleration mode), the speed of the engine is controlled by the torque on the generator. Basically, the generator power is adjusted so that the engine turns at the desired speed. Hence, by adjusting the generator speed, the engine can operate at a relatively constant speed while the vehicle is driven at different speeds.

In the Prius, the engine is limited from 0 to 4000 RPM. The motor is limited from a small negative RPM for reverse and up to 6000 RPM (~103 mph or 165 km/h). The generator is limited to ± 5500 RPM. The ring gear and sun gear each have 78 & 30 teeth respectively. The four planet gears each have 23 teeth. The final drive ratio is 3.93 and the wheel radius is 0.287 m. Hence, $\omega_e = 0.7222\omega_m + 0.2778\omega_g$.

The control strategy is as follows. For the given vehicle speed, and a desired output power determined by the drive cycle, or driver inputs, the desired operating point of the engine can be determined based on the maximum efficiency curve of the engine. From the vehicle speed and desired engine speed, the desired generator speed can then be calculated. The generator speed is regulated by the inverter through controlling the output power of the generator (either as a generator or as a motor). Motor torque is determined by looking at the difference between the total vehicle power demand and the engine torque that is delivered to the ring gear. The battery provides power to the motors along with the electricity generated by the engine.

Example Consider a planetary gear train based transmission with an engine (carrier) providing 100 kW at 2000 RPM optimum operating point. The ring gear has 72 teeth and the sun gear 30 teeth. The final drive ratio is 3.7865, and the wheel radius is 0.283 m. (i) For vehicle speed of 45 mph or 20.6 m/s, the power demand for heavy acceleration at this speed is 120 kW. Find the speed and power for each component, assuming no losses. (ii) For vehicle speed of 70 mph, or 32.7 m/s, when cruising, the power demand is 70 kW. Calculate the speed and power of each component.

Solution:

$$\omega_e = \frac{N_r}{N_r + N_s}\omega_r + \frac{N_s}{N_r + N_s}\omega_s = 0.706\omega_r + 0.294\omega_s$$

i) At 45 mph, the ring gear (same as motor speed) is calculated from the above to be 2632 RPM. Therefore, the sun gear (generator) speed needs to be 482 RPM in order to operate the engine at 2000 RPM.

$$T_c(\text{engine}) = P_{engine}/\omega_{engine(carrier)} = 477\ \text{Nm}$$

$$T_r(\text{Ring\_gear}) = \frac{N_r}{N_r + N_s}T_c = 0.706 * 477 = 337\ \text{Nm}$$

$$T_s(\text{generator}) = \frac{N_s}{N_r + N_s}T_c = 0.294 * 477 = 140\ \text{Nm}$$

$$P_c(\text{engine}) = 100\ \text{kW}$$

$$P_r(\text{Ring\_gear}) = T_r\omega_r = 337 * 2 * \pi * 2632/60 = 92.9\ \text{kW}$$

$$P_s(\text{generator}) = T_s\omega_s = 140 * 2 * \pi * 482/60 = 7.1\ \text{kW}$$

$$P_c(\text{engine}) = P_r + P_s$$

$$P_{vehicle} = 120\ \text{kW}$$

$$P_m(\text{motor}) = P_{vehicle} - P_r = 27.1\ \text{kW}$$

$$P_{bat} = P_m - P_s = 20\ \text{kW}$$

ii) At 70 mph, the ring gear (same as motor speed) is calculated from the above to equation be 4080 RPM. Therefore, the sun gear (generator) speed needs to be –2995 RPM in order to operate the engine at 2000 RPM.

$$T_c(\text{engine}) = P_{engine}/\omega_{engine(carrier)} = 477\,\text{Nm}$$

$$T_r(\text{Ring\_gear}) = \frac{N_r}{N_r + N_s} T_c = 0.706 * 477 = 337\,\text{Nm}$$

$$T_s(\text{generator}) = \frac{N_s}{N_r + N_s} T_c = 0.294 * 477 = 140\,\text{Nm}$$

$$P_c(\text{engine}) = 100\,\text{kW}$$

$$P_r(\text{Ring\_gear}) = T_r\omega_r = 337 * 2 * \pi * 4080/60 = 144\,\text{kW}$$

$$P_s(\text{generator}) = T_s\omega_s = 140 * 2 * \pi * (-2995)/60 = -44\,\text{kW}$$

$$P_c(\text{engine}) = P_r + P_s$$

$$P_{vehicle} = 70\,\text{kW}$$

$$P_m(\text{motor}) = P_{vehicle} - P_r = -74\,\text{kW}$$

$$P_{bat} = P_m - P_s = -30\,\text{kW}$$

10.4 Complex HEVs

Complex HEVs usually involve the use of planetary gear systems and multiple electric motors (in the case of four/all-wheel drive). One typical example is a four-wheel drive system that is realized through the use of separate drive axles, as shown in Figure 10.10. The generator in this system is used to realize the series operation as well as to control the engine operation condition for maximum efficiency. The two electric motors are used to realize all-wheel drive, and to provide better performance in regenerative braking. It may also enhance the vehicle stability control and antilock braking control by using the two electric motors.

10.4.1 GM Two-Mode Hybrid Transmission

The GM two-mode hybrid transmission, shown in Figures 10.11 and 10.12, is a typical complex hybrid system. It was initially developed by GM (Alison) in 1996 and later

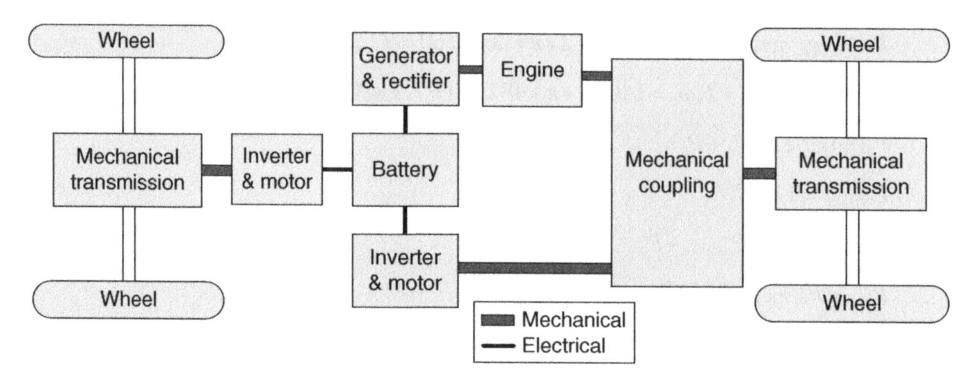

Figure 10.10 The electrical four-wheel drive system using a complex architecture.

Figure 10.11 The Chrysler Aspen Two-Mode Hybrid.

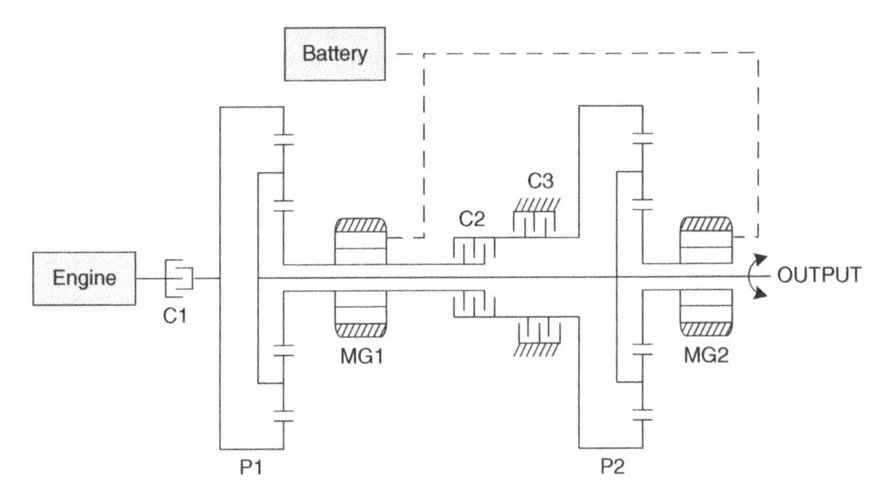

Figure 10.12 GM two-mode hybrid transmission.

advanced by General Motors, Chrysler, BMW, and Mercedes-Benz with a joint venture named Global Hybrid Cooperation in 2005. The GM two-mode hybrids use two planetary gear sets and two electric machines to realize two different operation modes – a high-speed mode, and a low-speed mode.

The GM Two-Mode hybrid electric powertrain (or transmission) is shown in Figure 10.12 [1, 2]. This powertrain consists of two planetary gear sets P1 & P2, two electric machines MG1 and MG2, and three clutches C1, C2, and C3. The powertrain is capable of providing electric continuous variable transmission (e-CVT) for both high-speed and low-speed operations, hence, two-mode. The two-mode concept can be referred to and compared to the Toyota and Ford hybrid electric vehicle powertrain whose operation is limited to only one mode. In principle, the two-mode operation can provide more flexibility for transmission control, can increase drivability, and can improve vehicle performance and fuel economy.

Operation principle of the two-mode powertrain: In the GM two-mode hybrid transmission, the engine is connected to the ring gear of planetary gear P1 through clutch C1. Electric machine MG1 is connected to the sun gear of P1. The carrier of P1 is connected to the final drive through the output shaft. MG2 is connected to the sun gear of planetary P2. The carrier of P2 is also connected to the output shaft. There is a dual-position clutch that connects either the ring gear of P2 to the ground, or the ring gear of P2 to the shaft of MG1.Through control of C2 and C3, different operating modes can be realized. The engine in this system can be kept at the best speed and torque combination to achieve the best fuel economy by controlling the input/output of the two electric machines. The engine may be stopped or idle during vehicle launch and backup, as well as at low power demand. In cruising conditions, the engine efficiency is further enhanced by cylinder deactivation, also known as active fuel management (GM) or the multi-displacement system (Chrysler). Note that this discussion is generic and may not be the same as those implemented in a real vehicle by the automobile manufacturers.

In the following derivations, ω is the angular velocity, T is torque, N is the number of teeth of a gear, and P is power. Subscript s stands for the sun gear, r for the ring gear, c for the planetary carrier, 1 for planetary gear set 1, 2 for planetary gear set 2, g for motor/generator one, or MG1, m for motor/generator two, or MG2, and out is for output or final drive.

Mode 0: Vehicle launch and backup: During vehicle launch and backup, the system is operating in motor alone mode (Mode 0). C2 is open and C3 engages to ground the ring gear of P2. In this mode, there are two possibilities for engine operation, either off or idle at cranking speed (approximately 800 RPM) by adjusting MG1 speed. MG1 torque is not transmitted to the final drive. MG2 provides the needed torque to launch the vehicle forward or backward. Figure 10.13 shows the power flow during launch and backup. The speed/torque relationships are:

$$\omega_{out} = \frac{N_{s2}}{N_{s2} + N_{r2}} \omega_m$$

$$T_{out} = \frac{N_{s2} + N_{r2}}{N_{s2}} T_m$$

In the final implementation, C1 was eliminated. Therefore, the engine is always connected to the ring gear of P1. Since the carrier of P1 is always connected to the final drive, MG1 needs to be controlled so that the engine is either at zero or at a certain speed:

$$\omega_g = \frac{N_{s1} + N_{r1}}{N_{s1}} \omega_{c1} - \frac{N_{r1}}{N_{s1}} \omega_e$$

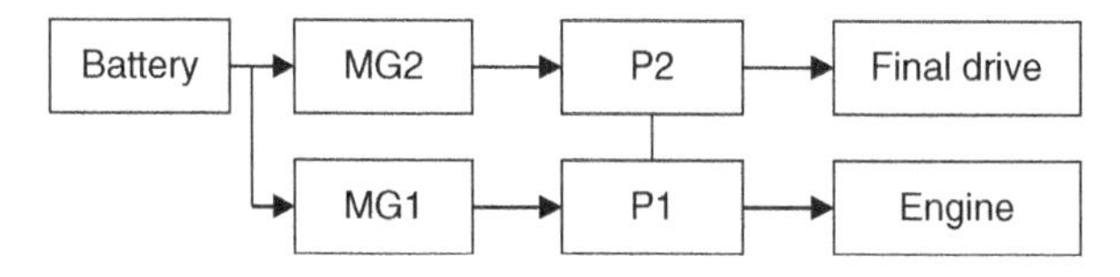

Figure 10.13 Power flow during launch and backup.

If the engine is maintained at 800 RPM without fuel injection, there are still friction losses but the engine can be ignited at any time without delay. If the engine is controlled to be at zero speed, then MG1 will need to be controlled so that the ring gear of P1 reaches 800 RPM before the engine can be started. In this case, the engine is controlled by cylinder deactivation.

Mode 1: Low range: Mode 1 is also called the low range, or low-speed mode. In this mode, C1 is engaged, C2 is open, and C3 is engaged. The second planetary gear works as a speed reduction gear for MG2. Figure 10.14 illustrates the mechanical connections of the transmission. The engine may be controlled by partial cylinder deactivation to further save fuel and reduce emissions based on vehicle power demand. The torque and speed relationships during steady state operations can be expressed as:

$$T_g = \frac{N_{s1}}{N_{r1}} T_e$$

$$\omega_{c1} = \frac{N_{r1}}{N_{s1} + N_{r1}} \omega_e + \frac{N_{s1}}{N_{s1} + N_{r1}} \omega_g$$

$$\omega_{out} = \omega_{c2} = \omega_{c1} = \frac{N_{s2}}{N_{s2} + N_{r2}} \omega_m$$

$$T_{out} = \frac{N_{r1} + N_{s1}}{N_{r1}} T_e + \frac{N_{s2} + N_{r2}}{N_{s2}} T_m$$

The different operations in Mode 1 can be described as follows:

1) Engine alone mode (CVT 1): MG2 is off (freewheel) and MG1 can be either in motoring mode or in generating mode. When MG1 is in a motoring mode, P1 acts as a speed coupling mechanism to couple the speed of the engine and MG1. When

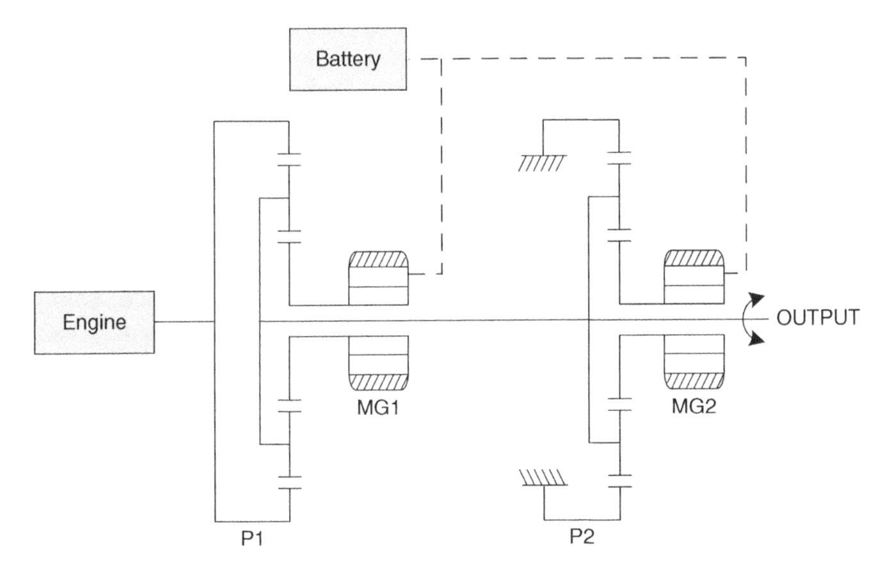

Figure 10.14 Low Range.

MG1 is in generating mode, engine power is split between the final drive and MG1, with power generated by MG1 charging the battery. Since the battery can quickly be charged fully, this mode is generally brief.

2) Combined mode (CVT 2): MG2 is turned on to assist the driving. P2 acts as a torque coupling mechanism to add the torque of the engine (P1 carrier portion) and MG2. If needed, both MG1 and MG2 can work in motoring mode to maximize the driving torque.
3) Power split mode (CVT 3): MG2 is in generating mode to charge the battery. MG1 can be either motoring or generating.

Mode 2: High range: Mode 2 is called the high range, or high-speed mode. C1 is engaged, C2 is engaged, but C3 is open. In this mode, the sun gear of P1 is connected to the ring gear of P2 through MG1, that is, MG1, S1, and R2 will have the same speed. Figure 10.15 shows the mechanical connections of the transmission in Mode 2. In this operating mode, the engine is generally kept at a constant speed to achieve the best fuel economy. MG1 and MG2 are controlled in either motoring or generating mode depending on the vehicle speed and power demand. The torque and speed relationships during steady-state operation are as follows:

$$T_g = \frac{N_{s1}}{N_{r1}} T_e + \frac{N_{r2}}{N_{s2}} T_m$$

$$\omega_{c1} = \frac{N_{r1}}{N_{s1} + N_{r1}} \omega_e + \frac{N_{s1}}{N_{s1} + N_{r1}} \omega_g$$

$$\omega_{c2} = \frac{N_{r2}}{N_{s2} + N_{r2}} \omega_g + \frac{N_{s2}}{N_{s2} + N_{r2}} \omega_m$$

$$\omega_{c2} = \omega_{c1}$$

$$\omega_{r2} = \omega_{s1} = \omega_g$$

$$T_{fd} = \frac{N_{r1} + N_{s1}}{N_{r1}} T_e + \frac{N_{s2} + N_{r2}}{N_{s2}} T_m$$

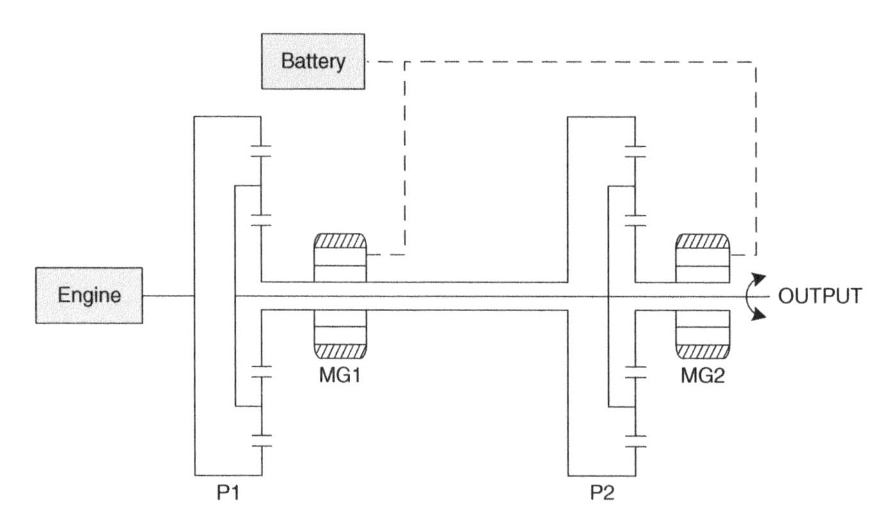

Figure 10.15 High range.

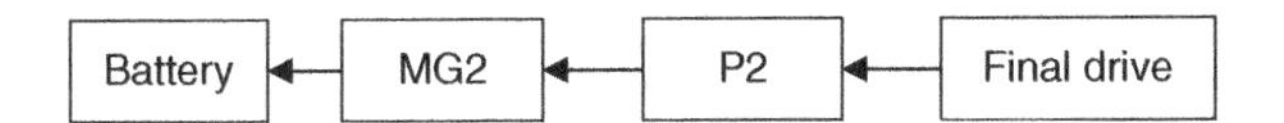

Figure 10.16 Power flow in regenerative braking.

Similar to Mode 1, the engine may be controlled by partial cylinder deactivation to further save fuel and reduce emission based on vehicle power demand.

Mode 3: Regenerative braking: During regenerative braking, C1 is open, C2 is open and C3 engages to ground the ring gear of P2 (Mode 3). The engine and MG1 are off or freewheel. MG2 provides the needed braking torque for the vehicle, and at the same time, stores regenerative braking energy in the onboard battery. Figure 10.16 shows the power flow during regenerative braking. The speed/torque relationship is given by:

$$\omega_m = \frac{N_{s2} + N_{r2}}{N_{s2}} \omega_{out}$$

$$T_m = \frac{N_{s2}}{N_{s2} + N_{r2}} T_{out}$$

Hydraulic/frictional braking may be controlled in coordination with regenerative braking to maximize the braking torque, and/or maintain vehicle stability and prevent wheel locking. In this case, MG2 only provides a portion of the braking torque.

Transitions between modes 0, 1, 2, and 3: In general, the transition is performed at a condition that can minimize mechanical disturbance to the overall vehicle system. The vehicle is usually launched by MG2 with the engine off (Mode 0). MG1 is turned on before transitioning to Mode 1, such that the engine speed reaches approximately 800 RPM. The transition from Mode 0 to Mode 1 is characterized by the engine turning on. This typically happens when the power demand reaches a certain limit such that MG2 is no longer capable of providing the needed torque. The power demand is a combination of vehicle speed, acceleration demand, vehicle load, and road conditions.

The transition from Mode 1 to Mode 2 happens when the sun gear of P1 and the ring gear of P2 reach the same speed. In other words, since the ring gear of P1 is grounded (zero speed), the transition from Mode 1 to Mode 2 will happen when the sun gear of P1 or MG1 reaches zero speed. Similarly, transition from Mode 2 to Mode 1 also happens when the speed of MG1 reaches zero.

The transition from Mode 1 to Mode 3, or Mode 2 to Mode 3 is triggered by the braking request from the driver (brake pedal is pressed).

Example 1 Both planetary gear sets have 30 teeth for the sun gear and 70 teeth for the ring gear. The engine is kept at 800 RPM in Mode 0, ramped up from 800 RPM to 2000 RPM in Mode 1, and kept at 2000 RPM in Mode 2. The wheel radius is 0.28 m. Vehicle speed V ranges from −40 km/h to 160 km/h. The final drive gear ratio (including axle) is 3.3.

Solution:

The final drive speed is a function of vehicle speed:

$$\omega_{out} = (V * 1000/3600/0.28) * 3.3 * 60/2\pi = 31.2V(\text{RPM})$$

In Mode 0, the engine is kept at 800 RPM and the speed of the ring gear of P2 is zero. Therefore:

$$\omega_m = \frac{N_{s2} + N_{r2}}{N_{s2}}\omega_{c2} = 3.33\omega_{out}$$

$$\omega_e = 800\,\text{RPM}$$

$$\omega_g = \frac{N_{s1} + N_{r1}}{N_{s1}}\omega_{c1} - \frac{N_{r1}}{N_{s1}}\omega_e = 3.33\omega_{out} - 2.33\omega_e$$

In Mode 1, the engine speed will ramp up from 800 RPM to 2000 RPM. Note that the engine can be turned on or kept idling. The engine on/off is determined by vehicle power demand. The speed relationships are the same as in Mode 0.

When the speed of the sun gear of P1 reaches zero, the vehicle will shift from Mode 1 to Mode 2. In Mode 2, the engine speed is kept at 2000 RPM. The speed relationships are:

$$\omega_e = 2000\,\text{RPM}$$

$$\omega_g = \frac{N_{s1} + N_{r1}}{N_{s1}}\omega_{c1} - \frac{N_{r1}}{N_{s1}}\omega_e = 3.33\omega_{out} - 2.33\omega_e$$

$$\omega_m = \frac{N_{s2} + N_{r2}}{N_{s2}}\omega_{c2} - \frac{N_{r2}}{N_{s2}}\omega_{r2} = 3.33\omega_{out} - 2.33\omega_g$$

Figure 10.17 shows the speeds of the system: engine, MG1, MG2, and final drive.

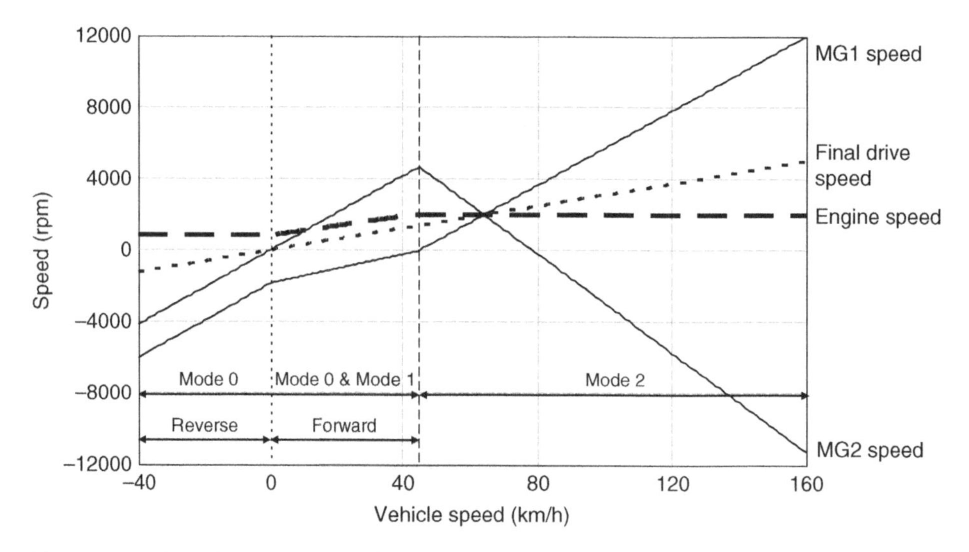

Figure 10.17 Speed relationships of the two-mode transmission in example 1.

Example 2 Planetary gear set 1 has 35 teeth for the sun gear and 65 teeth for the ring gear. Planetary gear set 2 has 30 teeth for the sun gear and 70 teeth for the ring gear. The engine is kept at 0 RPM in Mode 0; ramped up from 0 RPM to 3000 RPM in Mode 1, and kept at 3000 RPM in Mode 2. The wheel radius is 0.28 m. Vehicle speed V ranges from −40 km/h to 160 km/h. The final drive gear ratio (including axle) is 3.

Solution:

The final drive speed is a function of vehicle speed.

$$\omega_{out} = (V * 1000/3600/0.28) * 3.3 * 60/2\pi = 31.2V(\text{RPM})$$

In Mode 0, the engine is kept at 0 RPM and the speed of the ring gear of P2 is zero. Therefore:

$$\omega_m = \frac{N_{s2} + N_{r2}}{N_{s2}}\omega_{c2} = 3.33\omega_{out}$$

$$\omega_g = \frac{N_{s1} + N_{r1}}{N_{s1}}\omega_{c1} = 2.86\omega_{out}$$

In Mode 1, the engine speed will ramp up from 0 to 3000 RPM. Note that the engine can be turned on or kept idling. Engine on/off is determined by vehicle power demand. The speed relationships are:

$$\omega_m = \frac{N_{s2} + N_{r2}}{N_{s2}}\omega_{c2} = 3.33\omega_{out}$$

$$\omega_g = \frac{N_{s1} + N_{r1}}{N_{s1}}\omega_{c1} - \frac{N_{r1}}{N_{s1}}\omega_e = 2.86\omega_{out} - 1.86\omega_e$$

When the speed of the sun gear of P1 reaches zero, the vehicle will shift from Mode 1 to Mode 2. In Mode 2, engine speed is kept at 3000 RPM. The speed relationships are:

$$\omega_e = 3000\,\text{RPM}$$

$$\omega_m = \frac{N_{s2} + N_{r2}}{N_{s2}}\omega_{c2} - \frac{N_{r2}}{N_{s2}}\omega_{r2} = 3.33\omega_{out} - 2.33\omega_g$$

$$\omega_g = \frac{N_{s1} + N_{r1}}{N_{s1}}\omega_{c1} - \frac{N_{r1}}{N_{s1}}\omega_e = 2.86\omega_{out} - 1.86\omega_e$$

Figure 10.18 shows the speeds of the system: engine, MG1, MG2, and final drive.

10.4.2 Dual Clutch Hybrid Transmissions

There are a few variants of automatic transmissions such as automated manual transmission (AMT), continuous variable transmission (CVT), and dual clutch transmission (DCT). Each of these technologies has its own penetration levels in different regions of the world (North America, Europe, Asia). The advantages of DCT include high efficiency, low cost, and driving comfort. Conservative estimates peg DCT technology to be

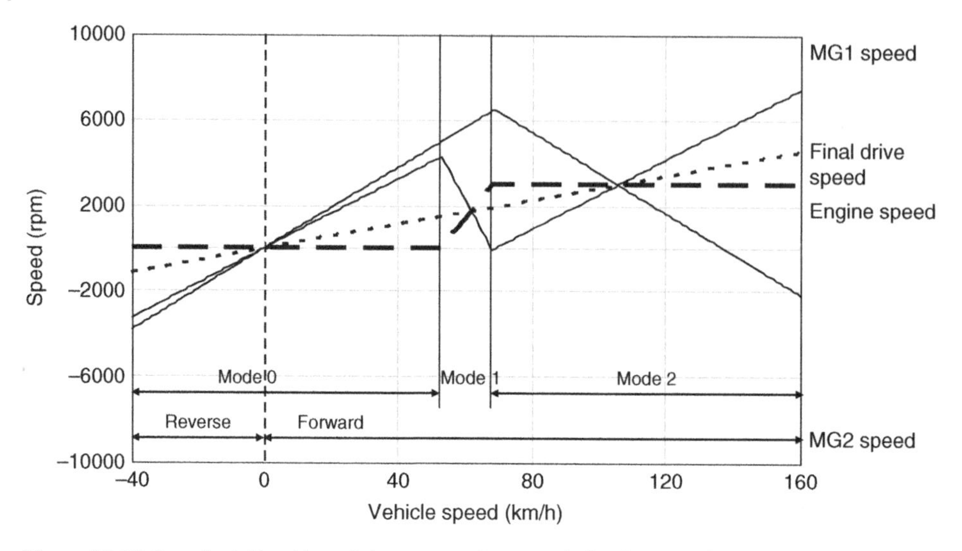

Figure 10.18 Speed relationships of the two-mode transmission in example 2.

Table 10.1 Qualitative comparison of automatic and manual transmissions.

Aspects	Automatic transmission	Manual transmission	Desired transmission
Cost	Expensive	Lower	Low
Efficiency	Moderate	High	High
Ease of use	Easy	Hard	Easy
Comfort	Good	Poor	Good

around 10% of the global market by 2015. Table 10.1 compares the advantages and disadvantages of automatic and manual transmission.

DCT technology is well suited for high torque diesel engines and high revving gas engines alike. Some of the major drivers for DCT include:

- flexible and software tunability
- gear ratio flexibility the same as that of manual lay-shaft transmissions allowing greater compatibility with any engine characteristics.

10.4.2.1 Conventional DCT Technology

A typical DCT architecture has a lay-shaft with synchronizers used for maximum efficiency. It also has launch clutches (either wet or dry) used with electronics along with mechanical or hydraulic actuation systems to achieve the automatic shifting. The lay-shaft transmissions yield an efficiency of 96% or better as compared to 85–87% efficiency of automatic transmissions [3, 4].

Figure 10.19 shows the diagram of a DCT based transmission. It is a typical setup used in the latest vehicle models with a DCT. It consists of two coaxial shafts each having the

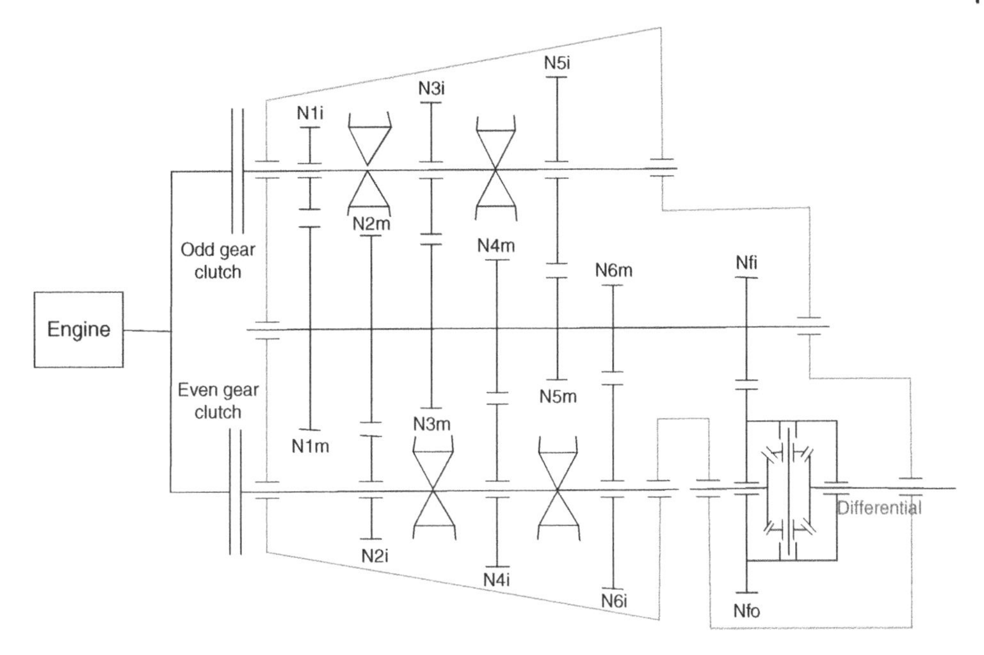

Figure 10.19 Dual clutch transmission. (Note: the reverse gear is omitted from the diagram).

odd and even gears. It is tantamount to having two transmissions, hence the name, dual clutch transmission.

In a DCT system, the two clutches are connected to two separate sets of gears. The odd gear set is connected to one of the clutches and the even gear set to the other clutch. It is necessary to pre-select the gears to realize the benefits of the DCT system. Accordingly, the off-going clutch is released simultaneously as the oncoming clutch is engaged. This gives an uninterrupted torque supply to the driveline during the shifting process. This pre-selection of gears can be implemented using either complicated controllers such as fuzzy logic or simple ones such as selections based on the next anticipated vehicle speed.

10.4.2.2 Gear Shift Schedule

Initially, when the vehicle starts, gear N1i is synchronized. Therefore, engine torque is transmitted to the final drive through gears N1i and N1m. The vehicle speed increases as the odd clutch engages. When the vehicle speed reaches a certain threshold, gear N2i is synchronized. As the even clutch engages (the odd clutch disengages), engine torque is shifted from gear N1i to N2i, so the engine torque is transmitted through gear N2i and N2m. As vehicle speed increases further, N3i is synchronized. Then the odd clutch would engage and the even clutch would disengage. This process will continue until the vehicle speed becomes stable (from N3i to N4i, from N4i to N5i, and from N5i to N6i).

During downshift, the process is reversed. For example, assume initially that N4i is synchronized and the even clutch is engaged. During downshift, N3i is synchronized before the even clutch opens. When the even clutch disengages and the odd clutch engages,

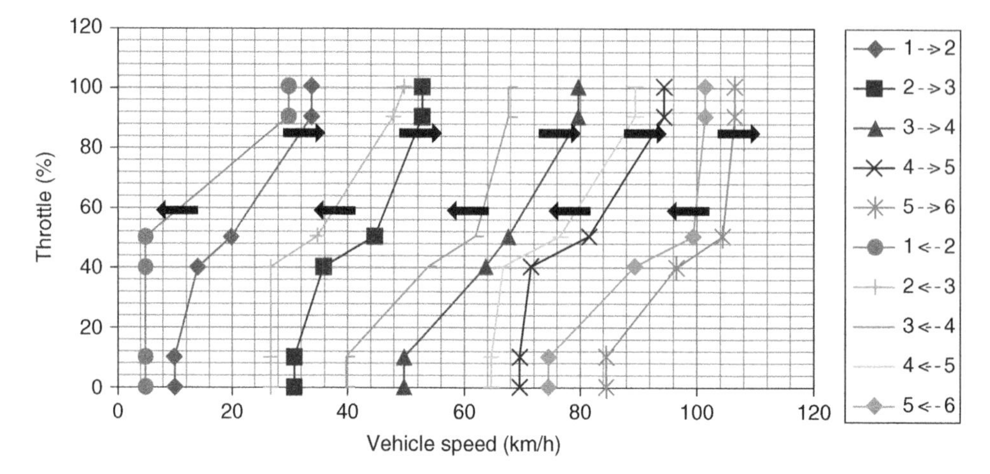

Figure 10.20 Gear shift schedule.

engine torque is transferred from N4i to N3i. Similarly, N2i would be synchronized before the even clutch engages.

Since all transitions in a DCT are managed by gear synchronizers and two clutches, there is no need for a torque converter in a DCT. The transitions (gear shifting and torque shifting) are very smooth. Control of the synchronizers and clutches or shift controller is computerized in the vehicle. The shift controller decides the upshifts or the downshifts of the transmission as per the gear shift schedule shown from left to right in Figure 10.20. This controller intelligently pre-selects the higher or the lower gear depending on the current and desired vehicle velocity.

10.4.2.3 DCT-based Hybrid Powertrain

The diagram for a DCT based hybrid powertrain is shown in Figure 10.21 [5]. The transmission is a six-speed AMT. The hybrid powertrain consists of two motors with each coupled mechanically onto the two shafts using a standard gear reduction. Due to the presence of the motor/generator, the vehicle can be reversed without the reverse gear. The odd shaft houses gears 1, 3, and 5 and the even shaft houses gears 2, 4, and 6. The two motors can also be operated as generators as needed by the hybrid control strategy.

10.4.2.4 Operation of DCT Based Hybrid Powertrain

The DCT based hybrid powertrain shown in Figure 10.21 has seven operating modes when the vehicle is in motion and one additional operating mode for standstill charging.

Motor alone mode: The vehicle is always launched in the motor only mode unless the battery's state of charge (SOC) is below the minimum level. In this mode, the gears are selected according to the shift logic controller. The vehicle operates in this mode up to a maximum speed defined by the controller, provided the SOC is greater than the minimum SOC for the battery as per the system design. Since the engine does not operate in this mode, the dual clutches are disengaged to prevent any backlash to the engine. Either motor can be used for the launch and backup of the vehicle. The equations for this mode are:

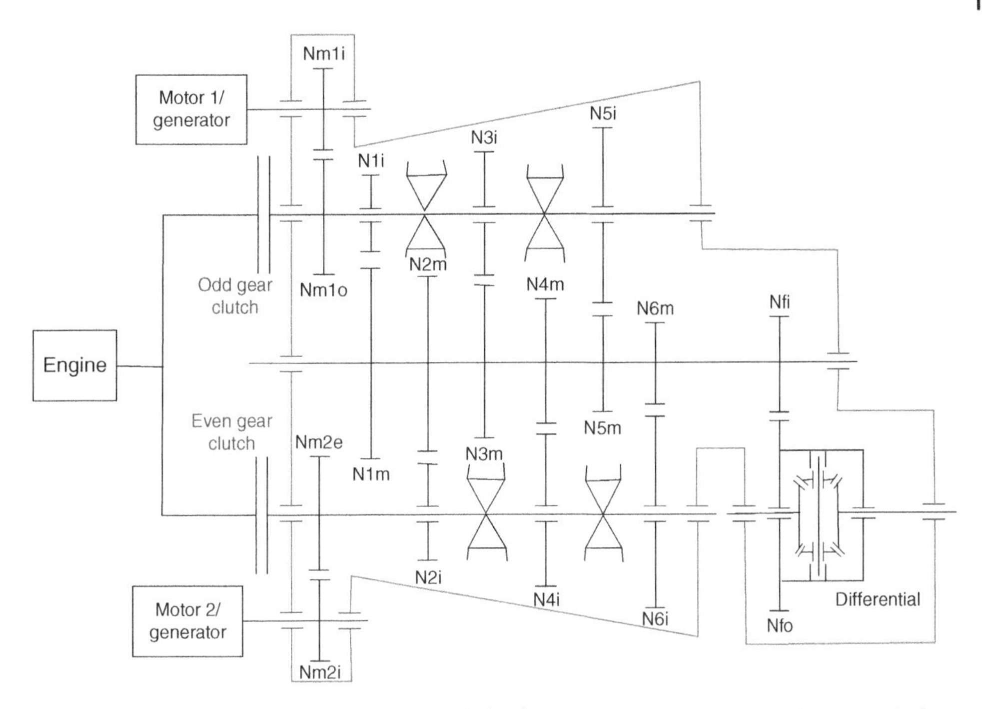

Figure 10.21 Hybrid powertrain based on dual clutch transmissions. Reverse gear is not needed because the motors can be used to back up the vehicle.

$$\omega_o = \frac{\omega_m}{i_g i_a i_m}$$

$$T_o = i_a i_g i_m \times T_m$$

Combined mode: This mode is selected when a high torque is required for situations such as sudden acceleration or climbing a grade. This mode is also selected if the vehicle speed becomes more than the maximum speed defined by the controller in the motor alone mode. Both the engine and the motor provide the propulsion power to the drive shaft. Depending on the vehicle speed, the transmission shift controller selects the proper dual clutch and the gears. The power flow is shown in Figure 10.22. The equations for this mode are:

$$\omega_o = \frac{\omega_m}{i_g i_a i_m} = \frac{\omega_e}{i_g i_a}$$

$$T_o = i_g i_a i_m T_m + i_g i_a T_e$$

Engine alone mode: This mode involves the engine as the only source of propulsion. The engine controller ensures that the engine transmits power to the lowest possible gear ratio such that the engine remains in the best efficiency window. The equations for this mode are:

$$\omega_o = \frac{\omega_e}{i_g i_a}$$

$$T_o = i_a i_g \times T_e$$

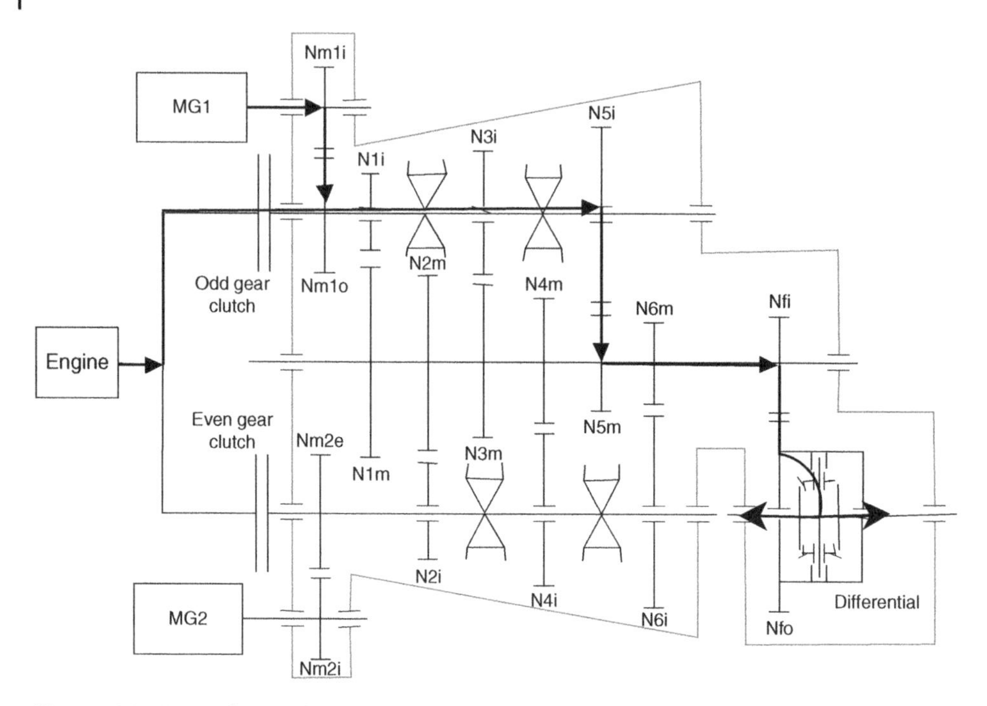

Figure 10.22 Power flow in the combined mode.

Regenerative braking mode: The motor is coupled to the output shaft through gears and it can function as a generator as well. It is used to recover the energy during braking to charge the battery. Depending on the current clutch that is used, the controller decides which motor is to be operated in this mode. In case the motor torque is not sufficient to slow down the vehicle, a conventional braking system is used to supplement the braking demand.

The equations for this mode are:

$$\omega_{in} = \frac{\omega_{\mathrm{m}}}{i_m i_g i_a}$$

$$T_{in} = -i_m i_a i_g \times T_m$$

Power split mode: This mode is used to charge the battery when the vehicle is in motion. The vehicle controller decides on this mode if the engine supplies more power than that required to drive the vehicle. The excess power is then used to charge the battery. The motor on the same lay-shaft that drives the output shaft is selected to act as the generator to charge the battery. The motor controller selects the correct motor depending on the shaft that is transmitting the power to the final drive. The equations for this mode are

$$\omega_o = \frac{\omega_{\mathrm{m}}}{i_a i_m i_g} = \frac{\omega_{\mathrm{e}}}{i_a i_g}$$

$$T_o = i_a i_g T_e - T_m$$

Standstill charge mode: This mode can be used to crank start the engine or charge the battery when the vehicle is in a standstill position. The controller opts for this mode when the battery SOC is lower than the minimum SOC level permitted by the design. This is the only operating mode when the engine is cranked and the vehicle is in a standstill position. Since the vehicle is not moving and no power is transmitted to the drive train, all the gears are disengaged for safety. The kinematic equations for this mode are

$$\omega_o = 0$$

$$T_e = T_m i_m$$

$$\omega_e = \frac{\omega}{i_m}$$

Series hybrid mode: This mode offers a very interesting option for the DCT based hybrid powertrain. The engine is operated in a region near its sweet spot (by adaptively changing the gear ratios) so that the torque generated from the engine is used by one of the motors to generate electricity. This electricity is then used by another motor on the other shaft to drive the vehicle. This, therefore, gives an option of having the DCT based hybrid powertrain operating as a series hybrid. The kinematic equations for this mode are

$$\omega_o = \frac{\omega_m}{i_a i_m i_g}$$

$$T_o = T_m i_m i_g i_a$$

10.4.3 Hybrid Transmission Proposed by Zhang, et al.

An alternate hybrid transmission was proposed to use one electric motor, a planetary gear set and four fixed gears to realize automated transmission and CVT for a parallel hybrid, as shown in Figure 10.23 [6]. The design is based on the concept of AMTs. It uses a combination of lay-shaft gearing and planetary gearing for power transmission. The lay-shaft gears on the input shaft and the motor shaft freewheel unless engaged by the shifter-synchronizer assemblies. The carrier of the planetary gear train is connected to the input shaft that picks up the engine torque. The sun is connected to the motor shaft if so engaged. One motor is used either as the driving assisting unit or as the generator in charging and regenerative braking operations. Mode switching and gear shifting are realized by shifters actuated by computer and controlled step motors as in an AMT. The hybrid system has five operating modes for vehicle driving and one standstill mode for emergency or convenience operations. The six operating modes and the related kinematics are described in the following.

Motor alone mode: The vehicle is always launched in the motor alone mode. In this mode, the motor shaft is engaged by the shifter and power is transmitted to the final drive by the motor gears. Vehicle backup is realized by reversing the motor rotation. All other gears freewheel in this mode. The operating parameters for the engine and the motor in this mode are related by the following equations:

$$\omega_o = \frac{\omega_m}{i_m}$$

$$T_o = i_m T_m$$

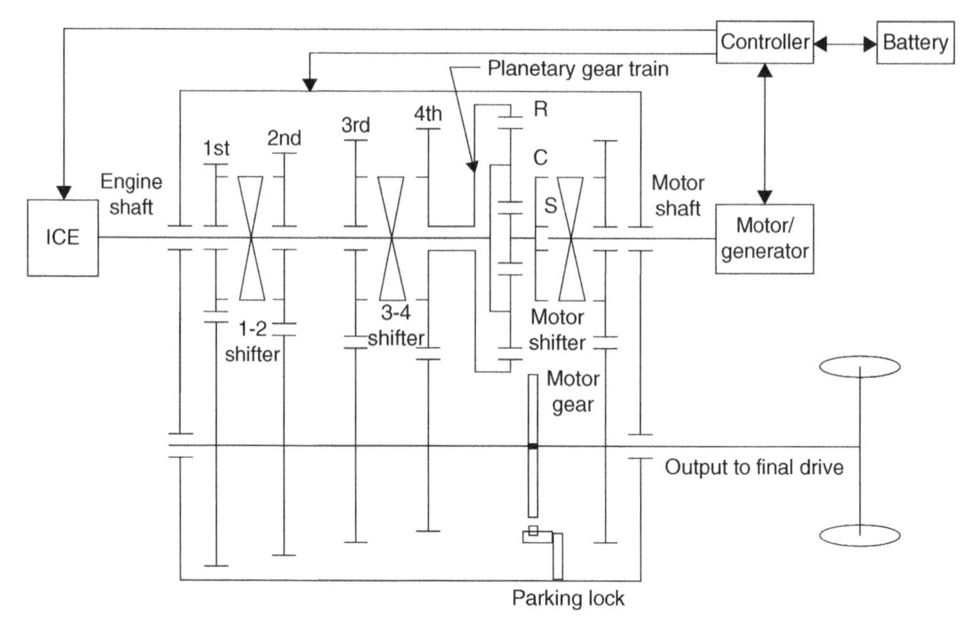

Figure 10.23 Hybrid transmission proposed by Zhang, *et al.*

where, ω_o and ω_m are the angular velocities of the output shaft and the motor respectively, T_o and T_m are the output torque and the motor torque respectively, i_m is the motor gear ratio.

Combined power mode: The combined mode is used when high power is required in situations such as accelerating and hill climbing. In this mode, the motor shaft and one of the lay-shaft pairs are engaged. One of the four available gears as shown in Figure 4.15 can be selected according to the vehicle operating condition. The motor and the engine operate at speeds that are mechanically linked by the gear ratios to drive the vehicle jointly, with the operating parameters related as follows:

$$\omega_o = \frac{\omega_e}{i_k} = \frac{\omega_m}{i_m}$$

$$T_o = i_k T_e + i_m T_m$$

where, ω_e and T_e are the angular velocity and the torque of the engine respectively, and i_k(with $k = 1, 2, 3, 4$) is the gear ratio of the lay-shaft gear pairs.

Engine alone mode: The engine alone mode is the most efficient mode for highway cruising. In this mode, a lay-shaft gear pair with a low gear ratio is engaged to transmit the engine torque to the output shaft with the motor shaft in neutral. The vehicle runs like a conventional vehicle in this mode. The operating parameters are linked by the lay-shaft gear ratio:

$$\omega_o = \frac{\omega_e}{i_k}$$

$$T_o = i_k T_e$$

Electric CVT mode: The electric CVT mode provides two degrees of freedom for vehicle operation control, which allows optimization of engine operation for best fuel economy. In this mode, the engine drives the vehicle and powers the generator for battery charging at the same time. The sun gear of the planetary gear train is coupled to the motor shaft by the shifter and the fourth lay shaft gear is coupled to the ring gear. The operating parameters of the system are governed by the characteristics of the planetary gear train:

$$\omega_m = \frac{N_s + N_r}{N_s}\omega_e - \frac{N_r}{N_s}i_4\omega_o$$

$$T_m = T_s = \frac{N_s}{N_s + N_r}T_e$$

$$T_o = T_r i_4 = \frac{N_r}{N_s + N_r}i_4 T_e$$

where N_r and N_s are the number of teeth of the ring gear and sun gear, respectively, and T_R and T_s are the ring gear torque and sun gear torque, which are distributed from the engine torque at a constant proportion. The output torque T_o and the angular velocity ω_o are determined by the vehicle driving condition. The engine torque and the torque provided to the generator are determined by optimizing the engine efficiency. In the electric CVT mode, the engine speed ω_e is optimized at the point for the highest efficiency corresponding to the required torque. The generator speed ω_m is controlled such that the engine speed and torque are optimized.

Energy recovery mode: In the energy recovery mode, the motor is coupled to the output shaft through the motor gear pair by the shifter and functions as a generator. The relations for the operating parameters are the same as those in the motor alone mode, with the power flow reversed.

Standstill mode: In this mode, the motor is engaged by the shifter to the sun gear and the parking locker is applied to lock the output shaft (ring gear). The lay-shaft gears freewheel. This mode can be used to crank start the engine or use the engine to charge a low battery at standstill. It can also be used as a generator for household electricity or other conveniences if a bidirectional power converter is provided. The parameters are:

$$\omega_e = \frac{N_s}{N_r + N_s}\omega_m$$

$$T_m = \frac{N_s}{N_r + N_s}T_e$$

10.4.4 Renault IVT Hybrid Transmission

In the Renault infinitely variable transmission (IVT), as shown in Figure 10.24, there are two electric motors MG1 and MG2, two planetary gears sets P1 and P2, but no clutches [7]. MG1 is connected to the sun gear of P1; MG2 is connected to the sun gear of P2; the engine is connected to the carrier of P2 as well as the ring gear of P1; the carrier of P1 and the ring gear of P2 are coupled together and connected to the final drive. The system is

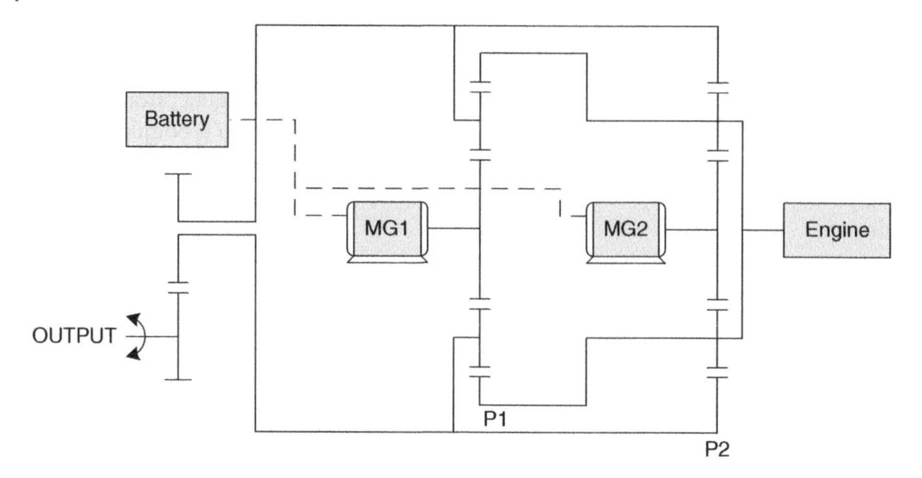

Figure 10.24 Renault two-mode transmission.

capable of providing infinitely variable transmission by controlling the two electric motors to match the vehicle speed while optimizing the operation of the engine.

Since there are no clutches, there is only one operating mode, and the steady state torque speed relationships are:

$$\omega_e = \omega_{c2} = \frac{N_{r2}}{N_{s2} + N_{r2}}\omega_o + \frac{N_{s2}}{N_{s2} + N_{r2}}\omega_{mg2}$$

$$\omega_o = \omega_{c1} = \frac{N_{r1}}{N_{s1} + N_{r1}}\omega_e + \frac{N_{s1}}{N_{s1} + N_{r1}}\omega_{mg1}$$

$$T_o = T_{r2} + T_{c1} = \frac{N_{r2}}{N_{s2} + N_{r2}}T_e + \frac{N_{r1} + N_{s1}}{N_{s1}}T_{mg1}$$

$$T_{mg2} = \frac{N_{s2}}{N_{s2} + N_{r2}}T_e$$

$$T_{mg1} = \frac{N_{r2}}{N_{r2}}T_e$$

10.4.5 Timken Two-Mode Hybrid Transmission

The Timken hybrid powertrain shown in Figure 10.25 is also a two-mode hybrid system [8]. The transmission contains two electric motors MG1and MG2, two planetary gears, P1 and P2, two clutches, C1 and C2, and two locks, B1 and B2.

The engine is connected to the ring gear of P1; MG1 is connected to the sun gear of P1 and via a clutch (C2) to the ring gear of P2; MG2 is connected to the sun gear of P2; the carrier of P2 is connected to the output shaft; the carrier of P1 is connected through C1 to the output shaft, or can be locked by B1. By controlling the two clutches and the two

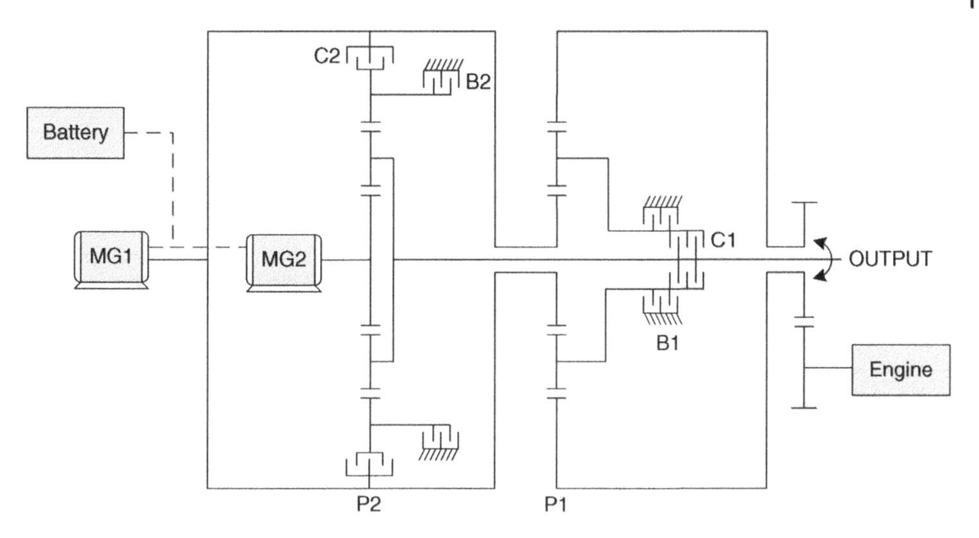

Figure 10.25 Timken two-mode transmission.

locks, the system can operate in high range or low range, based on vehicle operating conditions.

Mode 0: Launch and reverse: The vehicle can be launched by MG2. In this mode, B2 locks the ring gear of P2. MG2 torque is transferred through the sun gear of P2 to the carrier of P2. Since the sun gear of P1 is coupled to the ring gear of P2, the sun gear of P1 is also locked. In this case, the carrier of P1 needs to be locked by B1 so that the engine is stalled as well. The equations are:

$$\omega_{mg2} = \omega_{s2} = \frac{N_{s2}}{N_{s2} + N_{r2}} \omega_o$$

$$T_o = \frac{N_{s2} + N_{r2}}{N_{s2}} T_{mg2}$$

Mode 1: Low-speed operation: In this mode, B2 locks the ring gear of P2; C1 engages the engine, as shown in Figure 10.26. The operation of this mode is exactly the same as the GM two-mode hybrid.

Mode 2: High-speed operation: In this mode, C1 engages the carrier of P1; C2 engages MG1. The sun gear of P1, the ring gear of R2, and MG1 will have the same speed. This mode is also the same as the GM two-mode hybrid powertrain as shown in Figure 10.27.

Mode 4: Series operating mode: The powertrain can also operate in series mode as shown in Figure 10.28 by locking the carrier of P1 and the ring gear of P2. In this mode, engine power is delivered to MG1 through the sun gear of P1 (with the carrier locked). The electricity generated by MG1 will be delivered to MG2 which drives the sun gear of P2 which in turn drives the carrier of P2 with the ring gear locked. The torque-speed equations are:

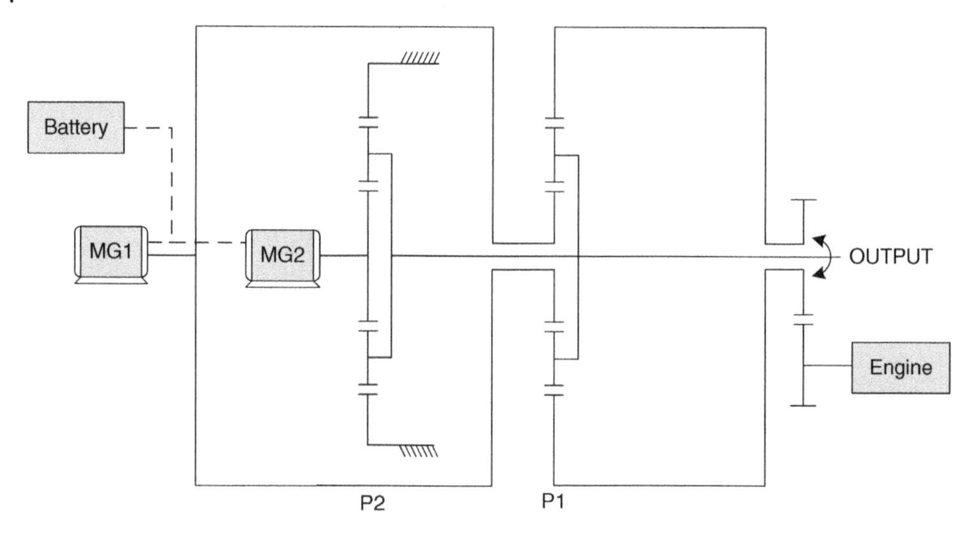

Figure 10.26 Low speed mode of the Timken two-mode transmission.

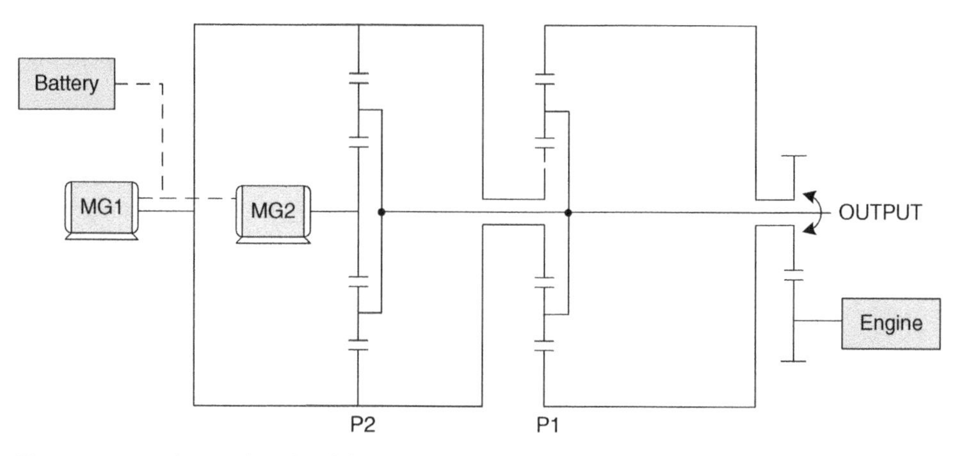

Figure 10.27 High-speed mode of the Timken two-mode transmission.

$$\omega_o = \omega_{c2} = \frac{N_{s2}}{N_{s2} + N_{r2}}\omega_{mg2}$$

$$\omega_{mg1} = -\frac{N_{r1}}{N_{s1}}\omega_e$$

$$T_o = \frac{N_{r2} + N_{s2}}{N_{s2}}T_{mg2}$$

$$T_{mg1} = \frac{N_{s1}}{N_{r1}}T_e$$

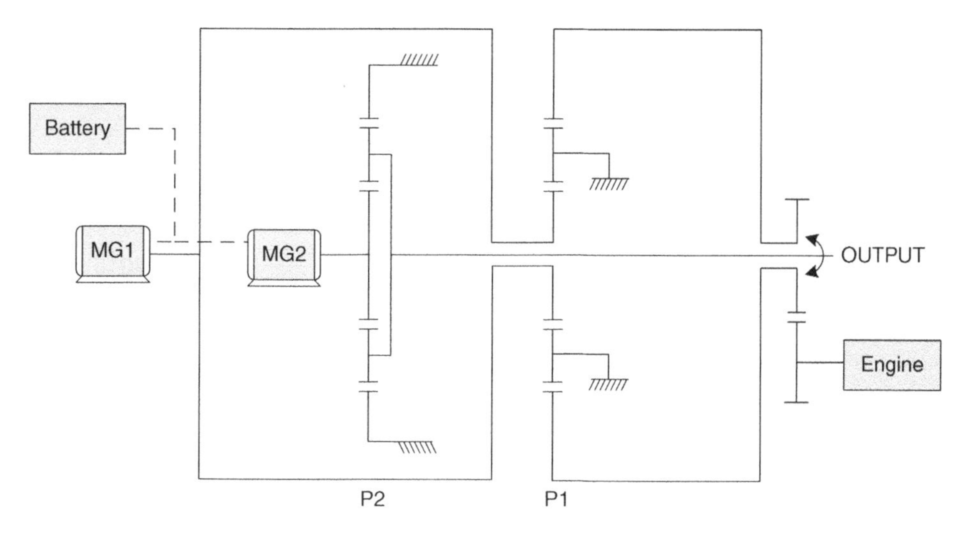

Figure 10.28 Series operating mode of the Timken two-mode transmission.

Mode transition: Similar to any other hybrid powertrain, the transition between different modes needs to happen at the moment when mechanical disturbance to the system can be minimized. For example, with the carrier of P1 locked to B1, the engine can be started by MG1. In order to engage the carrier of P1 with the final drive shaft, first B1 needs to be released, then the speed of MG1 will be controlled such that C1 will accelerate to the same speed as the final drive shaft, and then C1 will engage the carrier of P1 to the final drive shaft. Similarly, in order to engage the ring gear of P2 to the sun gear of P1 (and MG1), the sun gear speed of P1 and MG1 is first brought down to zero and then C2 is engaged.

10.4.6 Tsai's Hybrid Transmission

In the hybrid system proposed by Tsai, as shown in Figure 10.29, the transmission includes one electric motor, two clutches, two planetary gear sets, and two locks [9, 10].

As shown in Table 10.2, there are 14 different combinations of operating modes based on the different configurations of the clutches and locks, but there are only seven valid modes.

The speed/torque relationships can be written in two conditions, that is, C1 engaged or C2 engaged.

When C1 engages (B1 freewheel):

$$\omega_e = \frac{N_{s1}}{N_{s1}+N_{r1}}\omega_m + \frac{N_{r1}}{N_{s1}+N_{r1}}\omega_o$$

$$T_o = \frac{N_{r1}}{N_{s1}+N_{r1}}T_e$$

$$T_m = \frac{N_{s1}}{N_{s1}+N_{r1}}T_e$$

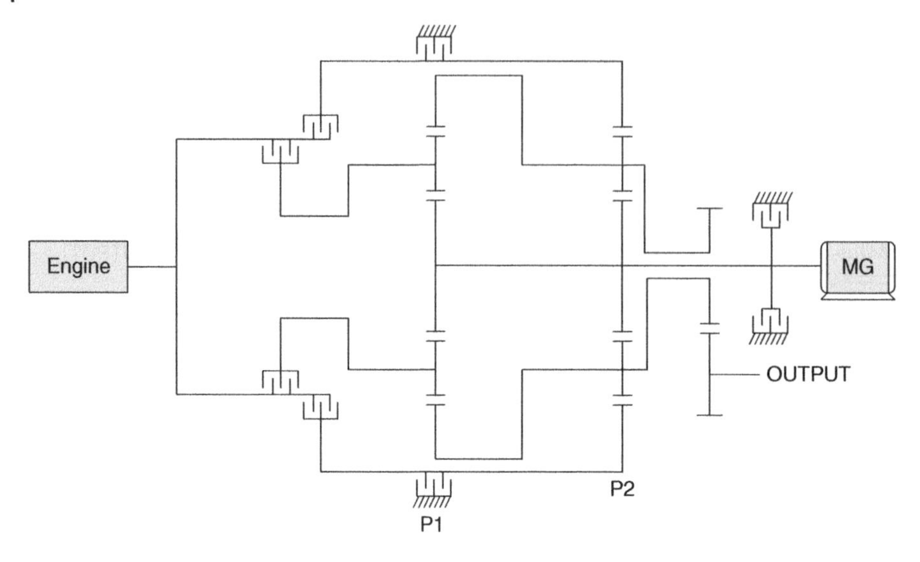

Figure 10.29 Multimode hybrid transmission proposed by Tsai, *et al.*

Table 10.2 Different combinations of operating modes in Tsai's hybrid transmission.

C1	C2	B1	B2	State	Additional modes
0	0	0	0	None	
0	0	0	1	None	
0	0	1	0	Motor alone	Engine idle, generator/regen
0	0	1	1	None	
0	1	0	0	Engine + motor	Motor, generator/regen
0	1	0	1	Engine alone	Motor stationary
0	1	1	0	None	
0	1	1	1	None	
1	0	0	0	Engine alone/CVT	Generator/charging
1	0	0	1	Engine + motor	Motor stationary
1	0	1	0	Engine + motor	Motor, generator/regen
1	0	1	1	None	
1	1	0	0	Engine + motor	Motor, generator/regen
1	1	0	1	None	
1	1	1	0	None	
1	1	1	1	None	

When C1 engages (B1 lock):

$$\omega_e = \frac{N_{s1}}{N_{s1} + N_{r1}}\omega_m + \frac{N_{r1}}{N_{s1} + N_{r1}}\omega_o$$

$$\omega_o = \frac{N_{s2}}{N_{s2} + N_{r2}}\omega_m$$

$$T_o = \frac{N_{r1}}{N_{s1} + N_{r1}}T_e + \frac{N_{s2} + N_{r2}}{N_{s2}}T_m$$

When C2 engages:

$$\omega_o = \frac{N_{s2}}{N_{s2} + N_{r2}}\omega_m + \frac{N_{r2}}{N_{s2} + N_{r2}}\omega_e$$

$$T_o = \frac{N_{s2} + N_{r2}}{N_{r2}}T_e = \frac{N_{s2} + N_{r2}}{N_{s2}}T_m$$

When C1 and C2 both engage:

$$\omega_o = \omega_m = \omega_e$$

$$T_o = T_m = T_e$$

10.4.7 Hybrid Transmission with Both Speed and Torque Coupling Mechanism

The hybrid configuration proposed by Ehsani, et al. in [11] uses one electric motor, three clutches, and two locks to achieve both speed coupling and torque coupling functions, as shown in Figure 10.30 [11].

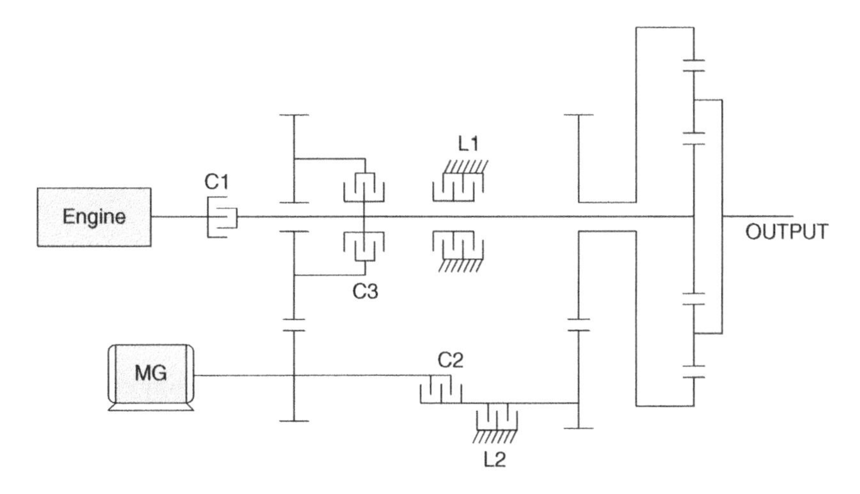

Figure 10.30 Hybrid transmission proposed in [11].

Mode 0: Vehicle launch and backup (motor alone mode), and regenerative braking. C1 open, C2 closed, C3 open, L1 closed, and L2 open. Only the motor torque is transmitted to the final drive. The sun gear of the planetary gear is locked. The torque/speed relationships are:

$$\omega_{out} = \frac{N_r}{N_s + N_r}\omega_m$$

$$T_{out} = \frac{N_s + N_r}{N_r}T_m$$

Mode 1: Engine alone mode. C1 closed, C2 open, C3 open, L1 open, L2 closed. In this mode, the motor is off – only the engine is transferring torque to the final drive. The ring gear is locked. The torque/speed relationships are:

$$\omega_{out} = \frac{N_s}{N_s + N_r}\omega_e$$

$$T_{out} = \frac{N_s + N_r}{N_s}T_e$$

Mode 2: Low-speed mode. C1 closed, C2 open, C3 close, L1 open, L2 closed. In this mode, the engine torque and the motor torque are added to provide the maximum drivetrain torque for acceleration needs. The ring gear is locked and the motor torque is added to the engine shaft. The torque/speed relationships are:

$$\omega_{out} = \frac{N_s}{N_s + N_r}\omega_e$$

$$\omega_m = \frac{N_1}{N_2}\omega_e$$

$$T_{out} = \frac{N_s + N_r}{N_s}\left(T_e + \frac{N_1}{N_2}T_m\right)$$

Mode 3: Combined and power split mode (CVT). C1 closed, C2 closed, C3 open, L1 open, L2 open. In this mode, the motor and the engine output are coupled to the planetary gear on the sun gear and ring gear, respectively. The output and input relationships are:

$$\omega_{out} = \frac{N_s}{N_s + N_r}\omega_e + \frac{N_r}{N_s + N_r}\omega_m$$

$$T_{out} = \frac{N_s + N_r}{N_s}T_e = \frac{N_s + N_r}{N_r}T_m$$

The vehicle could be running between Mode 2 and Mode 3 during highway cruising. Mode transition in this transmission is complicated. In order to reduce mechanical disturbance, the locks have to be engaged at zero speed and the clutches have to be engaged when the two sides of the gears have similar speeds.

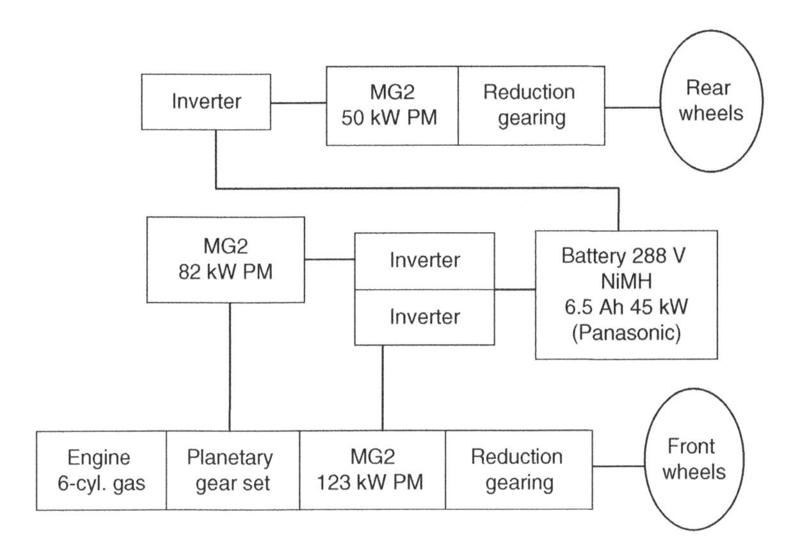

Figure 10.31 Schematics of electric four-wheel drive hybrid system.

10.4.8 Toyota Highlander and Lexus Hybrid, e-Four Wheel Drive

The Toyota Highlander and Lexus hybrid vehicles feature an electric four-wheel drive, or e-four, with the front wheels driven by a planetary gear based hybrid powertrain, and the rear wheels driven by an electric motor. The generalized schematics are shown in Figure 10.31.

In this scheme, the engine is connected to the carrier of the planetary gear set, the generator is connected to the sun gear, and the ring gear is connected to the final drive of the front axle. The total powertrain torque available in any driving condition is:

$$T_{out} = \frac{N_r}{N_s + N_r} T_e + \frac{N_1}{N_2} T_{m1} + \frac{N_3}{N_4} T_{m2}$$

The generator torque is:

$$T_g = \frac{N_s}{N_s + N_r} T_e$$

The power relationship, when neglecting losses, is:

$$P_{out} = P_{m1} + P_{m2} + P_r$$

$$P_e = P_r + P_g$$

$$P_{m1} + P_{m2} = P_g + P_B$$

A simplified version of the e-four is shown in Figure 10.32 with the goal of reducing the overall system cost [12]. In comparison to the above configurations, this configuration has a significant cost advantage and is simple to manufacture, because it does not involve any modification to the front axle design. However, this design does not allow the flexibility of engine speed control. Besides, the system is not very efficient during power split

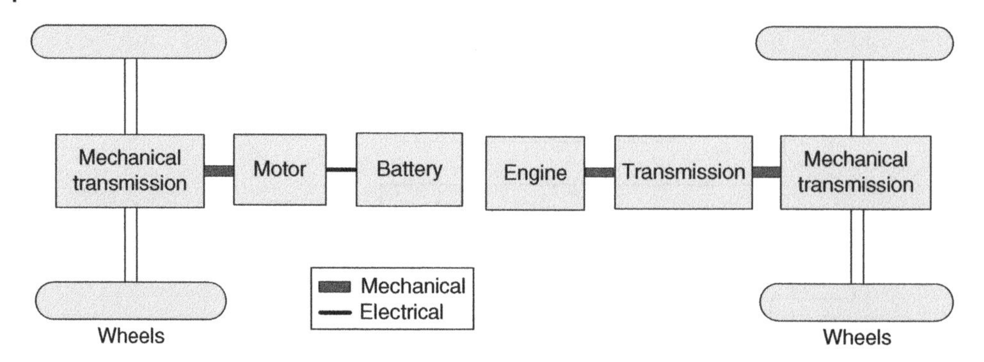

Figure 10.32 Hybrid powertrain with separate driving axles.

mode operation, because engine power needs to be transferred through the vehicle body to the rear axle and then to the electric motor. The total torque of the powertrain is:

$$T_{out} = k_e T_e + k_m T_m$$

where k_e and k_m are the gear ratios of the engine transmission and motor transmission, respectively.

10.4.9 CAMRY Hybrid

In the Camry hybrid, there are two planetary gear sets, as shown in Figure 10.33. The engine, generator, and planetary gear set 1 are configured the same way as in the Prius, that is, the engine is coupled to the carrier, and the generator is coupled to the sun gear.

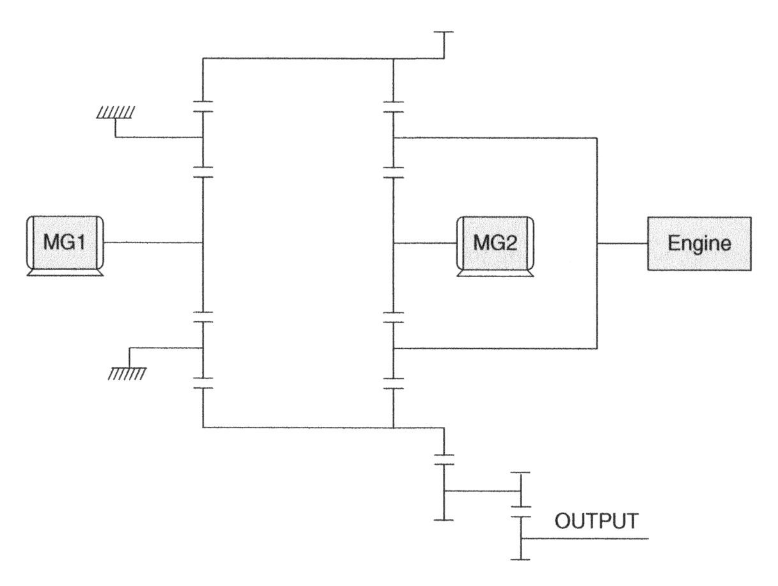

Figure 10.33 Toyota Camry hybrid transmission.

However, the ring gear is connected to a counter gear which is also connected to the ring gear of the motor speed reduction planetary gear. The motor is connected to the sun gear of planetary gear set 2, and the carrier of planetary gear set 2 is grounded. The multi-function gear connects the counter gear and the final drive.

The ring gear speed can be calculated from the vehicle speed:

$$\omega_r = \frac{N_2 N_4}{N_1 N_3} i_{fd} V$$

The engine and generator satisfy:

$$\omega_e = \frac{N_{r1}}{N_{r1} + N_{s1}} \omega_r + \frac{N_{s1}}{N_{r1} + N_{s1}} \omega_g$$

The motor and the ring gear satisfy:

$$\omega_r = -\frac{N_{s2}}{N_{r2}} \omega_m$$

The torque at the final drive is:

$$T_{out} = \left(\frac{N_{r2}}{N_{s2}} T_m + \frac{N_{r1}}{N_{r1} + N_{s1}} T_e \right) \frac{N_2 N_4}{N_1 N_3} i_{fd}$$

10.4.10 Chevy Volt Powertrain

The Chevy Volt from GM has been described as an extended-range electric vehicle or EREV. The exact powertrain configuration is not yet public. However, a number of sources have suggested that the Volt employs two electric motors and a planetary gear set, along with the engine and three clutches as shown in Figure 10.34 [13, 14]. The Volt is equipped with a 16 kWh lithium-ion battery pack, a 125 kW induction motor and a 1.4 liter, four-cylinder engine. The initial driving range of 25 to 50 miles (40–80 km) can be achieved by using energy from the onboard battery, and additional driving range can be achieved by using gasoline.

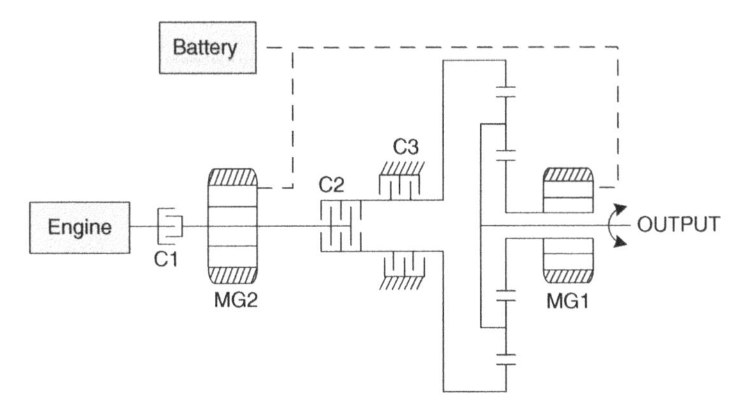

Figure 10.34 The Chevy Volt transmission.

In the Volt transmission, MG1 is the main drive motor, which is connected to the sun gear of the planetary system. The ring gear can be grounded through clutch C3, or connected to MG2 by engaging clutch C2. The carrier is connected to the final drive. The engine can be connected to MG1 through clutch C1. MG2 can be either connected to the engine to become a generator through C1, or connected to the ring gear to become a motor through C2.

The operation of the system can be described as follows:

Mode 1: Single motor driving mode. In this mode, MG1 drives the sun gear with the ring gear locked by C3. The torque of the motor is transmitted through the carrier to the final drive. The engine and MG2 are idle. All driving power is provided by the battery. This mode is suitable for launch, backup, low-speed driving, and cruising. The equation is:

$$T_{out} = \frac{N_s}{N_s + N_r} T_{MG2}$$

Mode 2: Dual motor driving mode. In this mode, MG2 is connected to the ring gear through C2, with C3 disengaged. The engine is idle with C1 open. Both motors receive power from the battery to drive the carrier which delivers torque to the final drive. The equations are:

$$\omega_{out} = \frac{N_s}{N_s + N_r}\omega_{MG2} + \frac{N_r}{N_s + N_r}\omega_{MG1}$$

$$T_{out} = \frac{N_s}{N_s + N_r} T_{MG2} = \frac{N_r}{N_s + N_r} T_{MG1}$$

Since the speed of the two motors is added, this can achieve a high cruising speed for the vehicle. Hence, the authors believe this mode is used for highway cruising with the battery only.

Mode 3: Extended driving range. In this mode, C1 is engaged, so the engine drives MG1 which is now a generator. C2 is open and C3 is engaged to lock the ring gear. Electricity generated by MG2 is delivered to MG1 through power inverters. Only MG1 is driving the final drive. Hence, the output torque expression is the same as in Mode 1.

$T_{out} = \frac{N_s}{N_s + N_r} T_{MG2}$ and $T_{MG1} = kT_e$, where k is the gear ratio between the engine and MG2.

Mode 4: With the engine turned on and C1 engaged, C2 is engaged as well (but C3 is open). Now, a portion of the engine torque can be transmitted to the ring gear to drive the carrier, which delivers the combined engine and MG1 torque to the final drive. Another portion of the engine torque is still used to drive MG2 to generate electricity. This mode is suitable for high speed and heavy accelerations. Thus:

$$kT_e = T_{MG1} + T_r$$

$$\omega_{out} = \frac{N_s}{N_s + N_r}\omega_{MG2} + \frac{N_r}{N_s + N_r}\omega_{MG1}$$

Regenerative braking mode: The engagement of the clutches is the same as in Mode 1. The only difference is that the wheels are now driving MG1 to generate electricity and at the same time generating the required braking torque to slow down the vehicle.

10.5 Non-Ideal Gears in the Planetary System

In the previous sections, the torque/speed relations were given under the assumption that the losses of transmission/gears are neglected, and the kinetic motion is also neglected. When gear losses are considered, the torque equations of a planetary gear set in steady state can be expressed as:

$$T_r - \xi_r \omega_r = \frac{N_r}{N_r + N_s}(T_c - \xi_c \omega_c)$$

$$T_s - \xi_s \omega_s = \frac{N_s}{N_r + N_s}(T_c - \xi_c \omega_c)$$

where ξ_c, ξ_r, and ξ_s stands for frictional loss of carrier, ring gear, and sun gear, respectively.

When gear losses are considered, the torque relationships of the Prius transmission are:

$$T_r + \xi_r \omega_r = \frac{N_r}{N_r + N_s}(T_e - \xi_c \omega_c)$$

$$T_g + \xi_s \omega_s = \frac{N_s}{N_r + N_s}(T_e - \xi_c \omega_c)$$

$$T_o = T_r + \frac{N_{m1}}{N_{m2}}(T_m - \xi_m \omega_m)$$

where ζ is the friction coefficient of each gear, and subscripts m – motor, r – ring gear, s – sun gear, c – carrier, g – generator, e – engine, and ζ_m is the motor output gear.

10.6 Dynamics of Planetary Gear Based Transmissions

When the dynamics of the transmission is considered, there will also be transients in the transmission. For any given rotational system, the rotational dynamics can be written as:

$$T_{in} = T_{out} + J\frac{d\omega}{dt}$$

In the following, the dynamics of the Toyota Prius planetary gear based hybrid transmission is further analyzed. The analysis of other systems should be very similar.

The inertia of the final drive shaft and axle are transferred to the output shaft of the transmission [2]:

$$J_{fd} = \frac{1}{G_a^2}J_a + J_{fd\_sh}$$

where the subscripts are: a – axle, sh – final drive shaft, G_a is the final drive gear ratio. The final drive and the ring gear are coupled directly to the motor shaft. Therefore, on the motor shaft, the total inertia is:

$$J_{ma} = J_m + J_r + J_{fd}$$

The sun gear is coupled to the generator. Therefore, the total generator shaft inertia is:

$$J_{gq} = (J_s + J_g)$$

The engine shaft inertia is the total of the engine crankshaft and the carrier:

$$J_e = J_{crank} + J_c$$

where J_{crank} is the crankshaft inertia. From generator to carrier:

$$J_{gc} = \frac{(N_r + N_s)N_r}{N_s^2}(J_g + J_s)$$

The generator shaft inertia can be transferred to the motor shaft and engine shaft. The equivalent inertias of the engine shaft and the motor shaft are:

$$J_{eq} = J_e + \frac{(N_r + N_s)^2}{N_s^2} J_{gq}$$

$$J_{mq} = J_{ma} + \frac{N_r^2}{N_s^2} J_{gq}$$

On the generator shaft:

$$T_g = \frac{N_s}{N_s + N_r}\{T_e - \xi_c \omega_c - J_{eq}\dot{\omega}_e - J_{gc}\dot{\omega}_m\} - \xi_s \omega_s$$

On the output shaft:

$$T_o = T_m - \xi_m \omega_m + \frac{N_r}{N_r + N_s}(T_e - \xi_c \omega_c) - \left(\frac{N_r}{N_r + N_s} J_{eq} + J_{gc}\right)\dot{\omega}_e - \left(\frac{N_r}{N_r + N_s} J_{gc} + J_{mq}\right)\dot{\omega}_m$$

This torque will drive the final drive shaft at a certain speed. Due to slip of the wheels, $\lambda = \frac{\omega r - V}{V}$, there exists a traction force $F_{fd} = mg\mu(\lambda)$, where $\mu(\lambda)$ is the traction coefficient. This traction force is to overcome the resistive force of the vehicle during driving:

$$F_{fd} = mg\mu(\lambda) = mg \sin\alpha + \frac{1}{2}C_D A_F \rho V^2 + mg(C_o + C_1 V^2) + m\frac{dV}{dt}$$

10.7 Conclusions

It is noted that most planetary gear-based HEVs, including Toyota, Ford, and GM two-mode hybrid, do not include a separate dedicated starter for the engine. The engine is started by one of the motor/generators in an appropriate condition. Due to the fact that the engine usually starts at the time when the drive needs more power, such as on acceleration, there is usually a "jerk" or "hiccup" because a portion of the motor torque is diverted to start the engine. The battery has limited power capability.

Another issue that many drivers have experienced is the weakness of the 14 V battery used to supply power to the vehicle's auxiliary power, such as for wipers, headlights, entertainment systems, power steering, and hydraulic compressor.

The authors feel that, if a starter–alternator is added, the "jerk" during acceleration can be eliminated because the engine may be started by the 14 V onboard battery. Besides, this starter can also be used to charge the 14 V battery when the engine is on which may also ease the burden on the 14 V battery.

In the case of the Toyota and Ford hybrid systems, an additional clutch between the engine and the planetary gear system could smooth the acceleration.

Control of these powertrains is complicated. Often an advanced control algorithm is needed to manage the system. Fuzzy logic, dynamic programming, and wavelet transforms are popular in the power management of complex hybrid vehicles.

References

1 A.G. Holmes, and M.R. Schmidt, Hybrid electric powertrain including a two-mode electrically variable transmission, US Patent US6478705 B1, Nov 12, 2002.
2 J.M. Miller, Hybrid Electric Vehicle Propulsion System Architectures of the e-CVT Type, *IEEE Transactions on Power Electronics*, Vol. 21, No. 3, May 2006, pp. 756–767.
3 M. Kulkarni, T. Shim, and Y. Zhang, Shift dynamics and control of dual-clutch transmissions, *Mechanism and Machine Theory* Vol. 42, No. 2, pp. 168–182, Feb 2007.
4 Y. Liu, D. Qin, H. Jiang, and Y. Zhang, A Systematic Model for Dynamics and Control of Dual Clutch Transmissions, *Journal of Mechanical Design*, 131, 061012, 2009.
5 A. Joshi, N. P. Shah and C. Mi, Modeling and Simulation of a Dual Clutch Hybrid Vehicle Powertrain, 5th IEEE Vehicle Power and Propulsion Conference, Dearborn, Michigan, USA, September 7–11, 2009.
6 Y. Zhang, H. Lin, B. Zhang, and C. Mi, Performance modeling of a multimode parallel hybrid powertrain, *Journal of Mechanical Design, Transactions of the ASME*, Vol. 128, No. 1, Jan 2006, pp. 79–80.
7 A. Villeneuve, Dual mode electric infinitely variable transmission, in Proc. SAE TOPTECH Meeting Continuously Variable Transmission., pp. 1–10. March 8–11, 2004, Detroit
8 X. Ai, T. Mohr, and S. Anderson, An electro-mechanical infinitely variable speed transmission, presented at the Proc. SAE Congress Expo, 2004.
9 L.W. Tsai, G.A. Schultz, and N. Higuchi, A Novel Parallel Hybrid Transmission, *Journal of Mechanical Design, Transactions of the ASME*, Vol.123, June 2001, pp. 161–168.
10 G.A. Schultz, L.W. Tsai, N. Higuchi, and I.C. Tong, Development of a Novel Parallel Hybrid Transmission, SAE 2001 World Congress, Detroit, Michigan March 2001.
11 M. Ehsani, Y. Gao, and A. Emadi, Modern Electric, Hybrid Electric, and Fuel Cell Vehicles: Fundamentals, Theory, and Design, Second Edition (Power Electronics and Applications Series)," CRC Press, 2009.
12 http://reviews.cnet.com/suv/2006-toyota-highlander-hybrid/1707-10868_7-31352761.html.
13 B.M. Conlon, P.J. Savagian, A.G. Holmes, and M.O. Harpster, Jr., Output split electrically-variable transmission with electric propulsion using one or two motors, US patent US2009/0082171 A1, filed Sept 10, 2007, and published March 26, 2009.
14 J.M. Amend, Charge up, Chevy volt rises above sound, fury of introduction, *Wards Autoworld*, November 2010.

Index

Automotive Power Transmission Systems, First Edition. Yi Zhang and Chris Mi.

Printed and bound by CPI Group (UK) Ltd, Croydon, CR0 4YY

16/04/2025

14658384-0002